Urs Eggli · Leonard E. Newton

Etymological Dictionary of Succulent Plant Names

Springer

*Berlin
Heidelberg
New York
Hong Kong
London
Milan
Paris
Tokyo*

Urs Eggli · Leonard E. Newton

Etymological Dictionary
of Succulent Plant Names

Springer

Dr. Urs Eggli
Sukkulenten-Sammlung Zürich
Mythenquai 88
8002 Zürich
Switzerland

e-mail:
urs.eggli@gsz.stzh.ch

Professor Dr. Leonard E. Newton
Department of Botany
Kenyatta University
P.O. Box 43844
Nairobi 00100
Kenya

e-mail:
lnewton@avu.org

ISBN   978-3-642-05597-3

Cataloging-in-Publication Data applied for

Bibliographic information published by Die Deutsche Bibliothek
Die Deutsche Bibliothek lists this publication in the Deutsche Nationalbibliografie; detailed bibliographic data are available in the Internet at <http://dnb.ddb.de>

This work is subject to copyright. All rights reserved, whether the whole or part of the material is concerned, specifically the rights of translation, reprinting, reuse of illustrations, recitation, broadcasting, reproduction on microfilm or in any other way, and storage in data banks. Duplication of this publication or parts thereof is permitted only under the provisions of the German Copyright Law of September 9, 1965, in its current version, and permission for use must always be obtained from Springer-Verlag. Violations are liable for prosecution under the German Copyright Law.

Springer-Verlag is a part of Springer Science+Business Media
springeronline.com

© Springer-Verlag Berlin Heidelberg 2010
Printed in Germany

The use of general descriptive names, registered names, trademarks, etc. in this publication does not imply, even in the absence of a specific statement, that such names are exempt from the relevant protective laws and regulations and therefore free for general use.

Cover design: design & production GmbH, 69121 Heidelberg, Germany

31/3150-WI- 5 4 3 2 1 0 - Printed on acid-free paper

Contents

Preface	VII
Abbreviations	IX
Introduction	XI
References	XVII
Dictionary	1

Preface

Names are important elements to handle the diversity of items in daily life – persons, objects, animals, plants, etc. Without such names, it would be difficult to attach information to such items and to communicate information about them, and names are usually used without giving them much thought.

This is not different for plants. When dealing with plants, however, it soon becomes apparent that the situation is somewhat more complex. Botanists use Latin names to bring order into the vast diversity, while everyday usage resorts to vernacular or "popular" names. As practical as these vernacular names are (it is not suggested that you should ask your greengrocer for a kilogram of *Solanum tuberosum* or *Musa paradisiaca* subsp. *sapientum*), their most important drawback is the fact that they vary widely, not only from one language to another but also from country to country, even from region to region within a large country. More importantly, vernacular names in any given language are usually only available for the plants growing locally, or for plants of some special importance, such as crops and vegetables, medicinal plants, or important garden plants. For all other plants, the Latin names used by botanists and other scientists have to be employed.

Such names often appear complicated or even awkward to the ears of those not accustomed to them. Names are best memorized when their meaning is apparent, but the ever diminishing general knowledge of the classical languages (and here especially Latin) makes the information on the meaning of these scientific names more and more inaccessible. This is the point where the present "Etymological Dictionary" tries to fill a gap. We have tried to give a complete alphabetical list, together with concise explanations, of all the currently accepted names of succulent plants (including cacti).

This task was greatly facilitated by the publication of the recently completed series "Illustrated Handbook of Succulent Plants" (Eggli & Hartmann, 2001–2003), and explanations of the etymologies of all names used in those volumes were originally part of the planned layout. For various reasons, etymological explanations were finally included only for the genus names, but several authors of the "Handbook" series had already supplied etymological explanations for the names of species in the groups they contributed. It was therefore an easy decision to continue this work, expand it to cover *all* accepted names of succulents (including cacti), and thus provide a standardised set of etymological explanations for this intriguing and highly interesting group of plants. It is our hope that the present etymological dictionary will help towards a better understanding of the scientific names of the plants covered – for hobby collectors, horticulturists and botanists alike.

Acknowledgements

The compilation of this etymological dictionary relied heavily on the data supplied by several authors to the "Illustrated Handbook of Succulent Plants" series. Our sincere thanks go to F. Albers, S. Arroyo-Leuenberger, C. C. Berg, A. Chautems, B. Descoings, S. Carter, P. Forster, G. Germishuizen, H. 't Hart (†), H.-D. Ihlenfeldt, E. van Jaarsveld, M. Kimnach, U. Meve, R. Moran, R. Nyffeler, G. D. Rowley, G. F. Smith, J. Thiede, and W. J. de Wilde. Our work to complete and standardise the etymological data, and especially our search for biographical information, was greatly supported by Gordon D. Rowley and Dieter J. Supthut.

Others who supplied much valuable information are M. B. Bayer, V. Gapon, M. J. & R. C. Kimberley, A. B. Pullen, L. Springate, and N. P. Taylor. Numerous colleagues have supplied further snippets of information, and we are grateful for all their help. Additional help was also provided by several authors of recently described taxa, as well as by living persons who were honoured with the name of a plant taxon. Their help is also greatly appreciated.

Finally, it is the pleasant duty of one of us (U.E.) to thank the director of the Sukkulenten-Sammlung Zürich, Dr. Thomas Bolliger, as well as the administration of Grün Stadt Zürich, for permission to use computing infrastructure for the etymological database that was developed during this project. A further word of sincere thanks goes to Springer Verlag, and especially to Dr. Jutta Lindenborn, who favoured the present project and thus enabled this volume to become a companion volume to the "Illustrated Handbook of Succulent Plants".

Nairobi / Zürich, January 2004   Leonard E. Newton & Urs Eggli

Abbreviations

Arab.	Arabian
C	Central
Comp.	Comparative
Dept.	Department
Dim.	Diminutive
Distr.	District
E	East, eastern
Engl.	English
esp.	especially
f.	female
fl.	(floruit) flourishing, active
Gen.	Genitive
Germ.	German
Gr.	Greek
ICBN	International Code of Botanical Nomenclature
IHSP	Illustrated Handbook of Succulent Plants
Ital.	Italian
Lat.	Latin
m.	male
MLat.	Medieval Latin
N	North, northern
n.	neuter
NE	Northeast, northeastern
NW	Northwest, northwestern
Pl.	Plural
Prov.	Province
RBG	Royal Botanic Gardens
RSA	Republic of South Africa
S	South, southern
SE	Southeast, southeastern
Span.	Spanish
Superl.	Superlative
SW	Southwest, southwestern
syn.	synonym
USA	United States of America
W	West, western

Introduction

Coverage

The taxa for which etymologies are supplied in this volume are those accepted in the volumes of the Illustrated Handbook of Succulent Plants (Eggli 2001–2003, Hartmann 2001 and Albers & al. 2002). In the case of the cacti, the list of accepted names was derived from The Cactus Family (Anderson 2001), which in turn is primarily based on the second edition of the CITES Cactaceae Checklist (Hunt 1999). The many synonyms, i.e. names that have gone out of use, are not included here. Deviations from the sources just cited are relatively minor and can be described as follows:

Illustrated Handbook of Succulent Plants (IHSP): With the exception of a couple of corrections of errors and the inclusion of a very few names erroneously not covered in these volumes, there are no deviations. A few of the etymologies for generic names in the IHSP have been slightly modified. In addition, names of doubtful or uncertain application were included with a short discussion and description in the main body of the text in the case of the Aizoaceae, but were listed separately and without explanation in the other four volumes. Consequently, the doubtful names in Aizoaceae are included in our list, unless the discussion in Hartmann (2001) suggests that they do not belong in the family, whilst those listed in the other volumes are omitted. Many new taxa have been described since the IHSP volumes appeared, and there have also been taxonomic changes for some existing taxa. As the main aim of this book is to explain names appearing in the Handbook, new combinations are ignored here (though this does not mean that the editors wish to discredit the publications concerned). If the newly published taxa are interpreted as having at least some standing (at the editors' discretion), etymologies have been included in this volume. Such taxa were always accepted when they were published by the authority / authorities who contributed the relevant Handbook treatment. The infraspecific taxa (esp. varieties) described as new in recent years are mostly unlikely to be of major taxonomic importance, and these are consequently ignored for this work.

Cactaceae: Deviations from the taxonomy presented by Anderson (2001) are again relatively few. The most obvious change is the recognition of the genera *Pierrebraunia*, *Sulcorebutia* and *Weingartia* (included in the synonymy of *Arrojadoa* and, respectively for the latter two, in *Rebutia*). At species level, changes are mostly due to research published subsequently to the compilation of Anderson's book, and usually concern names accepted by Anderson, but which are now recognized as synonyms. This is most notably the case for *Opuntia*. For several of the larger genera, additional infraspecific taxa have been accepted in comparison with the taxonomy by Anderson, again based on recent research. All these changes have been made with the goal of giving an even coverage of etymologies for all relevant taxa, and they reflect on-going research as well as (in some cases) a continuation of current usage.

This book covers almost exclusively names of naturally occurring taxa, but not of artificial hybrids or cultivars (garden selections, including selected variants from natural populations, hybrids, and chimaeras). The names included in this list are those whose formation and use are governed by the "International Code of Botanical Nomenclature" or ICBN (Greuter & al. 2000). A cultivar name consists of a botanical name followed by a cultivar epithet enclosed in single quotes, and the application of cultivar epithets is controlled by a separate code, the "International Code of Nomenclature for Cultivated Plants" (Trehane & al. 1995). Cultivar epithets must be words in a modern language, though before this rule was introduced in 1959 some Latin words were used as cultivar epithets, and their continued use is allowed. As they are in single quotes and are not italicised, they should not be confused with botanical epithets.

Sources of Information

As indicated in the Preface, many IHSP authors included etymologies in their manuscripts, and these contributions formed the basis of our list. When completing the list we found that many names were without obvious meaning or application, and in such cases reference was made to protologues, though some uncertainties remain. Secondary sources, especially for biographical data relating to commemorative names, included lists of field collectors (e.g. Dorr 1997; Gunn & Codd 1981), obituaries and archival material. The main sources are listed at the end of this Introduction.

Many nomenclatural authors give no indication of how they arrived at the name, and even the application of a descriptive epithet is not always obvious from the description of the plant. Very frequently a name commemorates a person about whom absolutely no information is given in the protologue, and in spite of much searching in contemporary literature some of these people remain as shadowy figures of the past, in some cases known only by a surname. The ICBN recommends that authors of new names should include in the protologue the explanation of a name whose meaning is not obvious (Rec. 60H). When observed, this would be a great help to readers. A further recommendation should be made that authors give some basic biographical information to identify persons commemorated in new plant names.

Nomenclature – The Science of Naming Plants

The use of Latin scientific names for plants goes back to medieval times, when Latin was the "lingua franca" of science. Early literature used "phrase names", which were essentially very brief Latin descriptions of the plants. In 1753 the Swedish naturalist Carl Linnaeus (1707-1778) published a list of plants known at that time, with a single-word "trivial name" for each species in each genus. This was quickly established as a system, still in use today, of giving each species a binomial, i.e. a combination of a generic name followed by a specific epithet. Additional epithets indicate infraspecific taxa, such as subspecies, varieties, or forms. Today, the formation and application of botanical names are governed by the "International Code of Botanical Nomenclature" or ICBN (see Greuter & al. 2000 for the current edition). This is a set of rules agreed at an international botanical conference, with the aim of ensuring stability in plant nomenclature. The conference is held every six years, and the ICBN is modified slightly at each conference as users find problems or loopholes in applying the rules and propose improvements.

Although often referred to as Latin names, scientific names include many words from classical Greek or from modern languages. Whatever the source, all names are treated as Latin with regard to grammar. Generic names are nouns (substantives). Epithets may be adjectives, participles treated as adjectives, or nouns. In Latin there are three genders for nouns, namely masculine, feminine and neuter, and in a binomial the epithet must agree with the generic name in gender. For this reason spelling variants occur, though they are essentially the same word, such as *albus*, *alba* and *album*. These notes on Latin are necessarily brief. For further information on Latin grammar, usage and vocabulary, readers are referred to the excellent book by Stearn (1992a).

Botanical names mostly have three derivations. There are descriptive names, referring to some distinctive character of the plant, geographical names, referring to the type locality or distribution of the taxon, and commemorative names, based on personal names of people usually associated with the plants in some way. In the case of epithets the ICBN specifies several standard terminations that indicate the nature of the meaning. For example '-ensis' (m., f.) or '-ense' (n.) means "coming from" and indicates geographical origin or distribution. The terminations '-anus', '-inus' and '-icus' are also used for geographical names, with appropriate gender endings. In substantival commemorative epithets, '-i' or '-ii' is for a man, '-ae' or '-iae' is for a woman, and '-orum' or '-iorum' is for two or more persons. However, there are exceptions, such as '-ae' for a man whose name ends in 'a', such as *rivae* for Riva. Adjectival commemorative epithets may end in '-anus', '-ana' or '-anum', according to the gender of the generic name. Two frequently used Greek terminations to remember are '-oides' and '-opsis', which mean "resembling" or "looking like".

When names are derived from languages other than Latin, the ICBN imposes some standardisation, such as the treatment of accents and umlauts. Complications occur when a personal name is changed. For example, the German ö is transcribed as oe in plant names, but when Schönland emigrated to an English-speaking country, he changed the spelling of his name to Schonland. Epithets based on both spellings have been published, and both are valid. Other cases of possible confusion arise from a lack of standardisation in geographical names, such as Migiurtina and Mijerteina for a Province in Somalia, giving the equally valid epithets *migiurtinus* and *mijerteinus*.

If a published name does not have the correct spelling as laid down in the ICBN, it is to be corrected, though without a change in authorship. Such corrections must be carried out with care, for the ICBN includes both mandatory rules, called Articles, and Recommendations. However, even some of the Recommendations attached to Article 60, dealing with orthography, are rendered mandatory by statements in other parts of the Article. Thus Art. 60.8 makes Rec. 60G mandatory, and Art. 60.11 makes Rec. 60C1 mandatory.

Frequent errors include '*afrus*' instead of '*afer*' (*afer*, *afra*, *afrum*) and '*-ferus*' instead of '*-fer*' (*-fer*, *-fera*, *-ferum*). Accordingly *Echinocereus stoloniferus* had to be corrected to *E. stolonifer*. As the ending '*-iorum*' is used to commemorate two or more persons, *Gasteria baylissiana* should really be *G. baylissiorum*, and *Anacampseros bayeriana* should be *A. bayeriorum*. The ICBN states that epithets should not include hyphens, unless the two words would normally stand alone (Art. 60.9). Therefore *purpureo-croceus* should be corrected to *purpureocroceus*, but *meyeri-johannis* is accepted. The formation of compound words can lead to errors if Latin and Greek words are mixed unwittingly. Also, connecting vowels can be chosen incorrectly. Thus *rubromarginata* had to be corrected to *rubrimarginata*. In addition to the publications by Brown (1956) and Stearn (1992a), a recent book by Radcliffe-Smith (1998) is an invaluable reference for aspiring authors wishing to avoid problems with compound words.

Apart from errors, there are many inconsistencies that represent a lack of standardisation, and it is not always clear from the ICBN if any automatic correction is required. In some cases, classical and mediaeval spellings of the same word are both allowed. For example the classical *silvaticus* and mediaeval *sylvaticus* have the same meaning and derivation but both are allowed (Art. 60.1, Ex. 1). Other cases of inconsistency include *bemarahaensis* and *bemarahensis*, both accepted at present, and the easily confused *litoralis* and *littoralis*. A confusing multiplicity of epithets is formed from quarci- and quartzi- in compound words referring to quartz, and from quarciti-, quartziti- and quarziti- in compound words referring to quartzite.

Some errors might go unnoticed for a long time. Several corrections were made tacitly in the Illustrated Handbook of Succulent Plants and other corrections will be found in this dictionary, but we did not scan all epithets to ensure complete standardisation. Attempting to correct or standardise all the questionable names would be to enter a minefield of uncertainties in interpreting some finer points of the ICBN. For example, we wonder how a plant named for the Bura region in Kenya became *Euphorbia buruana*, though it is treated in the IHSP as an "intentional latinisation" (Art. 60.7). Also, although the aim of the ICBN is to achieve stability in names, we are not sure that it is useful to make corrections to names that have been established for a considerable time and not so far questioned. We note that there are some precedents for this view. Technically, *Huernia* should probably have had to be corrected to *Heurnia*, but it is now the universally accepted spelling. Similar situations apply to the now ubiquitous *Mammillaria* (grammatically correct as *Mamillaria*) and *Pereskia* (for Peiresc, and thus also spelled *Peireskia* in some early literature).

A different kind of error is where the name is inappropriate because of a misunderstanding when the plant was named. An example is *Haworthia parksiana*, in which the word "Parks" was wrongly thought to be the name of a person. *Euphorbia lateriflora* was described from a herbarium specimen, in which the morphology was distorted during drying and mounting. In the case of *Cephalopentandra ecirrosa*, both the generic name and the specific epithet are erroneous. Details of these examples will be found in this dictionary. The ICBN does not allow correction of such errors (Art. 51) and so we must continue to use these names.

Pronunciation of botanical names tends to vary according to the nationality of the speaker, as is evident at any international conference involving plants. Stearn (1992b) commented that how they are pronounced matters little provided that they sound pleasant and are understood. Short names such as *Cereus* and *Sedum* are easy to say, but many people baulk at such monstrosities as *Cephalopentandra* and *Coleocephalocereus* and (for western tongues) *dzhavachischvilii*. The approach should be to break down the word into syllables and pronounce each syllable in turn until you can run off the whole word easily. Of the few rules to be remembered, it can be pointed out that in Latin every vowel should be pronounced separately and the stress falls on the penultimate syllable, thus A-lo-e, not A-loe, which is why it has been written as *Aloë* in some literature.

Layout and Content of the Alphabetical List

In this book we have listed generic names and epithets of species and infraspecific taxa. A family name is based on the name of the type genus, and needs no further explanation. Since the terminal spelling of most epithets is determined by the gender of the generic name, as explained above, variants of the same epithet will be found amongst plant names. In the list we have followed the usual convention adopted in dictionaries and have usually given the masculine version only, to avoid endless repetition of etymologies. Thus *albus* is given, but the same explanation applies to *alba* and *album*. Relating gender variants to the entry in the list should be easy, as in the example just given. The only exceptions in our list are where the link with the masculine spelling might not be obvious, such as '*afer*' (m.) for '*afra*' and '*afrum*' (in which case '*afra*' is the list entry). To determine which gender spelling is appropriate for a particular genus, readers are referred to the main six volumes of the Illustrated Handbook of Succulent Plants (Eggli & Hartmann, 2001–2003), or Anderson (2001) for cacti.

For each name, or each part of compound names, the original language, base word and meaning are given, followed where necessary by the application for the particular taxon. For commemorative names an effort has been made to give some basic biographical data for the persons concerned. These include title, commonly used forename, surname, dates of birth and (where applicable) death, nationality and occupation. Additional notes may include association with the taxon concerned, such as having collected the type specimen, with succulent plants in general, or with the author of the name.

Statistics and 'Awards'

The dictionary has 7006 terms with 8142 different explanations for 11439 taxa (572 genera, 9353 species and 1514 infraspecific taxa). Thus there is more than 20% "overlap", where the same explanation applies to two or more taxa. The most frequently used epithet for succulents is *grandiflorus* (21 taxa), with *pubescens* and *pulchellus* (16 each) sharing second place, followed by *robustus* (15). However, this is not a complete tabulation of names published for succulents as synonyms are excluded, and there might be other contenders for most frequently used overall.

The record for the person with the greatest number of commemorative names is held by Harry Hall, with 22 names (1 genus + 21 epithets), followed closely by Hans Herre (21 epithets). In third place is Neville Pillans, with 19 epithets. Again, synonyms are not included and so there could be other published commemorative names for these and other people, possibly even changing the overall league table. Also, many of the people featured in this dictionary have non-succulent plants, and even animals, named for them. One other record in succulent plant names is that Peter Bally has three commemorative epithets in one genus (*Euphorbia*).

Our nominations for the nicest names are *apicicephalius*, *kalisana* and *mallei*. The generic names *Calibanus* and *Prometheum* also have an interesting derivation. For the most inelegant name, our nomination is *ahremephianus*. Perhaps the award for the most unusual name goes to *mitejea*.

References

Albers, F. & Meve, U. (eds.) 2002. Illustrated Handbook of Succulent Plants: *Asclepiadaceae*. Berlin / Heidelberg (D) / New York (US): Springer-Verlag.

Anderson, E. F. 2001. The Cactus family. Portland (US: OR): Timber Press.

Barnhart, J. H. 1965. Biographical notes upon botanists. Boston (US: MA): G. K. Hall & Co. 3 vols.

Boerner, F. 1989. Taschenwörterbuch der botanischen Pflanzennamen. Berlin / Hamburg (D): Verlag Paul Parey.

Bossert, T. W. 1972. Biographical dictionary of botanists represented in the Hunt Institute Portrait Collection. Boston (US: MA): G. K. Hall & Co.

Brown, R. W. 1956. Composition of scientific words. Washington D.C. (US): Smithsonian Institution Press.

Brummitt, R. K. & Powell, C. E. 1992. Authors of plant names. Richmond (GB): Royal Botanic Gardens, Kew.

Chaudhri, M. N. & al. 1972. Index Herbariorum Part II (3): Collectors. Third instalment I-L. Regnum Vegetabile, 86.

Codd, L. E. & Gunn, M. 1985. Additional biographical notes on plant collectors in southern Africa. Bothalia 15: 631-654.

Desmond, R. 1994. Dictionary of British and Irish botanists and horticulturists. London (GB): Taylor & Francis Ltd. & The Natural History Museum.

Dorr, L. J. 1997. Plant collectors in Madagascar and the Comoro Islands. Richmond (GB): Royal Botanic Gardens, Kew.

Eggli, U. (ed.) 2001–2003. Illustrated Handbook of Succulent Plants: Monocotyledons; Dicotyledons; *Crassulaceae*. Berlin / Heidelberg (D) / New York (US): Springer-Verlag. 3 volumes.

Eggli, U. & Hartmann, H. E. K. (eds.) 2001–2003. Illustrated Handbook of Succulent Plants. Berlin / Heidelberg (D) / New York (US): Springer-Verlag. 6 volumes.

Friis, I. 1982. A list of botanical collectors in Ethiopia. Copenhagen (DK): Published by the author (cyclostyled).

Genaust, H. 1976. Etymologisches Wörterbuch der botanischen Pflanzennamen. [Ed. 1]. Basel (CH) / Stuttgart (D): Birkäuser Verlag.

Genaust, H. 1983. Idem, Ed. 2. Basel (CH) / Stuttgart (D) / Boston (US: MA): Birkhäuser Verlag.

Genaust, H. 1996. Idem, Ed. 3. Basel (CH) / Berlin (D) / Boston (US: MA): Birkhäuser Verlag.

Gillett, J. B. 1962. The history of the botanical exploration of the area of "The Flora of Tropical East Africa". Comptes Rendus IV Réunion AETFAT: 205-229.

Greuter, W. & al. (eds.) 2000. International Code of Botanical Nomenclature (St. Louis Code). Regnum Vegetabile vol. 138. Königstein (D): Koeltz Scientific Books.

Gunn, M. & Codd, L. E. 1981. Botanical exploration of Southern Africa. Cape Town (RSA): Balkema.

Haage, W. 1981. Kakteen von A bis Z. Leipzig / Radebeul (D): Neumann Verlag.

Hammer, S. A. 1993. The genus *Conophytum*. A conograph. Pretoria (RSA): Succulent Plant Publications.

Hammer, S. A. 1999. *Lithops* – Treasures of the Veld (Observations on the genus *Lithops* N. E. Br.). Hull Road (GB): British Cactus and Succulent Society.

Hammer, S. A. 2002. Dumpling and his wife: New views of the genus *Conophytum*. Norwich (GB): EAE Creative Colour Ltd.

References

Hartmann, H. E. K. (ed.) 2001. Illustrated Handbook of Succulent Plants. *Aizoaceae* A-E, *Aizoaceae* F-Z. Berlin / Heidelberg (D) / New York (US): Springer-Verlag. 2 volumes.

Hepper, F. N. & Neate, F. 1971. Plant collectors in West Africa. Regnum Vegetabile vol. 74.

Hunt, D. R. (ed.) 1999. CITES Cactaceae Checklist. Second Edition. Richmond (GB): Royal Botanic Gardens Kew & International Organization for Succulent Plant Study (IOS).

Jackson, W. P. U. 1990. Origins and meanings of names of South African plant genera. Rondebosch (RSA): UCT Ecolab.

Lanjouw, J. & Stafleu, F. A. 1954. Index Herbariorum Part II: Collectors. First instalment A-D. Regnum Vegetabile, 2.

Lanjouw, J. & Stafleu, F. A. 1957. Index Herbariorum Part II (2): Collectors. Second instalment E-H. Regnum Vegetabile, 9.

Mayr, H. 1998. Orchid names and their meanings. Vaduz (FL): A. R. G. Gantner Verlag.

Petschenig, M. & Skutsch, F. 1945. Der kleine Stowasser. Lateinisch-deutsches Wörterbuch. München (D): C. Freytag / Wien (A): Hölder-Pichler-Tempsky / Zürich (CH): Orell-Füssli.

Radcliffe-Smith, A. 1998. Three-language list of botanical name components. Richmond (GB): Royal Botanic Gardens, Kew.

Reynolds, G. W. 1950. The Aloes of South Africa. Johannesburg (RSA): The Aloes of South Africa Book Fund.

Rowley, G. D. 1997. A history of succulent plants. Mill Valley (US: CA): Strawberry Press.

Schubert, R. & Wagner, G. 1988. Pflanzennamen und botanische Fachwörter. Ed. 9. Leipzig / Radebeul (D): Neumann Verlag.

Seybold, S. 2002. Die wissenschaftlichen Namen der Pflanzen und was sie bedeuten. Stuttgart (D): Verlag Eugen Ulmer.

Stafleu, F. A. & Cowan, R. S. 1976–1988. Taxonomic literature. Ed. 2. Utrecht (NL): Bohn, Scheltema & Holkema. 7 vols.

Stafleu, F. A. & Mennega, E. A. 1992–2000. Taxonomic literature. Supplements to Ed. 2. Königstein (D): Koeltz Scientific Books. 6 vols.

Stearn, W. T. 1992a. Botanical Latin. Ed. 4. Newton Abbot (GB): David & Charles. xiv & 546 pp.

Stearn, W. T. 1992b. Stearn's dictionary of plant names for gardeners. London (GB): Cassel.

Voss, A. 1920. Botanisches Hilfs- und Wörterbuch. Berlin (D): Verlag von Paul Parey.

Wagenitz, G. 2003. Wörterbuch der Botanik. Ed. 2. Heidelberg / Berlin (D): Spektrum Akademischer Verlag / Gustav Fischer.

White, A. & Sloane, B. L. 1937. The Stapelieae. Ed. 2. Pasadena (US: CA): Abbey San Encino Press. 3 vols.

Wickens, G. E. 1982. Studies in the Flora of Arabia: III. A biographical index of plant collectors in the Arabian peninsula (including Socotra). Notes Roy. Bot. Gard. Edinburgh 40: 301-330.

Zimmer, G. F. 1912 (reprint 1946). A popular dictionary of botanical names and terms, with their English equivalents. London (GB): George Routledge & Sons.

A

aageodontus Gr. 'aages', hard; and Gr. 'odous, odontos', teeth. (*Aloe*)

abayensis For the occurrence in the Abay Gorge, Ethiopia. (*Orbea*)

abbreviatus Lat., shortened; (**1**) for the overall size. (*Senecio*) – (**2**) for the small plants, with few leaves and flowers. (*Phyllobolus*) – (**3**) for the very short internodes of the lateral branches. (*Mitrophyllum*) – (**4**) for the reduced leaf lamina. (*Ruschia*) – (**5**) for the short corolla tips. (*Ceropegia arabica* var.) – (**6**) application obscure. (*Uncarina*)

abchasicus For the occurrence in the region of Abchasia, W Caucasus, Georgia. (*Sedum*)

abdelkuri For the occurrence on the island Abd-El-Kuri off the coast of Socotra. (*Euphorbia*)

aberdeenensis For the occurrence near Aberdeen, Eastern Cape, RSA. (*Delosperma*)

abhaicus For the occurrence near the city of Abha, Asir Prov., Saudi Arabia. (*Aloe*)

aboriginus Lat., native, ancestral, aboriginal; because the type plants were growing on shell heaps made by the Florida aboriginal people. (*Harrisia*)

abramsii For Prof. LeRoy Abrams (1874–1956), US-American botanist at the Stanford University. (*Dudleya*)

abrotanifolius Lat. '-folius', -leaved; for the similarity of the leaves to those of *Artemisia abrotanum* (*Asteraceae*). (*Pelargonium*)

abruptus Lat., broken off, steep; application obscure. (*Octopoma*)

Absolmsia For Prof. Hermann M. C. L. F. zu Graf Solms-Laubach (1842–1915), German botanist. (*Asclepiadaceae*)

abyssi Gen. of Lat. 'abyssus', abyss; for the occurrence in a canyon near the Grand Canyon, Arizona, USA. (*Cylindropuntia*)

abyssicola Lat. 'abyssus', abyss; and Lat. '-cola', -dwelling. (*Aloe*)

abyssinicus Lat., Abyssinian, Ethiopian; for the occurrence there. (*Bulbine, Ceropegia, Delosperma, Euphorbia, Hypagophytum, Schlechterella*)

Acanthocalycium Gr. 'akanthos, akantha', thorn, prickle; and Gr. 'kalyx', calyx; for the spine-tipped scales on the receptacle and the spine-tipped perianth segments. (*Cactaceae*)

acanthocarpus Gr. 'akanthos, akantha', thorn, prickle; and Gr. 'karpos', fruit; for the spine-covered fruits. (*Cylindropuntia, Erythrina, Tetragonia*)

Acanthocereus Gr. 'akanthos, akantha', thorn, prickle; and *Cereus*, a genus of spiny columnar cacti. (*Cactaceae*)

acanthosetus Gr. 'akanthos, akantha', thorn, prickle; and Lat. '-setus', -bristled; for the bristle-like spines. (*Echinocereus pulchellus* ssp.)

acanthurus Gr. 'akanthos, akantha', thorn, prickle; and Gr. 'oura', tail; for the densely spined stems. (*Cleistocactus*)

acaulis Lat., stemless; (**1**) for the growth form. (*Furcraea, Opuntia*) – (**2**) for the growth form, but erroneously applied. (*Senecio*)

acervatus Lat., in heaps; for the cushion-forming growth habit. (*Euphorbia*)

acetosus Lat., sour; for the acid taste of the leaves. (*Pelargonium*)

achabensis For the occurrence on the Farm Achab, Northern Cape, RSA. (*Conophytum*)

Acharagma Gr. 'a-, an-', without; and Gr. 'charagma', groove; because the tubercles of the plant body are without a groove, contrasting with the related genus *Escobaria*. (*Cactaceae*)

achirasensis For the occurrence at Achiras, Prov. Córdoba, Argentina. (*Gymnocalycium monvillei* ssp.)

acicularis Lat., needle-shaped; (**1**) for the end spine of the leaves. (*Agave*) – (**2**) for the spination. (*Echinocereus engelmannii* var.)

aciculatus Lat., marked with fine irregular streaks; (**1**) for the petal colouration. (*Pelargonium*) – (**2**) application obscure. (*Opuntia*)

acidus Lat., acid, sour-tasting; for the taste of the fruits. (*Eulychnia*)

acifer Lat. 'acus', needle; and Lat. '-fer', carrying; for the spination. (*Echinocereus polyacanthus* ssp.)

acinacifolius Lat. 'acinaces', oriental scimitar; and Lat. '-folius', -leaved. (*Gasteria*)

acinaciformis Lat. 'acinaces', oriental scimitar; and Lat. '-formis', -shaped; for the leaf shape. (*Carpobrotus, Crassula*)

acinacispinus Lat. 'acinaces', oriental scimitar; and Lat. '-spinus', -spined. (*Gymnocalycium catamarcense* ssp.)

ackermannii For Georg Ackermann (fl. 1824), German cactus grower and importer. (*Disocactus*)

acklinicola Lat. '-cola', -dwelling; and for the occurence on Acklin Island, Bahamas. (*Agave*)

acocksii For John P. H. Acocks (1911–1979), South African plant ecologist, Botanical Survey Officer for the Botanical Research Institute, RSA, and a very active plant collector. (*Delosperma, Ruschia*)

aconcaguensis For the occurrence in the Aconcagua valley, C Chile. (*Eriosyce curvispina* var.)

acranthus Gr. 'akros', pointed, on the top; and Gr. 'anthos', flower; for the position of the flowers near the stem tips. (*Haageocereus*)

acre Lat. 'acer, acris, acre', sharp, acute, acid; for the sour-tasting sap of the plant in the mornings. (*Sedum*)

Acrodon Gr. 'akros', pointed, on the top; and Gr. 'odous, odontos', tooth; for the acute apically dentate leaves. (*Aizoaceae*)

acropetalus Gr. 'akros', pointed, on the top; and Gr. 'petalon', petal; for the spine-like mucro at the petal tips. (*Sedum*)

Acrosanthes Gr. 'akros', pointed, on the top; and Gr. 'anthos', flower; for the terminal flower above a dichotomous branching. (*Aizoaceae*)

acrosepalus Gr. 'akros', pointed, on the top; and Gr. 'sepalon', sepal. (*Lampranthus*)

acrotrichus Gr. 'akros', pointed, on the top; and Gr. 'thrix, trichos', hair; for the fibrous leaf tips. (*Dasylirion*)

actinocarpus Gr. 'aktis, aktinos', ray, rayed, star-like; and Gr. 'karpos', fruit; for the stellately spreading fruits. (*Sedum*)

actinocladus Gr. 'aktis, aktinos', ray, rayed, star-like; and Gr. 'klados', branch; for the branches, which radiate from the stem tip. (*Euphorbia*)

actites Gr. 'aktites', a watcher; for the 'lookout' locality high above the surrounding countryside. (*Plectranthus*)

acuispinus Lat. 'acus', needle; and Lat. '-spinus', -spined; for the end spine of the leaves. (*Agave cantala* var.)

aculeatangulus Lat. 'aculeatus', prickly, pointed; and Lat. 'angulus', angle; for the small prickles on the stem angles. (*Cissus quadrangularis* var.)

aculeatus Lat., prickly, pointed; (**1**) for the spiny stems. (*Euphorbia, Pereskia*) – (**2**) for the prickly stems. (*Adenia*) – (**3**) for the spiny leaf surface. (*Aloe*) – (**4**) for the slender stylar head. (*Cynanchum*)

acultzingensis For the occurrence at Acultzingo, Veracruz, Mexico. (*Mammillaria haageana* ssp.)

acuminatus Lat., pointed; (**1**) for the shape of the stem segments. (*Pseudorhipsalis*) – (**2**) for the pointed leaf tips. (*Cheiridopsis, Delosperma, Dudleya, Haworthia magnifica* var., *Hereroa, Machairophyllum, Ruschia*) – (**3**) for the petal tips. (*Drosanthemum*) – (**4**) perhaps for the corolla lobes. (*Crassula colorata* var., *Stapelia*)

acutangulus Lat. 'acutus', pointed, acute; and Lat. 'angulus', angle, corner; (**1**) for the sharply ribbed stems. (*Caralluma*) – (**2**) for the triquetrous leaves with sharp margins and keel. (*Ruschia*)

acutifolius Lat. 'acutus', pointed, acute; and Lat. '-folius', -leaved. (*Crassula tetragona* ssp., *Dischidia, Drosanthemum, Echeveria, Galenia, Lampranthus, Lenophyllum, Sesuvium, Stomatium*)

acutilobus Lat. 'acutus', pointed, acute; and Lat. 'lobus', lobe; for the pointed corolla lobes. (*Quaqua*)

acutipetalus Lat. 'acutus', pointed, acute; and Lat. 'petalum', petal. (*Khadia, Schwantesia*)

acutisepalus Lat. 'acutus', pointed, acute; and Lat. 'sepalum', sepal. (*Pseudosedum*)

acutissimus Superl. of Lat. 'acutus', pointed, acute; for the leaves. (*Aloe*)

acutus Lat., pointed, acute; (**1**) for the leaf shape. (*Villadia*) – (**2**) for the shape of the leaf tip. (*Aloinopsis*) – (**3**) for the petal tips. (*Conophytum*) – (**4**) for the acute tip of the shrivelled fruit. (*Elaeophorbia*)

adamsonii For Prof. Robert S. Adamson (1885–1965), British botanist, 1923–1950 professor at Cape Town University, RSA. (*Trachyandra*)

Adansonia For Michel Adanson (1727–1806), French surgeon and naturalist who discovered the "Baobab" around 1750 in Senegal. (*Bombacaceae*)

adansonii As above. (*Nolana*)

additus Lat., joined to, added; for the frequent presence of an additional flower in the otherwise three-flowered inflorescence. (*Antimima*)

addoensis For the occurrence at Addo, Eastern Cape, RSA. (*Senecio scaposus* var.)

adelae For Madame Adèle Le Chartier (fl. 1908), perhaps an acquaintance of the French botanist and physician Raymond Hamet, without further data. (*Kalanchoe*)

adelaidensis For the occurrence at Adelaide, Eastern Cape, RSA. (*Delosperma*, *Haworthia coarctata* var.)

adelmarii For Dr. Adelmar F. Coimbra Filho (fl. 1983), Brazilian zoologist in Rio de Janeiro. (*Cereus*)

adenensis For the occurrence near Aden, S Yemen. (*Caralluma*, *Euphorbia balsamifera* ssp.)

Adenia Gr. 'aden', gland; for the gland(s) almost always present on the leaves. (*Passifloraceae*)

Adenium Perhaps for the city of Aden, Yemen; or from Gr. 'aden', gland. (*Apocynaceae*)

adenochilus Gr. 'aden', gland; and Gr. 'chilos', lip; for the lipped nectary glands. (*Euphorbia*)

adenotrichus Gr. 'aden', gland; and Gr. 'thrix, trichos', hair; i.e. glandular-hairy. (*Rosularia*)

adeyanus For Mrs. Monica Adeya Masinde (*1937), Kenyan social worker for health and nutrition, mother of the Kenyan botanist Dr. Patrick Siro Masinde. (*Cissus*)

adigratanus For the occurrence at Adigrat, Tigre Prov., Ethiopia. (*Aloe*)

adjuranus For the occurrence in the area inhabited by the Somali Adjuran people. (*Euphorbia*)

adoensis For the occurrence near Adua, Ethiopia. (*Coccinia*)

adolfofriedrichii For Adolfo M. Friedrich (1897–1987), German photographer emigrating 1925 to Brazil and settling 1930 in Paraguay, war photographer and cactus collector during the Chaco War between Paraguay and Bolivia. (*Echinopsis*)

adolphi For Prof. Heinrich Gustav Adolf Engler (1844–1930), German botanist in Berlin and director of the Berlin Botanical Garden and Museum. (*Sedum*)

adpressifolius Lat. 'adpressus', appressed; and Lat. '-folius', -leaved; for the deflexed leaves lying on the stem. (*Euphorbia beharensis* var.)

Adromischus Gr. 'hadros', thick, sturdy; and Gr. 'mischos', flower stalk; for the stout pedicels. (*Crassulaceae*)

adscendens Lat., ascending; for the growth habit. (*Caralluma*, *Cotyledon*, *Harrisia*)

aduncus Lat., hooked; for the apically recurved leaf tips. (*Lampranthus*)

adustus Lat., blackened, scorched; for the blackish spines. (*Echinocereus*)

Aeollanthus Gr. 'Aiolos', God of the Winds; and Gr. 'anthos', flower (Jackson 1990). (*Lamiaceae*)

Aeonium Gr. 'aionion', everliving plant; for the succulent nature and the assumed longevity of the plants. (*Crassulaceae*)

aequatorialis Lat., equatorial, equator-; for the occurrence in Ecuador (= Span., equator. (*Opuntia*)

aequilaterus Lat. 'aequus', equal; and Lat. 'latus, lateris', side; for the triquetrous leaves. (*Carpobrotus*)

aequoris Lat., of the plains; for the habitat. (*Euphorbia*)

aereus Lat., made from copper, bronze or brass; for the petal colour. (*Delosperma*)

aerocarpus Gr. 'aer', air; and Gr. 'karpos', fruit; for the ballon-like wind-dispersed fruits (or from Lat. 'aes, aeris', verdigris, an alloy of copper (bronze); for the fruit colour). (*Eriosyce*)

aeruginosus Lat., verdigris-coloured; for the colour of the branches. (*Euphorbia*)

Aeschynanthus Gr. 'aischyne', shame; and Gr. 'anthos', flower; for the predominantly red flowers. (*Gesneriaceae*)

aestiflorens Lat. 'aestas', summer; and Lat. '-florens', -flowering. (*Conophytum minusculum* ssp.)

aestivalis Lat., pertaining to the summer; for the flowering time. (*Pelargonium*)

aestivus Lat., pertaining to the summer; presumably for the flowering time. (*Lampranthus*)

Aethephyllum Gr. 'aethes', unusual, uncommon; and Gr. 'phyllon', leaf; for the pinnatifid leaves, which are uncommon in the family. (*Aizoaceae*)

aethiopicus For the occurrence in Africa ('Aethiopia' in classical use means Africa, usually South Africa). (*Sansevieria*)

aethiops Gr., sunburnt, black, scorched; for the black spination. (*Cereus*)

aetnensis For the occurrence on Mt. Aetna, Sicily, Italy. (*Sedum*)

affinis Lat., allied to; for the relationship to previously known taxa. (*Aloe, Ceropegia, Delosperma, Dudleya abramsii* ssp., *Echeveria, Galenia, Lampranthus*)

afra Lat. 'afer, afra, afrum', African. (*Portulacaria*)

africanus Lat., for the occurrence in Africa. (*Adenia wightiana* ssp., *Aloe, Ceropegia, Galenia, Kedrostis, Schlechterella, Sterculia*)

Afrovivella From Lat. 'Africa', Africa, for the occurrence there; and in allusion to the similarly rosette-producing former genus *Sempervivella* (now *Sedum, Crassulaceae*). (*Crassulaceae*)

Agave Gr. 'Agave', daughter of Kadmos and sister of Semele in Gr. mythology, also the mother of Pentheus, who she murdered in an outburst of fury; also Gr. 'agavos', stately, noble, illustrious; for the stately nature of many species, but also for the ferocious leaf margin teeth present in many species. (*Agavaceae*)

agavoides Gr. '-oides', resembling; and for the genus *Agave* (*Agavaceae*). (*Ariocarpus, Echeveria*)

aggeria Lat. 'agger', heap, mound; and Gr. suffix '-ia', having the characteristic; for the growth habit. (*Grusonia*)

agglomeratus Lat., collected into a mass; for the group-forming growth. (*Grusonia*)

aggregatus Lat., aggregated, clustered; (**1**) for the offsetting rosettes. (*Orostachys malacophylla* var.) – (**2**) for the crowded branches. (*Euphorbia*) – (**3**) for the numerous flowers clustered in the leaf axils. (*Sinningia*) – (**4**) for the congested inflorescences. (*Ruschia*)

agnetae For Agnes Roggen (fl. 1975), wife of P. A. Roggen, who was a good friend of the describing author. (*Parodia concinna* ssp.)

agninus Lat., of a lamb; for the small teeth along the leaf margins, and comparing the gaping leaves of a pair with a lamb's mouth. (*Stomatium*)

agnis Gen. of Lat. 'agnus', lamb; for Agnese Battista († 2002), stillborn daughter of the Italian succulent plant collector Luigi Battista. (*Haworthia*)

agudoensis For the occurrence at Agudo, Rio Grande do Sul, Brazil. (*Rhipsalis*)

aguirreanus For Ing. Gustavo Aguirre Benavides (*1915), Mexican cactus specialist and cultivator in Parras de la Fuente, Coahuila, Mexico. (*Acharagma*)

ahmarensis For the occurrence at Al Ahmar, Yemen. (*Aloe*) – (**2**) For the occurrence in the Ahmar Mts., Somalia. (*Ceropegia*)

ahremephianus For the initials RMF of Roger M. Ferryman (fl. 2002), English photographer and cactus collector. (*Copiapoa*)

Aichryson Gr. 'aei', always; and Gr. 'chrysos', gold; for the flower colour. (*Crassulaceae*)

aitonis Perhaps for William Aiton (1731–1793), British gardener and botanist at Kew. (*Mesembryanthemum*)

Aizoanthemum Gr. 'anthemon', flowering plant, flower; and for the similarity to the genus *Aizoon*. (*Aizoaceae*)

aizoides Gr. '-oides', resembling; and for the genus *Aizoon* (*Aizoaceae*). (*Aichryson, Senecio talinoides* ssp.)

Aizoon Gr., an everliving plant. (*Aizoaceae*)

aizoon As above. (*Aeonium, Phedimus, Prometheum*)

ajgal From the vernacular Berber name for the

Dragon Tree in the region of Jbel Imzi, Morocco. (*Dracaena draco* ssp.)
aktites Gr., coast-dweller; for the coastal occurrence. (*Agave*)
alacriportanus For the occurrence near Porto Alegre, Rio Grande do Sul, Brazil (Lat. 'portus alacer' = Port. 'Porto Alegre', merry port). (*Parodia*)
alamosanus For the occurrence at Los Alamos and the Alamos Mts., Sonora, Mexico. (*Ferocactus, Sedum*)
alamosensis For the occurence near Los Alamos, Sonora, Mexico. (*Stenocereus*)
alatus Lat., winged; (**1**) for the ridges on the caudex. (*Myrmecodia*) – (**2**) for the winged branches. (*Euphorbia, Pseudorhipsalis, Ruschia*) – (**3**) for the winged rhachis. (*Erythrophysa*) – (**4**) for the winged corolla-angles. (*Echeveria*) – (**5**) for the winged seeds. (*Bulbine*) – (**6**) application obscure. (*Crassula*)
albanensis For the occurrence near Albany, Eastern Cape, RSA. (*Rhombophyllum*)
albanicus As above. (*Mestoklema*)
albatus Lat., clothed white; for the flower colour. (*Juttadinteria*)
albens Lat., becoming white, whitened; for the flower colour. (*Drosanthemum*)
albersianus For Prof. Dr. Focke Albers (*1940), German botanist in Münster, working with the karyology and morphology of Asclepiadaceae and Geraniaceae. (*Cibirhiza*)
albersii As above. (*Quaqua*)
albertensis For the occurrence near Prince Albert, Western Cape, RSA. (*Euphorbia*)
albertii For Albert von Regel (1845–1908), Swiss-born Russian physician and botanist of German descent in St. Petersburg; oldest son of the botanist Eduard A. Regel. (*Sedum*)
albertiniae For the occurrence near Albertinia, Western Cape, RSA. (*Crassula capensis* var.)
albertisii For L. M. d'Albertis (1841–1901), Italian botanist. (*Myrmecodia*)
albescens Lat., becoming white; (**1**) for the leaf colour. (*Agave*) – (**2**) for the spine colour. (*Mammillaria decipiens* ssp.)

albiarmatus Lat. 'albus', white; and Lat. 'armatus', armed; for the white spinatation. (*Mammillaria coahuilensis* ssp.)
albicans Lat., whitish; for the spination. (*Mammillaria*)
albicaulis Lat. 'albus', white; and Lat. '-caulis', -stemmed; for the whitish-blue stem colour. (*Cereus*)
albicephalus Lat., 'albus', white; and Gr. 'kephale', head; for the white cephalium. (*Melocactus, Micranthocereus*)
albiceps Lat. 'albus', white; and Lat. '-ceps', headed; for the white appearance of the stem tips caused by the closely-set areoles. (*Browningia*)
albicolumnaris Lat. 'albus', white; and Lat. 'columnaris', column-shaped; for the shortly cylindrical white-spined plant bodies. (*Frailea pygmaea* ssp.)
albicolumnarius Lat. 'albus', white; and Lat. 'columnarius', having columns; for the shortly cylindrical white-spined plant bodies. (*Escobaria*)
albicomus Lat. 'albus', white; and Lat. 'coma', hair, mane; for the white hair-like spination. (*Mammillaria*)
albidiflorus Lat. 'albidus', whitish; and Lat. '-florus', -flowered. (*Anacampseros, Eriosyce chilensis* var.)
albidior Comp. of Lat. 'albidus', white, whitish; for the pale glaucous leaves. (*Agave mitis* var.)
albido-opacus Lat. 'albidus', whitish; and Lat. 'opacus', dull, not shining; for the flower colour. (*Umbilicus*)
albidus Lat., whitish; (**1**) for the wax-covered leaves. (*Machairophyllum*) – (**2**) for the flower colour. (*Aloe, Dischidia*)
albiflorus Lat. 'albus', white; and Lat. '-florus', -flowered. (*Aloe, Arrojadoa, Beschorneria, Conophytum, Drosanthemum, Dudleya, Mammillaria, Polymita, Rebutia, Rhadamanthus, Tylecodon, Villadia*)
albiflos Lat. 'albus', white; and Lat. 'flos', flower. (*Haemanthus*)
albifuscus Lat. 'albus', white; and Lat. 'fuscus', brown; application not explained. (*Frailea gracillima* ssp.)

albilanatus Lat. 'albus', white; and Lat. 'lanatus', woolly; for the white wool-like spination. (*Mammillaria*)
albinota Lat. 'albus', white; and Lat. 'nota', mark; for the white-spotted leaf surface. (*Rabiea*)
albipilosus Lat. 'albus', white; and Lat. 'pilosus', hairy; for the white-hairy corolla. (*Brachystelma, Tridentea marientalensis* ssp.)
albipollinifer Lat. 'albus', white; Lat. 'pollen', pollen; and Lat. '-fer, -fera, -ferum', -carrying. (*Euphorbia*)
albiporcatus Lat. 'albus', white; and Lat. 'porcatus', ridged, ribbed (from Lat. 'porca', ridge between two furrows); for the stem architecture. (*Cissus*)
albipunctus Lat. 'albus', white; and Lat. 'punctum', dot, spot; for the white-spotted leaf surface. (*Rabiea*)
albisaetacens Lat. 'albus', white; and Lat. 'saetacens', with bristles; for the spination of the stem segments. (*Tunilla*)
albiseptus Lat. 'albus', white; and Lat. 'septum', partition, cross-wall; for the disposition of the white corolla lobes. (*Ceropegia*)
albisetatus Lat. 'albus', white; and Lat. '-setatus', -bristled; for the sometimes almost white spination. (×*Haagespostoa*)
albispinus Lat. 'albus', white; and Lat. '-spinus', -spined. (*Haageocereus*)
albissimus Lat., whitest (Superl. of Lat. 'albus', white). (*Avonia*)
albisummus Lat. 'albus', white; and Lat. 'summus', highest (Superl. of 'superus', high); for the tall white-hairy stems. (*Pilosocereus*)
albivenius Lat. 'albus', white; and Lat. 'vena', vein; for the leaves. (*Ipomoea*)
alboareolatus Lat. 'albus', white; and Lat. 'areolatus', with areoles. (*Gymnocalycium*)
albocastaneus Lat. 'albus', white; and Lat. 'castaneus', chestnut-brown; for the flower colour. (*Orbea*)
albomarginatus Lat. 'albus', white; and Lat. 'marginatus', margined; (**1**) for the leaves of the primary rosettes. (*Sedum*) – (**2**) for the white leaf margins. (*Agave*)
albopectinatus Lat. 'albus', white; and Lat. 'pectinatus', comb-like, pectinate; for the spination. (*Rebutia*)
alboroseus Lat. 'albus', white; and Lat. 'roseus', rose-coloured; for the petals. (*Conicosia pugioniformis* ssp., *Leipoldtia, Stomatium*)
alboruber Lat. 'albus', white; and Lat. 'ruber, rubra, rubrum', red; for the bicoloured petals. (*Antimima*)
albostriatus Lat. 'albus', white; and Lat. 'striatus', striped. (×*Duvaliaranthus*)
albovestitus Lat. 'albus', white; and Lat. 'vestitus', clothed; for the heavy bloom on the tepals. (*Aloe*)
albovillosus Lat. 'albus', white; and Lat. 'villosus', villous, hairy; for the hairy leaves and cyathia. (*Euphorbia gueinzii* var.)
albovirens Lat. 'albus', white; and Lat. 'virens', green; for the colour of the branches. (*Euphorbia pillansii* var.)
Albuca Lat. 'albucus', "Asphodel" (*Asphodelus sp.*, from Lat. 'albus', white, for the white flowers); for the similarity of some species to Asphodel. (*Hyacinthaceae*)
albus Lat., white; for the flower colour. (*Astridia, Bulbine, Crassula, Gibbaeum, Oscularia, Pelargonium, Sedum*)
alcahes From the vernacular name of the plants in Baja California, Mexico. (*Cylindropuntia*)
alcicornis Lat. 'alces', elk; and Lat. 'cornu', horn; (**1**) for the intricate branching pattern. (*Euphorbia*) – (**2**) probably for the inflorescences. (*Crassula*)
aldabrensis For the occurrence on Aldabra Island in the W Indian Ocean. (*Aloe*)
alensis For the occurrence in the Sierra del Alo, Mexico. (*Pilosocereus*)
alexanderi For Edward J. Alexander (1901–1985), US-American botanist and curator of the herbarium at the New York Botanical Garden. (*Sedum*) – (**2**) For W. B. Alexander (fl. 1921), who first collected this taxon. (*Tephrocactus*)
alexandrii For Dr. Richard Chandler Prior (formerly Alexander) (1809–1902), British botanist. (*Sterculia*)
alfredii For Alfred Hance, son of the British

botanist Henry F. Hance (1827–1886). (*Sedum*) – (**2**) For Alfred Razafindratsira (fl. 2001), Madagascan plant collector and owner of a succulent plant nursery. (*Aloe, Euphorbia*)

Algastoloba Intergeneric hybrid name that combines the names of the involved genera *Aloe, Gasteria* and *Astroloba*. (*Aloaceae*)

algidus Lat., cold; for the occurrence in cold Siberia. (*Rhodiola*)

algoensis For the occurrence in Algoa Park, near Port Elizabeth, Eastern Cape, RSA. (*Lampranthus*)

aliciae For Miss Alice Pegler (1861–1929), teacher and amateur botanist in the Eastern Cape, RSA. (*Plectranthus*) – (**2**) For Mlle. Alice Leblanc (fl. 1908), an intimate acquaintance of the the French botanist and physician Raymond Hamet. (*Orostachys*)

alidae For Alida P. Zonneveld-van Leeuwen (fl. 1986), wife of the Dutch *Crassulaceae* specialist Dr. B. Zonneveld. (*Sempervivum*)

aliwalensis For the occurrence near Aliwal North, Eastern Cape, RSA. (*Delosperma*)

alko-tuna From the local vernacular name of the plants in Bolivia, from Quechua 'alko', dog, and Span. 'tuna', fruit of *Opuntia* spp. (*Opuntia*)

allantoides Gr., sausage-shaped; for the leaf shape. (*Sedum*)

allionii For Carlo Allioni (1728–1804), Italian physician and botanist. (*Sempervivum globiferum* ssp.)

alloplectus Gr. 'allos', other or another; and Gr. 'plektos', plaited, twisted; application obscure. (*Plectranthus*)

allosiphon Gr. 'allos', other or another; and Gr. 'siphon', tube; for the floral tube that is said to distinguish the taxon from related species. (*Parodia*)

Alluaudia For Charles Alluaud (1861–1949), French entomologist and natural history collector of independent means, working in Africa and Madagascar, often in association with the Natural History Museum, Paris. (*Didiereaceae*)

Alluaudiopsis Gr. '-opsis', similar to; and for the genus *Alluaudia* (*Didiereaceae*). (*Didiereaceae*)

Aloe From the Gr. ('aloe'), Arabian ('alloch') and Hebrew ('ahalim') names for the plants. (*Aloaceae*)

aloides Gr. '-oides', resembling; and for the genus *Aloe* (*Aloaceae*). (*Nananthus*)

aloifolius Lat. '-folius', -leaved; for the similarity of the leaves to those of *Aloe* (*Aloaceae*). (*Yucca*)

Aloinopsis Gr. '-opsis', similar to; and for the genus *Aloe* (*Aloaceae*). (*Aizoaceae*)

alonsoi For Alonso García Luna (*1980), Mexican discoverer of the taxon. (*Turbinicarpus*)

alooides Gr. '-oides', resembling; and for the genus *Aloe* (*Aloaceae*). (*Aloe, Bulbine, Dudleya saxosa* ssp., *Senecio*)

alpester Ital. 'alpestre', alpine (from Lat. 'Alpes', Alps); for the high-altitude distribution. (*Crassula, Rosularia, Sedum*)

alpinus Lat., alpine; for the high-altitude distribution. (*Brachystelma, Ectotropis, Esterhuysenia*)

alsinefolius Lat. '-folius', -leaved; for the similarity of the leaves to those of *Alsine* (now *Minuartia*, "Sandwort" etc., *Caryophyllaceae*). (*Sedum*)

alsinoides Gr. '-oides', resembling; and for the former genus *Alsine* (now *Minuartia*, "Sandwort" etc., *Caryophyllaceae*). (*Crassula pellucida* ssp.)

alsius Lat., frosty; for the cold northerly habitat. (*Rhodiola*)

Alsobia Gr. 'alsos', forest; and perhaps Gr. 'sobe', horse tail; for the epiphytic growth and the pendent stems. (*Gesneriaceae*)

alstonii For Captain Edward G. Alston (fl. 1891–1917), farm manager and botanical collector in the Northern Cape, RSA. (*Adromischus, Avonia quinaria* ssp., *Cephalophyllum, Crassula, Hoodia*)

alternans Lat., alternating; (**1**) for the alternately arranged leaflets of the leaves. (*Pelargonium*) – (**2**) application obscure. (*Kalanchoe*)

alternicolor Lat. 'alternans', alternating; and Lat. 'color', colour; for the variegated branches. (*Euphorbia aggregata* var.)

alticola Lat. 'altus', high; and Lat. '-cola', inhabiting; for the occurrence at high altitudes. (*Aloe megalacantha* ssp., *Crassula*, *Delosperma*, *Kalanchoe*, *Khadia*)

altigenus Lat. 'altus', high; and Lat. 'genus', birth, origin; for the high-altitude habitat. (*Ruschia*)

altissimus Comp. of Lat. 'altus', high; (**1**) for the tall stems. (*Browningia*) − (**2**) for the comparatively tall leaves. (*Haworthia angustifolia* var.)

altistylus Lat. 'altus', high; and Lat. 'stylus', style. (*Lampranthus*)

altus Lat., high; (**1**) for the high-altitude occurrence. (*Sempervivum*) − (**2**) for the comparatively tall bodies formed from the fused leaves of a pair. (*Conophytum bilobum* ssp.)

alversonii For Andrew H. Alverson (1845–1916), British-born mineral prospector in California, later cactus collector, propagator and dealer. (*Escobaria*)

amabilis Lat., loveable, pleasing; (**1**) for the general appearance. (*Cheiridopsis*, *Lampranthus*, *Rhodiola*) − (**2**) and for Mabel Grande (fl. 2002), pseudonym of a well-known US-American Mesemb specialist. (*Phyllobolus*)

amajacensis For the occurrence in the valley of the Río Amajac and near the village of Santa Maria Amajac, Hidalgo, Mexico. (*Mammillaria*)

amanensis For the occurrence in the Amanus Mts. in S Turkey. (*Rosularia sempervivum* ssp.)

amaniensis For the occurrence at Amani in the Usambara region, Tanzania. (*Senecio*)

amarantoides Gr. '-oides', resembling; and for the genus *Amaranthus* (*Amaranthaceae*). (*Cistanthe*)

amarifontanus Lat. 'amarus', bitter; and Lat. 'fontanus', relating to a spring; for the occurrence at Bitterfontein, RSA. (*Euphorbia*)

amatolicus For the occurrence in the Amatola Mts., Eastern Cape, RSA. (*Crassula sediflora* var., *Schizoglossum*)

amatymbicus Probably erroneously formed from the locality Tambukiland, from where the taxon was described. (*Pelargonium*)

amazonicus For the occurrence in the Amazonas region, Brazil. (*Pseudorhipsalis*) −

(**2**) For the occurrence in the Amazonas region, Peru. (*Praecereus euchlorus* ssp.)

ambarivatoensis For the occurrence at Ambarivato in the Ankarana Reserve, Madagascar. (*Euphorbia*)

ambatoensis For the occurrence in the Sierra Ambato, Prov. Catamarca, Argentina. (*Gymnocalycium*)

ambatofinandranae For the occurrence near Ambatofinandrahana, Madagascar. (*Euphorbia stenoclada* ssp.)

ambigens Lat., doubting ('ambigere', to doubt); application obscure. (*Aloe*)

ambiguus Lat., doubtful; (**1**) because the basionym author was uncertain about the generic placing. (*Plectranthus*) − (**2**) because the taxon was seen as intermediate between *Claytonia* and *Calandrinia* when first published. (*Cistanthe*) − (**3**) because the placement within one of the sections of the genus was unresolved. (*Drosanthemum*) − (**4**) application obscure. (*Sinocrassula*)

amboensis For the occurrence in the former Amboland (now Ovamboland) in N Namibia. (*Neorautanenia*)

ambohipotsiensis For the Ambohipotsi Section of the Tsimbazaza Botanical Garden, Antananarivo, Madagascar, where the plants were found growing. (*Euphorbia*)

amboinicus From the occurrence on Amboina Island, Moluccas. (*Plectranthus*)

ambolensis For the occurrence in the Ambolo territory, SE Madagascar. (*Kalanchoe*)

ambongensis For the occurence in the valley of the Ambongo River, Madagascar. (*Pachypodium*)

ambovombensis For the occurrence near the town of Ambovombe, S Madagascar. (*Ceropegia*, *Cynanchum*, *Euphorbia*)

ambroseae For Mrs. M. N. Ambrose (fl. 1960), who collected plants in Southern Rhodesia (now Zimbabwe) and Moçambique for L. C. Leach. (*Euphorbia*)

amecaensis For the occurrence near Ameca, México, Mexico. (*Disocactus speciosus* fa.)

amecamecanus For the purported type locality Amecameca, México, Mexico (*Sedum*)

amerhauseri For Mr. Helmut Amerhauser (*1941), Austrian cactus collector, *Gymno*-

calycium specialist and co-founder of the Austrian "Arbeitsgemeinschaft Gymnocalycium". (*Frailea, Gymnocalycium*)

americanus Lat., American; for the occurrence on the American continent. (*Agave*)

amethystinus Lat., amethyst-coloured; (**1**) for the leaf-colour. (*Graptopetalum*) − (**2**) for the occurrence near Brejinho das Ametistas, Bahia, Brazil. (*Melocactus bahiensis* ssp.)

amiclaeus From the ancient town of Amiclae [Amyclae], present-day Monticelli (near Napoli), Italy. (*Opuntia*)

amicorum Gen. Pl. of Lat. 'amicus', friend, i.e. of friends; (**1**) for the friends on the Mountain Club of Kenya expedition when the taxon was discovered. (*Aloe*) − (**2**) because the taxon was discovered together by four friends. (*Conophytum tantillum* ssp.) − (**3**) for the two plant-hunting friends John J. Lavranos and Len E. Newton, who collected the type material of the taxon. (*Euphorbia*) − (**4**) perhaps because the taxon was discovered by several friends of the South African botanist Louisa Bolus. (*Ruschia*)

amilis Probably from Lat. 'amor', love; according to the protologue for the most-beloved daughter of Carlos Spegazzini, who died shortly before the publication of the taxon. (*Portulaca*)

ammak From the local vernacular name of the plants in Arabia (perhaps, though unlikely, from Arab. 'ammaq', to deepen ?). (*Euphorbia*)

ammophilus Gr. 'ammos', sand; and Gr. 'philos', friend; for the preferred sandy habitat. (*Crassula, Opuntia*)

amoenus Lat., beautiful, pleasing. (*Antimima, Echeveria, Lampranthus*)

ampanihensis For the occurrence near the town of Ampanihy, S Madagascar. (*Cynanchum*)

ampanihyensis As above. (*Euphorbia decaryi* var.)

Amphibolia Gr. 'amphibolos', ambiguous, doubtful; for the doubtful generic placement of the type species. (*Aizoaceae*)

amphibolius Gr. 'amphibolos', ambiguous, doubtful; for the doubtful generic placement of the taxon. (*Lampranthus*)

Amphipetalum Gr. 'amphi-', around, double, of two kinds; and Gr. 'petalon', petal; for the two distinct petal types. (*Portulacaceae*)

amphoralis Lat., pertaining to amphoras; for the amphora-shaped corolla. (*Echeveria*)

amplectens Lat., clasping; for the tubular stem-clasping leaf sheaths. (*Aspazoma*)

amplexicaulis Lat. 'amplexus', clasping; and Lat. '-caulis', -stemmed; for the amplexicaul (i.e. clasping the stem) leaves. (*Sedum*)

ampliatus Lat., enlarged, increased; (**1**) for the subglobose leaves. (*Ruschia*) − (**2**) for the broadly inflated corolla. (*Ceropegia*)

ampliflorus Lat., 'amplus', wide, large; and Lat. '-florus', -flowered. (*Senecio*)

ampliphyllus Lat., 'amplus', wide, large; and Gr. 'phyllon', leaf; for the large leaves. (*Euphorbia*)

amstutziae For Mrs. E. Amstutz (fl. 1956), botanist working for several years in Oroya, Peru, and discoverer of the taxon. (*Browningia*)

amudatensis For the occurrence at Amudat, Upe County, Uganda. (*Aloe*)

amydros Gr., indistinct, dim, obscure; for the undistinguished small plants. (*Monanthes polyphylla* ssp.)

Anacampseros Gr. 'anakamptein', to bring back, and Gr. 'eros', love; from the alleged aphrodisiac properties of the plant to which the name was originally applied by Plinius and Plutarch, now thought to have been a *Sedum*. Alternatively from Gr. 'anakampto', to bend down; and Gr. '-eros', capable of; and relating to the reflexed ripening fruits of some taxa. (*Portulacaceae*)

anacampseros Gr. 'anakamptein', to bring back, and Gr. 'eros', love; antique Gr. name of the plant, because of its alleged aphrodisiac properties. Alternatively from Gr. 'anakampto', to bend down; and Gr. '-eros', capable of; and relating to the procumbent stems. (*Hylotelephium*)

anacanthus Gr. 'an-', without; and Gr. 'akantha', thorn, spine; for the absence of spines on the stems. (*Opuntia*)

anagensis For the occurrence in the Anaga Mts., Tenerife, Canary Islands. (*Monanthes*)

analalavensis For the occurrence near Analalava, N Madagascar. (*Euphorbia*)

analavelonensis For the occurrence near Analavelona, Prov. Toliara, Madagascar. (*Euphorbia*)

ancashensis For the occurrence in Prov. Ancash, Peru. (*Armatocereus mataranus* ssp.)

anceps Lat., two-edged; (**1**) for the two-angled young stems. (*Acrosanthes*) – (**2**) for the two-angled stems. (*Erepsia*)

ancistrophorus Gr. 'ankistron', hook, barb; and Gr. '-phoros', carrying; for the hooked spines. (*Echinopsis*)

andegavensis For the occurrence near Angers (Lat. Andecaves), Région Anjou, France. (*Sedum*)

andersonianus For Prof. Dr. Edward ("Ted") F. Anderson (1932–2001), US-American botanist and *Cactaceae* specialist. (*Mammillaria perezdelarosae* ssp.)

andersonii As above. (*Opuntia, Turbinicarpus schmiedickeanus* ssp.)

andinus For the occurrence in the S American Andes. (*Furcraea, Melocactus, Sedum*)

andohahelensis For the occurrence in the Massif de l'Andohahela, S Madagascar. (*Aloe*)

andongensis For the occurrence at Pungo Andongo, Angola. (*Aloe*)

andreae For Wilhelm Andreae (1895–1970), German owner of a brewery, cactus specialist and horticulturist in Bensheim, honorary member of the Deutsche Kakteen-Gesellschaft. (*Gymnocalycium*)

andreaeanus For Dieter Andreae (fl. 1961), German cactus horticulturist in Otzberg-Lengfeld, Germany, son of Wilhelm Andreae. (*Eriosyce, Huernia*)

andrefandrovanus For the Andrefandrova Section of the Tsimbazaza Botanical Garden, Antananarivo, Madagascar, where the plant was growing. (*Euphorbia*)

andringitrensis For the occurrence in the Andringitra Mts., Madagascar. (*Aloe, Cynanchum*)

androsaceus For the similarity of the cushion-like growth to that of the genus *Androsace* (*Primulaceae*). (*Antimima*)

Aneilema Gr. 'a-, an-', without; and Gr. 'eilema', involucre; because the inflorescences are without conspicuous subtending bracts. (*Commelinaceae*)

anemoniflorus For the genus *Anemone* (*Ranunculaceae*); and Lat. '-florus', -flowered. (*Duvalia*)

anethifolius Lat. '-folius', -leaved; for the similarity of the leaves to those of *Anethum graveolens* ("Dill", *Apiaceae*). (*Pelargonium*)

angavokeliensis For the occurrence on Mt. Angavokely, C Madagascar. (*Cynanchum*)

angelae For Mrs. Angeles G. Lopez de Kiesling († 1985), former wife the Argentinian botanist Roberto Kiesling. (*Gymnocalycium*)

angelensis For the occurrence on Isla Angel de la Guarda, Baja California, Mexico. (*Mammillaria*)

angelesiae For Mrs. Angeles G. Lopez de Kiesling († 1985), former wife of the Argentinian botanist Roberto Kiesling. (*Echinopsis*)

angelica For Mrs. Angelique Wallace (fl. 1934), wife of a former Chief Engineer of the South African Railways. (*Aloe*)

angelicae For Angelika Rusch (fl. 1923), wife of Ernst J. Rusch and a friend of the Dinters. (*Conophytum*)

anglicus Lat., English; for the occurrence in England. (*Sedum*)

angolensis For the occurrence in Angola. (*Aloe, Chortolirion, Glossostelma, Huernia verekeri* var., *Monadenium, Raphionacme, Stapelia schinzii* var., *Synadenium, Tavaresia*)

angrae For the occurrence near Angra Pequena (Lüderitz Bay), Namibia. (*Euphorbia*)

anguinus Lat., snake-like; for the habit of the stems. (*Cleistocactus baumannii* ssp.)

angularis Lat., angled; for the angularly winged branches. (*Euphorbia*)

anguliger Lat. 'angulus', angle, corner; and Lat. '-ger, -gera, -gerum', carrying, bearing; for the shape of the stem segments. (*Epiphyllum*)

angulipes Lat. 'angulus', angle, corner; and Lat. 'pes', foot; for the angled peduncles. (*Gibbaeum*)

angustatus Lat., narrowed; for the leaf shape. (*Stathmostelma*)

angustiarum Gen. of Lat. 'angustiae', narrow place, narrow pass; for the occurrence in canyons. (*Agave*)

angustiflorus Lat. 'angustus', narrow; and Lat. '-florus', -flowered; for the narrowly elongated cyathia. (*Euphorbia*)

angustifolius Lat. 'angustus', narrow; and Lat. '-folius', -leaved. (*Acrosanthes, Bulbine, Corpuscularia, Echeveria, Fockea, Haworthia, Kalanchoe, Myrmecodia, Odontophorus, Pedilanthus tithymaloides* ssp., *Pterodiscus, Sedum, Stomatium, Tetragonia*)

angustilobus Lat. 'angustus', narrow; and Lat. 'lobus', lobe; for the corolla lobes. (*Duvalia, Echidnopsis*)

angustipetalus Lat. 'angustus', narrow; and Lat. 'petalum', petal; for the petals. (*Corpuscularia, Pelargonium*)

angustissimus Superl. of Lat. 'angustus', narrow; for the comparatively narrow leaves. (*Aspidoglossum, Yucca*)

angustus Lat., narrow; (**1**) because the leaves are narrow compared with the typical variety. (*Monadenium invenustum* var.) − (**2**) perhaps for the leaf shape. (*Hylotelephium, Rhodiola*)

anhuiensis For the occurrence in Anhui Prov., China. (*Sedum*)

anisitsii For Prof. Dániel J. Anisits (1856–1911), Hungarian pharmacist and plant and animal collector, 1883–1907 in Asunción, Paraguay, 1909–1911 working as bacteriologist in Berlin, supplied cactus specimens to K. Schumann. (*Gymnocalycium*)

anivoranoensis For the occurrence near Anivorano, NE Madagascar. (*Aloe*)

ankarafantsiensis For the occurrence near Ankarafantsika, Madagascar. (*Euphorbia viguieri* var.)

ankaranensis For the occurrence in the Falaise d'Ankarana, NW Madagascar. (*Aloe*)

ankaratrae For the occurrence at Ankaratra, Madagascar. (*Euphorbia duranii* var.)

ankarensis For the occurrence at the rocks of Ankara, NW Madagascar. (*Euphorbia*)

ankazobensis For the occurrence near Ankazobe, Prov. Antananarivo, Madagascar. (*Euphorbia*)

ankirihitrensis For the occurrence near Ankirihitra, W Madagascar. (*Cyphostemma*)

ankoberensis For the occurrence at Ankober, Ethiopia. (*Aloe*)

annae For Anna S. Schchia (fl. 1969), botanical explorer in the Caucasus region, Georgia. (*Sempervivum*)

annamariae For Mrs. Annamarie Braus Ross, wife of Erich Ross who sponsored the expedition that discovered the taxon. (*Euphorbia*)

annamicus For the occurrence in the Annam region, Vietnam. (*Kalanchoe*)

annianus For Anni Lau (fl. 1981), wife of the succulent plant collector Alfred B. Lau. (*Mammillaria*)

annularis Lat., ring-shaped, arranged in a circle; for the arrangement of the nectary glands. (*Delosperma*)

annulatus Lat., with a ring (from Lat. 'annulus', ring); (**1**) for the pattern produced by the leaf scars on the lateral branches. (*Vanzijlia*) − (**2**) application obscure. (*Hoodia pilifera* ssp.)

annuus Lat., annual; (**1**) for the often annual growth habit. (*Bulbine*) − (**2**) for the annual growth habit. (*Dorstenia, Sedum*)

anomalus Gr. 'anomales, anomalos', abnormal, uneven; (**1**) for the abnormal characteristics when compared to *Ornithogalum* (*Hyacinthaceae*). (*Drimia*) − (**2**) for the aberrant morphological features of the taxon. (*Agave, Drosanthemum*) − (**3**) for the charactistics when compared to the former genus *Stylophyllum* (*Crassulaceae*). (*Dudleya*) − (**4**) application obscure. (*Crassula atropurpurea* var., *Miraglossum*)

anoplia Gr., without armour (Gr. 'hopla', armour); for the lack of spines. (*Euphorbia*)

Anredera Unresolved. (*Basellaceae*)

ansamalensis For the occurrence at Ansamala, S Madagascar. (*Cynanchum*)

antandroi For the occurrence at Antandroi in SW Madagascar; or for the occurrence in the territory of the Antandroi tribe. (*Aloe, Senecio*)

antanimorensis For the occurrence at Antanimora, Madagascar. (*Aloe acutissima* var.)

Antegibbaeum Lat. 'ante', before; and for the

genus *Gibbaeum* (*Aizoaceae*), for which the genus was thought to be a predecessor in evolution. (*Aizoaceae*)

antemeridianus Lat. 'ante', before; and Lat. 'meridianus', pertaining to midday, noon; for the flowers opening in the morning. (*Lampranthus*)

antennifer Lat. 'antenna', antenna (insects); and Lat. '-fer, -fera, -ferum', -carrying; for the very narrow corolla lobes. (*Ceropegia*)

anteojoensis For the occurrence at the base of Mt. Anteojo in the Sierra de la Madera, Coahuila, Mexico. (*Cylindropuntia*)

anteuphorbium Gr. 'anti', against; and Gr. 'euphorbion', spurge (cf. *Euphorbia*); from the supposed value as antidote against spurge poison. (*Senecio*)

antezanae For Lucio Antezana (fl. 1953), Bolivian agronomist and director of the Forestry Division of Oruro, who assisted the Bolivian botanist M. Cárdenas on one of his trips. (*Echinopsis*)

anthonyanus For Dr. Harald E. Anthony (fl. 1950) of Englewood, New Jersey, USA, who first flowered the taxon. (*Selenicereus*)

anthonyi For A. W. Anthony (fl. 1896), US-American botanical collector. (*Dudleya*)

Anthorrhiza Gr. 'anthos', flower; and Gr. 'rhiza', root; for the root-derived spines around the flowers. (*Rubiaceae*)

antidysentericus Gr. 'anti', against; and Gr. 'dysenterikos', dysentery; for the use of the tubers as a remedy against dysentery. (*Pelargonium*)

antillanus For the occurrence in the Greater Antilles. (*Furcraea, Opuntia*)

antillarum As above. (*Agave*)

Antimima Gr. 'antimimos', imitating; for the superficial similarity of the type species to *Argyroderma necopinum* (now *A. octophyllum*). (*Aizoaceae*)

antiquorum Lat., of the ancients; because the taxon was known and used medicinally in ancient times. (*Euphorbia*)

antisyphiliticus Gr. 'anti', against; and Lat. 'syphiliticus', gonorrhoeic; for the supposed use of the latex. (*Euphorbia*)

antoinii For F. Antoine (1815–1886), Austrian horticulturist. (*Myrmecodia platytyrea* ssp.)

antoniae For Margery S. Anthony (*1924), US-American botanist. (*Cylindropuntia*)

antonii For Dr. Antony Vincent Hall (*1936), English-born botanist at the University of Cape Town, RSA. (*Lampranthus*)

antsingiensis For the occurrence at Antsingy, W Madagascar. (*Euphorbia moratii* var.)

antsingyensis As above. (*Aloe*)

antsiranensis For the occurrence in the region of Antsiranana (Diego Suarez), N Madagascar. (*Sarcostemma*)

antso From the local vernacular name of the plants in Madagascar. (*Euphorbia*)

anulatus Lat., with a ring; from Lat. 'an[n]ulus', ring; application obscure. (*Hoya*)

aoracanthus Gr. 'aor', sword; and Gr. 'akantha', thorn, spine; for the fierce spination. (*Tephrocactus*)

apachensis For the occurrence near the Apache Trail in Arizona, USA. (*Echinocereus*)

apamensis For the occurrence near San Lorenzo Apam, Hidalgo, Mexico. (*Mammillaria wiesingeri* ssp.)

Apatesia Gr. 'apatesis', deception; for the possible confusion with the genus *Hymenogyne* (*Aizoaceae*). (*Aizoaceae*)

apedicellatus Gr. 'a-', without; and Lat. 'pedicellatus', pedicellate; for the sessile flowers. (*Agave*)

apertiflorus Lat. 'apertus', open, uncovered; and Lat. '-florus', -flowered; for the widely opening flowers. (*Sedum eriocarpum* ssp.)

apertus Lat., open, uncovered; (**1**) for the widely opening flowers. (*Villadia*) – (**2**) because the filamentous staminodes do not cover the flower centre completely. (*Erepsia*) – (**3**) probably for the flower shape. (*Tromotriche*)

apetalus Gr. 'a-, an-', without; and Gr. 'petalon', petal; for the seemingly absent but in reality very inconspicuous petals. (*Dorotheanthus*)

aphyllus Lat., leafless. (*Crassula, Euphorbia*)

apicicephalius Lat. 'apex, apicis', tip; and Gr. 'kephale', head; for the cephalia formed at the stem tips. (*Cephalocereus*)

apiciflorus Lat. 'apex, apicis', tip; and Lat. '-florus', -flowered; for the flowers, which appear predominantly at the stem tips. (*Corryocactus*)

apiculatus Lat., apiculate, ending abruptly in a short point; (**1**) for the leaf tips. (*Braunsia*) – (**2**) for the petal tips. (*Villadia cucullata* ssp.)

aplocaryoides Gr. '-oides', resembling; and for the genus *Aplocarya* (now *Nolana*, *Nolanaceae*). (*Nolana*)

Apodanthera Gr. 'apodos', without a foot; and Gr. 'antheros', flowering; for the sessile flowers. (*Cucurbitaceae*)

apodanthus Gr. 'apodos', without a foot; and Gr. 'anthos', flower; for the sessile flowers. (*Mesembryanthemum*)

apoleipon Gr., missing link; because the taxon is thought to be the hitherto unknown progenitor of *Sedum sexangulare* (*Crassulaceae*). (*Sedum*)

appalachianus For the occurrence in the Appalache Mts., USA. (*Talinum*)

apparicianus For Appariício Pereira Duarte (1910–1984), Brazilian botanist at the Rio de Janeiro Botanical Garden. (*Euphorbia*)

appendiculatopsis Gr. '-opsis', similar to; and for its likeness to the non-succulent *Cynanchum appendiculatum*. (*Cynanchum*)

appendiculatus Lat., appendiculate, with small projections; (**1**) for the stipules. (*Pelargonium*) – (**2**) for the appendages on the ventral side of the carpels. (*Aeonium*)

applanatus Lat., flattened; (**1**) for the flatly spreading growth form. (*Euphorbia*) – (**2**) for the flattened rosettes. (*Agave*)

appressus Lat., appressed, pressed together; for the imbricately arranged leaves. (*Corpuscularia*)

approximatus Lat., approximate, neighbouring; for the close relationship with other taxa. (*Ruschia*)

Aptenia Gr. 'apten', unfledged, unable to fly, i.e. unwinged; for the wingless fruits of the type species. (*Aizoaceae*)

apurimacensis For the occurrence in Dept. Apurimac, Peru. (*Opuntia*)

aquaticus Lat., aquatic; for the habitat. (*Crassula*)

aquosus Lat., watery; relating to the Mexican vernacular name "Tuna de Agua" = water tuna. (*Pereskiopsis*)

arabicus For the occurrence in Arabia. (*Caralluma, Ceropegia, Raphionacme, Sarcostemma*)

arachnacanthus Gr. 'arachne', spider; and Gr. 'akanthos', spine, thorn; for the radiating radial spines. (*Echinopsis ancistrophora* ssp.)

arachnoides Gr., like a spider (Gr. 'arachne', spider); for the cobweb-like hairs. (*Anacampseros*)

arachnoideus Lat., from Gr. 'arachnoides', like a spider; (**1**) for the cobweb-like hairs at the leaf tips. (*Sempervivum*) – (**2**) for the slender delicate central spines. (*Mammillaria prolifera* ssp.) – (**3**) for the spidery spines on the leaf margins and keels. (*Haworthia*) – (**4**) for the narrow corolla lobes spreading like the legs of a spider. (*Caralluma*)

aragonii For Manuel Aragón (fl. 1901), director of the Dirección General de Estadística, Costa Rica. (*Stenocereus*)

arahaka From the vernacular name of the plants in Madagascar. (*Euphorbia*)

araliaceus For the resemblance to species of *Aralia* (*Araliaceae*). (*Steganotaenia*)

aralioides Gr. '-oides', resembling; and for the genus *Aralia* (*Araliaceae*). (*Cissus*)

aramberri For the occurrence near Aramberri, Nuevo León, Mexico. (*Mammillaria winterae* ssp.)

arancioanus For Prof. Gina Arancio (fl. 1995), Chilean botanist and curator at the herbarium at La Serena, Chile. (*Cistanthe*)

araneifer Lat. 'araneum', cobweb, or Lat. 'aranea', spider; and Lat. '-fer, -fera, -ferum', -carrying; perhaps for the indumentum of the leaves; or for the shape of the corona. (*Aspidoglossum*)

araneus Lat., spidery; for the soft dense cobweb-like leaf prickles. (*Haworthia arachnoidea* var.)

araucanus For the occurrence in the region of the Araucarias in S Argentina. (*Pterocactus*)

araysianus For the occurrence at Jebel al Arays, S Yemen. (*Orbea*)

arborescens Lat., becoming tree-like (from

Lat. 'arbor', tree). (*Aloe, Crassula, Kalanchoe, Monadenium, Othonna, Sansevieria, Stoeberia*)

arboreus Lat., tree-like (from Lat. 'arbor', tree). (*Aeonium, Euphorbia atrocarmesina* ssp., *Leptocereus*)

arboricola Lat. 'arbor', tree; and Lat. '-cola', -dwelling; for the epiphytic habit. (*Echinopsis*)

arboriformis Lat. 'arbor', tree; and Lat. '-formis', -shaped; for the growth habit. (*Mestoklema*)

arbusculoides Gr. '-oides', resembling; and for the similarity to *Tetragonia arbuscula* (*Aizoaceae*). (*Tetragonia*)

arbusculus Lat., small tree (Dim. of Lat. 'arbor', tree); for the growth form. (*Cylindropuntia, Euphorbia, Tetragonia*)

arbuthnotiae For Miss Isobel A. Arbuthnot (1870–1963), Irish immigrant to RSA in 1888 and herbarium assistant at the Bolus Herbarium, later at the Compton Herbarium. (*Lampranthus*)

arcei For Lucio Arce (fl. 1956), Bolivian agronomist and student of the Bolivian botanist M. Cárdenas. (*Opuntia*)

arceuthobioides Gr. '-oides', resembling; and for the genus *Arceuthobium* (*Viscaceae*). (*Euphorbia*)

archboldianus For R. Archbold (1907–1976), American philanthropist and zoologist. (*Myrmecodia*) – (**2**) For Mrs. Archbold (fl. 1937), a mentor of botanical expeditions. (*Hoya*)

archerae For Mrs. Jacoba (Kowie) M. N. Archer (fl. 1966), without further data. (*Lithops pseudotruncatella* ssp.)

archeri For Joseph Archer (1871–1954), Englishman, emigrated to RSA in 1890 as railway worker, became station master in Matjiesfontein, and succulent plant collector, 1921–1939 curator of Karoo Garden, Whitehill, RSA. (*Drosanthemum, Haworthia marumiana* var., *Ruschia, Tanquana*) – (**2**) For Philip G. Archer (*1922), British accountant and succulent plant enthusiast resident in Kenya 1950–1974. (*Aloe, Echidnopsis, Huernia*)

archiconoideus Gr. 'arche', beginning, past time; to avoid a homonym vs. *Maihueniopsis conoidea*, and because the taxon was discovered earlier than the contender. (*Maihueniopsis*)

arctotoides Gr. '-oides', resembling; and for the genus *Arctotis* (*Asteraceae*); for the similarly laciniate leaves. (*Othonna auriculifolia* var.)

arcuatus Lat., curved like a bow; for the stems. (*Ophionella*)

arduus Lat., steep, ascending, difficult, sudden; for the ascending stems or their marked constrictions. (*Armatocereus*)

arebaloi For Francisco Arebalo (fl. 1955), Bolivian plant collector who collected the type. (*Echinopsis*)

arenaceus Lat., sandy; for the appearance of the spines, which look as if covered by loose sand. (*Sulcorebutia*)

arenarius Lat., sand-; for the sandy habitat. (*Brachystelma, Ceropegia, Cistanthe, Cynanchum, Lampranthus, Opuntia polyacantha* var., *Plinthus, Sedum, Sempervivum globiferum* ssp., *Stapelianthus*)

arenicola Lat. 'arena', sand; and Lat. '-cola', inhabiting; (**1**) for the occurrence on a sandstone outcrop. (*Plectranthus*) – (**2**) for the preferred sandy habitat. (*Aloe, Euphorbia virosa* ssp., *Lampranthus, Parakeelya, Quaqua, Rhadamanthus*)

Arenifera Lat. 'arena', sand; and Lat. '-fer, -fera, -ferum', -carrying; for leaf surfaces covered with sand. (*Aizoaceae*)

arenosus Lat., sandy; for the preferred habitat. (*Brownanthus, Lampranthus, Stapelia*)

areolatus Lat., areolate (from Lat. 'areola', small space); for the roughened and patterned surface of the tuber. (*Anthorrhiza*)

arequipensis For the occurrence near Arequipa, Dept. Arequipa, Peru. (*Neoraimondia*)

arfakianus For the type locality in the Arfak Mts., Irian Jaya. (*Myrmephytum*)

argenteo-maculosus Lat. 'argenteus', silvery; and Lat. 'maculosus', spotted; for the white-spotted leaves. (*Haworthia pygmaea* var.)

argenteus Lat., silvery; (**1**) for the greyish tomentose young plant parts. (*Tylosema*) – (**2**) for the spination. (*Cylindropuntia imbricata* var., *Thelocactus conothelos* ssp.) – (**3**) for the silvery appearance of the densely hairy leaves. (*Antimima*) – (**4**) for the silver-grey leaf colour. (*Lampranthus*)

argenticauda Lat. 'argenteus', silvery; and Lat. 'cauda', tail; for the peduncle, which is covered with large silvery bracts. (*Aloe*)

argentinensis For the occurrence in Argentina. (*Cereus, Gymnocalycium pflanzii* ssp., *Portulaca*)

argentinus As above. (*Trianthema*)

argillosus Lat., full of clay, clayey; possibly for the leaf colour, described as "dirty green". (*Lampranthus*)

Argyroderma Gr. 'argyros', silver; and Gr. 'derma', skin; for the leaf colour. (*Aizoaceae*)

Aridaria From Lat. 'aridus', dry; for the arid habitat of the taxon. (*Aizoaceae*)

aridicola Lat. 'aridus', dry; and Lat. '-cola', -dwelling; for the habitat. (*Ceropegia, Pelargonium*)

aridimontanus Lat. 'aridus', dry; and Lat. 'montanus', -mountain; for the dry mountain where the type was collected. (*Trachyandra*)

aridus Lat., dry; for the preferred habitat. (*Bursera, Euphorbia, Mammillaria petrophila* ssp., *Quaqua*)

Ariocarpus Gr. 'aria', "Whitebeam" (*Sorbus aria*); and Gr. 'karpos', fruit; or perhaps incorrect formation from Gr. 'erion', wool, because the fruits arise from the woolly crown of the plants. (*Cactaceae*)

aristatus Lat., awned; (**1**) for the awn-like leaf tips. (*Aloe, Haworthia, Pelargonium*) – (**2**) for the awn-like petal tips. (*Erepsia, Villadia*)

aristolochioides Gr. '-oides', resembling; and for the genus *Aristolochia* ("Dutchmen's Pipe", *Aristolochiaceae*). (*Ceropegia*)

aristulatus Dim. of Lat. 'aristatus', awned; for the awn-like leaf tips. (*Antimima*)

arizonicus For the occurrence in Arizona, USA. (*Agave, Cistanthe parryi* var., *Dudleya pulverulenta* ssp.)

arkansanus For the occurrence in Arkansas, USA. (*Yucca*)

armandii For Armand Rakotozafy (fl. 1964), Madagascan teacher and discoverer of the taxon. (*Ceropegia*)

armatissimus Superl. of Lat. 'armatus', armed; for the prominent marginal teeth of the leaves. (*Aloe*)

Armatocereus Lat. 'armatus', armed; and *Cereus*, a genus of columnar cacti; for the fierce spination of some taxa. (*Cactaceae*)

armatus Lat., armed; (**1**) for the stout stiffly pointed tubercles of the stems. (*Quaqua*) – (**2**) for the stout spination. (*Echinocereus engelmannii* var., *Echinocereus reichenbachii* ssp., *Eriosyce curvispina* ssp., *Opuntia*)

armeniacus Lat., apricot-coloured, pertaining to the apricot (from Gr. 'Armenia', Armenia, which is the supposed origin of the apricot tree); for the similarity of the leaf body to a halved apricot. (*Conophytum maughanii* ssp.)

armenus For the occurrence in (Turkish) Armenia. (*Sempervivum*)

armianus From the initials of Anthony R. Mitchell (*1938), English horticulturist in RSA for many years, and a specialist of *Conophytum* (*Aizoaceae*). (*Conophytum, Eriospermum, Othonna, Portulacaria, Tylecodon sulphureus* var.)

armillatus Lat., adorned with arm bangles; for the spination that shows dark bands encircling the plant, giving it the appearance of a racoon's tail. (*Mammillaria*)

armstrongii For William Armstrong (1901–?), collector of succulent plants in Humansdorp, RSA. (*Gasteria nitida* var.)

arnostianus For Dr. Arnost Janousek (fl. 1986), Czech cactus hobbyist. (*Parodia*)

arnotii For David Arnot (1821–1894), son of an English settler in RSA, collected plants for Kew in 1858. (*Brachystelma, Stapelia, Talinum*)

arnottianus For George Arnott Walker (1799–1868), Scottish botanist. (*Ceropegia*)

aromaticus Lat. 'aromaticus', aromatic, spicy (originally Gr. 'aromatikos'); for the scent, which is unusual in the genus. (*Kalanchoe*)

arrabidae For Don Francisco Antonio de Arrábida (fl. 1821–1831), Brazilian bishop and co-editor of Vellozo's "Flora Fluminensis". (*Pilosocereus*)

Arrojadoa For Dr. Miguel Arrojado Lisboa (fl. 1920), superintendent of the "Estrada de Ferro Central do Brasil" and botanical collector. (*Cactaceae*)

arteagensis For the occurrence in the Arteaga Cañón, Coahuila / Nuevo León border region, Mexico. (*Echinocereus parkeri* ssp.)

Arthraerva Gr. 'arthron', joint; for the jointed branches and the similarity to the genus *Aerva* (*Amaranthaceae*). (*Amaranthaceae*)

Arthrocereus Gr. 'arthron', joint; and *Cereus*, a genus of columnar cacti; for the jointed columnar stems. (*Cactaceae*)

arthurolfago For Dr. Arthur Tischer (1895–2000), German Mesemb specialist and founding member of the International Organisation for Succulent Plant Study; and for Rolf Rawé (fl. 1970s, 1980s), German nurserymen and Mesemb enthusiast in RSA; and Lat. '-ago' from Lat. 'agere', to perform, achieve; in honour of two important specialists of the genus *Conophytum*. (*Conophytum lithopsoides* ssp.)

articulatus Lat., jointed; (**1**) for the rhizomes with alternating thick and slender parts. (*Pelargonium*) – (**2**) for the stems. (*Pectinaria, Psilocaulon, Senecio, Tephrocactus*)

artus Lat., close, pressed together; for the densely packed cushion-like growth form. (*Bergeranthus*)

artvinensis For the occurrence near Artvin, Turkey. (*Sempervivum*)

arubensis For the occurrence on the island of Aruba, West Indies. (*Agave*)

arvernensis For the occurrence in the Auvergne, France. (*Sempervivum tectorum* var.)

asarifolius Lat. '-folius', -leaved; and for the similarity to the leaves of *Asarum europaeum* ("Asarabacca", *Aristolochiaceae*). (*Pelargonium*)

asbestinus Lat., asbestos-; application obscure. (*Aizoon*)

ascendens Lat., ascending, climbing. (*Alluaudia*)

ascensionis For the occurrence near Ascensión, Nuevo León, Mexico. (*Mammillaria glassii* ssp.)

Asclepias For Asklepios, the ancient Gr. deity of medicine. (*Asclepiadaceae*)

aselliformis Lat. 'asella', literally: little donkey, root of German 'Assel', woodlice; and Lat. '-formis', having the form of; for the appearance of the areoles. (*Pelecyphora*)

asemus Gr. 'asemos', indistinct, indistinguished; for the few distinguishing characters. (*Haworthia monticola* var.)

ashtonii For H. Ashton (fl. 1922, 1932), who collected the type of the taxon. (*Delosperma*)

asparagoides Gr. '-oides', resembling; and for the genus *Asparagus* (*Asparagaceae*). (*Myrsiphyllum*)

Aspazoma Gr. 'aspazomai', to embrace; for the tubular stem-clasping leaf sheaths. (*Aizoaceae*)

asper Lat., rough; (**1**) for the rough stems. (*Anredera*) – (**2**) for the rough leaf surface. (*Cheiridopsis, Erepsia, Hereroa, Ruschia*) – (**3**) for the papillate corolla. (*Huernia*)

aspericaulis Lat. 'asper', rough; and Lat. '-caulis', -stemmed. (*Euphorbia*)

asperiflorus Lat. 'asper', rough; and Lat. '-florus', -flowered. (*Pectinaria articulata* ssp.)

asperifolius Lat. 'asper', rough; and Lat. '-folius', -leaved. (*Aloe*)

asperispinus Lat. 'asper', rough; and Lat. '-spinus', -spined; for the somewhat tomentose-roughened spines. (*Escobaria missouriensis* ssp.)

asperrimus Superl. of Lat. 'asper', rough; for the leaf surface. (*Agave*)

asperulus Lat., a little asperous, somewhat rough (Dim. of Lat. 'asperus', rough); (**1**) for the stem surface. (*Delosperma*) – (**2**) for the papillate leaves. (*Peperomia*)

asphodeloides Gr. '-oides', resembling; and for the genus *Asphodelus* ("Asphodel", *Asphodelaceae*). (*Bulbine*)

Aspidoglossum Gr. 'aspis, aspidos', shield; and Gr. 'glossa', tongue; for the shape of the corona segments. (*Asclepiadaceae*)

Aspidonepsis Gr. 'aspis, aspidos', shield; and Gr. 'anepsia', cousin; for the close relationship with the genus *Aspidoglossum* (*Asclepiadaceae*). (*Asclepiadaceae*)

aspillagae For Don Manuel Aspillaga (fl. 1929), Chilean owner of the Hacienda where the taxon was discovered. (*Eriosyce*)

assumptionis For the occurrence at Asunción, Paraguay. (*Opuntia*)

assurgens Lat., becoming erect; for the growth habit. (*Leptocereus*)

assyriacus Lat., Assyrian; for the occurrence there. (*Sedum*)

astephanus Gr. 'a-, an-', without; and Gr. 'stephanos', garland, wreath; application obscure. (*Dischidia*)

asterias Lat. 'asterias', starfish; for the body appearance. (*Astrophytum*)

asterias From Gr. 'aster', star; for the flower shape. (*Stapelia*)

asthenacanthus Gr. 'asthenes', weak, feeble; and Gr. 'akantha', thorn, spine; for the minute spines. (*Euphorbia*)

Astridia For Astrid Schwantes (fl. 1926), wife of the German Mesemb specialist Prof. G. Schwantes. (*Aizoaceae*)

astrispinus Gr. 'aster, astros', star; and Lat. '-spinus', -spined; for the star-like branching of the spinescent peduncles. (*Euphorbia stellispina* var.)

Astroloba Gr. 'aster, astros', star; and Gr. 'lobos', lobe; for the stellately spreading perianth lobes. (*Aloaceae*)

astrophorus Gr. 'aster, astros', star; and Gr. '-phoros', carrying; for the star-like radiating sterile peduncles. (*Euphorbia*)

Astrophytum Gr. 'aster, astros', star; and Gr. 'phyton', plant; for the body shape. (*Cactaceae*)

astyanactis For Astyanax, son of Mr. Allah Kouamé, engineer at the Eaux et Forêts of Ivory Coast, who accompanied the expedition on which the taxon was discovered. (*Dorstenia*)

asuntapatensis For the occurrence in the mountain range of Asunta Pata, Prov. J. Bautista Saavedra, Dept. La Paz, Bolivia. (*Lepismium*)

atacamensis For the occurrence in the region of the Chilean Atacama Desert. (*Copiapoa*) – (**2**) For the occurrence at higher altitudes of the Chilean Atacama region. (*Echinopsis, Maihueniopsis*)

atherstonei For Dr. William G. Atherstone (1814–1898), English medical practitioner and naturalist settling in RSA with his parents in 1820. (*Ipomoea*)

atlanticus Lat. 'Atlanticus', from the Atlas Mountains. (*Sempervivum*)

atoto Perhaps from the local vernacular name of the plants on the islands of the Pacific. (*Euphorbia*)

atra Lat. 'ater, atra, atrum', black; for the colour of the old leaves. (*Vlokia*)

atratus Lat., blackened, dark; (**1**) for the black stems. (*Ruschia*) – (**2**) for the dark leaf coloration upon flowering. (*Sedum*)

atrichocarpus Gr. 'a-', without; Gr. 'trichos', hair; and Gr. 'karpos', fruit; for the glabrous fruits. (*Jatropha schweinfurthii* ssp.)

atrispinus Lat. 'ater, atra, atrum', black; and Lat. '-spinus', -spined. (*Euphorbia, Opuntia*)

atrocarmesinus Lat. 'ater, atra, atrum', black; and Lat. 'carmesinus', crimson; for the inflorescence colour. (*Euphorbia*)

atroflorus Lat. 'ater, atra, atrum', black; and Lat. '-florus', -flowered; for the dark cyathia. (*Euphorbia*)

atrofuscus Lat. 'ater, atra, atrum', black; and Lat. 'fuscus', brown; for the leaf colour. (*Haworthia magnifica* var.)

atropatanus Lat. 'Atropatene', Azerbaijan in NW Iran; for the occurrence there. (*Sempervivum*)

atropes Lat. 'ater, atra, atrum', black; and Lat. 'pes', foot; perhaps for the spine colouration. (*Opuntia*)

atropilosus Lat. 'ater, atra, atrum', black; and Lat. 'pilosus', hairy; for the dark hairs covering the receptacle. (*Selenicereus*)

atropurpureus Lat. 'ater, atra, atrum', black; and Lat. 'purpureus', purple; (**1**) for the leaf colour. (*Echeveria*) – (**2**) for the leaf colour under dry conditions. (*Crassula*) – (**3**) for the colour of some flower parts. (*Schizoglossum*) – (**4**) for the inner face of the corolla lobes. (*Tylecodon*) – (**5**) for the colour of the nectar glands. (*Euphorbia*)

atrorubens Lat. 'ater, atra, atrum', black; and Lat. 'rubens', red; for the colour of the corona segments. (*Schizoglossum bidens* ssp.)

atrosanguineus Lat. 'ater, atra, atrum', black; and Lat. 'sanguineus', blood-red; for the flower colour. (*Huerniopsis*)

atrovirens Lat. 'ater, atra, atrum', black; and Lat. 'virens', green; for the dark green leaves. (*Agave*)

atroviridis Lat. 'ater, atra, atrum', black; and Lat. 'viridis', green; for the colour of the plant bodies. (*Eriosyce crispa* ssp.)

atrox Lat., atrocious; for the vicious spines. (*Euphorbia*)

atsaensis For the occurrence at Atsa Pass, Tibet. (*Rhodiola*)

attastoma Gr. 'stoma', mouth; and for the ant genus *Atta*; for the horned cyathial glands, which resemble the mouth-parts of the ants. (*Euphorbia*)

attenuatus Lat., drawn out, attenuate, tapered, weakened; (**1**) for the slender stems. (*Brachystelma, Ceropegia*) – (**2**) for the stems that become more slender apically. (*Caralluma adscendens* var.) – (**3**) for the leaf shape. (*Agave, Dudleya, Haworthia, Haworthia reticulata* var., *Juttadinteria, Monsonia, Sedum jurgensenii* ssp.) – (**4**) for the sepals. (*Pelargonium*) – (**5**) application obscure. (*Drosanthemum*)

attonsus Lat. 'at-', towards; and Lat. 'tonsus', shaven, having become glabrous; for the lack of the typical tuft of hairs or bristles at the leaf tip. (*Trichodiadema*)

atuntsuensis For the occurrence at A-tun-tsu, Yunnan, China. (*Rhodiola*)

auberi For Pedro A. Auber (fl. before 1825, 1843), French-born Spanish director of the Botanical Garden La Habana, Cuba. (*Opuntia*)

aubrevillei For Prof. André Aubréville (1897–1982), eminent French botanist, Africa specialist and long-time director of the Laboratoire de Phanérogamie at the Muséum d'Histoire Naturelle in Paris. (*Kalanchoe*)

aucampiae For Miss Juanita Aucamp (fl. 1929), who collected plants on her farther's farm near Postmasburg, RSA. (*Lithops*)

augustinus For the occurrence near St. Augustin near Toliara (Tuléar), Madagascar. (*Aloe descoingsii* ssp.)

aurantiacus Lat., orange; for the flower colour. (*Aloe haworthioides* var., *Disocactus, Fenestraria rhopalophylla* ssp., *Matucana, Opuntia, Portulaca, Pterodiscus, Thelocactus conothelos* ssp.)

aurasensis For the occurrence in the Auras Mts., Lüderitz-Süd, Namibia. (*Antimima*)

auratus Lat., golden; for the predominant spine colour. (*Eriosyce*)

aureicentrus Lat. 'aureus', yellow; and Lat. 'centrum', centre; for the colour of the central spines. (*Parodia*)

aureiceps Lat. 'aureus', yellow; and Lat. '-ceps', -headed; for the spination. (*Mammillaria rhodantha* ssp.)

aureiflorus Lat. 'aureus', yellow; and Lat. '-florus', -flowered. (*Matucana, Pereskia, Rebutia*)

aureilanatus Lat. 'aureus', yellow; and Lat. '-lanatus', -woolly; for the radial spines. (*Mammillaria*)

aureispinus Lat. 'aureus', yellow; and Lat. '-spinus', -spined. (*Haageocereus pseudomelanostele* ssp., *Mammillaria rekoi* ssp., *Opuntia, Pilosocereus*)

aureopurpureus Lat. 'aureus', yellow; and Lat. 'purpureus', purple; for the petal coloration. (*Drosanthemum*)

aureospinus Lat. 'aureus', yellow; and Lat. '-spinus', -spined. (*Myrmecodia*)

aureoviridiflorus Lat. 'aureus', yellow; Lat. 'viridis', green; and Lat. '-florus', -flowered. (*Euphorbia*)

aureus Lat., yellow, golden yellow; (**1**) for the spine and flower colour. (*Coleocephalocereus*) – (**2**) for the flower colour. (*Aeonium, Agave, Corryocactus, Echinopsis, Lampranthus, Opuntia, Quaqua, Trichodiadema*)

auriazureus Lat. 'aureus', yellow; and Lat. 'azureus', azure, deep blue; for the blue epidermis and the golden yellow spination. (*Micranthocereus*)

auriculatus Lat., with small ears, auriculate; for the leaf shape. (*Kalanchoe nyikae* ssp.)

auriculifolius Lat. 'auricula', small ear (also the plant *Primula auricula*); and Lat. '-folius', -leaved; with leaves like an Auricula, or auriculate leaves. (*Othonna*)

auriflorus Lat. 'aureus', yellow; and Lat. '-florus', flowered. (*Conophytum*)

aurilanatus Lat. 'aureus', yellow; and Lat. 'lanatus', woolly; for the yellow cephalium wool. (*Pilosocereus aurisetus* ssp.)

aurisetus Lat. 'aureus', yellow; and Lat. '-setus', bristly; for the spination of the stems. (*Pilosocereus*)

auritus Lat., eared, auriculate; for the sometimes auriculate leaves. (*Pelargonium*)

aurusbergensis For the occurrence on the Aurusberg in the Sperrgebiet of Namibia. (*Crassula*, *Tylecodon*)

ausensis For the occurrence near Aus, Namibia. (*Crassula*, *Juttadinteria*)

australianus For the occurrence in Australia. (*Grahamia*)

australis Lat., southern; (**1**) for the occurrence in Australia. (*Brachychiton*, *Hoya*, *Marsdenia*, *Sarcostemma viminale* ssp.) – (**2**) for the occurrence in New Zealand. (*Cordyline*, *Disphyma*) – (**3**) for the occurrence in S Peru. (*Haageocereus*) – (**4**) for the relatively southern distribution in relation to the bulk of the genus. (*Brachystelma*, *Echeveria*, *Pterocactus*, *Sedum*) – (**5**) for the relatively southern distribution in relation to the other subspecies. (*Dudleya attenuata* ssp., *Sedum erythrospermum* ssp., *Zaleya galericulata* ssp.)

austricola Lat. 'auster', south, southern; and Lat. '-cola', inhabiting; for the occurrence in the very S of RSA. (*Drosanthemum*, *Lampranthus*)

austrinus Lat. 'austrinus', southern; for the occurrence in the SE USA. (*Opuntia*)

austroarabicus Lat. 'auster', south; and Lat. 'arabicus', Arabian; for the occurrence in S Saudi Arabia. (*Aloe*)

Austrocactus Lat. 'auster', south; and Lat. 'cactus', cactus; for the occurrence in S South America. (*Cactaceae*)

Austrocylindropuntia Lat. 'auster', south; Lat. 'cylindrus', cylinder; and for the genus *Opuntia*; for the cylindrical stem segments and the occurrence in South America. (*Cactaceae*)

autumnalis Lat., autumn-; for the flowering time. (*Drosanthemum*)

avasmontanus For the occurrence in the Auas Mts. (Lat. 'montanus', mountain-), Namibia. (*Euphorbia*)

avellanidens Lat. 'avellaneus', hazelnut-brown; and Lat. 'dens', tooth; for the colour of the leaf margin teeth. (*Agave*)

avia Lat., desolate place, wilderness; for the habitat. (*Yucca angustissima* var.)

Avonia Perhaps from Lat. 'avus', grandfather; for the white 'old'-looking stipular scales. (*Portulacaceae*)

awashensis For the occurrence in the Awash National Park, Ethiopia. (*Euphorbia*)

awdelianus For the occurrence at Bilad Awdeli, S Yemen. (*Caralluma*)

axthelmianus For Mr. Axthelm (fl. 1923), without further data. (*Ruschia*)

ayopayanus For the occurrence in Prov. Ayopaya, Dept. Cochabamba, Bolivia. (*Corryocactus*, *Parodia*)

ayresii For Thomas Ayres (1828–1913), British agriculturist and naturalist, living in RSA for most of his adult life. (*Sesuvium*)

aytacianus For Dr. Zeki Aytaç (fl. 1994), Turkish botanist in Ankara. (*Sedum*)

aztatlensis For the occurrence near San Miguel Aztatla, Oaxaca, Mexico. (*Echeveria longissima* var.)

Aztekium For the similarity of the body structure of this Mexican genus to Aztec sculptures. (*Cactaceae*)

azulensis For the occurrence at Pedra Azul, Minas Gerais, Brazil. (*Pilosocereus*)

azureus Lat., azure, deep blue; for the body colour. (*Melocactus*, *Opuntia*)

B

babatiensis For the presumed occurrence at Babati, Northern Prov., Tanzania. (*Aloe*)

babiloniae For the occurrence on the Babilonstoringsberge, Western Cape, RSA, where the type and only specimen was found. (*Erepsia*)

baccatus Lat., baccate, berry-like; for the fleshy fruits. (*Yucca*)

baccifer Lat. 'bacca', berry; and Lat. '-fer, -fera, -ferum', carrying; for the fruits. (*Rhipsalis*)

bachelorum MLat. 'baccalaureus', bachelor, apprentice, unmarried, single; for the unbranched growth. (*Conophytum*)

backebergianus For Curt Backeberg (1894–1966), German export merchant, later horticulturist and *Cactaceae* specialist. (*Mammillaria*)

backebergii As above. (*Echeveria chiclensis* var., *Echinopsis*)

badius Lat., reddish-brown; for the leaf colour. (*Haworthia mirabilis* var.)

badspoortensis For the occurrence at Badspoort near Calitzdorp, Western Cape, RSA. (*Crassula*)

Baeriopsis Gr. '-opsis', similar to; and for the genus *Baeria* (*Asteraceae*). (*Asteraceae*)

baeseckei For the German botanist and collector (Paul ?) Baesecke (fl. 1913, Namibia). (*Anacampseros*)

baeticus Lat., from Baetia, which was a region of the Roman Empire in S Spain; for the occurrence there. (*Sedum hirsutum* ssp.)

baga From the local vernacular name of the plants in W Africa. (*Euphorbia*)

bagamoyensis For the occurrence in the Bagamoyo Distr., Tanzania. (*Sansevieria*)

bagshawei For Dr. Arthur W. G. Bagshawe (1871–1950), British botanist, attached to the Anglo-German Kenya-Uganda Boundary Commission in 1904–1905. (*Stathmostelma welwitschii* var.)

bahamanus For the occurrence on the Bahamas. (*Agave*)

bahamensis As above. (*Pedilanthus tithymaloides* ssp.)

bahiensis For the occurrence in the State of Bahia, Brazil. (*Cissus, Discocactus, Leocereus, Melocactus, Pereskia, Pierrebraunia*)

baileyi For Liberty H. Bailey (1858–1954), US-American horticultural botanist, mainly at Cornell University where he founded the Bailey Hortorium. (*Yucca*) – (**2**) For Mr. Vernon Bailey (fl. 1906). (*Echinocereus reichenbachii* ssp.)

bainesii For J. Thomas Baines (1820–1875), English artist and explorer, esp. in S Africa. (*Corallocarpus, Cyphostemma*)

baioensis For the occurrence on Baio Mt., NE Kenya. (*Euphorbia*)

bakeri For John G. Baker (1834–1920), British botanist at Kew. (*Aloe*) – (**2**) For William Baker (fl. 1993), Californian nurseryman. (*Echeveria*) – (**3**) For Dr. Marc A. Baker (fl. 1989), US-American botanist at the Arizona State University. (*Opuntia*)

balansanus For Benedict Balansa (1825–1891), French commercial plant collector in Paraguay, Uruguay and Oceania, and a friend of A. Bonpland. (*Peperomia*)

baldianus For J. Baldi (fl. 1905), an acquaintance and sponsor of Carlos Spegazzini. (*Echinopsis, Gymnocalycium*)

baldratii For Isaia Baldrati (fl. 1930), Italian collector in Eritrea. (*Orbea*)

baleensis For the occurrence in the Bale Region, Ethiopia. (*Euphorbia, Sedum*)

balfourii For Sir Isaac Bayley Balfour (1853–1922), Scottish botanist in Edinburgh. (*Sedum*)

baliolus Lat., streaked, blotched; application obscure. (*Euphorbia*)

ballii For John S. Ball (fl. 1964), Zimbabwean forestry officer. (*Aloe*)

ballsii For Edward K. Balls (1892–1984), English botanist collecting in Asia Minor, Greece and S America. (*Echeveria, Sempervivum marmoreum* ssp.)

ballyanus For Dr. Peter R. O. Bally (1895–1980), Swiss botanist at the Coryndon Museum, Nairobi, widely travelling in E Africa, and resident in Kenya from the 1930s. (*Ceropegia, Euphorbia*)

ballyi As above. (*Adenia, Aloe, Echidnopsis, Euphorbia, Kalanchoe*)

balonensis For the occurrence at the Balonne River, New South Wales, Australia. (*Parakeelya*)

balsameus Lat., balsamic; for the scent of the latex observed by the discoverer. (*Euphorbia gariepina* ssp.)

balsamifer Lat. 'balsamum', balm, resin; and Lat. '-fer, -fera, -ferum', -carrying; (**1**) for the resinous scent of the plant. (*Aeonium*) – (**2**) perhaps for the reportedly non-toxic latex. (*Euphorbia*)

balsamifluus Lat. 'balsamum', balm, resin; and Lat. 'fluere', to flow; for the viscid nature of the plants. (*Nolana*)

balsasensis For the occurrence near Balsas, valley of the Río Marañón, Prov. Amazonas, Peru. (*Armatocereus rauhii* ssp.)

bambusiphilus Gr. 'philos', friend; for the occurrence in a bamboo-forest. (*Mammillaria xaltianguensis* ssp.)

banae For R. Bana (fl. 1993), officer of the Service des Eaux et Forêts, Madagascar. (*Euphorbia*)

baradii For Dr. Gerald (Jerry) S. Barad (*1923), US-American gynaecologist and succulent plant enthusiast, specialist on the pollination of asclepiads, 1990–1993 president of the Cactus and Succulent Society of America. (*Caralluma, Conophytum klinghardtense* ssp., *Euphorbia*)

barbadensis For the occurrence on Barbados (Lesser Antilles). (*Agave*)

barbatus Lat., bearded; (**1**) for the dense hair cover of the plants. (*Plectranthus*) – (**2**) for the bearded leaf margins. (*Crassula*) – (**3**) for the bristles on the leaf tip. (*Trichodiadema*) – (**4**) for the papillae around the fissure between the fused leaves of a pair. (*Conophytum obscurum* ssp.) – (**5**) for the fringed (ciliate) perianth segments. (*Mammillaria*) – (**6**) for the papillate corolla. (*Huernia*)

barberae For [Mrs. F. W.] Mary E. Barber (née Bowker) (1818–1899), English writer, painter and naturalist in RSA, whose parents emigrated to RSA in 1820. (*Aloe, Brachystelma, Tetradenia*)

barbertonicus For the occurrence at Barberton, Mpumalanga (formerly Transvaal), RSA. (*Senecio*)

barbeyi For William Barbey (1842–1914), Swiss philanthropist and botanist in Geneva. (*Cotyledon, Sedum*)

barbicollis Lat. 'barba', beard; and Lat. 'collum', collar; for the tuft of hairs at the base (neck) of the nectary glands. (*Euphorbia*)

barbulatus Lat., with a little beard; for the somewhat hairy rosettes. (*Sempervivum*)

bargalensis For the occurrence at Bargal, Somalia. (*Aloe*)

bariensis For the occurrence in the Bari region, Somalia. (*Euphorbia*)

baringoensis For the occurrence in the Baringo Distr., Kenya. (*Euphorbia heterospina* ssp.)

barkerae For Miss Winsome F. Barker (1907–1994), South African botanist for the National Botanic Gardens, 1957–1972 Curator of the Compton Herbarium, Kirstenbosch. (*Drosanthemum*)

barklyi For Sir Henry Barkly (1815–1898), English Governor in RSA from 1870–1877, and keen naturalist. (*Ceropegia africana* ssp., *Crassula, Mesembryanthemum,* ×*Orbelia, Pelargonium, Tavaresia*)

barnardii For Dr. Keppel Harcourt Barnard (1887–1964), British marine biologist, lived in RSA from 1911. (*Ruschia*) – (**2**) For W. G. Barnard (fl. 1934–1939), Stock Inspector in RSA and plant collector, esp. in Sekukuniland, made useful notes on native lore and plant uses. (*Euphorbia*)

barnhartii For Dr. John H. Barnhart (1871–1949), US-American botanist and bibliographer at the New York Botanical Garden. (*Euphorbia*)

baronii For Rev. Richard Baron (1847–1907), British botanist and missionary in Madagascar 1872–1907. (*Pachypodium, Senecio*)

barorum Gen. Pl., perhaps for Rev. Richard Baron (1847–1907) and family, British botanist and missionary in Madagascar 1872–1907. (*Senecio*)

barrancensis For the occurrence in a deep barranca (Span., narrow valley or gorge). (*Agave inaequidens* ssp.)

barringtonensis For the occurrence on Barrington Island (= Isla Santa Fé) in the Galápagos. (*Opuntia echios* var.)

barrydalensis For the occurrence near Barrydale, Western Cape, RSA. (*Piaranthus*)

barthelowanus For Capt. Benjamin Barthelow (fl. 1911), in whose company Dr. J. N. Rose cruised the waters of Baja California on U.S. Steamer "Albatross". (*Echinocereus*)

bartramii For Edwin B. Bartram (1878–1964), US-American bryologist and plant collector. (*Graptopetalum*)

basalticus For the occurrence on basaltic lava. (*Crassula*)

Basella From the local Indian Malayalam name of the plant. (*Basellaceae*)

basiclavicaulis Lat. (originally Gr.) 'basis', base, bottom; Lat. 'clava', club, cudgel; and Lat. '-caulis', -stemmed; for the basally thickened stems. (*Dioscorea*)

basilaris Lat., basal; because new stem segments are produced predominantly from the base of the plant. (*Opuntia*)

basuticus For the occurrence in Basutoland (now Lesotho). (*Delosperma*)

batallae For María A. Batalla Zepeda (fl. 1973), Mexican botanist. (*Sedum*)

batesianus For John T. Bates (1884–1966), British trolley bus conductor in London and succulent plant enthusiast, who collaborated with N. E. Brown and others.. (*Gasteria*, *Haworthia marumiana* var.)

batesii Probably for Henry W. Bates (1825–1892), British entomologist, botanist and natural history explorer. (*Sedum*)

battandieri For Prof. Jules A. Battandier (1848–1922), French botanist and explorer of Algeria, professor at the medical school of Alger. (*Sedum*)

batteniae For Mrs. Auriol Batten (née Taylor, *1918), well-known South African botanical artist and teacher. (*Albuca*)

baueri For Dr. Ralf Bauer (*1968), German dentist and specialist on epiphytic cacti. (*Epiphyllum*)

baumannii For Mr. Nap. Baumann (fl. 1844), horticulturist at Mulhouse and Bolwillers (France). (*Cleistocactus*) – (**2**) For a Mr. Baumann (fl. 1969), without further data. (*Espostoa*)

baumii For Hugo Baum (1867–1950), German horticulturist and botanist in Rostock, collecting on the expedition led by P. van der Keller 1899–1900 into the interior of Huila Prov., Angola, and travelling 1925 in Mexico. (*Erythrina*, *Jatropha*, *Mammillaria*)

baviaanus For the occurrence in the Baviaanskloof, RSA. (*Huernia brevirostris* ssp.)

baxaniensis For the occurrence at a place "Baxan" [Baján ?] in Mexico. (*Acanthocereus*)

baxteri For Mr. W. D. Baxter (fl. 1914), plant collector in RSA. (*Machairophyllum*)

baxterianus For Edgar Baxter (1903–1967), US-American sociologist and enthusiastic student and collector of cacti. (*Mammillaria petrophila* ssp.)

Bayerara For M. Bruce Bayer (*1935), South African agricultural entomologist, succulent plant enthusiast and gardener, and former curator of the Karoo National Botanic Gardens, Worcester, RSA, specialist on *Haworthia*; and suffix '-ara', indicating plurigeneric hybrids. (*Aloaceae*)

bayeri For M. Bruce Bayer (*1935), South African agricultural entomologist, plant enthusiast and gardener, and former curator of the Karoo National Botanic Gardens, Worcester, RSA, and specialist on *Haworthia*. (*Euphorbia*, *Gasteria brachyphylla* var., *Haworthia*, *Huernia*, *Tylecodon*)

bayerianus As above. (*Quaqua*) – (**2**) For Waldemar F. Bayer (1903–1985), South African magistrate who discovered the taxon, and his son M. Bruce Bayer, co-discoverers, the latter a well-known South African agricultural entomologist and plant enthusiast and gardener, who pursued it. (*Anacampseros*)

baylissianus For Lt. Colonel Roy D. Bayliss (1909–1994) and Mrs. R. Bayliss, English plant collectors in RSA. (*Gasteria*)

baylissii For Roy D. Bayliss (1909–1994), English motor specialist, became Lt. Colonel during World War II service, emigrating to Africa 1947, living variously in Kenya, Zambia and RSA, from 1973 collector for the Botanical Research Institute, Pretoria, RSA. (*Euphorbia*, *Haworthia angustifolia* var., *Lampranthus*, *Stapelia*, *Tromotriche*)

Baynesia For the occurrence in the Baynes Mts., NW Namibia, which in turn are named for Maudsley Baynes, English explorer who first investigated the area in 1911. (*Asclepiadaceae*)

bayrianus For Alfred Bayr (1905–1970), Austrian bank director in Linz, cactus hobbyist and former president of the Austrian Cactus Society GÖK. (*Gymnocalycium*)

Beaucarnea For Monsieur Beaucarne (fl. 1861), Belgian succulent plant grower and notary from Eename near Audenarde, who first collected flowers of *Beaucarnea recurvata*. (*Nolinaceae*)

beaufortensis For the occurrence near Beaufort, Beaufort West Distr., Western Cape, RSA. (*Ruschia, Stomatium*)

beauverdii For Gustave Beauverd (1867–1942), Swiss botanist and curator of the Boissier herbarium, Geneva. (*Kalanchoe, Sedum*)

beccarii For Dr. Odoardo Beccari (1843–1920), Italian botanist and ant-plant specialist. (*Myrmecodia, Myrmephytum*)

bedinghausii For H. J. Bedinghaus (fl. 1860s), Belgian horticulturist in whose garden the taxon was discovered. (*Furcraea*)

beetzii For Dr. Beetz (fl. 1922), geologist for the Diamond Company in present-day Namibia. (*Stoeberia*)

begardii For M. Begard (fl. 1984), chief gardener at Tsimbazaza Botanical Garden, Madagascar. (*Euphorbia primulifolia* var.)

Begonia For Michel Begon (1638–1710), French Governor of Santo Domingo and promotor of botany. (*Begoniaceae*)

beguinii For Abbé Beguin (fl. 1896), of Brignoles, France, who produced many fine *Aloe*- and *Gasteria*-hybrids, which were sold by the Haage & Schmidt nursery in Germany. (×*Gasteraloe, Turbinicarpus*)

beharensis For the occurrence at Behara, Madagascar. (*Euphorbia, Kalanchoe*)

beillei For Prof. Dr. Lucien Beille (1862–1946), French pharmacist, physician and botanist, professor at the University of Bordeaux and director of the botanical garden. (*Euphorbia*)

Beiselia For Karl-Werner Beisel (fl. 1979), German horticultural grower of succulents. (*Burseraceae*)

beiselii As above. (*Mammillaria karwinskiana* ssp.)

bekinolensis For the occurrence on Mt. Bekinoly, C Madagascar. (*Cynanchum gerrardii* ssp.)

belavenokensis For the occurrence near Belavenoka, Taolanaro Prov., Madagascar. (*Aloe*)

bellatulus Diminutive of Lat. 'bellus', beautiful. (*Aloe*)

bellavistensis For the occurrence near Bellavista, Dept. Amazonas, Peru. (*Melocactus*)

bellidiflorus Lat. '-florus', -flowered; and for *Bellis perennis* ("Daisy"; *Asteraceae*); for the superficially similar flowers. (*Acrodon*)

bellidiformis Lat. '-formis', -shaped; and for *Bellis perennis* ("Daisy"; *Asteraceae*); for the superficially similar flowers. (*Dorotheanthus*)

bellus Lat., pleasant, beautiful. (*Agave toumeyana* ssp., *Aloe, Drosanthemum, Echeveria, Graptopetalum, Hoya lanceolata* ssp., *Lithops karasmontana* ssp., *Mammillaria nunezii* ssp., *Opuntia, Sansevieria, Sedum*)

bemarahaensis For the occurrence in the Bemaraha region, W Madagascar. (*Euphorbia*)

bemarahensis As above. (*Cissus microdonta* fa., *Delonix leucantha* ssp., *Euphorbia moratii* var.)

beneckei For Stephan (Etienne) Benecke (1808–1879), German merchant emigrating to Mexico, founded 1875 the Camara Nacional de Comercio. (*Mammillaria, Stenocereus*)

bengalensis For the occurrence in the Bengal region (NE India, Bangladesh). (*Dischidia*)

benguellensis For the occurrence in Benguel[l]a Prov., Angola. (*Dorstenia, Sesamothamnus*)

benguetensis For the occurrence in Benguet Prov., Luzon, Philippines. (*Hoya*)

bensonii For Prof. Dr. Lyman Benson (1909–1993), US-American botanist at Pomona College, California, and student of cacti. (*Opuntia*)

bentii For Theodore Bent (1851 or 1852–1897), British archaeologist and explorer, especially in Africa and Arabia, died from malaria caught during an expedition to Africa. (*Echidnopsis*, *Kalanchoe*)

berchtii For Dr. C. A. Ludwig Bercht (*1945), Dutch cactus hobbyist and *Gymnocalycium* specialist. (*Gymnocalycium*)

berevoanus For the occurrence near Berevo, W Madagascar. (*Aloe*)

Bergeranthus For Alwin Berger (1871–1931), German botanist and succulent plant specialist and long-time curator of the Hanbury Garden at La Mortola, Italy; and Gr. 'anthos', flower. (*Aizoaceae*)

bergeri For Alwin Berger (1871–1931), German botanist and succulent plant specialist and long-time curator of the Hanbury Garden at La Mortola, Italy. (*Euphorbia*, *Kalanchoe*, *Sedum*)

bergerianus As above. (*Stapelia schinzii* var.)

Bergerocactus As above. (*Cactaceae*)

berghiae For Mrs. J. Bergh (fl. 1937), without further data. (*Lampranthus*)

bergianus For Prof. Dr. Cornelis C. Berg (*1934), Norwegian botanist. (*Dorstenia*)

bergii For Peter Jonas Bergius (1730–1790), Swedish physician and botanist. (*Euphorbia*)

bergioides Gr. '-oides', similar to; and for the genus *Bergia* (*Elatinaceae*). (*Crassula*)

berillonianus For Dr. Edgar Berilloni (fl. 1913); without further data. (*Sedum*)

berlandieri For Jean L. Berlandier (1805–1851), botanical traveller of Belgian origin in Mexico, later working as physician without ever having studied this profession. (*Echinocereus*)

bernadetteae For Mrs. Bernadette Castillon (fl. 2000); expert cultivator of Madagascan succulents. (*Aloe*)

bernalensis For the occurrence on Cerro Bernal, Tamaulipas, Mexico. (*Graptopetalum paraguayense* ssp.)

berorohae For the occurrence near the village of Beroroha, S Madagascar. (*Euphorbia*)

beroticus For the occurrence near the Rio Bero, Angola. (*Euphorbia*)

bertemariae For Berte Marie Ulvester (fl. 2000), wife of Dr. Maurizio Dioli, Italian veterinary officer in Ethiopia. (*Aloe*)

berthelotii For Sabin Berthelot (1794–1880), French Consul in Santa Cruz de Tenerife and explorer of the flora of the Canary Islands. (*Euphorbia*)

bertinii For Pierre Bertin (1800–1891), plant collector in Patagonia, without further data. (*Austrocactus*)

bertramianus For Paul Bertram (fl. 1920), German clergyman and cactus hobbyist in Erfurt, Germany. (*Echinopsis*)

Beschorneria For Friedrich W. C. Beschorner (1806–1873), German physician and botanist, director of the Institute of Public Assistance and the Lunatic Asylum at Owinsk, Poland. (*Agavaceae*)

beswickii For Mr. Beswick (fl. 1922), successful grower of succulent plants in Queenstown, Eastern Cape, RSA. (*Khadia*)

bethencourtianus Probably for Jean (Juan) de Béthencourt, Norman knight who in 1402 discovered the Canary Islands. (*Aichryson*)

betiformis Lat. 'beta', beetroot, mangold; and Lat. '-formis', having the form of; for the caudex. (*Cyphostemma*)

betsileensis For the occurrence at Betsileo, Toliara Prov., Madagascar. (*Aloe*)

bettinae For Bettina Hoover (fl. 1965), wife of the US-American botanist Robert F. Hoover. (*Dudleya abramsii* ssp.)

beukmanii For Mr. C. Beukman (fl. 1935, 1940), schoolteacher and amateur botanist in Bonnievale, Western Cape, RSA, field collector and grower of succulents. (*Haworthia mirabilis* var.)

bevilaniensis For the occurrence at Bevilany, Madagascar. (*Euphorbia milii* var.)

bhidei For R. K. Bhide (fl. 1902, 1911), Indian botanist and keeper of the herbarium at the College of Science, Poona, India. (*Kalanchoe*)

bhupinderianus For Mrs. Bhupinder Kour Sarkaria († before 2002), wife of the Indian medical doctor and succulent plant enthusiast J. Sarkaria. (*Caralluma*)

biaculeatus Lat. 'bi-', two; and Lat. 'aculeatus', prickly, spiny; for the paired stipular spines. (*Euphorbia*)

bianoensis For the occurrence on the Biano Plateau, Shaba Prov., Zaïre. (*Monadenium*)

bicarinatus Lat. 'bi-', two; and Lat. 'carinatus', keeled; (**1**) for the pair of keeled leaves. (*Conophytum*) – (**2**) for the occasionally doubly keeled leaves. (×*Astroworthia*) – (**3**) for the 2-keeled calyx. (*Sarcozona*)

bicolor Lat., two-coloured; (**1**) for the spination. (*Cereus*) – (**2**) for the yellow or pink flowers. (*Portulaca*) – (**3**) for the two-coloured flowers. (*Adromischus, Agave, Drosanthemum, Echeveria, Gasteria, Lampranthus, Thelocactus*) – (**4**) for the mixture of red buds and greenish-white flowers on the inflorescences. (*Aloe marlothii* var.) – (**5**) for the yellow flowers with a white ring around the throat. (*Pachypodium rosulatum* fa.)

bicomitum Lat. 'bi-', two; and Lat. 'comitor', accompany; commemorating G. W. Reynolds and N. R. Smuths who travelled together in search of plants. (*Aloe*)

bicornis Lat. 'bi-', two; and Lat. 'cornu', horn; most probably for the calyx with two lobes much longer than the others. (*Psilocaulon*)

bidens Lat. 'bi-', two-; and Lat. 'dens, dentis', tooth; for the segments of the corona. (*Schizoglossum*)

bieblii For Wolfgang Biebl (fl. 1995), German cactus collector. (*Pygmaeocereus*)

bifidus Lat. 'bi-', two-; and Lat. '-fidus', -divided; for the often branched inflorescences. (*Echeveria*)

biflorus Lat. 'bi-', two-; and Lat. '-florus', -flowered. (*Aspidoglossum*)

bifolius Lat. 'bi-', two-; and Lat. '-folius', -leaved. (*Whiteheadia*)

biformis Lat. 'bi-', two; and Lat. '-formis', -shaped; (**1**) for the two different stem forms (terete and leaflike flattened). (*Disocactus*) – (**2**) for two forms of leaves. (*Antimima*)

bigelovii For John M. Bigelow (1804–1878), US-American surgeon and botanist. (*Brandegea, Cylindropuntia, Nolina*)

biharamulensis For the occurrence in the Biharamulo Distr., Tanzania. (*Euphorbia*)

bihendulensis For the occurrence at Bihendula, Somalia. (*Echidnopsis*)

Bijlia For Mrs. D. van der Bijl (fl. 1930–1937), South African naturalist, nurserywoman in Great Brak River, Western Cape, field collector and founder and first president of the South African Succulent Society. (*Aizoaceae*)

bijliae As above. (*Machairophyllum, Ruschia, Stapelia*)

bikitaensis For the occurrence at Bikita Mine, SE Zimbabwe. (*Brachystelma*)

bilobatus Lat. 'bi-', two; and Lat. 'lobatus', lobed; for the outer appendix of the corona. (*Hoya*)

bilobus Lat., two-lobed; for the fused but two-lobed leaf pairs. (*Conophytum*)

bilocularis Lat. 'bi-', two; and Lat. 'locularis', with locules or cavities; for the 2-celled ovaries. (*Euphorbia candelabrum* var.)

binus Lat., two by two; for two different types of leaf pairs. (*Antimima, Braunsia*)

biolleyi For Paul A. Biolley (1862–1908), Swiss teacher and naturalist in Costa Rica. (*Weberocereus*)

bipapillatus Lat. 'bi-', two; and Lat. 'papillatus', papillate; for the filaments. (*Ruschia*)

bipartitus Lat. 'bi-', two; and Lat. 'partitus', -parted; application obscure. (*Kalanchoe*)

biplanatus Lat. 'bi-', two; and Lat. 'applanatus', flattened; for the leaves with both faces flat. (*Crassula*)

bisetosus Lat. 'bi-', two; and Lat. 'setosus', bristly-hairy; for the two bristles at the base of the lower spines of each areole. (*Opuntia*)

bisinuatus Lat. 'bi-', two-; and Lat. 'sinuatus', sinuate; for the shape of the staminal corona segments. (*Cynanchum*)

bispinosus Lat. 'bi-', two-; and Lat. 'spinosus', spiny; for the spines arranged in pairs. (*Pachypodium*)

bitataensis For the occurrence at Bittata Rocks, S Ethiopia. (*Euphorbia*)

bituminosus Lat., full of asphalt or pitch; for the scent of the plants. (*Aichryson*)

blaauwianus For A. F. H. Blaauw (1903–1978), Dutch succulent plant enthusiast. (*Parodia concinna* ssp.)

blackbeardianus For Miss Gladys I. Blackbeard (1891–1975), gardener and nature lover near Grahamstown, RSA, and succulent plant enthusiast. (*Haworthia bolusii* var.)

blackburniae For Mrs. H. Blackburn (fl. 1936), wife of the station master at Calitzdorp, Western Cape, RSA. (*Gibbaeum, Haworthia*)

blainei For Blaine Tree Welsh (fl. 1980), who discovered the taxon. (*Sclerocactus spinosior* ssp.)

blakeanus For Dr. Sidney F. Blake (1892–1959), US-American botanist at the Department of Agriculture, Washington D.C. and specialist for *Asteraceae*. (*Pereskiopsis*)

blakei For Stanley Thatcher Blake (1911–1973), Australian botanist. (*Plectranthus*)

blandus Lat., mild, pleasing, kind; (**1**) for the nature of the plants. (*Lampranthus*) – (**2**) for the flowers. (*Conophytum*)

bleckiae For Mary Bellerue-Bleck (1933–1999), US-American horticulturist and succulent plant specialist, for some time co-owner of Abbey Garden Nursery, and 1983-1990 curator of the succulent plant collection at the Johannesburg City Botanical Garden. (*Tylecodon*)

bleo From the local vernacular name for the plants in Colombia. (*Pereskia*)

blepharantherus Gr. 'blepharis', eye lash; and Gr. 'anthera', anther; for the ciliate anthers. (*Brachystelma*)

blepharophyllus Gr. 'blepharis', eye lash; and Gr. 'phyllon', leaf; for the ciliate leaf margins. (*Rosularia, Sedum*)

blissii For Mr. Bliss (fl. 1911) of Orpington, England, who raised this hybrid in his garden. (*Agave*)

blochmaniae For Mrs. I. M. Blochman (fl. 1896) who discovered the taxon. (*Dudleya*)

Blossfeldia For Harry Blossfeld (*1913), German botanist and plant collector in S America; son of Robert Blossfeld. (*Cactaceae*)

blossfeldianus For Robert Blossfeld († 1945), German horticulturist in Potsdam and later Lübeck; father of Harry Blossfeld. (*Kalanchoe, Mammillaria*)

blossfeldiorum Lat. Gen. Pl. of Blossfeldius; for father Robert and son Harry Blossfeld, German horticulturist and botanist / plant collector, respectively. (*Espostoa*)

blyderiverensis For the occurrence at the Blyde River, Mpumalanga, RSA. (*Huernia quinta* var.)

blyderivierensis As above. (*Aloe minima* var.)

bocasanus For the occurrence in the Sierra de Bocas, San Luis Potosí, Mexico. (*Mammillaria*)

bocensis For the occurrence at Las Bocas, Sonora, Mexico. (*Mammillaria*)

bodenbenderianus For Dr. Bodenbender (fl. 1928), German mineralogist in Argentina. (*Gymnocalycium*)

bodenghieniae For Mlle. Bodenghien (fl. 1987), Belgian who travelled with F. Malaisse in Zaïre. (*Monadenium*)

bodleyae For Mrs. Elise Bodley van Wyk (1922–1997); South African botanical illustrator painting all known *Tylecodon* species. (*Tylecodon*)

boehmeri For L. Boehmer (fl. 1890s?), botanical collector in Japan. (*Orostachys*)

boehmianus For R. Böhm (fl. 1888), a pioneer biochemist. (*Adenium obesum* ssp.)

boelderlianus For Rudolf Bölderl (fl. 1988), German cactus enthusiast and *Mammillaria* specialist in München, Germany. (*Mammillaria*)

bogneri For Josef Bogner (*1939), curator at the Munich botanical gardens and specialist in *Araceae*. (*Begonia, Kalanchoe*)

bohlei For Bernhard Bohle (fl. 2001), German cactus hobbyist and specialist on Brazilian cacti. (*Pilosocereus*)

boinensis For the occurrence in the former Iboina or Boina Region (now Mahajanga = Majunga) in NW Madagascar. (*Euphorbia*)

boisii For Mr. Bois (fl. 1914), assistant at the Natural History Museum Paris. (*Kalanchoe*)

boissieri For Prof. Pierre Edmond Boissier (1810–1885), eminent Swiss botanist in Geneva and traveller. (*Euphorbia*)

boiteaui For Pierre L. Boiteau (1911–1980), French botanist in Madagascar and curator of the Botanical Garden at Antananarivo. (*Aloe, Euphorbia*)

boivinianus For Louis H. Boivin (1808–1852), French botanist and plant collector, esp. active on the islands in the Indian Ocean. (*Delonix, Talinella*)

boivinii As above. (*Momordica*)

bojeri For Wenceslas Bojer (1797–1856), Czech-born naturalist and explorer who settled on Mauritius. (*Crassocephalum*)

bokei For Prof. Norman H. Boke (1913-1996), US-American botanist, specialising in developmental anatomy of cacti. (*Epithelantha*)

boldinghianus For Isaäc Boldingh (1879–1938), Dutch botanist active in Indonesia and working on the flora of the Dutch West Indies after retirement. (*Agave*)

boldinghii As above. (*Opuntia*)

boleanus For the occurrence in the Bole valley, Ethiopia. (*Huernia*)

bolivarii For Dr. Cándido Bolívar y Pieltain (fl. 1970), professor at the Escuela Nacional de Ciencias Biológicas, Mexico. (*Bursera*)

bolivianus For the occurrence in Bolivia. (*Cumulopuntia*, *Lepismium*)

boliviensis As above. (*Furcraea*)

bollei For Carl A. Bolle (1821–1909), German dendrologist and ornithologist in Berlin and avid plant collector. (*Aichryson*)

bolusiae For Dr. H. M. Louisa Bolus (née Kensit) (1877–1970), South African botanist and Mesemb specialist; daughter-in-law of Harry Bolus. (*Conophytum*, *Ruschia*, *Stomatium*)

bolusianus For Harry Bolus (1834–1911), English-born South African banker and botanist, emigrated to RSA in 1850. (*Ipomoea*)

bolusii As above. (*Euphorbia*, *Haworthia*, *Pleiospilos*)

bombycinus Lat., silky, silk-like; for the glassy silk-like radial spines. (*Mammillaria*)

bombycopholis Gr. 'bombyx', silk, silkworm moth; and Gr. 'pholidos', reptile scale; for the silky wool-like covering of the bud scales. (*Senecio*)

bommeljei For Mr. Cornelis Bommeljé (fl. 1968), Dutch succulent plant hobbyist. (*Parodia tabularis* ssp.)

bonatzii For Hans-Joachim Bonatz (fl. 1991), German cactus collector in Berlin. (*Echinocereus adustus* ssp., *Turbinicarpus*)

bondanus For the occurrence near Bonda Mission, Zimbabwe. (*Aloe cameronii* var.)

bongolavensis For the occurrence near Bongolava, Madagascar. (*Euphorbia*)

boninensis For the occurrence on Bonin Island (Ogasawara-Shoto), Japan. (*Sedum uniflorum* ssp.)

bonkerae For Mrs. Frances Bonker (fl. 1932), Pasadena, California, USA. (*Echinocereus*)

bonnafousii For Mr. Bonnafous (fl. 1916) [perhaps Dr. Victor Bonafous, French physician in Marseille], a friend of the French physician and botanist R. Hamet. (*Hylotelephium*)

bonnieae For Bonnie Brunkow (fl. 1997), US-American cactus enthusiast who discovered the taxon. (*Echinopsis*, *Maihueniopsis*)

bonnieri Perhaps for Prof. Dr. Gaston Bonnier (1853–1922), French botanist, professor of botany in Paris 1887–1922. (*Sedum*)

bonplandii For Dr. Aimé J. A. Bonpland (Goujaud) (1773–1858), French explorer and botanist, accompanied Humboldt on his travels, later working as physician in Prov. Misiones, Argentina. (*Harrisia*, *Opuntia*)

booleanus For George Boole Hinton (*1990), great grandson of the well-known Mexican collector George B. Hinton (1882–1942), and son of G. S. Hinton. (*Sedum*, *Turbinicarpus mandragora* ssp.)

boolii For Herbert W. Bool (fl. 1953), US-American plant enthusiast in Phoenix, and one of the founders of the Desert Botanical Garden of Arizona. (*Mammillaria*)

boomianus For Dr. Boudewijn K. Boom (1903–1980), Dutch horticultural botanist at Wageningen University. (*Discocactus zehntneri* ssp.)

Boophane Gr. 'bouphonos', killing cattle; for the possibly poisonous nature of some of its species. (*Amaryllidaceae*)

booysenii For Mr. W. A. Booysen (fl. 1968), South African farmer near Sutherland, Northern Cape, on whose farm the type of the taxon was collected. (*Dorotheanthus*)

boranae For the occurrence in the region inhabited by the Boran ethnic group, Kenya. (*Kalanchoe*)

boranensis As above. (*Euphorbia*, *Raphionacme*) – (**2**) For the occurrence near Borana Awraja, Ethiopia. (*Pelargonium*)

borcherdsii For Dr. W. M. Borcherds (fl. 1929), without further data. (*Schwantesia*)

borchersii For Dr. Ph. Borchers (fl. 1932), German privy councellor in Bremen and 1932 leader of a German-Austrian expedition to the Peruvian Andes. (*Oroya*)

bordenii For the collector T. E. Borden (fl. 1906), without further data. (*Hoya*)

borealis Lat., northern; (**1**) for the distribution in the N of Madagascar. (*Operculicarya*) – (**2**) for the distribution in relation to other species. (*Lampranthus*) – (**3**) for the distribution in relation to the other subspecies. (*Conophytum lithopsoides* ssp., *Ferocactus fordii* ssp., *Khadia*, *Pectinaria articulata* ssp., *Sedum obtusatum* ssp.) – (**4**) for the distribution in North Africa. (*Tetragonia*)

borinquensis Probably for the occurrence on the island of Borinque, off Puerto Rico. (*Opuntia*)

borissovae For Antonina G. Borissova (1903–1970), Russian botanist working with *Crassulaceae*. (*Sedum*, *Sempervivum*)

borschii For Fred J. Borsch (fl. 1944), who cultivated the taxon. (*Sedum*)

borthii For Hans Borth (*1925), Austrian librarian, alpinist and plant collector. (*Gymnocalycium*)

borzianus For Prof. Antonino Borzi (1852–1921), Italian botanist in Messina and Palermo, 1893–1921 director of the botanical garden at Palermo. (*Moringa*)

boscawenii For Lieut.-Col. Mildmay Thomas Boscawen (1892–1958), English military officer, became a sisal grower in Tanzania after the first world war, where he developed a fine garden of ornamental plants. (*Aloe*)

bosscheanus For Mr. L. Van den Bossche of Tirlemont (fl. 1910), Belgian horticulturist in whose gardens the first specimen of this species grown from seed flowered. (*Faucaria*)

bosseri For Jean M. Bosser (*1922), French botanist and agronomical engineer, and director of ORSTOM in Antananarivo, Madagascar. (*Aloe*, *Ceropegia*, *Euphorbia*, *Odosicyos*, *Seyrigia*)

bosserianus As above. (*Delosperma*)

bothae For the occurrence at Botha Ridge, Eastern Cape, RSA. (*Euphorbia*)

botijae For the occurrence in the Quebrada Botija, Antofagasta Prov., Chile. (*Senecio*)

botryoides Gr. 'botrys', bunch, raceme; and Gr. '-oides', like; for the racemose inflorescences. (*Umbilicus*)

botswanicus For the occurrence in Botswana. (*Jatropha*)

bottae For Paolo Emilio Botta (1802–1870), who travelled in Arabia in 1837. (*Euphorbia*)

bougheyi For Prof. Arthur S. Boughey (fl. 1958) of the University College of Rhodesia and Nyasaland, who encouraged L. C. Leach to study succulent Euphorbias, and who collected the type of the taxon. (*Euphorbia*)

bourgaeanus For Eugène Bourgeau (1815–1877), French botanical traveller and collector. (*Euphorbia*)

bourgaei As above. (*Sedum*)

bouvetii For Georges Bouvet (1850–1929), French pharmacist and botanist and director of the Botanical Garden at Angers, France. (*Kalanchoe*)

bouvieri For Prof. Bouvier (fl. 1916), French botanist and Professor at the Natural History Museum of Paris. (*Rhodiola himalensis* ssp.)

bovicornutus Lat. 'bos, bovis', ox, cow; and Lat. 'cornutus', horned; for the leaf margin teeth. (*Agave*)

Bowiea For James Bowie (1789–1869), English horticulturist and botanical collector in S Africa. (*Hyacinthaceae*)

bowiea As above. (*Aloe*)

bowieanus As above. (*Eriospermum*)

bowkeri For James Henry Bowker (1822–1900), naturalist and government official in RSA. (*Ceropegia*, *Pelargonium*)

boyce-thompsonii For Colonel William Boyce Thompson (1869–1930), US-American business men, mining magnate and philanthropist, founder of the Boyce-Thompson Southwestern Arboretum at Superior, Arizona. (*Echinocereus*)

boylei For F. Boyle (fl. 1892), without further data. (*Aloe*)

boyuibensis For the occurrence near Boyuibe, Prov. Cordillera, Dept. Santa Cruz, Bolivia. (*Echinopsis*)

bozsingianus For Franz Bozsing (1912–1990), Austrian *Gymnocalycium* specialist. (*Melocactus*)
braceanus For Mr. Brace (fl. 1904), US-American botanical collector on the Bahamas (?). (*Agave*)
brachiatus Lat., branching like arms; for the opposite branching pattern. (*Euphorbia*)
brachyacanthus Gr. 'brachys', short; and Gr. 'akanthos', thorn, spine. (*Opuntia sulphurea* ssp.)
brachyandrus Gr. 'brachys', short; and Gr. 'aner, andros', man, [botany] stamens. (*Lampranthus*)
brachyanthus Gr. 'brachys', short; and Gr. 'anthos', flower. (*Gymnocalycium monvillei* ssp., *Pistorinia*)
brachycalyx Gr. 'brachys', short; and Gr. 'kalyx', calyx. (*Lewisia*)
brachycaulos Gr. 'brachys', short; and Gr. 'kaulos', stem. (*Monanthes*)
Brachycereus Gr. 'brachys', short; and *Cereus*, a genus of columnar cacti; for the short stems of this columnar cactus. (*Cactaceae*)
Brachychiton Gr. 'brachys', short; and Gr. 'chiton', covering; for the short covering around the seed. (*Sterculiaceae*)
brachycladus Gr. 'brachys', short; and Gr. 'klados', branch; for the short stem segments. (*Opuntia basilaris* var.)
brachylobus Gr. 'brachys', short; and Gr. 'lobos', small lobe, leaflet; for the short corolla segments. (*Kalanchoe*)
brachypetalus Gr. 'brachys', short; and Gr. 'petalon', petal. (*Corryocactus*, *Crassula pellucida* ssp.)
brachyphyllus Gr. 'brachys', short, and Gr. 'phyllon', leaf. (*Crassula decumbens* var., *Euphorbia*, *Gasteria*)
brachypus Gr. 'brachys', short; and Gr. 'pous', foot; for the short or absent stem. (*Agave brittoniana* ssp.)
brachystachys Gr. 'brachys', short; and Gr. 'stachys', spike; for the inflorescences. (*Aloe*)
brachystachyus Gr. 'brachys', short; and Gr. 'stachys', spike; for the inflorescences. (*Crassula*)
Brachystelma Gr. 'brachys', short; and Gr. 'stelma', crown, garland, wreath; for the nature of the corona. (*Asclepiadaceae*)
brachytrichion Gr. 'brachys', short; and Gr. 'trichion', small hair; for the pubescent-hairy spine surface. (*Mammillaria*)
bracteatus Lat., bearing (conspicuous) bracts. (*Antimima*, *Erepsia*, *Grahamia*, *Kalanchoe*, *Pedilanthus*, *Sansevieria*)
bracteolatus Lat., bearing (conspicuous) bracteoles. (*Brachystelma*)
bracteosus Lat., with many bracts; for the conspicuous bracts. (*Agave*, *Anthorrhiza*, *Pachyphytum*)
bradei For Dr. Alexander C. Brade (1881–1971), German botanist settling 1910 in Brazil, working as building engineer and farmer, and 1928–1952 as biologist. (*Cipocereus*, *Hatiora epiphylloides* ssp., *Weberocereus*)
bradtianus For George M. Bradt (fl. 1896), editor of "The Southern Florist & Gardener" of Louisville, Kentucky, USA. (*Grusonia*)
bradyi For Major L. G. Brady (fl. 1960), US-American who discovered the taxon. (*Pediocactus*)
brakdamensis For the occurrence near Brakdam, Namaqualand, Northern Cape, RSA. (*Euphorbia*, *Ruschia*)
brandbergensis For the occurrence in the Brandberg area, C Namibia. (*Euphorbia monteiri* ssp., *Lithops gracilidelineata* ssp.)
branddraaiensis For the occurrence at Branddraai, Mpumalanga, RSA. (*Aloe*)
Brandegea For Townsend S. Brandegee (1843–1925), US-American botanist. (*Cucurbitaceae*)
brandegeei As above. (*Echinocereus*, *Mammillaria*)
brandhamii For Dr. Peter E. Brandham (*1937), British plant geneticist at the Jodrell Laboratory, Kew, England, with a strong interest in *Aloaceae*. (*Aloe*)
brandtii For Friedrich ('Fred') Brandt (fl. 1980), Superintendent of the Huntington Botanical Gardens, USA. (*Echeveria colorata* fa.)
Brasilicereus For the genus *Cereus* (*Cactaceae*) and its occurrence in Brazil. (*Cactaceae*)

brasiliensis For the occurrence in Brazil. (*Brasiliopuntia, Echinopsis, Parodia mammulosa* ssp., *Pilosocereus, Pseudoacanthocereus*) – (**2**) For the erroneously presumed occurrence in Brazil. (*Pterodiscus*)

Brasiliopuntia From the close relationship to the genus *Opuntia*, and the occurrence in Brazil. (*Cactaceae*)

brassii For Leonard J. Brass (1900–1971), Australian botanist widely travelling in Asia. (*Euphorbia, Myrmecodia*)

braunii For Carl P. J. G. Braun (1870–1935), German pharmacist and botanist, for many years employed in Tanzania. (*Sansevieria*) – (**2**) For Otto Braun (fl. 1956), agronomist in Bolivia. (*Cereus*) – (**3**) For Dr. Pierre Braun (*1959), German agronomist and specialist on Brazilian cacti. (*Melocactus, Tacinga*)

brauniorum For Dr. Pierre J. Braun (*1959), German agronomist and specialist on Brazilian cacti, and his wife Beate. (*Pierrebraunia*)

Braunsia For Dr. Hans H. J. C. Brauns (1857–1929), German physician and entomologist in Willowmore, RSA. (*Aizoaceae*)

braunsii As above. (*Chasmatophyllum, Stomatium*) – (**2**) For Dr. R. Brauns (fl. 1915), who provided material of the taxon. (*Euphorbia*)

bravoae For Dr. Helia Bravo Hollis (1901–2001), Mexican botanist and cactus specialist at the Universidad Nacional Autónoma de México. (*Mammillaria hahniana* ssp.)

bravoanus For Ventura Bravo (fl. 1954), a friend of the botanist E. Sventenius from Gomera (Canary Islands). (*Euphorbia*) – (**2**) For Dr. Helia Bravo Hollis (1901–2001), Mexican botanist and cactus specialist at the Universidad Nacional Autónoma de México. (*Ariocarpus, Opuntia*)

breekpoortensis For the occurrence at Breekpoort, Steinkopf Distr., Northern Cape, RSA. (*Ruschia*)

breviaculeatus Lat. 'brevis', short; and Lat. 'aculeatus', prickly; for the short stipular spines. (*Euphorbia greenwayi* ssp.)

breviarticulatus Lat. 'brevis', short; and Lat. 'articulatus', jointed; for the short branch segments. (*Euphorbia*)

brevibracteatus Lat. 'brevis', short; and Lat. 'bracteatus', provided with bracts. (*Ruschia*)

brevicalyx Lat. 'brevis', short; and Lat. 'calyx', calyx. (*Pachypodium densiflorum* var.)

brevicarpus Lat. 'brevis', short; and Gr. 'karpos', fruit. (*Antimima, Aridaria*)

brevicaulis Lat. 'brevis', short; and Lat. '-caulis', stem. (*Pachypodium, Talinum*)

brevicollis Lat. 'brevis', short; and Lat. 'collum', neck; most probably for the very short pedicels. (*Antimima*)

brevicorolla Lat. 'brevis', short; and Lat. 'corolla', corolla. (*Kalanchoe aromatica* var.)

breviculus Dim. of Lat. 'brevis', short, i.e. somewhat short. (*Haworthia reinwardtii* var.)

brevicymus Lat. 'brevis', short; and Lat. 'cymus', cyme; for the inflorescences. (*Ruschia*)

breviflorus Lat. 'brevis', short; and Lat. '-florus', -flowered. (*Aloe excelsa* var., *Eulychnia, Fouquieria splendens* ssp., *Pistorinia*)

brevifolius Lat. 'brevis', short; and Lat. '-folius', -leaved. (*Aloe, Aloe deltoideodonta* var., *Crassula, Dicrocaulon, Drosanthemum, Dudleya blochmaniae* ssp., *Hereroa, Machairophyllum, Pachyphytum, Portulaca, Ruschia, Sedum, Sesuvium, Talinum, Yucca*)

brevihamatus Lat. 'brevis', short; and Lat. 'hamatus', hooked; for the short hooked central spines. (*Parodia alacriportana* ssp., *Sclerocactus*)

brevilabra Lat. 'brevis', short; and Lat. 'labrum', (upper) lip; for the short upper lip of the corolla. (*Dauphinea*)

brevilobus Lat. 'brevis', short; and Lat. 'lobus', lobe; (**1**) for the short free petal tips. (*Caralluma arachnoidea* var., *Stapeliopsis*) – (**2**) for the short anther appendages. (*Glossostelma*)

brevipedatus Lat. 'brevis', short; and Lat. 'pedatus', pedate; application obscure. (*Parakeelya*)

brevipedicellatus Lat. 'brevis', short; and Lat. 'pedicellatus', pedicellate. (*Brachystelma, Sarcostemma*)

brevipes Lat. 'brevis', short; and Lat. 'pes', foot; (**1**) for the caudex shape. (*Dioscorea sylvatica* var.) – (**2**) for the short pedicels. (*Dudleya, Ruschia*)

brevipetalus Lat. 'brevis', short; and Lat. 'petalum', petal. (*Agave, Aichryson, Delosperma, Erepsia, Sempervivum*)

brevipilus Lat. 'brevis', short; and Lat. 'pilus', hair; for the shortly pubescent leaves. (*Sempervivum*)

breviramus Lat. 'brevis', short; and Lat. 'ramus', branch. (*Euphorbia*)

brevirostris Lat. 'brevis', short; and Lat. 'rostrum', beak; application obscure. (*Huernia*)

brevis Lat., short; (**1**) for the short branches. (*Drosanthemum, Euphorbia*) – (**2**) for the short leaves and calyx tyube. (*Conophytum*) – (**3**) for the short corona lobes. (*Marsdenia*) – (**4**) for the short segments of the staminal corona. (*Aspidoglossum*)

breviscapus Lat. 'brevis', short; and Lat. 'scapus', scape; for the short inflorescences. (*Aloe*)

brevisepalus Lat. 'brevis', short; and Lat. 'sepalum', sepal. (*Delosperma*)

brevispicatus Lat. 'brevis', short; and Lat. 'spicatus', spike-bearing; for the short flower spikes. (*Tetradenia*)

brevispinus Lat. 'brevis', short; and Lat. '-spinus', -spined. (*Agave, Armatocereus, Echinocereus enneacanthus* ssp., *Lepismium*)

brevistaminus Lat. 'brevis', short; and Lat. '-staminus', pertaining to the stamens; for the short stamens. (*Lampranthus*)

brevistylus Lat. 'brevis', short; and Lat. 'stylus', style. (*Corryocactus*)

brevitortus Lat. 'brevis', short; and Lat. 'tortus', twisted; for the short twisted branches. (*Euphorbia*)

brevitubulatus Lat. 'brevis', short; and Lat. 'tubulatus', tubular; for the short corolla tube. (*Brachystelma*)

bridgesii For Dr. Thomas Bridges (1807–1865), English-born plant collector in Chile, Peru, Bolivia and California, introducing many cacti to Europe. (*Copiapoa, Echinopsis*)

Brighamia For William T. Brigham, explorer and collector in Hawaii 1864–1865. (*Campanulaceae*)

brinkmanianus For J. Brinkman (1950–1994), Dutch biologist and artist. (*Dorstenia cuspidata* var.)

briquetii For Dr. John I. Briquet (1870–1931), Swiss botanist in Geneva, and director of the Geneva Botanical Garden 1896–1931. (*Kalanchoe*)

brissemoretii For Dr. Brissemoret (fl. 1925), French pharmacologist. (*Sedum*)

bristolii For Barkley Bristol (fl. 1934), US-American from Nogales, Arizona, who first found the taxon. (*Echinocereus*)

britteniae For Lilian L. Britten (1886–1952), South African botanist at Rhodes University, RSA. (*Corpuscularia, Faucaria, Ruschia*)

brittonianus For Dr. Nathaniel L. Britton (1859–1934), US-American botanist, founder and first director of the New York Botanical Garden, and cactus specialist. (*Agave*)

brittonii As above. (*Dudleya*)

broadwayi For Walter E. Broadway (1888–1922), US-American botanist and naturalist on Tobago Island. (*Melocactus*)

bromfieldii For Mr. H. Bromfield (fl. 1933); without further data. (*Lithops*)

brookeae For Miss Winifred M. Brooke (fl. 1949), who collected plants in Bolivia. (*Cleistocactus*)

brookii For Herbert A. Brook (fl. 1907), Registrar of the Bahamas, for his valuable aid during N. L. Britton's exploration of these islands. (*Harrisia*)

broomii For Dr. Robert Broom (1866–1951), Scottish physician and palaeontologist emigrating to RSA in 1896. (*Aloe, Crassula barbata* ssp., *Rhinephyllum*)

broussonetii For Prof. Pierre M. A. Broussonet (1761–1807), French botanist and zoologist in Montpellier, collector in Morocco and the Canary Islands. (*Euphorbia*)

Brownanthus For Dr. Nicholas E. Brown (1849–1934), English botanist at Kew specializing in African succulents; and Gr. 'anthos', flower. (*Aizoaceae*)

brownianus For Dr. Nicholas E. Brown (1849–1934), English botanist at Kew specializing in African succulents. (*Brachystelma*)

brownii As above. (*Cheiridopsis, Conophytum ectypum* ssp., *Lampranthus, Raphionacme*)

Browningia For W. E. Browning (fl. 1920), former director of the Instituto Inglés at Santiago, Chile, who "was the friend of all Americans who visited Santiago". (*Cactaceae*)

bruceae For Miss Eileen A. Bruce (1905–1955), British botanist at Kew who worked on African plants and also worked in RSA 1946–1952. (*Brachystelma*)

bruchii For Dr. Carlos Bruch (fl. 1923), cactus collector in Córdoba, Argentina, and contemporary of Carlos Spegazzini. (*Echinopsis, Gymnocalycium*)

brunellii For H. E. Brunell (fl. 1952), botanist in Göteborg, Sweden. (*Euphorbia*)

brunneodentatus Lat. 'brunneus', brown; and Lat. 'dentatus', toothed; for the brown marginal teeth of the leaves. (*Aloe*)

brunneostriatus Lat. 'brunneus', brown; and Lat. 'striatus', striate; for the striate leaves. (*Aloe*)

brunnescens Lat., brownish; for the colour of the spines and the tomentum of the areoles. (*Rebutia*)

brunneus Lat., brown; (**1**) for the epidermis colour. (*Conophytum*) – (**2**) for the flowers with their brownish inside. (*Agave*)

brunnthaleri For Josef Brunnthaler (1871–1914), Austrian botanist and conservator of the Botanical Museum of the Vienna University, collected in RSA 1909–1910. (*Delosperma*)

brunonianus Lat. 'Bruno', Brown (personal name), for Dr. Robert Brown (1773–1858), Scottish botanist and plant collector, circumnavigated Australia 1801–1803, later Keeper of Botany at the British Museum (Natural History), London. (*Sarcostemma viminale* ssp.)

Brunsvigia Honouring the House of Braunschweig [Brunswick]-Lüneburg. (*Amaryllidaceae*)

brunsvigiifolius Lat. '-folius', -leaved; for the similarity of the leaves to those of some species of *Brunsvigia* (*Amaryllidaceae*). (*Bulbine*)

bruynsii For Dr. Peter V. Bruyns (*1957), South African mathematician and succulent plant botanist. (*Aloe, Bulbine, Conophytum, Euphorbia, Haworthia, Scopelogena*)

bryantii For Edward G. Bryant (fl. 1918–1932), mining engineer in RSA and plant collector. (*Stomatium*)

bryoniifolius Lat. '-folius', -leaved; and for the similarity to the leaves of *Bryonia* ("White Bryony", etc.; *Cucurbitaceae*). (*Cucumella*)

bubalinus Lat., buff, also an African Buffalo; for the occurrence by the Buffels River, Western Cape, RSA. (*Euphorbia*)

bubonifolius Lat. '-folius', -leaved; and for the similarity to the leaves of *Bubon* (syn. of *Athamanta*; *Apiaceae*). (*Pelargonium*)

buchananii For John Buchanan (1821–1903), Scottish clergyman, resident in RSA 1861–1877. (*Aloe, Brachystelma, Dorstenia*)

bucharicus For the occurrence in the then Emirate of Bukhara in present-day Uzbekistan / Tadzhikistan. (*Pseudosedum*)

buchlohii For Prof. Günther Buchloh (*1923), German botanist (bryologist) in Stuttgart, collecting higher plants with Prof. Rauh in Madagascar in 1961. (*Aloe*)

buchnerianus For Mr. Buchner (fl. 1894), who collected the taxon. (*Aeollanthus*)

buchtienii For Otto Buchtien, who collected plants in Bolivia in 1932 and 1934. (*Cleistocactus*)

buchubergensis For the occurrence on the Buchuberge in Namibia. (*Antimima*)

buderianus For Prof. Dr. Buder (fl. 1933), Breslau, Poland, who prepared the drawings for K. von Poellnitz's synopsis of *Anacampseros*. (*Avonia recurvata* ssp.)

buekii For Heinrich W. Buek (1796–1878), German physician and botanist in Hamburg. (*Thelocactus tulensis* ssp.)

buenekeri For Rudolf Heinrich Büneker (fl. 1922), cactus collector of German descent in Rio Grande do Sul, Brazil, brother-in-law of Leopoldo Horst, father of Rudi W. Büneker. (*Gymnocalycium*) – (**2**) For Mr. F. Bueneker (fl. 1961), cactus collector in Brazil, of German descent. (*Parodia alacriportana* ssp.) – (**3**) For Rudi W. Büneker (fl. 1987), Brazilian cactus collector of German descent, son of R. H. Büneker. (*Frailea*)

buettneri For Prof. Dr. Oscar A. R. Büttner (1858–1927), German botanist, head of a research station in Togo 1890–1891, later professor in Berlin. (*Aloe*)

buhrii For Elias A. Buhr (fl. 1971), farmer in the Northern Cape, RSA. (*Aloe*)

buiningianus For Albert F. H. Buining (1901–1976), Dutch public servant and cactus enthusiast and expert on Brazilian cacti. (*Frailea*)

buiningii As above. (*Parodia, Uebelmannia*)

bukobanus For the occurrence near Bukoba, Tanzania. (*Aloe*)

bulbicaulis Lat. 'bulbus', bulb; and Lat. 'caulis', stem; for the bulbous base of the plants. (*Aloe*)

bulbifer Lat. 'bulbus', bulb; and Lat. '-fer, -fera, -ferum', -bearing. (*Portulaca, Sedum*)

bulbillifer Lat. 'bulbilla', small bulb, bulbil; and Lat. 'fer, -fera, -ferum', -bearing; for the bulbils that develop on the inflorescences. (*Aloe*)

Bulbine Lat., an onion-like plant (from Lat. 'bulbus', bulb). (*Asphodelaceae*)

bulbinifolius For the genus *Bulbine* (*Asphodelaceae*); and Lat. '-folius', -leaved. (*Senecio*)

bulbispinus Lat. 'bulbus', bulb; and Lat. '-spinus', -spined; (**1**) for the basally enlarged stipular spines. (*Euphorbia*) – (**2**) for the basally enlarged spines. (*Grusonia*)

bulbocalyx Lat. 'bulbus', bulb; and Lat. 'calyx', calyx; for the funnel-shaped flowers with narrowed mouth and spreading perianth segments. (*Eriosyce*)

bulbosus Lat., bulbous, tuberous (from Lat. 'bulbus', bulb, tuber); for the tuberous rootstock. (*Bulbine, Ceropegia, Sinningia*)

bullockii For Arthur A. Bullock (1906–1980), British botanist at Kew, and specialist for *Asclepiadaceae*. (*Aloe*)

bullulatus From Lat. 'bullula', small bladder; for the tuberculate leaf surface. (*Astroloba*)

bumammus Gr. prefix 'bu-', huge, great; and Lat. 'mamma', brest, teat; for the very large tubercles of the plant bodies. (*Coryphantha elephantidens* ssp.)

bupleurifolius Lat. '-folius', -leaved; and for the similarity of the leaves to species of the genus *Bupleurum* ("Hare's Ear"; *Apiaceae*). (*Euphorbia*)

bupleuroides Gr. '-oides', similar to; and for the genus *Bupleurum* ("Hare's Ear"; *Apiaceae*). (*Rhodiola*)

burchardii For Oscar Burchard (1863–1949), German botanist. (*Caralluma*)

burchellii For William J. Burchell (1781–1863), British naturalist and explorer, collected widely in RSA 1811–1815, visited Brazil 1825. (*Aizoon, Brachystelma, Lithops lesliei* ssp., *Rhipsalis*)

burdettii For Anthony F. M. Burdett (fl. 2000); collected plants in Malawi, without further data. (*Sansevieria*)

burgeri For Dr. William C. Burger (*1932), US-American botanist, 1961–1965 at the Addis Abeba University, Ethiopia. (*Ceropegia, Euphorbia*) – (**2**) For Mr. S. Burger (fl. 1962), without further data. (*Trichodiadema*) – (**3**) For Willem Burger (fl. 1967), South African farmer in Namaqualand, Northern Cape, on whose farm the type of the taxon was collected. (*Conophytum*)

burgersfortensis For the occurrence near Burgersfort, Mpumalanga, RSA. (*Aloe*)

burkei For Joseph Burke (1812–1873), English plant collector, travelled 1840–1842 with the German explorer Karl Zeyher in RSA. (*Raphionacme*)

burmanicus For the occurrence in Burma (now Myanmar). (*Sansevieria*)

burmannii For Nicolaus L. Burman (1743–1793), Dutch physician and botanist in Amsterdam, son of Johannes Burman. (*Euphorbia*)

burnatii For Emile Burnat (1828–1920), Swiss engineer, industrialist, magistrate and amateur botanist. (*Sempervivum montanum* ssp.)

burrageanus For Guy H. Burrage, steamship commander for the 1911 collecting trip of the US-American botanist J. N. Rose. (*Cylindropuntia alcahes* var.)

burragei As above. (*Fouquieria*)

burrensis For its occurrence in the Serranias del Burro, Coahuila, Mexico. (*Echinocereus reichenbachii* ssp.)

burrito Span. 'burro', donkey; the plant is locally named 'cola de burro' = donkey's tail, for its pendent densely leafy stems. (*Sedum*)

Bursera For Joachim Burser (1593–1649), German physician and botanist. (*Burseraceae*)

burtoniae Probably for Helen M. Rousseau Burton (née Kannemeyer) (1878–1973), South African amateur naturalist and collector. (*Delosperma, Ruschia*)

buruanus For the occurrence in the Bura region, Kenya. (*Euphorbia*)

busseanus For Dr. W. Busse (fl. 1902), German agricultural officer in Tanzania. (*Sesamothamnus*)

bussei As above. (*Aloe, Euphorbia*)

buxbaumianus For Prof. Dr. Franz Buxbaum (1900–1979), Austrian botanist, high school teacher and specialist of cactus morphology. (*Coleocephalocereus*)

buysianus For T. G. ("Buys") Wiese (fl. 1987), South African farmer and succulent plant enthusiast in the Northern Cape, RSA. (*Conophytum reconditum* ssp.)

bwambensis For the occurrence in the Bwamba Forest, Uganda. (*Euphorbia*)

bylesianus For Ronald S. Byles (fl. 1957), English cactus hobbyist. (*Pygmaeocereus*)

byrnesii For Edward M. Byrnes (fl. 1905), grower of J. N. Rose's collection of *Crassulaceae* in Washington. (*Echeveria secunda* fa.)

C

cabrae For the plant collector Cabra-Michel (fl. 1903). (*Glossostelma*)

cabrerae For Prof. Dr. Angel L. Cabrera (1908–1999), Argentinian botanist of Spanish origin, founder of the Argentinian botanical society. (*Cistanthe, Echinopsis*)

cabuya From the vernacular name of the plants in Costa Rica. (*Furcraea*)

cacalioides Gr. '-oides', resembling; and for the genus *Cacalia* (*Asteraceae*). (*Othonna, Tylecodon*)

cacozela Unknown, perhaps from a local vernacular name for the plants on the Bahamas. (*Agave*)

cactiformis Lat., for the resemblance to cacti. (*Larryleachia*) – (**2**) Lat., for the resemblance to succulent climbing cacti. (*Cissus*)

cactus Lat. 'cactus', cactus; for the similarity to some cacti. (*Euphorbia*)

caducifolius Lat. 'caducus', falling, caducous; and Lat. '-folius', -leaved; for the quickly deciduous leaves. (*Euphorbia*)

caducus Lat., falling, caducous; for the early caducous leaves. (*Sedum*)

caerulans Lat., becoming blue; for the bluish-green branches. (*Euphorbia*)

caerulescens Lat., becoming blue; for the bluish-green mature branches. (*Euphorbia*)

caeruleus Lat., blue; for the flower colour. (*Anthorrhiza, Raphionacme, Sedum*)

caesius Lat., light blue; (**1**) for the body colour. (*Melocactus curvispinus* ssp.) – (**2**) for the leaf colour. (*Pachyphytum*) – (**3**) perhaps erroneously for the leaf colour. (*Aloe striatula* var.) – (**4**) application obscure, perhaps for the leaves, which are dark green when fresh but grey-green when dry. (*Tetragonia*)

caespitosus Lat., cespitose, tufted; for the growth form. (*Cephalophyllum, Delosperma, Dudleya, Duvalia, Echinopsis maximiliana* ssp., *Lampranthus, Mila, Sedum*)

caffer Lat. 'caffer, caffra, caffrum', from the old name Caffraria for S Africa. (*Brachystelma, Pelargonium, Talinum, Tinospora*)

cahum From the vernacular name of the plants in S Mexico. (*Furcraea*)

caineanus For the occurrence in the valley of the Río Caine, Potosí / Cochabamba, Bolivia. (*Browningia, Echinopsis*)

cairicus For the occurrence near Cairo, Egypt. (*Ipomoea*)

cajalbanensis For the occurrence in the Sierra de Cajálbana, Cuba. (*Agave*)

cajasensis For the occurrence near Cajas, Prov. Mendez, Dept. Tarija, Bolivia. (*Echinopsis*)

cakilifolius For the genus *Cakile* ("Sea Rocket"; *Brassicaceae*); and Lat. '-folius', -leaved. (*Othonna*)

calamiformis Lat. 'calamus', reed; and Lat. '-formis', -shaped; (**1**) for the reed-like branches. (*Euphorbia*) – (**2**) for the narrowly cylindrical (reed-like) leaves. (*Cylindrophyllum*)

Calamophyllum Gr. 'kalamos', quill, reed; and Gr. 'phyllon', leaf; for the slender cylindrical leaves. (*Aizoaceae*)

calandrus Gr. 'kalos', beautiful; and Gr. 'aner, andros', man, [botany] stamen; for the beautifully coloured stamens. (*Leipoldtia*)

calcairophila French 'calcaire', lime, limestone; and Gr. 'philos', friend; for the ecological preference. (*Aloe*)

calcaratus Lat. 'calcar', a spur; (**1**) for the spur-like lobes of the stems. (*Hylocereus*) – (**2**) for the spurred leaves. (*Lampranthus, Sedum*) – (**3**) for the spurred sepals. (*Sedum celatum* fa.) – (**4**) for the prominently spurred cyathia. (*Pedilanthus*)

calcareus Lat., chalky, limy; because the taxon grows on limestone. (*Gunniopsis, Haworthia mirabilis* var., *Ruschia, Sempervivum, Titanopsis*)

calcaricus Lat. 'calx, calcis', limestone; for the occurrence on limestone outcrops. (*Talinum*)

calcicola Lat. 'calx, calcis', limestone; and Lat. '-cola', inhabiting. (*Beschorneria, Dudleya, Ruschia, Sedum*)

calcirupicola Lat. 'calx, calcis', limestone; Lat. 'rupes', steep rocks; and Lat. '-cola', -dwelling. (*Cereus jamacaru* ssp.)

calculus Lat., pebble; for the shape of the fused leaf pair. (*Conophytum*)

calderanus For the occurrence near Caldera, N Chile. (*Copiapoa*)

calderoniae For Graciela Calderón de Rzedowski (*1931), Mexican botanist and wife of Jerzy Rzedowski. (*Echeveria*)

caledonicus For the occurrence near Caledon, Western Cape, RSA. (*Pelargonium*)

Calibanus Named for Shakespeare's monster Caliban from the play 'The Tempest'; perhaps for the massive caudex. (*Nolinaceae*)

calidicola Lat. 'calidus', hot; and Lat. '-cola', inhabiting; for the occurrence in low altitude hot dry valleys. (*Euphorbia cooperi* var.)

calidophilus Lat. 'calidus', hot; and Gr. 'philos', friend; for the preference for hot sites. (*Aloe*)

californicus For the occurrence on the Baja California peninsula (Mexico). (*Euphorbia, Portulaca*) – (**2**) For the occurrence in the State of California, USA. (*Cylindropuntia*)

calipensis For the occurrence near Calipan, Puebla, Mexico. (*Coryphantha*)

calitzdorpensis For the occurrence near Calitzdorp, Little Karoo, Western Cape, RSA. (*Delosperma, Huernia guttata* ssp.)

calliantholilacinus Gr. 'kallos', beauty; Gr. 'anthos', flower; and Lat. 'lilacinus', lilac-coloured; for the beautiful flower colour. (*Echinopsis*)

callianthus Gr. 'kallos', beauty; and Gr. 'anthos', flower. (*Rhodiola*)

callichromus Gr. 'kallos', beauty; and Gr. 'chroma', colour; for the flowers. (*Echinopsis, Hylotelephium*)

callichrous Gr. 'kallos', beauty; and Gr. '-chrous', -coloured; for the flowers. (*Sedum*)

callifer Lat. 'callus', callus, hardened thickening; and Lat. '-fer, -fera, -ferum', -carrying; for the small hard protrusions on the leaves. (*Ruschia*)

Callisia Gr. 'kallos', beauty. (*Commelinaceae*)

calmallianus For the occurrence at Calmallí, Baja California, Mexico. (*Cylindropuntia*)

calochlorus Gr. 'kalos' / 'kallos', beautiful; and Gr. 'chloros', green; for the glossy green epidermis. (*Coryphantha, Echinopsis, Gymnocalycium*)

caloderma Gr. 'kalos' / 'kallos', beautiful; and Gr. 'derma', skin; for the variegation of the branches. (*Euphorbia*)

calodontus Gr. 'kalos' / 'kallos', beautiful; and Gr. 'odous, odontos', tooth; for the teeth at the leaf margins. (*Agave*)

caloglossus Gr. 'kalos' / 'kallos', beautiful; and Gr. 'glossa', tongue; for the remarkable tongue-like corona segments. (*Fanninia*)

caloruber Gr. 'kalos' / 'kallos', beautiful; and Lat. 'ruber, rubra, rubrum', red; for the flower colour. (*Echinopsis obrepanda* ssp.)

calvatus Lat., made bald; for the absence of a tuft of hairs or bristles on the leaf tip. (*Trichodiadema*)

calvus Lat., bald, hairless, glabrous; (**1**) for the glabrous plants. (*Gunniopsis*) – (**2**) because the stems are spiny but without hairs. (*Espostoa*)

calycinus From Gr. 'calyx', cup, cover; (**1**) for the calyx-like perianth lobes of the female flower. (*Synadenium*) – (**2**) for the persistent sepals. (*Talinum*) – (**3**) for the sepals, which enlarge after flowering. (*Delosperma*) – (**4**) for the well-developed calyx. (*Cistanthe, Drosanthemum, Hereroa, Monadenium orobanchoides* var., *Tetragonia, Zeuktophyllum*) – (**5**) for the corolla form. (*Hoya*)

calycosus Lat., with a conspicuous calyx. (*Echeveria*)

Calymmanthium Gr. 'kalymma', covering; and Gr. 'anthos', flower; for the envelope of vegetative tissue surrounding the flower buds. (*Cactaceae*)

calyptratus Lat., provided with a kalyptra (from Gr. 'kalyptra', covering, woman's hat); probably for the perianth remains, which cover the ripening fruits. (*Parakeelya*)

Calyptrotheca Gr. 'kalyptra', covering, woman's hat; and Gr. 'theke', a case; for the dehiscence of the capsule, which opens from the base with 6 slits, the top falling off like a lobed cap. (*Portulacaceae*)

camachoi For Carlos Camacho (fl. 1933), Chilean agronomical engineer, who first collected the taxon. (*Maihueniopsis*)

camarguensis For the occurrence near Camargo, Prov. Cinti, Dept. Chuquisaca, Bolivia. (*Echinopsis*)

camdeboensis For the Khoi name 'Camdebo', meaning green elevations, and given to the region near Aberdeen, Eastern Cape, RSA, where the taxon is found. (*Monsonia*)

cameronii For Kenneth J. Cameron (± *1862), Scottish, 1890–± 1903 planter in Malawi for the African Lakes Corporation. (*Aloe, Euphorbia, Synadenium*)

camforosma Lat. 'camphora', camphor; and Gr. 'osme', scent; application obscure. (*Aizoon*)

camilla For Camilla R. Huxley-Lambrick (*1952), British botanist and ant-plant specialist. (*Anthorrhiza*)

camillei For Mr. Camille (fl. 1895), without further data. (*Zaleya*)

campanulatus Lat., bell-shaped; for the flower shape. (*Brachystelma, Ceropegia, Cotyledon, Fouquieria splendens* ssp., *Hesperaloe, Huernia, Kalanchoe*)

campanuliflorus Lat. 'campanula', small bell; and Lat. '-florus', -flowered. (*Pseudosedum*)

camperi For Manfredo Camperio (fl. 1894), resident in Eritrea. (*Aloe*)

campestris Lat., pertaining to plains; for the habitat. (*Crassula, Ruschia, Yucca*)

campii For Walter Camp (fl. ± 1990), US-American discoverer of the taxon. (*Cylindropuntia*)

campos-portoanus For Paulo de Campos Porto (*1889), Brazilian botanist and long-time director of the Botanical Garden Rio de Janeiro. (*Rhipsalis*)

camptotrichus Gr. 'kamptos', curved, with a bend; and Gr. 'thrix, trichos', hair; for the bristle-like intertwined spination. (*Mammillaria decipiens* ssp.)

canaliculatus Lat., channelled, grooved; (**1**) for the leaves. (*Phyllobolus, Sansevieria, Senecio*) – (**2**) for the deeply channelled leaves. (*Echeveria*)

canariensis For the occurrence on the Canary Islands. (*Aeonium, Aizoon, Euphorbia*)

canarinus Lat., canary-yellow; for the flower colour. (*Aloe*)

canatlanensis For the occurrence near Canatlán, Durango, Mexico. (*Coryphantha recurvata* ssp.)

cancellatus Lat., grided, barred, latticed; for the striped corolla. (*Ceropegia*)

candelabrum Lat., candlestick; (**1**) for the branching pattern, which is similar to a candelabrum. (*Euphorbia*) – (**2**) for the inflorescences. (*Ceropegia, Dudleya*)

candelaris From Lat., 'candela', candle; for the candelabriform growth. (*Browningia*)

candelilla From the local vernacular name "Candelilla"; Dim. of Span. 'candela', candle; for the colourful tubular flowers. (*Cleistocactus*)

candens Lat., being glossy white, glossy; for the pale papillate and thus glittering leaves. (*Drosanthemum*)

candicans Lat., becoming pure white; (**1**) for the bract colour. (*Aloe deltoideodonta* var.) – (**2**) for the flower colour. (*Echinopsis*)

candidus Lat., pure white; (**1**) for the densely white-farinose leaves. (*Dudleya*) – (**2**) for the spination. (*Mammilloydia*) – (**3**) for the flower colour. (*Disocactus ackermannii* fa., *Lampranthus, Schlumbergera microsphaerica* ssp.)

candollei For Prof. Augustin P. de Candolle (1778–1841), Swiss botanist in Geneva and well known author of numerous important botanical publications, including the famous illustrated "Histoire des Plantes Grasses". (*Sedum*)

canelensis For the occurrence in the Sierra Canelo, Chihuahua, Mexico. (*Mammillaria*)

canescens Lat., becoming grey, greyish; (**1**) for the silver-grey indumentum covering the whole plant. (*Sinningia*) – (**2**) for the colour of the bark. (*Jatropha*)

canigueralii For Father Juan Cañigueral (fl. 1961), priest at the Recoleta in Sucre, Bolivia. (*Sulcorebutia*)

caninus Lat., pertaining to dogs; perhaps for the aroma of the leaves. (*Plectranthus*)

canis Lat. 'canis', dog; for Theo Campbell-Barker (fl. 2002), who discovered the taxon; in allusion to the dog's barking and thus a jocular rendering of the family name 'Barker'. (*Aloe*)

cannellii For Ian C. Cannell (*1937), Zimbabwean civil engineer, working for the Rhodesian / Zimbabwean Ministry of Roads and becoming Provincial Roads Engineer, who travelled and collected with L. C. Leach and others. (*Aloe, Euphorbia, Monadenium*)

canonotatus Lat. 'canus', greyish-white; and Lat. 'notatus', marked; for the whitish triangle at the nodes of many shoots. (*Ruschia*)

cantabricus For the occurrence in the Cordillera Cantábrica, Spain. (*Sempervivum*)

cantabrigiensis For the Cambridge Botanic Garden, where the taxon was found growing. (*Kalanchoe*)

cantala Perhaps from a local Asian vernacular name for the plants. (*Agave*)

cante For Cante, a Mexican charitable foundation in San Miguel de Allende operating a botanical garden. (*Echeveria*)

cantelovii For Herbert C. Cantelow (fl. 1941), US-American businessman in the Pacific Coast Shipping Industry, who discovered the taxon. (*Lewisia*)

canus Lat., grey; (**1**) for the grey hair cover. (*Brachystelma*) – (**2**) for the leaf colour. (*Pleiospilos compactus* ssp.)

capensis For the occurrence at the Cape of Good Hope, or in the former Cape Province, RSA. (*Commiphora, Crassula, Fockea, Kedrostis, Othonna, Stenostelma*) – (**2**) For the occurrence at Cabo San Lucas, Baja California, Mexico. (*Agave, Mammillaria, Yucca*)

capillaceus Lat., hair-like, thread-like; for the slender stems and pedicels. (*Lampranthus*)

capillaris Lat., pertaining to hair; application obscure. (*Drosanthemum*)

capillensis For the occurrence near the locality Capilla del Monte, Prov. Córdoba, Argentina. (*Gymnocalycium*)

capitatus Lat., capitate, provided with a head; (**1**) for the head-like inflorescence. (*Aloe, Bulbine, Monadenium*) – (**2**) for the flower heads, which are arranged in secondary heads. (*Coulterella*)

capitellus Lat., small head; for the inflorescences. (*Crassula*)

capmanambatoensis For the occurrence at Cap Manambato, N Madagascar. (*Aloe, Euphorbia*)

capornii For A. St. Clair Caporn (fl. 1915), who collected the type specimen. (*Ruschia*)

capricornis Lat., goat-horned (from Lat. 'caper, capra', goat; and Lat. 'cornu', horn); for the twisted spines. (*Astrophytum*)

capsaintemariensis For the occurrence at Cap Sainte Marie, S Madagascar. (*Euphorbia*)

capuronianus For René P. R. Capuron (1921–1971), French botanist and specialist in forest trees. (*Euphorbia viguieri* var.)

capuronii As above. (*Euphorbia, Senecio*)

caput-aureus Lat. 'caput', head; and Lat. 'aureus', golden; for the inflorescence colour. (*Euphorbia*)

caput-medusae Lat., medusa head; (**1**) for the branches radiating from the thickened body and resembling the snake-head of the Gorgon in Greek mythology. (*Euphorbia*) – (**2**) for the spreading long and very slender tubercles of the plant bodies. (*Digitostigma*) – (**3**) for the tangled leaves resembling the snake-head of the Gorgon in Greek mythology. (*Bulbine*)

caput-viperae Lat. 'caput', head; and Lat. 'vipera', snake; for the appearance of the plant. (*Pseudolithos*)

caracassanus For the occurrence near Caracas, Venezuela. (*Opuntia*)

Caralluma From Arabian 'qarh al-luhum', wound in the flesh, abscess; for the floral odour of some taxa. (*Asclepiadaceae*)

carambeiensis For the occurrence near Carambei, Paraná, Brazil. (*Parodia*)

cardenasianus For Dr. Martin Cárdenas (1899–1973), Bolivian botanist and student of cacti. (*Echinopsis ancistrophora* ssp., *Gymnocalycium spegazzinii* ssp., *Portulaca, Sulcorebutia*)

cardinalis Lat., cardinal-red; for the flower colour. (*Sinningia*)

cardiophyllus Gr. 'kardia', heart; and Gr. 'phyllon', leaf; for the leaf shape. (*Jatropha*)

cardiospermus Gr. 'kardia', heart; and Gr. 'sperma', seed; for the seed shape. (*Opuntia*)

caribaeicola Lat. 'Caribae [Insulae]', Caribbean [Islands]; and Lat. '-cola', inhabiting; for the occurrence. (*Agave*)

caribaeus Lat., pertaining to the Caribbean region; for the origin. (*Cylindropuntia*)

Carica Lat. 'Carica' (add 'ficus', fig), Karyan Fig (the fig originally came from Karya in Asia Minor); for the fruits, which are compared with the Karyan figs. (*Caricaceae*)

caricus Lat., from Karya, a region in Asia Minor (SW Turkey). (*Sedum eriocarpum* ssp.)

carinans Lat., becoming keeled; for the lower face of the leaves. (*Hereroa*)

carinatus Lat., carinate, keeled; (**1**) for the acutely angular stems. (*Caralluma adscendens* var.) – (**2**) for the keeled leaves. (*Gasteria*) – (**3**) for the ribs on flowers and fruits. (*Leptocereus*) – (**5**) application obscure. (*Aspidoglossum, Stenostelma*)

carmenae For Prof. Carmen González Castañeda (fl. 1953), wife of the Mexican engineer and cactus enthusiast M. Castañeda. (*Mammillaria*)

carmenensis For the occurrence on Isla del Carmen, Gulf of California, Mexico. (*Ferocactus diguetii* var.)

carminanthus Lat. 'carmineus', carmine; and Gr. 'anthos', flower. (*Gymnocalycium*)

carmineus Lat., carmine; for the flower colour. (*Echeveria*)

carminiflorus Lat. 'carmineus', carmine; and Lat. '-florus', -flowered. (*Haageocereus pseudomelanostele* ssp.)

Carnegiea For Andrew Carnegie (1835–1919), Scottish-born US-American industrialist and philanthropist, founder of the Carnegie Institution of Washington. (*Cactaceae*)

carnegiei As above. (*Sedum*)

carneifolia Lat. 'carneus', meat-, fleshy; and Lat. '-folius', -leaved; for the somewhat succulent leaves. (*Peperomia*)

carnerosanus For the occurrence at Paso de Carneros, Mpio. Saltillo, Coahuila, Mexico. (*Yucca*)

carneus Lat., meat- fleshy; for the flower colour. (*Aloe, Mammillaria, Pelargonium auritum* var., *Pelargonium*)

carnicolor Lat. 'caro, carnis', flesh; and Lat. 'color', colour; for the colour of the flowers and/or leaves. (*Echeveria*)

carnosus Lat., fleshy; (**1**) for the fleshy roots. (*Ceropegia*) – (**2**) for the succulent stems. (*Orbea, Pelargonium*) – (**3**) for the succulent leaves. (*Codonanthe, Didelta, Glottiphyllum, Heliophila, Hoya, Nolana, Othonna, Phylohydrax*)

caroli For Dr. Carl A. Lückhoff (1914–1961), South African physician, naturalist and painter. (*Conophytum*) – (**2**) For Dr. Charles F. Juritz (fl. 1922), in whose garden near Cape Town, RSA, the plant flowered. (*Ruschia*)

caroli-henrici For Prof. Karl Heinz Rechinger (*fil.*) (1906–1998), Austrian botanist in Vienna, and editor of 'Flora Iranica'. (*Pelargonium, Sedum*)

caroli-linnaei For Dr. Carl von Linné [Linnaeus] (1707–1778), Swedish botanist, physician and zoologist, founder of modern plant systematics and nomenclature. (*Melocactus*)

caroli-schmidtii For Karl Schmidt († 1919), owner of the Haage & Schmidt nursery in Erfurt, Germany. (*Cheiridopsis*)

carolineae For Caroline Wheeler (née Jones) (1960–2000), wife of Charlie Wheeler, Kenya, both active in the conservation of Kenya's environment. (*Aloe*)

carolinensis For the occurrence near Carolina, Prov. San Luis, Argentina. (*Gymnocalycium andreae* ssp.) – (**2**) For the occurrence near Carolina, Mpumalanga, RSA. (*Delosperma, Khadia*)

Carpanthea Gr. 'karpos', fruit; and Gr. 'anthos', flower; probably because the open fruits look like star-shaped flowers. (*Aizoaceae*)

carpathicus For the occurrence in the Carpathian Mountains (Ukraine, Romania). (*Sempervivum montanum* ssp.)

carpianus For Bernard Carp, living in RSA from 1948 as a nurserymen and plant explorer. (*Conophytum*)

carpii As above. (*Stoeberia*)

Carpobrotus Gr. 'karpos', fruit; and Gr. 'brotos', edible; for the juicy edible fruits. (*Aizoaceae*)

carretii For a Mr. Carret (fl. 1898), without further data. (*Mammillaria*)

carrissoanus For Luis W. Carrisso (1886–1937), Portuguese botanist working on the flora of Angola. (*Ceraria, Portulaca*)

carrizalensis For the occurrence near Carrizal Alto, Chile. (*Eriosyce crispa* var.)

Carruanthus Gr. 'anthos', flower; and for the occurrence in the Karoo region, RSA. (*Aizoaceae*)

carsonii For Mr. Carson, collector with the Swedish Rhodesia-Congo Expedition 1911–1912. (*Glossostelma*)

cartagensis For the occurrence near Cartago, Costa Rica. (*Epiphyllum*)

carterae For Miss Beatrice Orchard Carter (1889–1939), South African botanical artist at the Bolus Herbarium, Cape Town, RSA. (*Delosperma*)

carterianus For Mrs. Susan Carter Holmes (*1933), English botanist at RBG Kew, and specialist on *Euphorbia* and *Aloe* in Tropical Africa. (*Euphorbia*)

cartilagineus Lat., cartilaginous, firm but flexible; for the leaf appendages. (*Orostachys*)

cartwrightianus For Alfred Cartwright (fl. 1918) of the British Consular Service at Guayaquil, Ecuador. (*Armatocereus*)

carunculifer Lat. 'caruncula', caruncle [a seed appendage]; and Lat. '-fer, -fera, -ferum', -carrying; for the presence of an obvious caruncle on the seeds. (*Euphorbia*)

caryophyllaceus Lat. 'caryophyllon', cloves; or for the similarity to the flowers of some *Caryophyllaceae* (Pink Family). (*Adromischus*)

Caryotophora Gr. 'karyon, karyotos', nut; and Gr. '-phoros', carrying; for the nut-like fruits. (*Aizoaceae*)

cashelensis For the occurrence at Cashel, Zimbabwe. (*Huernia longituba* ssp.)

cassythoides Gr. '-oides', resembling; and for the genus *Cassytha* (*Lauraceae*). (*Euphorbia*)

castaneus Lat., pertaining to the chestnut; (**1**) for the body colour. (*Frailea*) – (**2**) for the spine colour. (*Eriosyce subgibbosa* var.) – (**3**) for the flower colour. (*Aloe*) – (**4**) for the spiny fruits resembling the closed casks of the chestnut. (*Eulychnia*)

castellae For Manuel T. Castellá (fl. 1974), Mexican cactus hobbyist and president of the Sociedad Mexicana de Cactología, who first collected the taxon. (*Peniocereus*)

castellanosii For Prof. Dr. Alberto Castellanos (1897–1968), Argentinian botanist in Córdoba and assistant to Carlos Spegazzini, from 1955 onwards in Brazilian exile. (*Gymnocalycium*)

castello-paivae For Barão do Castello de Paiva (fl. 1859), Portuguese officer. (*Aeonium*)

castellorum Gen. Pl. of Lat. 'castellum', castle, fortress; for the occurrence on historic fortress mountains. (*Aloe*)

castillonii For Mr. Jean-Bernard Castillon (fl. 2001), professor at the University of La Réunion. (*Euphorbia*)

catamarcensis For the occurrence in Prov. Catamarca, Argentina. (*Gymnocalycium*)

cataphractus Lat., armoured; for the colour pattern of the plant body, which makes it appear to be armoured with platelets. (*Frailea*)

cataphyllaris Lat. 'cataphyllum', cataphyll, lower bract-like leaf; for the reduced bract-like leaves. (*Ceropegia*)

cataphyllatus Lat. 'cataphyllum', cataphyll, lower bract-like leaf; for the lower leaves. (*Bulbine*)

cataractarum Gen. Pl. of Lat. 'cataracta', waterfall; for the occurrence near waterfalls. (*Euphorbia*)

catarinensis For the occurrence in the State of Santa Catarina, Brazil (*Parodia alacriportana* ssp.)

catavinensis For the occurrence near Cataviña, Baja California, Mexico. (*Cylindropuntia ganderi* var.)

catenatus Lat., chained; for the small root tubers produced in chains. (*Monadenium*)

catengianus For the occurrence near Catengue, Angola. (*Aloe*)

catenulatus Dim. of Lat. 'catenatus', chained; for the chain-like succession of stem segments. (*Rhipsalis pacheco-leonis* ssp.)

caterviflorus Lat. 'caterva', crowd; and Lat. '-florus', -flowered. (*Euphorbia*)

catharticus Lat., cathartic; the root was used as an emetic. (*Jatropha*)

cathcartensis For the occurrence near Cathcart, Eastern Cape, RSA. (*Brachystelma*)

catingicola Lat. '-cola', -dwelling; and for the occurence in the Brazilian Caatinga vegetation. (*Discocactus heptacanthus* ssp., *Pilosocereus*)

catorce For the occurrence near Real de Catorce, San Luis Potosí, Mexico. (*Sedum*)

cattimandoo From the local name 'katti', knife; and 'jemado', medicine; because the latex was used in local medicines. (*Euphorbia*)

caucasicus For the occurrence in the Caucasus. (*Hylotelephium, Sempervivum*)

caudatus Lat., with a tail; – (**2**) – (**2**) for the caudate tips of the leaflets. (*Entandrophragma*) – (**3**) for the calyx lobes. (*Phyllobolus*) – (**4**) for the corolla segments. (*Orbea*) – (**5**) for the corolla segments and the anther appendices. (*Hoya*) – (**6**) for the tips of the stigma lobes. (*Lampranthus, Ruschia*)

caulerpoides Gr. '-oides', resembling; and for the green alga *Caulerpa clavifera*. (*Portulaca*)

caulescens lat., stem-forming. (*Oscularia, Sansevieria, Senecio scaposus* var.)

Caulipsolon Anagram of *Psilocaulon* (*Aizoaceae*), where the type species was formerly placed. (*Aizoaceae*)

caulopodius Gr. 'kaulos', stem; and Gr. 'podion', foot; for the numerous stems from the base of the plant. (*Monadenium ellenbeckii* fa.)

cauticola Lat. 'cautes', rock, cliff; and Lat. '-cola', -dwelling. (*Hylotelephium*)

Cavanillesia For Antonio J. Cavanilles (1745–1804), Spanish clergyman and botanist in Paris and later in Madrid. (*Bombacaceae*)

caylae For Monsieur Cayla (fl. 1936), Governor-General of Madagascar. (*Pelargonium*)

caymanensis For the occurrence on the Cayman Islands, British West Indies. (*Consolea millspaughii* ssp.)

ceciliae For Mrs. Evelyn Cecil (née Alicia Margaret Amherst, later Lady Rockley) (1865–1941), English flower painter, gardener, author and botanical collector. (*Glossostelma*)

cedarbergensis For the occurrence on the Cedarberg, Western Cape, RSA. (*Oscularia, Ruschia*)

cedrimontanus Lat. 'cedrus', cedar; and Lat. 'montanus', mountain-; for the occurrence at the Cedarberg, Western Cape, RSA. (*Stapelia*)

cedrorum Gen. Pl. of Lat. 'cedrus', cedar; for the renowned private botanical garden "Les Cèdres" of Julien Marnier-Lapostolle in S France. (*Euphorbia, Senecio*)

cedrosensis For the occurrence on Cedros Island off the coast of Baja California, Mexico. (*Cylindropuntia*)

Ceiba From Span. 'ceiba, ceibo', some sort of american cotton tree; probably of Haitian origin with roots in Tupi-Guarani 'iba', tree. (*Bombacaceae*)

celatus Lat., hidden; (**1**) for the late discovery because of its hidden habitat. (*Euphorbia*) – (**2**) application obscure. (*Sedum*)

celiae Anagram for 'Alice', for Mlle. Alice Leblanc (fl. 1910, 1913), an intimate acquaintance of the French botanist and physician Raymond Hamet. (*Sedum*)

celosioides Gr. '-oides', resembling; and for the genus *Celosia* (*Amaranthaceae*). (*Cistanthe*)

celsianus For Jean-François Cels (1810–1888), well-known French gardener and grower of cacti and orchids (together with his brother Auguste Cels). (*Oreocereus*)

centralifer Lat. 'centralis', central; and Lat. '-fer, -fera, -ferum', -carrying; for the presence of central spines. (*Mammillaria compressa* ssp.)

centrocapsula Lat. 'centrum', centre; and Lat. 'capsula', capsule; for the position of the fruit capsule in the centre of the spinescent inflorescence. (*Ruschia*)

cepaceus Lat. 'cepa', bulb, onion; i.e. with a bulb. (*Bulbine*)

cepaea The ancient Latin name of the plant; Lat., a plant with purslane-like leaves, from Gr. 'kepos', garden, perhaps for the similarity of the leaves with those of the cultivated garden purslane. (*Sedum*)

cephaliomelanus Gr. 'kephale', head; here:

cephalium; and Gr. 'melas, melano-', black; for the dark cephalium hairs. (*Facheiroa*)

Cephalocereus Gr. 'kephale', head; and for *Cereus*, a genus of columnar cacti; for the cephalium formed by these columnar cacti. (*Cactaceae*)

Cephalocleistocactus Gr. 'kephale', head; and for the genus *Cleistocactus*; for the relationship and the lateral pseudocephalium. (*Cactaceae*)

cephalomacrostibas Gr. 'kephale', head; Gr. 'makros', large; and Gr. 'stibas', bed, cushion, padding; for the large and near the stem tips almost confluent areoles. (*Echinopsis*)

Cephalopentandra Gr. 'kephale', head; Gr. 'penta', five; and Gr. 'aner, andros', male; for the cluster of stamens, erroneously stated to be 5 in the protologue. (*Cucurbitaceae*)

cephalophorus Gr. 'kephale', head; and Gr. '-phoros', carrying; (**1**) for the compact flower heads. (*Senecio*) − (**2**) for the capitate inflorescences. (*Aloe*, *Cistanthe*)

Cephalophyllum Gr. 'kephale', head; and Gr. 'phyllon', leaf; for the tufted leaves of some species. (*Aizoaceae*)

Ceraria Either Gr. 'keras', horn; or Lat. 'cera', wax; application obscure. (*Portulacaceae*)

ceratophyllus Gr. 'keratos', (forked) horn, antler; and Gr. 'phyllon', leaf; for the leaves, which resemble stag's antlers. (*Pelargonium*)

Ceratosanthes Gr. 'keratos', (forked) horn, antler; and Gr. 'anthos', flower; application obscure. (*Cucurbitaceae*)

ceratosepalus Gr. 'keratos', (forked) horn, antler; and Gr. 'sepalon', sepal; for the hornlike extension surmounting the tips of the sepals. (*Trianthema*)

cerdanus For the occurrence near Cerda, Prov. Cornelio Saavedra, Dept. Potosí, Bolivia. (*Echinopsis*)

cerealis Lat., pertaining to Ceres, the Roman deity of plant growth; but also for the occurrence at Ceres, Western Cape, RSA. (*Drosanthemum*)

cerebellus Dim. of Lat. 'cerebrum', brain; for Dr. Peter V. Bruyns (*1957), South African mathematician and botanist; also alluding to Lat. 'bellum', war, and Lat. 'bellus', beautiful. (*Conophytum ernstii* ssp.)

cereiformis Lat. 'cereus', candle (also the name of a genus of columnar *Cactaceae*); and Lat. '-formis', -shaped; for the shape of the branches. (*Echidnopsis*, *Euphorbia*)

cereoides Gr. '-oides', resembling; and for the genus *Cereus* (*Cactaceae*); (**1**) for the bristly stems. (*Rhipsalis*) − (**2**) for the densely leaf-covered stems, which from a distance resemble the spiny stems of a small columnar cactus. (*Peperomia*)

ceresianus For the occurrence near Ceres, Western Cape, RSA. (*Ruschia*)

Cereus Lat., made of wax, candle; for the erect columnar growth of many species. (*Cactaceae*)

cereusculus Dim. of *Cereus*; for the dwarf bristly stems of the juvenile form. (*Rhipsalis*)

cerifer Lat. 'cera', wax; and Lat. '-fer, -fera, -ferum', -carrying; for the pruinose stems. (*Adenia gummifera* var.)

cerinthoides Gr. '-oides', resembling; and for the genus *Cerinthe* (*Boraginaceae*). (*Tradescantia*)

ceriseus From Lat. 'cerasus', cherry (cf. French 'cerise'); for the flower colour. (*Lampranthus*)

Cerochlamys Gr. 'keros', wax, wax candle; and Gr. 'chlamys', cloak; for the wax cover of the leaves. (*Aizoaceae*)

Ceropegia Gr. 'keros', wax, wax candle; and Gr. 'pegynai', assemble, unite; perhaps for the chandelier-like inflorescences of some species. (*Asclepiadaceae*)

cerralboa For the occurrence on Isla Cerralvo (Cerralbo), Baja California, Mexico. (*Mammillaria*)

cerralvensis For the occurrence on Isla Cerralvo, Baja California, Mexico. (*Dudleya nubigena* ssp.)

cerulatus Lat., a little waxy (from Lat. 'cera', wax); for the leaf surface. (*Agave*)

cervifolius Lat. 'cervus', stag; and Lat. '-folius', -leaved; because "the form of the leaflets with the irregular lobes resembles the antlers of a stag". (*Commiphora*)

chabaudii For John A. Chabaud (fl. 1905), plant grower in Port Elizabeth, Eastern Cape, RSA. (*Aloe*)

chacalapensis For the occurrence near Chacalapa, Oaxaca, Mexico. (*Stenocereus*)

chachapoyensis For the occurrence near Chachapoyas, Prov. Amazonas, Peru. (*Corryocactus*)

chacoanus For the occurrence in the Chaco region, Paraguay / Argentina. (*Cleistocactus baumannii* ssp., *Echinopsis rhodotricha* ssp.)

chacoensis For the occurrence in the Chaco region of Bolivia. (*Gymnocalycium*)

chaffeyi For Dr. Elswood Chaffey (fl. 1915–1923), US-American plant collector in Mexico. (*Escobaria dasyacantha* ssp., *Opuntia*)

chalaensis For the occurrence near Chala, Arequipa, Peru. (*Echinopsis*, *Haageocereus*)

chalumnensis For the occurrence at Chalumna, Eastern Cape, RSA. (*Haworthia reinwardtii* fa.)

chamaecereus Gr. 'chamai', low, prostrate, on the ground; and for *Cereus*, a genus of columnar cacti; for the growth and size of the plants. (*Echinopsis*)

chamelensis For the occurrence near Chamela, Jalisco, Mexico. (*Agave*)

chanetii For L. Chanet (fl. 1908), French botanical collector. (*Orostachys*)

chapalensis For the occurrence near Lake Chapala, Jalisco / Michoacán, Mexico. (*Echeveria*)

chapototii For Dr. Chapotot (fl. 1915), French physician in Lyon. (*Kalanchoe*)

charadzeae For Anna L. Charadze [Kharadze] (1905–1971), Georgian botanist. (*Sempervivum*)

charazaniensis For the occurrence near Charazani, Prov. Bautista Saavedra, Dept. La Paz, Bolivia. (*Corryocactus*)

charleswilsonianus For Charles Wilson (fl. 1997), succulent plant enthusiast in Pretoria, RSA. (*Euphorbia*)

Chasmanthera Gr. 'chasma', cleft; and Lat. 'anthera', anther. (*Menispermaceae*)

Chasmatophyllum Gr. 'chasma', cleft; and Gr. 'phyllon', leaf; for the distinct gap between the leaves of a pair. (*Aizoaceae*)

chauveaudii For Dr. Gustav L. Chauveaud (1859–1933), French botanist in Paris. (*Sedum*)

chauviniae For Marie von Chauvin (fl. 1920), German naturalist and Mesemb enthusiast. (*Conophytum*)

chavena From the local vernacular name 'Chaveña' or 'Chaveño' for the plants in Mexico. (*Opuntia*)

chazaroi For Miguel Cházaro Basañez (fl. 2002), Mexican botanist specializing in *Crassulaceae* and parasitic plants. (*Echeveria*)

Cheiridopsis Gr. 'cheiris, cheiridis', sleeve; and Gr. '-opsis', similar to; for the sleeve-like sheath of the short leaf-pair basally surrounding the long leaf-pair in the resting period. (*Aizoaceae*)

chelidonius MLat. *Chelidonium*, the traditional medieval name for *Ranunculus ficaria* ("Celandine"; *Ranunculaceae*), which has similar leaves. (*Pelargonium*)

chende From the local vernacular name of the plants in Mexico. (*Polaskia*)

chenopodioides Gr. '-oides', resembling; and for the genus *Chenopodium* (*Chenopodiaceae*). (*Tetragonia*)

cheranganiensis For the occurrence on the Cherangani Hills, Kenya. (*Aloe*)

chersinus From Gr. 'xersos', desert country; for its habitat. (*Euphorbia*)

cherukondensis For the occurrence near Cherukonda, Andhra Pradesh, India. (*Kalanchoe*)

chevalieri For Auguste Chevalier (1873–1956), French botanist and professor at the Académie des Sciences, Paris. (*Kalanchoe*, *Monadenium*)

chiangii For Fernando Chiang (*1943), botanist in Mexico (?). (*Hesperaloe funifera* ssp.)

chiapensis For the occurrence in the Mexican state of Chiapas. (*Agave*)

chicamochaensis For the occurrrence in the region of Chicamocha, Dept. Santander, Colombia. (*Melocactus schatzlii* ssp.)

chichensis For the occurrence in Prov. Nor Chichas, Dept. Potosi, Bolivia. (*Cumulopuntia*)

chichipe From the local vernacular name of the plant in Mexico. (*Polaskia*)

chiclensis For the occurrence at Chicla, Dept. Lima, Peru. (*Echeveria*)

chihuahuensis For the occurrence in the Mexican state of Chihuahua. (*Echeveria, Escobaria, Opuntia, Sedum*)

chilensis For the occurrence in Chile. (*Carica, Carpobrotus, Eriosyce, Neowerdermannia*)

chilianensis For the occurrence at Qilinshan, Gansu Prov., China. (*Sedum erici-magnusii* ssp.)

chiloensis From the erroneous assumption that the taxon is native to the island of Chiloe, off the coast of S Chile. (*Echinopsis*)

chilonensis For its occurrence at Chilón, Dept. Santa Cruz, Bolivia. (*Echeveria*)

chimanimanensis For the occurrence on Mt. Chimanimani, Zimbabwe. (*Kalanchoe velutina* ssp.)

chinensis For the occurrence in China. (*Hylotelephium sieboldii* var., *Sedum subtile* ssp.) – (2) For the erroneously presumed occurrence in China, possibly a corruption of 'guineensis'. (*Sansevieria*)

chingtungensis For the occurrence in the Ching-tung Region, Yunnan, China. (*Sedum*)

chionocephalus Gr. 'chion', snow; and Gr. 'kephale', head; for the white-woolly stem apices. (*Mammillaria*)

chiotilla From the local vernacular name of the fruits of the plants in S Mexico. (*Escontria*)

chiquitanus For the occurrence in Prov. Chiquitos, Dept. Santa Cruz, Bolivia. (*Frailea, Gymnocalycium*)

Chirita From 'cheryta', the Hindustani name for the gentian. (*Gesneriaceae*)

chisoensis For the occurrence in the Chisos Mts., Big Bend National Park, Texas, USA. (*Echinocereus*)

chisosensis As above. (*Opuntia*)

chitralicus For the occurrence in the Chitral Distr., Pakistan. (*Rosularia adenotricha* ssp.)

chloracanthus Gr. 'chloros', green; and Gr. 'akantha', spine, thorn; for the spiny leaves. (*Haworthia*)

chloranthus Gr. 'chloros', green; and Gr. 'anthos', flower. (*Aloe, Brachystelma, Echinocereus viridiflorus* ssp., *Umbilicus*)

chlorocarpus Gr. 'chloros', green; and Gr. 'karpos', fruit. (*Browningia*)

chloropetalus Gr. 'chloros', green; and Gr. 'petalon', petal. (*Sedum*)

Chlorophytum Gr. 'chloros', yellowish-green, pale green; and Gr. 'phyton', plant; for the leaves of some taxa. (*Anthericaceae*)

chloroticus Lat., yellowish green; for the stem colour. (*Opuntia*)

chlorozonus Gr. 'chloros', green; and Gr. 'zone', zone, band; for the colour and markings of the corolla. (*Brachystelma*)

chludowii For a Mr. Chludow (fl. 1896), without further data. (×*Gasteraloe beguinii* nvar.)

choananthus Gr. 'choanos', funnel; and Gr. 'anthos', flower. (*Tromotriche*)

choapensis For the occurrence in the valley of the Río Choapa, C Chile. (*Eriosyce curvispina* var.)

chocoensis For the occurrence in Chocó Prov., W Colombia. (*Pseudorhipsalis amazonica* ssp.)

cholla From the widespread Mexican vernacular name "Cholla" for Cylindropuntias. (*Cylindropuntia*)

chontalensis For the occurrence in the region inhabited by the Chontal Indians in Oaxaca, Mexico. (*Selenicereus*)

chordifolius Lat. 'chorda', cord; and Lat. '-folius', -leaved. (*Senecio talinoides* ssp.)

chortolirioides Gr. '-oides', resembling; and for the genus *Chortolirion* (*Aloaceae*). (*Aloe*)

Chortolirion Gr. 'chortos', feeding place; and Gr. 'leirion', lily; for the preferred habitat in grassland. (*Aloaceae*)

chotaensis For the occurrence in the valley of the Río Chota, Cajamarca, Peru. (*Cleistocactus*)

chrisocruxus For Chris Barnhill (fl. 2001), US-American *Conophytum* enthusiast; and

Lat. 'crux, crucis', cross; for the discoverer of the taxon and for the cross-shaped marking on the surface. (*Conophytum*)

chrisolus For Chris Rodgerson (*1958), English self-employed electrician and nurseryman, succulent plant enthusiast and author specialising in *Conophytum* and *Adromischus*, who discovered the taxon during one of his field trips in RSA; and Lat. 'solus', solitary, alone; because the plants do not offset. (*Conophytum*)

christianeae For Christiane Peckover (fl. 1993), wife of the South African succulent plant enthusiast Ralph Peckover. (*Brachystelma*)

christianii For H. Basil Christian (1871–1950), South African agriculturist and amateur botanist, emigrated to Zimbabwe in 1911 and established a large private garden in 1914, which is now the Ewanrigg National Park. (*Aloe*)

christophoranum For Christopher (Chris) Hemming (fl. 1968), US-American botanist with the Desert Locust Survey, without further data. (*Pelargonium*)

chrysacanthion Gr. 'chrysos', gold; and Gr. 'akanthion', small spine, small thorn; for the colour of the delicate spination. (*Parodia*)

chrysacanthus Gr. 'chrysos', gold; and Gr. 'akantha', spine, thorn. (*Anthorrhiza, Ferocactus, Pilosocereus*)

chrysanthemifolius For the genus *Chrysanthemum* (*Asteraceae*, formerly including the 'Chrysanthemums' of the trade); and Lat. '-folius', -leaved. (*Rhodiola*)

chrysanthus Gr. 'chrysanthes', with (golden-) yellow flowers. (*Agave, Echidnopsis, Echinopsis, Prometheum*)

chryseus Lat. (from Gr. 'chryseos'), golden, gold-coloured; for the spine colour. (*Haageocereus pseudomelanostele* ssp.)

chrysicaulus Gr. 'chrysos', golden; and Gr. 'kaulos', stem. (*Sedum*)

chrysocardius Gr. 'chrysos', gold; and Gr. 'kardia', heart; for the yellow centre of the flowers. (*Selenicereus*)

chrysocarpus Gr. 'chrysos', gold; and Gr. 'karpos', fruit; for the yellow spines covering the ripening fruits. (*Stenocereus*)

chrysocentrus Gr. 'chrysos', gold; and Gr. 'kentron', centre; for the flower colour. (*Echinocereus engelmannii* var.)

chrysocephalus Gr. 'chrysos', gold; and Gr. 'kephale', head; for the golden spined cephalia. (*Cephalocleistocactus*)

chrysochete Gr. 'chrysos', gold; and Gr. 'chaite', long hair; for the yellow bristly spination. (*Echinopsis*)

chrysoglossus Gr. 'chrysos', gold; and Gr. 'glossa', tongue; for the yellow long and dense inflorescences. (*Agave*)

chrysoleucus Gr. 'chrysos', golden; and Gr. 'leukos', white; for the flower colour of many plants in this variable species. (*Monilaria*)

chrysophthalmus Gr. 'chrysos', gold; and Gr. 'ophthalmos', eye; for the golden-eyed flowers. (*Phyllobolus*)

chrysostachys Gr. 'chrysos', gold; and Gr. 'stachys', spike; for the yellow inflorescences of the type material. (*Aloe*)

chrysostele Gr. 'chrysos', gold; and Gr. 'stele', pillar, column; for the appearance of the plants. (*Pilosocereus*)

chrysostephanus Gr. 'chrysos', gold; and Gr. 'stephanos', wreath; for the golden yellow corona. (*Orbea*)

chrysus Gr. 'chrysos', gold; for the petal colour. (*Drosanthemum*)

chuhsingensis For the occurrence in the Chuhsing Region, Yunnan, China. (*Sedum*)

churchillianus For Sir Winston Churchill (1874–1965), British statesman, on the occasion of his official visit to Belgium in 1945. (*Sedum*)

churinensis For the occurrence in the valley of the Río Churin, Dept. Lima, Peru. (*Weberbauerocereus*)

cibdelus Gr. 'kibdela', false; perhaps for the similarity to *Euphorbia indecora*. (*Euphorbia*)

Cibirhiza Lat. 'cibus', nourishment, food; and Gr. 'rhiza', root; for the edible root tubers. (*Asclepiadaceae*)

cicatricosus Lat., scarred (from Lat. 'cicatrix', scar); (**1**) for the conspicuous scars left by the fallen pedicels. (*Caralluma*) – (**2**) for the tuberculate stems. (*Senecio*)

ciferrii For Raffaele Ciferri (1897–1964), Italian mycologist. (*Portulaca*)

ciliaris Lat., fringed, ciliate; for the fringe of cilia on the leaf bases. (*Aloe*)

ciliatus Lat., ciliate (from Lat. 'cilium', eyelash); (**1**) for the hairiness of the plants. (*Ceropegia, Plectranthus*) – (**2**) for the leaves. (*Echeveria setosa* var.) – (**3**) for the leaf margins. (*Aeonium, Crassula, Jatropha, Sedum trullipetalum* var., *Trachyandra*) – (**4**) for the cilia at the leaf base. (*Brownanthus*) – (**5**) for the petal tips. (*Monsonia*) – (**6**) for the corolla margins. (*Orbea, Rhytidocaulon*) – (**7**) for the hairy corolla lobes. (*Echidnopsis sharpei* ssp.)

cilicicus For the occurrence in Cilicia, S Turkey. (*Sedum*)

ciliosus Lat., ciliate; for the leaves. (*Sedum radiatum* ssp., *Sempervivum*)

cimiciodorus For *Cimex*, a genus of Bedbugs; and Lat. 'odorus', scented; for the bedbug-like scent of the flowers. (*Ceropegia*)

cinctus Lat., encircled, girdled; (**1**) for the colour pattern on the corolla. (*Quaqua*) – (**2**) for the purple margins of the petals. (*Ruschia*)

cinerascens Lat., becoming ash-grey; for the spines. (*Copiapoa, Echinocereus*)

cinereus Lat., ash-grey; (**1**) for the colour of the bark. (*Jatropha*) – (**2**) for the body colour. (*Copiapoa*) – (**3**) for the leaf colour. (*Namibia*)

cinnabari Lat., cinnabar-red; for the colour of the stem exudate. (*Dracaena*)

cinnabarinus Lat., cinnabar-red; for the flower colour. (*Disocactus, Echinopsis*)

cinnamomifolius Lat. '-folius', -leaved; for the similarity of the leaves to those of *Cinnamomum* ("Cinnamon"; *Lauraceae*). (*Hoya*)

Cintia For the occurrence in Prov. Nor Cinti, Dept. Chuquisaca, Bolivien. (*Cactaceae*)

cintiensis For the occurrence in the provinces of Nor-Cinti and Sud-Cinti, Dept. Chuquisaca, Bolivia. (*Weingartia*)

Cipocereus For *Cereus*, a genus of columnar cacti, and the occurrence in the Serra do Cipo, Minas Gerais, Brazil. (*Cactaceae*)

cipolinicola Lat. '-cola', -dwelling; for the occurrence on Cipolin limestones in Madagascar. (*Aloe capitata* var.)

Circandra Gr. 'kirkos', ring, circle; and Gr. 'aner, andros', male; for the ring of short stamens that surround the gynoeceum. (*Aizoaceae*)

circinatus Lat., round like a circle; for the root tuber. (*Brachystelma*)

ciribe Unknown, probably from a local vernacular name for the plants on Baja California, Mexico. (*Cylindropuntia bigelovii* var.)

cismontanus Lat., this side of the mountain; for the occurrence on the Pacific hillside slopes of the foothills of the S Californian Coastal Ranges. (*Nolina*)

Cissus Lat., from Gr. 'kissos', ivy; name used by Linné for an extra-European genus of climbing / twining plants. (*Vitaceae*)

Cistanthe For the genus *Cistus* (*Cistaceae*); and Gr. 'anthos', flower; for the superficially similar flowers. (*Portulacaceae*)

citreus Lat., lemon-like; for the lemon-yellow flower colour. (*Aloe*)

citriformis Lat. 'citrus', lemon; and Lat. '-formis', -shaped; for the lemon-shaped leaves. (*Senecio*)

citrinus Lat., lemon-yellow; for the flower colour. (*Aloe, Astridia, Echeveria racemosa* var., *Kalanchoe, Lampranthus, Umbilicus*)

Citrullus For the similarity of the fruits of some species to those of *Citrus* ("Orange", "Lemon" etc.; *Rubiaceae*). (*Cucurbitaceae*)

clandestinus Lat., hidden; for the sessile cyathia obscured by the leaves. (*Euphorbia*)

clarae For Mme. Claire Schaijes (fl. 1990), wife of Michel Schaijes, a colleage of the Belgian botanist F. Malaisse. (*Monadenium*)

clarkei For Paul Clarke (*1958), English management consultant resident in Kenya 1985–2001, and keen mountaineer, who discovered the plants on a remote mountain. (*Aloe*)

classenii For George A. Classen (1915–1982), Russian-born geologist, resident in Kenya from 1948, collected plants whilst travelling professionally as a hydrologist. (*Aloe, Euphorbia*)

clausenii For Prof. Robert T. Clausen (1911–

1981), US-American botanist and *Sedum* specialist. (*Sedum*)

clausus Lat., closed; perhaps for the very dense ("closed") shrubby growth habit. (*Eberlanzia*)

clava Lat., club; for the tapering club-shaped stems. (*Euphorbia*)

clavarioides Gr. '-oides', resembling; and for the genus *Clavaria* (a genus of club-shaped fungi); for the similarly shaped branches. (*Euphorbia, Maihueniopsis*)

clavatus Lat., club-shaped; (**1**) for the body shape. (*Coryphantha, Echinopsis, Eriosyce subgibbosa* ssp.) – (**2**) for the shape of the stem segments. (*Grusonia, Rhipsalis*) – (**3**) for the leaf shape. (*Crassula, Dorotheanthus, Sedum*) – (**4**) for the shape of the ovary. (*Ruschia*)

clavellatus Dim. of Lat. 'clavatus', club-shaped; for the leaf shape. (*Disphyma*)

clavellinus From the Dim. of Lat. 'clava', club; for the shape of the stem segments. (*Cylindropuntia molesta* var.)

claviceps Lat. 'clava', club; and Lat. '-ceps', -headed; for the shape of the plant body. (*Parodia schumanniana* ssp.)

clavicoronus Lat. 'clava', club; and Lat. 'corona', corona; for the club-shaped segments of the staminal corona. (*Stapelia*)

claviferens Lat. 'clava', club; and Lat. '-ferens', -carrying; for the club-shaped trichomes covering the plant bodies. (*Conophytum bilobum* ssp.)

claviflorus Lat. 'clava', club; and Lat. '-florus', -flowered; for the club-shaped flowers. (*Aloe*)

clavifolius Lat. 'clava', club; and Lat. '-folius', -leaved. (*Adromischus cristatus* var., *Jordaaniella, Othonna, Sedum*)

claviger Lat. 'clava', club; and Lat. '-ger, -gera, -gerum', -carrying; (**1**) for the club-shaped branches. (*Euphorbia*) – (**2**) for the appendix on the sepals. (*Portulaca*) – (**3**) for the clavate hairs in the corolla tube. (*Huernia*)

clavilobus Lat. 'clava', club; and Lat. 'lobus', lobe; for the club-shaped corona lobes. (*Ceropegia*)

clavipes Lat. 'clava', club; and Lat. 'pes, pedis', foot; for the club-shaped pedicels. (*Delosperma*)

clavispinus Lat. 'clava', club; and Lat. '-spinus', -spined; for the basally thickened central spines. (*Cleistocactus*)

claytonioides Gr. '-oides', resembling; and for the genus *Claytonia* (*Portulacaceae*). (*Peperomia*)

cleistanthus Gr. 'kleistos', closed; and Gr. 'anthos', flower; for the appearance of the flowers at anthesis. (*Dischidia*)

Cleistocactus Gr. 'kleistos', closed; and Lat. 'cactus', cactus; for the "closed" tubular flowers. (*Cactaceae*)

cleistogamus Gr. 'kleistos', closed; and Gr. 'gamos', wedding [i.e. referring to the flowers]; for the flowers that do not open (cleistogamous flowers). (*Rhipsalis baccifera* ssp.)

Cleretum Gr. 'kleros', fate, chance, lot; application obscure. (*Aizoaceae*)

climaxanthus Gr. 'climax', ladder, staircase; and Gr. 'anthos', flower; perhaps for the position of the flowers, which appear scattered over the length of the stems. (×*Haagespostoa*)

clivicola Lat. 'clivus', slope of a hill; and Lat. '-cola', -dwelling. (*Agave duplicata* ssp., *Euphorbia*)

clivorum Gen. Pl. of Lat. 'clivus', slope of a hill; for the habitat preference. (*Mitrophyllum*)

cloeteae For Miss. F. Cloete (fl. 1929), without further data. (*Delosperma*)

closianus For Prof. Dr. Dominique Clos (1821–1908), French physician and botanist in Toulouse. (*Crassula*)

cluytioides Gr. '-oides', resembling; and for the genus *Cluytia* (*Euphorbiaceae*). (*Jatropha lagarinthoides* var.)

coahuilensis For the occurrence in the state of Coahuila, Mexico. (*Grahamia, Mammillaria, Yucca*)

coalcomanensis For the occurrence in the Coalcomán region in Michoacán, Mexico. (*Pedilanthus*)

coarctatus Lat., crowded together; for the leaf arrangement. (*Haworthia*)

coccineus Lat., deep red; (**1**) for the red inflorescences. (*Crassula, Crassula perfoliata* var., *Monadenium, Rhodiola*) – (**2**) for the deep red bracts. (*Euphorbia gossypina* var.) – (**3**) for the red flowers. (*Echeveria, Echinocereus, Lampranthus*)

Coccinia From Lat. 'coccineus', deep red; for the red fruits. (*Cucurbitaceae*)

cochabambensis For the occurrence in Dept. Cochabamba, Bolivia. (*Cereus, Echinopsis, Opuntia*)

cochal From the local vernacular name "cochal" of the plant in Baja California, Mexico. (*Myrtillocactus*)

cochenillifer French 'cochenille', the cochineal insect; and Lat. '-fer, -fera, -ferum', -carrying. (*Opuntia*)

cochleatus Lat., spoon-like; for the leaf shape. (*Dischidia*)

cockerellii For Prof. Theodor D. A. Cockerell (1866–1948), British-born US-American naturalist, Professor of Zoology at the University of Colorado. (*Sedum*)

cocui From the local vernacular name of the taxon in Venezuela. (*Agave*)

coddii For Dr. Leslie E. W. Codd (1908-1999), celebrated botanist in RSA, and long-time director of the Botanical Research Institute, Pretoria. (*Brachystelma*)

Codonanthe Gr. 'kodon', bell; and Gr. 'anthos', flower. (*Gesneriaceae*)

codonanthus Gr. 'kodon', bell; and Gr. 'anthos', flower; for the corolla shape. (*Brachystelma*)

coeganus For the occurrence near Coega, Eastern Cape, RSA. (*Orthopterum*)

coelestis Lat., sky-blue; for the flower colour. (*Nolana*)

coeruleus Lat., blue; (**1**) for the leaf colour. (*Pachyphytum*) – (**2**) for the flower colour. (*Plectranthus*) – (**3**) for the presumed flower colour. (*Pterodiscus*)

coetzeei For B. J. Coetzee (*1943), South African botanist and ecologist at the Kruger National Park. (*Bulbine*)

cognatus Lat., related; for the similary to other taxa. (*Aspidonepsis, Opuntia*)

coimasensis For the occurrence near Las Coimas, Prov. Aconcagua, C Chile. (*Eriosyce senilis* ssp.)

coleae For Miss Edith Cole (1859–1940), Englishwoman who collected plants during a botanical expedition led by E. Lort-Phillips into N Somalia 1894–1895. (*Crassula volkensii* ssp.)

Coleocephalocereus Gr. 'koleos', sheath, scabbard; Gr. 'kephale', head; and for *Cereus*, a genus of columnar cacti; for the long lateral cephalium formed by these columnar cacti. (*Cactaceae*)

coleorum For Prof. Desmond T. Cole (*1922), South African professor of Bantu languages and *Lithops* specialist, and his wife and coworker Naureen A. Cole (*1935). (*Lithops*)

colimanus For the occurrence in the Mexican state of Colima. (*Agave*)

collenetteae For Mrs. I. Sheila Collenette (*1927), English plant collector, esp. in Asia and Arabia. (*Aloe*)

colliculinus Lat. 'colliculus', small hill; i.e. from small hills, for the habitat. (*Euphorbia*)

colliculosus Lat., covered with little rounded elevations; for the rough leaves. (*Thompsonella*)

colligatus Lat. 'cum-, con-', together with; and Lat. 'ligatus', joined; for the long and strictly erect calyx lobes at fruiting time, which are often clasping and appear to be joined together. (*Crassula*)

collinsii For G. N. Collins (fl. 1906), collected 1906 in Mexico, without further data. (*Mammillaria karwinskiana* ssp.)

collinus Lat., hill-; for the preferred hilly habitat. (*Aloe, Drosanthemum, Galenia, Portulaca*)

collomiae For Mrs. R. E. Collom (fl. 1924), who discovered the taxon. (*Dudleya saxosa* ssp.)

colocynthis Gr. 'kolokynthe' / Lat. 'colocynthis', "Colocynth", i.e. the vernacular name of a cucurbitaceous plant with large round and/or bitter fruits grown for its medicinal value. (*Citrullus*)

colombianus For the occurrence in Colombia. (*Acanthocereus, Portulaca*)

coloradensis For the occurrence in the drainage of the Colorado River, California, USA. (*Cylindropuntia acanthocarpa* var.)

coloratus Lat., coloured; (**1**) probably because the plants are flushed with red under dry conditions. (*Crassula*) – (**2**) for the partly red-coloured leaves. (*Echeveria*) – (**3**) for the leaf coloration. (*Agave*) – (**4**) for the brightly coloured spination. (*Ferocactus gracilis* ssp.)

coloreus Lat., coloured; for the spination. (*Maihueniopsis*)

colosseus From Lat. 'colossus', giant, colossus; for the habit of the plants. (*Cereus lamprospermus* ssp.)

colubrinus Lat., like a snake; (**1**) for the snake-like longitudinal markings on the branches. (*Euphorbia*) – (**2**) for the narrowly cylindrical stem segments. (*Opuntia*)

columbarius Lat., pertaining to pigeons; for the shape of the flower in lateral view likened to a pigeon alighting. (*Impatiens*)

columbianus For the occurrence in British Columbia, Canada. (*Lewisia*, *Opuntia*) – (**2**) For the occurrence in Colombia. (*Mammillaria*)

columbiensis As above. (*Epiphyllum*)

columella Lat., column, pillar; for the columnar body formed by the congested leaves. (*Crassula*, *Peperomia*)

columna-trajani Lat. 'columna', column; i.e. "Trajan's column"; for the stately habit of the plant. (*Cephalocereus*)

columnaris Lat., columnar; for the growth-form. (*Browningia*, *Crassula*, *Euphorbia*, *Fouquieria*, *Notechidnopsis*, *Parodia*)

Columnea For Fabio Colonna (Fabius Columna) (1567–1640), Italian botanist. (*Gesneriaceae*)

comacephalus Lat. 'coma', hair, mane; and Gr. 'kephale', head; for the dense spination at the body apex. (*Matucana*)

comarapanus For the occurrence near Comarapa, Prov. Valle Grande, Dept. Santa Cruz, Bolivia. (*Cereus*, *Echinopsis*, *Parodia*)

comaru Corruption of the the Bushman vernacular name "Kambroe" for these plants. (*Fockea*)

Commiphora Gr. 'kommi', gum, a substance used by the ancient Egyptians in preserving mummies; and Gr. '-phoros', -carrying; for the balm-like scented resin. (*Burseraceae*)

commixtus Lat., mixed up; (**1**) because the taxon at first sight seems to combine characters of several genera. (*Sedum*) – (**2**) perhaps because the taxon was previously known under an illegitimate name, or because of its occurrence mixed in dense thickets. (*Aloe*)

commutans Lat., changing; for the variability of the taxon. (*Parodia*)

commutatus Lat., changed, changing. (*Crassula rupestris* ssp., *Orbea*, *Portulaca*)

comollii For Prof. Giuseppe Comolli (1780–1849), Italian botanist and agronomist in Pavia. (*Sempervivum*)

comosus Lat., brush-like (from Lat. 'coma', hair, mane); (**1**) for the growth form. (*Alluaudia*, *Chlorophytum*) – (**2**) for the tuft of bracts at the inflorescence tips. (*Aloe*) – (**3**) for the large crowded bracts of the inflorescences. (*Euphorbia*)

compactus Lat., compact; (**1**) for the growth habit. (*Antimima*, *Crassula*, *Cynanchum*, *Duvalia caespitosa* var., *Leipoldtia*, *Pleiospilos*) – (**2**) for the massive caudex and the absence of visible branches. (*Beaucarnea*) – (**3**) for the compact rosettes. (*Pachyphytum*, *Sedum*) – (**4**) for the compactly arranged leaves. (*Peperomia nivalis* var., *Trianthema*) – (**5**) for the compact interwoven spination. (*Coryphantha*) – (**6**) for the inflorescences. (*Brachychiton*, *Synadenium*)

complanatus Lat., flattened into a plane; for the shape of the peduncle. (*Ruschia*)

complexus Lat., complex; for the complex arrangement of cymes. (*Euphorbia*)

compositus Lat., composite; application obscure. (*Parakeelya*)

compressicaulis Lat. 'compressus', compressed, flat; and Lat. '-caulis', -stemmed; for the irregularly sculptured stems. (*Echeveria*)

compressus Lat., compressed, flat; (**1**) for the low growth habit. (*Cephalophyllum*) – (**2**) for the closely arranged tubercles of the plant body. (*Mammillaria*) – (**3**) for the leaves. (*Antimima*, *Oscularia*, *Sedum*) – (**4**) for the distichous ('laterally compressed') leaf arrangement. (*Aloe*)

comptonianus For Prof. Robert H. Compton

(1886–1979), British botanist in RSA, Harold Paterson Professor at Cape Town University and second director of the National Botanical Garden, Kirstenbosch. (*Haworthia emelyae* var.)

comptonii As above. (*Aloe, Anacampseros, Conophytum, Crassula namaquensis* ssp., *Cylindrophyllum, Drosanthemum, Gibbaeum, Lithops, Oscularia, Rabiea, Rhinephyllum*)

comptus Lat., neat, adorned; (**1**) for the pretty flowers. (*Brachystelma*) – (**2**) application obscure. (*Piaranthus*)

conaconensis For the occurrence near the railway station Cona-Cona on the line from Cochabamba to Oruro, Prov. Arque, Dept. Cochabamba, Bolivia. (*Echinopsis*)

concarpus Lat. 'cum, con-', together with; and Gr. 'karpos', fruit; for the basally united follicles. (*Sedum*)

concavus Lat., concave; (**1**) for the upper face of the leaves. (*Delosperma*) – (**2**) for the top of the fused leaf pair. (*Conophytum*) – (**3**) for the top of the ovary. (*Bergeranthus, Drosanthemum, Hereroa*)

concinnus Lat., neat, pretty, elegant. (*Antimima, Huernia, Melocactus, Parodia, Sansevieria*)

concordans Lat., agreeing, harmonising; for the union of the genera *Ophthalmophyllum* (where the taxon was previously classified) and *Conophytum*. (*Conophytum*)

condensatus Lat., condensed, densely arranged; for the inflorescences. (*Pseudosedum*)

condensus Lat., condensed; for the leaves, which are fused in pairs. (*Antimima*)

confertiflorus Lat. 'confertus', crowded, pressed together, dense; and Lat. '-florus', -flowered. (*Agave, Sedum*)

confertifolius Lat. 'confertus', crowded, pressed together, dense; and Lat. '-folius', -leaved. (*Portulaca*)

confertus Lat., crowded, pressed together, dense; for the inflorescences. (*Pelargonium*)

confinalis Lat. 'confinis', bordering; for the occurrence along the ridges bordering Moçambique and Mpumalanga (RSA). (*Euphorbia*)

confinis Lat., bordering, related with; (**1**) for the intermediate position between other species of the genus. (*Eriosyce*) – (**2**) application obscure. (*Nolana*)

confluens Lat., confluent; perhaps for the medusoid habit with branches converging at the main stem tip. (*Euphorbia*)

conformis Lat., having the same form. (*Stapelia macowanii* var.)

confusus Lat., confused; (**1**) because the taxon was confused with others. (*Cephalophyllum, Quaqua, Sedum*) – (**2**) as the taxon was previously unknown. (*Aloe*)

congdonii For Joseph W. Congdon (fl. 1880), US-American attorney and plant collector in San Francisco. (*Lewisia*) – (**2**) For Colin Congdon (fl. 1994), British manager of a tea estate in Tanzania and amateur naturalist. (*Aloe*)

congensis For the occurrence in the Congo Distr. in N Angola. (*Sesuvium*)

congestiflorus Lat. 'congestus', congested, crowded; and Lat. '-florus', -flowered. (*Apodanthera, Caralluma, Euphorbia*)

congestus Lat., congested, crowded; (**1**) for the compact growth habit. (*Phyllobolus*) – (**2**) perhaps for the caudex. (*Cyphostemma*) – (**3**) for the compactly arranged leaves. (*Astroloba, Crassula, Delosperma, Peperomia*) – (**4**) perhaps for the appearance of the plants. (*Cylindropuntia*) – (**5**) for the crowded tubercles of the stems. (*Monadenium stapelioides* var.) – (**6**) for the densely crowded flowers. (*Agave*) – (**7**) application obscure. (*Ruschia*)

conglomeratus Lat., clustered; for the clump-forming habit of the plants. (*Copiapoa*)

congregatus Lat., clustered, aggregated; for the growth habit. (*Argyroderma*)

Conicosia Gr. 'konikos', conical; for the shape of the top of the fruit. (*Aizoaceae*)

conicus Lat. / Gr., conical; for the shape of the style. (*Hymenogyne*)

conifer Lat. 'conus', cone; and Lat. '-fer, -fera, -ferum', -carrying; for the cone-like appearance of young inflorescences. (*Aloe*)

coniflorus Lat. 'conus', cone; and Lat. '-florus', -flowered; for the cuneate inner perianth segments. (*Selenicereus*)

conjunctus Lat., joined, related; perhaps for the almost completely united corolla lobes. (*Orbeanthus*)

conjungens Lat., joined; because all branches arise from a common base. (*Opuntia*)

connatus Lat., connate; (**1**) for the basally fused leaves of a pair. (*Crassula*) – (**2**) for the bracteoles. (*Octopoma*) – (**3**) for the united spur-lobes of the involucral bracts. (*Pedilanthus*) – (**4**) for the usually united petal tips. (*Aspidoglossum*)

connivens Lat., connivent; (**1**) for the ascending leaves. (*Crassula tetragona* ssp.) – (**2**) for the position of the corolla lobes. (*Ceropegia fimbriata* ssp., *Pelargonium*)

conoideus Lat., conical; (**1**) for the shape of the plant body. (*Melocactus, Neolloydia*) – (**2**) for the narrowly conical operculum of the fruits. (*Portulaca*)

Conophytum Gr. 'konos', cone, pine cone; and Gr. 'phyton', plant; for the conical body of many taxa, formed from the fused leaves of a pair. (*Aizoaceae*)

conothelos Gr. 'konos', cone, pine cone; and Gr. 'thele', tubercle; for the shape of the tubercles of the plant body. (*Thelocactus*)

conrathii For Paul Conrath (1861–1931), Bohemian-Austrian naturalist and chemist. (*Ceropegia*)

consanguineus Lat., related by blood; for the close relationships to other taxa. (*Haworthia mirabilis* var.)

consobrinus Lat., cousin; for the relationship with *Euphorbia schimperi*. (*Euphorbia*)

Consolea For Michelangelo Console (1812–1897), Italian botanist at the Botanical Garden Palermo, Italy. (*Cactaceae*)

conspicuus Lat., conspicuous; (**1**) for the attractive plant bodies. (*Mammillaria haageana* ssp.) – (**2**) for the showy inflorescence. (*Lampranthus, Sansevieria*) – (**3**) Welwitsch reported it as easily seen from a ship at sea. (*Euphorbia*)

constanceae For Konstanze Zimmermann (fl. 1996), wife of the German physician and amateur botanist Norbert Zimmermann. (*Schwantesia*)

constrictus Lat., constricted; (**1**) for the fruits constricted in the middle. (*Yucca*) – (**2**) for the constricted capsule lid. (*Portulaca*)

contortus Lat., contorted; for the branches. (*Euphorbia*)

convexus Lat., convex; for the upper leaf face. (*Lampranthus*)

convolvuloides Gr. '-oides', resembling; and for the genus *Convolvulus* ("Bindweed"; *Convolvulaceae*); for the twining growth and the leaf shape. (*Ceropegia*)

conzattianus For Prof. Cassiano Conzatti (1862–1951), Mexican botanist of Italian origin in Oaxaca. (*Disocactus ackermannii* ssp.)

conzattii As above. (*Jatropha, Portulaca, Sedum*)

cooperi For Thomas Cooper (1815–1913), English plant collector working for W. W. Saunders and collecting in S Africa 1859–± 1862, father-in-law of N. E. Brown. (*Adromischus, Aloe, Crassula exilis* ssp., *Delosperma, Euphorbia, Haworthia, Ledebouria, Orbea*)

copalensis For the occurrence near Copala, Sinaloa, Mexico. (*Sedum*)

copiapinus For the occurrence near Copiapó, Chile. (*Euphorbia, Tetragonia*)

Copiapoa For the occurrence of some of the species near the city of Copiapó, Chile. (*Cactaceae*)

copiosus Lat., copious, abundant; for the abundantly produced flowers. (*Mestoklema, Oscularia, Ruschia*)

copleyae For Mrs. Gwen Copley (fl. 1940), wife of Hugh Copley, Chief Fisheries Officer, Kenya, who had the type plant in her Nairobi garden. (*Ceropegia crassifolia* var.)

coptonogonus Gr. 'koptos', chopped small, bruised; and Gr. 'gonia', corner, margin; application obscure. (*Stenocactus*)

coquimbanus For the occurrence near Coquimbo, Chile. (*Copiapoa, Echinopsis*)

coquimbensis As above. (*Cistanthe*)

corallicola Lat. 'corallum', coral; and Lat. '-cola', -dwelling; for the occurrence on coral-derived rocks. (*Consolea*)

coralliflorus Lat. 'corallum', coral; and Lat. '-florus', -flowered; for the flower colour. (*Lampranthus*)

coralliformis Lat. 'corallum', coral; and Lat. '-formis', -shaped; perhaps for the appearance of the plants with compressed leaves on erect shoots. (*Mesembryanthemum*)

corallinus Lat., coral-red, like a coral; (**1**) for the appearance of the succulent stems. (*Brownanthus*) – (**2**) for the leaf colour during dry periods. (*Crassula*) – (**3**) for the flower colour. (*Aloe*)

Corallocarpus Lat. 'corallum' / Gr. 'kouralion', coral; and Gr. 'karpos', fruit; for the coral-red fruits. (*Cucurbitaceae*)

coralloides Gr. 'kouralion', coral; and Gr. '-oides', like; for the habitat on coral rocks. (*Portulaca*)

corallothamnus Lat. 'corallum' / Gr. 'kouralion', coral; and Gr. 'thamnos', shrub; for the coral-like branching. (*Euphorbia mauritanica* var.)

cordatus Lat., heart-shaped; for the leaf shape. (*Crassula*)

corderoyi For Justus Corderoy (1832–1911), English miller and succulent plant cultivator at Blewbury near Didcot, Berkshire (now Oxfordshire). (*Duvalia*)

cordifolius Lat. 'cor, cordis', heart; and Lat. '-folius', -leaved. (*Anredera, Aptenia, Crassula, Schizoglossum*)

cordiformis Lat. 'cor, cordis', heart; and Lat. '-formis', -shaped; for the cordate leaves. (*Tylecodon*)

Cordyline Gr. 'kordyle', club, pestle; for the club-like roots of some taxa. (*Dracaenaceae*)

coriarius Lat., tanner; for its use in tanning leather. (*Psilocaulon*)

cormifer Lat. 'cormus', trunk, tuberous thickening; and Lat. '-fer, -fera, -ferum', -carrying; for the thickened rootstock. (*Sedum*)

corniculatus Lat., with a horn-like appendage, or curved like a horn; (**1**) perhaps for the spination. (*Euphorbia*) – (**2**) for the leaves, which are often incurved like the horns of a bull. (*Cephalophyllum*) – (**3**) for the long lobes of the staminal corona. (*Stenostelma*)

cornifer Lat. 'cornus', horn; and Lat. '-fer, fera, ferum', carrying, bearing; for the hooked central spines. (*Coryphantha*)

corniger Lat. 'cornus', horn; and Lat. '-ger, -gera, -gerum', -carrying; for the horn-like process of the petal tip. (*Cyphostemma*)

cornutus Lat., horned; for the erect corona segments. (*Piaranthus decorus* ssp.)

coronatus Lat., crowned; (**1**) for flowers that appear several together from near the stem apex. (*Echinopsis*) – (**2**) for the auriculate corona segments, which form a crown. (*Marsdenia*) – (**3**) for the membranous wing surrounding the operculum. (*Portulaca umbraticola* ssp.) – (**4**) for the ring of horns on top of the fruit. (*Tetragonia*)

coronopifolius For the genus *Coronopus* ("Swine Cress", "Wart Cress"; *Brassicaceae*); and Lat. '-folius', -leaved. (*Heliophila*)

corotilla From the local vernacular name of the plants in Peru. (*Cumulopuntia*)

Corpuscularia Lat. 'corpusculum', a small body; and Lat. '-arius', having the possession; for the clustered growth and the firm thick leaves. (*Aizoaceae*)

correllii For Dr. Donovan S. Correll (1908–1983), US-American botanist. (*Echinocereus viridiflorus* ssp.)

corrigioloides Gr. '-oides', resembling; and for the genus *Corrigiola* (*Caryophyllaceae*). (*Parakeelya*)

corroanus For Anibal Corro (fl. 1952), a friend of the Bolivian botanist Martín Cárdenas. (*Samaipaticereus*)

corrugatus Lat., corrugated, wrinkled; (**1**) for the surface of the stem segments. (*Tunilla*) – (**2**) for the leaves, which appear wrinkled from the rough tubercles. (*Astroloba*) – (**3**) for the apically wrinkled leaves. (*Conophytum roodiae* ssp.)

Corryocactus For T. A. Corry (fl. ± 1918), Chief Engineer of the Ferrocarril del Sur in Peru, who helped J. N. Rose during his Peruvian expedition. (*Cactaceae*)

cortusifolius Lat. '-folius', -leaved; and for the similarity to the leaves of *Cortusa* (*Primulaceae*). (*Pelargonium*)

corumbensis For the occurrence at Corumba in the Brazilian state of Mato Grosso do Sul. (*Jacaratia*)

corymbifer Lat., bearing corymbs [flat-headed inflorescences]. (*Senecio*)

corymbosus Lat., corymbose; for the arrangement of the cymes in a flat head (corymb). (*Euphorbia, Sedum*)

coryne Gr., club, pestle; for the stem shape of seedling plants. (*Stetsonia*)

corynephyllus Gr. 'koryne', club, pestle; and Gr. 'phyllon', leaf; for the leaf shape. (*Sedum*)

Coryphantha Gr. 'koryphe', summit, tuft; and Gr. 'anthos', flower; for the flowers appearing near the apex of the plant bodies. (*Cactaceae*)

costafolius Lat. 'costa', rib; and Lat. '-folius', -leaved; for the rib-shaped narrow leaves. (*Dudleya cymosa* ssp.)

costaricensis For the occurrence in Costa Rica. (*Epiphyllum, Hylocereus*)

costatus Lat., ribbed; for the ribbed fruit capsules. (*Ruschia*)

cotacajensis For the occurrence near Cotacajes, Prov. Ayopaya, Dept. Cochabamba, Bolivia. (*Echinopsis*)

Cotyledon Lat., "Pennywort" (*Umbilicus rupestris*), from Gr. 'kotyledon', cup, hollowed; because Pennywort was part of the genus originally. (*Crassulaceae*)

cotyledon For the similarity of the leaf rosettes to some species of *Echeveria* (*Crassulaceae*), then still known under the generic name *Cotyledon*. (*Lewisia*) – (**2**) For the leaves, which are similar to those of *Umbilicus rupestris* ("Pennywort", *Crassulaceae*), in medieval times known as *Cotyledon*, from Gr. 'kotyledon', cup, hollowed. (*Peperomia*)

cotyledonis Presumably from the resemblance to *Cotyledon* (*Crassulaceae*). (*Crassula, Senecio*) – (**2**) For the leaves, which are similar to those of *Umbilicus rupestris* ("Pennywort", *Crassulaceae*), in medieval times known as *Cotyledon*, from Gr. 'kotyledon', cup, hollowed. (*Pelargonium*)

couesii For Dr. Elliott Coues (1842–1899), US-American naturalist, physician and philosopher. (*Agave parryi* var.)

Coulterella For John M. Coulter (1851–1928), US-American botanist. (*Asteraceae*)

coursii For G. Cours (fl. 1965), French botanist. (*Cissus*)

coxii For G. Cox (fl. ± 1850), traveller in Argentina. (*Austrocactus*)

cradockensis For the occurrence near Cradock, Eastern Cape, RSA. (*Ruschia*)

craibii For Prof. William G. Craib (1882–1933), British botanist in Aberdeen, Edinburgh and Kew. (*Kalanchoe*) – (**2**) For Charles Craib (fl. 2001), enthusiastic amateur botanist in Johannesburg, RSA. (*Aloe*)

craigianus For Dr. Robert T. Craig (1902–1986), US-American dentist and specialist on the genus *Mammillaria* (*Cactaceae*). (*Echeveria*)

craigii As above. (*Sedum*)

crassicaulis Lat. 'crassus', thick; and Lat. 'caulis', stem. (*Echinopsis, Euphorbia francoisii* var., *Hoya, Monsonia, Pelargonium, Sceletium*)

crassicylindricus Lat. 'crassus', thick; and Lat. 'cylindricus', cylindrical; for the massive stem segments. (*Cumulopuntia*)

crassifolius Lat. 'crassus', thick; and Lat. '-folius', -leaved. (*Antimima, Ceropegia, Codonanthe, Columnea, Disphyma, Nematanthus, Osteospermum, Tapinanthus*)

crassigibbus Lat. 'crassus', thick; and Lat. 'gibba', swelling, gibbosity; for the bulging body tubercles. (*Parodia*)

crassihamatus Lat. 'crassus', thick; and Lat. 'hamatus', hooked; for the massive hooked central spines. (*Sclerocactus uncinatus* ssp.)

crassipedicellatus Lat. 'crassus', thick; and Lat. 'pedicellatus', pedicellate. (*Cynanchum*)

crassipes Lat. 'crassus', thick; and Lat. 'pes', foot; (**1**) for the large root tuber. (*Ipomoea*) – (**2**) for the thickened root and stem. (*Euphorbia*) – (**3**) for the thickened stems. (*Pelargonium*) – (**4**) application obscure. (*Aloe*)

crassirostratus Lat. 'crassus', thick; and Lat. 'rostratus', beaked; for the fruit shape. (*Kedrostis*)

crassisepalus Lat. 'crassus', thick; and Lat. 'sepalum', sepal; (**1**) for the thickly fleshy sepals. (*Ruschia*) – (**2**) for the thickly fleshy outer perianth segments ('sepals'). (*Cipocereus*)

crassiserpens Lat. 'crassus', thick; and Lat. 'serpens', creeping; for the thick stems. (*Cleistocactus*)

crassispinus Lat. 'crassus', thick; and Lat. '-spinus', -spined; (**1**) for the strong spination. (*Maihueniopsis*) – (**2**) for the strong leaf marginal teeth. (*Agave salmiana* ssp.)

crassissimus Superl. of Lat. 'crassus', thick; for the very thick leaves. (*Senecio*)

crassiusculus Dim. of Lat. 'crassus', thick; i.e. somewhat thickened, for the leaves. (*Cyphostemma*)

Crassocephalum Lat. 'crassus', thick; and Gr. 'kephale', head. (*Asteraceae*)

Crassula Dim. of Lat. 'crassus', thick; for the predominantly succulent leaves. (*Crassulaceae*)

crassulae Dim. of Lat. 'crassus', thick; for the succulent leaves and stems; or for the wrong assumption that its host is a species of *Crassula*. (*Viscum*)

crassularia For the genus *Crassula* and Lat. '-arius', pertaining to; perhaps for the superficial similarity. (*Sedum*)

crassulifolius For the genus *Crassula* (*Crassulaceae*; from Lat. 'crassus', thick); and Lat. '-folius', -leaved; for the leaves resembling those of *Crassula*. (*Nolana, Senecio*)

crassuloides Gr. '-oides', resembling; and for the genus *Crassula* (*Crassulaceae*). (*Delosperma*)

crassulus Dim. of Lat. 'crassus', thick; for the succulent leaves and stems, and perhaps suggesting a similarity to some species of *Crassula* (*Crassulaceae*). (*Tradescantia*)

crassus Lat., thick; (**1**) for the succulent stems. (*Opuntia, Plectranthus, Ruschia, White-sloanea*) – (**2**) for the thicker stems. (*Drosanthemum, Euphorbia enterophora* ssp.) – (**3**) for the thick and almost terete leaves. (*Delosperma, Hereroa*) – (**4**) application obscure. (*Malephora*)

crateriformis Lat. 'crater', bowl, vessel for mixing water and wine; and Lat. '-formis', -shaped; for the shape of the calyx tube. (*Argyroderma*)

creber Lat. 'creber, crebra, crebrum', pressed together, numerous; (**1**) for the leaves. (*Lampranthus*) – (**2**) for the dense inflorescences. (*Aspidoglossum*)

crebrifolius Lat. 'creber, crebra, crebrum', pressed together, numerous; and Lat. '-folius', -leaved; for the leaf arrrangement. (*Dudleya cymosa* ssp.)

creethae For Miss Creeth (fl. 1912), who collected the taxon in West Australia. (*Parakeelya*)

cremersii For Georges Cremers (*1936), French botanist. (*Aloe, Euphorbia*)

Cremnophila Gr. 'kremnos', cliff, slope; and Gr. 'philos', friend; for the habitat. (*Crassulaceae*)

cremnophilus Gr. 'kremnos', cliff, slope; and Gr. 'philos', friend; for the habitat. (*Albuca, Aloe, Bulbine, Crassula, Melocactus oreas* ssp.)

crenatus Lat., crenate, scalloped; (**1**) for the stem margins. (*Epiphyllum, Lepismium*) – (**2**) for the crisped leaf margins. (*Kalanchoe, Monadenium*)

crenulatus Lat., crenulate, i.e. with small rounded teeth; (**1**) for the stems. (*Caralluma*) – (**2**) for the leaf margins. (*Crassula, Echeveria, Rhodiola*)

creticus For the occurrence on Crete, Greece. (*Sedum, Sedum litoreum* var.)

cretinii For Monsieur Cretin (fl. 1916), assistant physician and good friend of the French physician and botanist Raymond Hamet. (*Rhodiola*)

crinitus Lat., with tufts of weak hairs; (**1**) for the hair-like stipules. (*Jatropha*) – (**2**) for the hair-like radial spines. (*Mammillaria*)

crispatulus Lat., minutely crisped or curled; for the leaf margins. (*Talinum*)

crispatus Lat., crisped or curled; (**1**) for the wavy margins of the stem segments. (*Rhipsalis*) – (**2**) for the wavy ribs of the plant body. (*Stenocactus*) – (**3**) for the spination. (*Sulcorebutia*) – (**4**) for the wavy leaf margins. (*Mesembryanthemum*)

crispus Lat., crisped, curled; (**1**) for the spination. (*Eriosyce*) – (**2**) for the leaf margins. (*Euphorbia, Ipomoea, Monadenium*)

cristatus Lat., crested; (**1**) for the crested undulate leaf tip. (*Adromischus*) – (**2**) for the fruits with four wings on top and convoluted ridges between them. (*Tetragonia*)

crithmifolius For *Crithmum maritimum* (*Apiaceae*); and Lat. '-folius', -leaved; for the leaf shape. (*Cyphostemma, Pelargonium*)

crithmoides Gr. '-oides', resembling; and for the genus *Crithmum* (*Apiaceae*). (*Sesuvium*)

croceiflorus Lat. 'croceus', saffron-yellow; and Lat. '-florus', -flowered; for the yellow margins of the perianth segments. (*Cleistocactus baumannii* ssp.)

croceus Lat., saffron-yellow; for the flower colour. (*Bulbine, Drosanthemum, Malephora*)

croizatii For Dr. Léon C. M. Croizat (1894–1982), botanist of French origin, became US-American citizen 1929, from 1947 living in Venezuela, and specialist of *Euphorbiaceae*. (*Euphorbia*)

cronemeyerianus For Gustav Cronemeyer († before 1922), curator of La Mortola Botanical Garden, who 1889 published the first plant list of the garden. (*Delosperma*)

croucheri For J. Croucher (fl. 1869), English gardener, first head gardener at Kew and later at the Peacock collection, and succulent plant specialist. (*Gasteria*)

cruciatus Lat., like a cross; (**1**) for the decussate leaves. (*Glottiphyllum*) – (**2**) for the shape of the top of the fused leaf pair, with the sides bowed inwards. (*Conophytum ectypum* ssp.)

cruciformis Lat. 'crux, crucis', cross; and Lat. '-formis', -shaped; for the appearance of the cross-section of the four-angled stems. (*Lepismium*)

cruciger Lat. 'crux, crucis', cross; and Lat. '-ger, -gera, -gerum', carrying, bearing; for the usually 4 central spines. (*Mammillaria*)

crundallii For Albert H. Crundall (1889–1975), British-born bank official and amateur botanist widely travelling in RSA and developing a garden of rare plants in Pretoria. (*Kalanchoe*)

cryptanthus Gr. 'kryptos', hidden, covered; and Gr. 'anthos', flower; for the small flowers. (*Mesembryanthemum*)

cryptocarpus Gr. 'kryptos', hidden, covered; and Gr. 'karpos', fruit; (**1**) for the small fruits hidden by the apical wool of the stem. (*Yavia*) – (**2**) for the small fruits hidden by the leaves. (*Plinthus*)

cryptocaulis Gr. 'kryptos', hidden, covered; and Lat. 'caulis', stem; for the underground stems. (*Euphorbia*)

cryptoflorus Gr. 'kryptos', hidden, covered; and Lat. '-florus', -flowering; for the flowers, which are hidden by the bracts. (*Aloe*)

cryptopetalus Gr. 'kryptos', hidden, covered; and Gr. 'petalon', petal. (*Portulaca*)

cryptopodus Gr. 'kryptos', hidden, covered; and Gr. 'podos', foot; (**1**) for the underground bulb. (*Hyacinthus*) – (**2**) because the flower bases are covered by the large bracts. (*Aloe*)

cryptospinosus Gr. 'kryptos', hidden, covered; and Lat. 'spinosus', spiny; for the minute spines. (*Euphorbia*)

crystalenius Unknown, probably from Lat. 'crystallum', crystal; application obscure and according to some sources perhaps referring to the edible fruits. (*Opuntia*)

crystallinus Lat., crystalline; (**1**) for the densely papillate stems and leaves. (*Trianthema*) – (**2**) for the appearance of the leaves, which are densely covered with translucent papillae. (*Galenia, Mesembryanthemum, Tetragonia*)

cubensis For the occurrence on Cuba. (*Escobaria, Opuntia, Portulaca*)

cubicus Lat., cubic, cube-shaped; for the square body shape. (*Conophytum*)

cubiformis Lat., having the shape of a cube; for the body shape. (*Pseudolithos*)

cucullatus Lat., hooded; for the corolla shape. (*Cynanchum, Villadia*)

Cucumella Dim. of Lat. 'cucumis', cucumber, gherkin; for the relatively small fruits. (*Cucurbitaceae*)

cucumerinus Lat., like a cucumber; for the resemblance of the branches to cucumbers. (*Euphorbia*)

Cucumis Lat. 'cucumis', cucumber, gherkin; i.e. the ancient name for the plant. (*Cucurbitaceae*)

Cucurbita Lat., gourd. (*Cucurbitaceae*)

cuencaensis For the occurrence near Cuenca, Ecuador. (*Echeveria*)

cuencamensis For the occurrence near Cuencamé, Durango, Mexico. (*Coryphantha durangensis* ssp.)

cufodontii For Prof. Giorgo (Georg) Cufodontis (1896–1974), Austrian botanist of Italian-Greek descent in Vienna. (*Ceropegia*)

cuija From the local vernacular name of the plants in San Luis Potosí, Mexico. (*Opuntia engelmannii* var.)

cuixmalensis For the occurrence at Rancho de Cuixmala, Jalisco, Mexico. (*Peniocereus*)

cultratus Lat., shaped like the blade of a knife; for the leaf shape. (*Crassula, Dudleya*)

cultriformis Lat. 'culter', knife; and Lat. '-formis', -shaped; for the leaf shape. (*Crassula atropurpurea* var.)

Cummingara For David M. Cumming (*1942), Scottish medical laboratory technician, amateur botanist and plant breeder, lived in Australia from 1962 and established Silky Oaks nursery, moved to RSA in 1994; plus the suffix '-ara', indicating plurigeneric hybrids. (*Aloaceae*)

cummingii For David M. Cumming (*1942), Scottish medical laboratory technician, amateur botanist and plant breeder, lived in Australia from 1962 and established Silky Oaks nursery, moved to RSA in 1994. (*Brachystelma*)

cumulatus Lat., in a heap, piled up; for the crowded stems. (*Euphorbia*)

Cumulopuntia Lat. 'cumulus', heap, pile; for the growth form and the close relationship to the genus *Opuntia* ("Prickly Pear"). (*Cactaceae*)

cundinamarcensis For the occurrence in Prov. Cundinamarca, Colombia. (*Agave*)

cuneatus Lat., wedge-shaped; (**1**) for the shape of the stem segments. (*Rhipsalis*) – (**2**) for the leaf-shape. (*Aeonium, Cotyledon, Euphorbia, Jatropha, Othonna*)

cuneneanus For the occurrence near the Cunene River, Angola. (*Euphorbia*)

cupreatus Lat., coppery; (**1**) for the leaf colour. (*Conophytum pellucidum* ssp.) – (**2**) for the colour of the leaf marginal teeth. (*Agave*)

cupreiflorus Lat. 'cupreus', coppery, copper-coloured; and Lat. '-florus', -flowered. (*Conophytum*)

cupressoides Gr. '-oides', resembling; and for the genus *Cupressus* ("Cypres"; *Cupressaceae*); for the similarly appressed leaves. (*Sedum*)

cupreus Lat., coppery, copper-coloured; for the flower colour. (*Jordaaniella*)

cupricola Lat. 'cuprum', copper; and Lat. '-cola', -inhabiting; for the occurrence on copper-rich soils. (*Monadenium*)

cuprispinus Lat. 'cuprum', copper; and Lat. '-spinus', -spined; for the spine colour. (*Euphorbia*)

cupularis Lat., having a small tube or cask; for the shape of the cyathium. (*Synadenium*)

cupulatus Lat., cup-shaped; (**1**) for the arrangement of the bracteoles. (*Ruschia*) – (**2**) for the cup-shaped corona. (*Brachystelma*)

curassavicus For the occurrence on Curaçao. (*Opuntia*)

curcas From a local vernacular name. (*Jatropha*)

curocanus For the occurrence near the Curoca Drift in the Moçamedes district of Angola. (*Euphorbia*)

currorii For Dr. A. B. Curror of HMS Water-Witch, active around 1840. (*Cyphostemma, Hoodia*)

currundayensis For the occurrence at the Cerro Currunday, Prov. La Libertad, Peru. (*Matucana aurantiaca* ssp.)

curtophyllus Lat., 'curtus', short; and Gr. 'phyllon', leaf; for the short leaves. (*Cephalophyllum, Drosanthemum*)

curtus Lat., short; for the short leaf sheaths. (*Ruschia*)

curvatus Lat., curved; (**1**) for the curved branches. (*Adenia globosa* ssp.) – (**2**) for the downwards-curved leaves. (*Bulbine latifolia* var.) – (**3**) for the downwards-curved peduncles. (*Kleinia*)

curviandrus Lat. 'curvus', curved, bent; and Gr. 'aner, andros', man, [botany] stamens; for the upwards curved stamens. (*Pelargonium*)

curviflorus Lat. 'curvus', curved, bent; and Lat. '-florus', -flowered; for the incurved petals. (*Lampranthus*)

curvifolius Lat. 'curvus', curved, bent; and Lat. '-folius', -leaved. (*Lampranthus*)

curviramus Lat. 'curvus', curved, bent; and Lat. 'ramus', branch; for the U-curved branches. (*Euphorbia*)

curvispinus Lat. 'curvus', curved, bent; and Lat. '-spinus', -spined. (*Eriosyce, Frailea, Melocactus, Parodia*)

curvospinus Lat. 'curvus', curved, bent; and Lat. '-spinus', -spined. (*Opuntia*)

curvulus Dim. of Lat. 'curvus', curve; for the slightly curved corolla. (*Kalanchoe*)

cuscutoides Gr. '-oides', resembling; and for the genus *Cuscuta* ("Dodder"; *Convolvulaceae / Cuscutaceae*); for the fine slender stems. (*Schizobasis*)

cuspidatus Lat., cuspidate, ending in a sharp point; for the leaf tips. (*Dorstenia, Echeveria, Sedum*)

cussackianus For Mr. Cussack (fl. 1895), plant collector in Australia. (*Trianthema*)

Cussonia For Pierre Cusson (1727–1783), French physician, botanist and mathematician in Montpellier. (*Araliaceae*)

cussonioides Gr. '-oides', resembling; and for the genus *Cussonia* ("Cabbage Tree"; *Araliaceae*); for the similar branching. (*Euphorbia*)

cuzcoensis For the occurrence near Cuzco, Peru. (*Echinopsis, Weberbauerocereus*)

cyananthus Gr. 'kyanos', dark blue; and Gr. 'anthos', flower; for the flower colour. (*Stephania*)

cyaneus Lat., dark blue; (**1**) for the leaf colour. (*Haworthia decipiens* var.) – (**2**) application obscure. (*Hylotelephium*)

cyanophyllus Gr. 'kyanos', dark blue; and Gr. 'phyllon', leaf; for the leaf-colour. (*Plectranthus*)

Cyanotis Gr. 'kyanos', dark blue; and Gr. 'ous, otos', ear; for the flowers. (*Commelinaceae*)

cyathiformis Lat. 'cyathus' / Gr. 'kyathos', cup; and Lat. '-formis', -shaped; (**1**) for the shape of the receptacle. (*Eberlanzia*) – (**2**) for the shape of the calyx tube. (*Lampranthus*)

Cyclantheropsis Gr. '-opsis', similar to; and for the genus *Cyclanthera* (*Cucurbitaceae*). (*Cucurbitaceae*)

cyclophyllus Gr. 'kyklos' (Lat. 'cyclus'), circle; and Gr. 'phyllon', leaf. (*Matelea, Othonna, Portulaca*)

cycniflorus Gr. 'kyknos' / Lat. 'cygnus', swan; and Lat. '-florus', -flowered; for the shape of the corolla. (*Ceropegia*)

cylindraceus Lat., cylindrical; (**1**) for the shape of the plant body. (*Ferocactus*) – (**2**) for the dense cylindrical inflorescence. (*Plectranthus*)

cylindratus Lat., cylindrical; for the shape of the fused leaf pair. (*Conophytum*)

cylindricus Lat., cylindrical; (**1**) for the shape of the stems. (*Austrocylindropuntia, Echinocereus viridiflorus* ssp., *Euphorbia, Sulcorebutia*) – (**2**) for the leaf shape. (*Calamophyllum, Othonna, Sansevieria, Senecio talinoides* ssp.)

cylindrifolius Lat. 'cylindricus', cylindrical; and Lat. '-folius', -leaved; for the terete leaves. (*Euphorbia*)

Cylindrophyllum Gr. 'kylindros', cylinder; and Gr. 'phyllon', leaf; for the leaf shape. (*Aizoaceae*)

Cylindropuntia Lat. 'cylindrus', cylinder; and for the genus *Opuntia* ("Prickly Pear"); for the cylindrical stem segments. (*Cactaceae*)

cymatopetalus Gr. 'kyma, kymatos', wave; and Gr. 'petalon', petal; for the wavy-erose petal tips. (*Sedum*)

cymbifer Lat. 'cymba', boat; and Lat. '-fer, -fera, -ferum', -carrying; for the shape of the cyathia. (*Pedilanthus*)

cymbifolius Lat. 'cymba', boat; and Lat. '-folius', -leaved; for the leaf shape. (*Kalanchoe, Ruschia*)

cymbiformis Lat. 'cymba', boat; and Lat. '-formis', -shaped; for the leaf shape. (*Corpuscularia, Crassula, Haworthia*)

cymbosepalus Lat. 'cymba', boat; and Lat. 'sepalum', sepal; for the concave boat-shaped sepals. (*Talinum*)

cymifer Lat. 'cyma', young sprout of cabbage, also cyme, a type of branched inflorescence; and Lat. '-fer, -fera, -ferum', carrying; for the inflorescence. (*Drosanthemum*)

cymochilus Gr. 'kyma', offset, child; and Gr. 'cheilos', lip, margin; perhaps for the branching nature, forming chains of segments. (*Opuntia*)

cymosus Lat., with many shoots; for the cymose inflorescences. (*Cistanthe, Crassula, Dudleya, Galenia, Ruschia, Synadenium*)

Cynanchum Gr. 'kynos', dog; and Gr. 'anchein', to choke; for the toxicity of the plants. (*Asclepiadaceae*)

Cyphostemma Gr. 'kyphos', swelling, hump; and Gr. 'stemma', garland, crown; for the nectary consisting of four hump-like glands. (*Vitaceae*)

cyprius Lat., Cyprian; for the occurrence on Cyprus. (*Sedum*)

Cypselea Gr. 'kypsele', bee-hive; for the fruit shape. (*Aizoaceae*)

cypseloides Gr. '-oides', resembling; and for the genus *Cypselea* (*Aizoaceae*). (*Trianthema*)

cyrenaicus For the occurrence in the Cyrenaica, NE Libya (*Sedum*)

Cyrtanthus Gr. 'kyrtos', curved; and Gr. 'anthos', flower; for the curved flower tube. (*Amaryllidaceae*)

cyrtophyllus Gr. 'kyrtos', curved; and Gr. 'phyllon', leaf; for the distal leaf halves, which are rolled back. (*Aloe*)

D

dabenorisanus For the occurrence in the Dabenoris mountain range, Northern Cape, RSA. (*Aloe*)

dactylifer Gr. 'daktylos', finger; and Lat. '-fer, -fera, -ferum', -carrying; (**1**) for the finger-like stem segments. (*Cumulopuntia*) – (**2**) for the finger-like petal appendages. (*Echeveria*)

dactylopsis Gr. 'daktylos', finger; and Gr. '-opsis', looking like; for the leaf shape. (*Cotyledon orbiculata* var.)

daigremontianus For Monsieur & Madame Daigremont (fl. 1914), French collectors of *Crassulaceae*. (*Kalanchoe*, *Sedum*)

dalettiensis For the occurrence near Daletti, Ethiopia. (*Euphorbia*)

damannii For Pater Damann, catholic missionary in Angola. (*Ceropegia*)

damaranus For the occurrence in Damaraland, Namibia. (*Euphorbia*)

dammannianus For the horticultural establishment of Dammann & Co. (fl. 1899) in Napoli, Italy. (*Echidnopsis*)

dandyanus For James E. Dandy (1903–1976), British botanist. (*Sedum obtusipetalum* ssp.)

dangeardii For Prof. Pierre A. Dangeard (1862–1947), French botanist. (*Kalanchoe velutina* ssp.)

danguyi For Paul A. Danguy (1862–1942), French botanist in Paris. (*Xerosicyos*)

darbandensis For the occurrence at Dar Banda, Central African Republic. (*Euphorbia*)

darrahianus For Charles Darrah († 1903), English succulent plant enthusiast in Manchester, whose widow presented his large plant collection to the City of Manchester to form a public collection. (*Opuntia*)

darwinii For Charles R. Darwin (1809–1882), British naturalist and evolutionary biologist. (*Maihueniopsis*)

dasyacanthus Gr. 'dasys', dense, rough, shaggy; and Gr. 'akantha', spine. (*Echinocereus*, *Escobaria*, *Euphorbia*, *Mammillaria laui* ssp.)

dasylirioides Gr. '-oides', resembling; and for the genus *Dasylirion* (*Nolinaceae*). (*Agave*)

Dasylirion Gr. 'dasys', dense, rough, shaggy; and Gr. 'leirion', lily; presumably for the long and untidy appearance of the leaves. (*Nolinaceae*)

dasyphyllus Gr. 'dasys', dense, rough, shaggy; and Gr. 'phyllon', leaf; (**1**) for the pubescent leaves. (*Pelargonium*, *Sedum*) – (**2**) for the papillate leaves. (*Antimima*)

datylio From the local vernacular name of the plant in Mexico. (*Agave*)

dauanus For the occurrence by the Dawa Parma River, Kenya. (*Euphorbia*)

Dauphinea For the occurrence near Fort Dauphin (= Tolanaro), SE Madagascar. (*Lamiaceae*)

dauphinensis As above. (*Talinella*)

davidbramwellii For Dr. David Bramwell (*1942), British botanist and director of the Jardín Canario in Tafira Alta, Gran Canaria. (*Aeonium*)

davisii For Colonel Jefferson Davis (fl. 1856), US-American military and Secretary of War. (*Cylindropuntia*) – (**2**) For Peter H. Davis (1918–1992), botanist in Edinburgh and editor of the 9-volume 'Flora of Turkey'. (*Rosularia*, *Sempervivum*) – (**3**) For Allie R. Davis (fl. 1939), US-American cactus dealer in Marathon (Texas). (*Echinocereus viridiflorus* ssp.)

davyanus For Dr. Joseph Burtt Davy (1870–1940), British botanist working in RSA 1903–1919, Chief of Division of Botany, Dept. of Agriculture, RSA. (*Aloe greatheadii* var.)

davyi As above. (*Delosperma*, *Euphorbia*, *Miraglossum*)

dawei For M. T. Dawe (fl. 1906), British forester in Uganda and Curator of the Entebbe Botanical Garden. (*Aloe*, *Euphorbia*, *Sansevieria*)

dawsonii For Dr. Elmer Yale Dawson (1918–1966), US-American botanist specialising in algae, Curator of Cryptogams at the US National Museum, also with strong interest in cacti, died by drowning in the Red Sea whilst collecting algae. (*Melocactus curvispinus* ssp.)

dealbatus Lat., whitened, chalked; for the leaves. (*Dracophilus*)

deamii For Charles C. Deam (1865–1953), US-American forester and self-taught botanist from Indiana, collected cacti in Guatemala. (*Opuntia*)

debilis Lat., weak; (**1**) for the growth habit. (*Ceropegia linearis* ssp., *Euphorbia ephedroides* var., *Lampranthus*, *Sedum*) – (**2**) for the leaves. (*Agave*)

debilispinus Lat. 'debilis', weak; and Lat. '-spinus', -spined. (*Euphorbia*)

deboeri For Dr. Hindrik W. de Boer (1885–1970), Dutch food chemist, director of a food investigation laboratory, and succulent plant specialist. (*Lithops villetii* ssp.)

debranus For the occurrence at Debre Berhan, Ethiopia. (*Aloe*)

debreczyi For Dr. Zsolt Debreczy (fl. 2002), Hungarian cactus specialist. (*Opuntia*)

decaisneanus For Prof. Dr. Joseph Decaisne (1807–1882), Belgian botanist in Paris and student of *Asclepiadaceae*. (*Ceropegia*, *Orbea*)

decandrus Gr. 'deka', ten; and Gr. 'aner, andros', man, [botany] stamen. (*Zaleya*)

Decarya For Raymond Decary (1891–1973), French financial administrator, botanist and plant collector in Madagascar 1916–1944. (*Didiereaceae*)

decaryi As above. (*Delonix*, *Euphorbia*, *Operculicarya*, *Pachypodium*, *Senecio*, *Stapelianthus*, *Uncarina*, *Xerosicyos*)

deceptor Lat., swindler, impostor; because of the earlier confusion of the plants with *Crassula deltoidea*. (*Crassula*)

deceptus Lat., deceptive; for the camouflaged appearance in habitat. (*Euphorbia*)

deciduus Lat., deciduous; (**1**) for the deciduous branches. (*Euphorbia*) – (**2**) for the deciduous leaves. (*Ceropegia*, *Cissus*, *Crassula*, *Drosanthemum*, *Phyllobolus*)

decipiens Lat., deceiving; (**1**) for the superficial resemblance to another taxon. (*Agave*, *Brachystelma*, *Haworthia*, *Huerniopsis*, *Mammillaria*, *Portulaca pilosa* ssp., *Tylecodon*, *Yucca*) – (**2**) for the similarity of the sterile plants to species of *Sedum*. (*Sedum*)

declinatus Lat., bent or curved downwards or forwards; for the horizontally spreading inflorescences (*Yucca*)

decliviticola Lat. 'declivis', slope; and Lat. '-cola', -dwelling. (*Euphorbia*)

decoratus Lat., decorated; for the prominent leaf markings. (*Conophytum uviforme* ssp.)

decorsei For Dr. J. Decorse, French botanist and entomologist, collecting in Madagascar 1898–1900. (*Aloe*, *Euphorbia*, *Sarcostemma*)

decorticans Lat., with peeling bark. (*Copiapoa*, *Portulaca*)

decorus Lat. 'decorus', graceful, noble. (*Aeonium*, *Erythrina*, *Piaranthus*, *Rauhia*, *Trichodiadema*)

decumbens Lat., decumbent; (**1**) for the growth habit. (*Aloe gracilis* var., *Coleocephalocereus fluminensis* ssp., *Crassula*, *Haageocereus*, *Matelea*, *Opuntia*, *Ruschia*, *Tetragonia*) – (**2**) for the decumbent inflorescence. (*Echeveria*)

decurrens Lat., decurrent; for the leaf base. (*Ruschia*)

decurvans Lat., becoming down-curved; for the branches of the inflorescence. (*Ruschia*)

decurvatus Lat., decurved; for the downwards curving branches. (*Phyllobolus*)

decurvus Lat., decurved; for the orientation of the inflorescences. (*Aloe*)

dedzanus For the occurrence on Dedza Mountain, Malawi. (*Aloe cameronii* var., *Euphorbia*)

defectus Lat., faulty, failure; for the lack of staminodes. (*Antimima*)

deficiens Lat., wanting, lacking; (**1**) for the low number of petals. (*Stomatium*) – (**2**) application obscure. (*Kalanchoe*)

deflersianus For Albert Deflers (1841–1921), French ranger and later botanist, travelling on the Arabian Peninsula in 1887. (*Ceropegia aristolochioides* ssp., *Orbea*)

deflersii As above. (*Senecio*)

deflexus Lat., deflexed; for the disposition of the shoots. (*Lampranthus*)

defoliatus Lat., with leaves fallen; (**1**) for the habit of flowering from leafless tubers. (*Sinningia*) – (**2**) for the rapidly withering leaves. (*Aridaria noctiflora* ssp.)

deformis Lat., misshapen; perhaps for the short almost sessile inflorescences. (*Haemanthus*)

deherdtianus For Cyril De Herdt (*1931) Belgian nurseryman and director of the well-known De Herdt nursery in Antwerpen, Belgium. (*Mammillaria*)

deightonii For Frederick C. Deighton (1903–1992), British mycologist, 1926–1949 at the Dept. of Agriculture, Njala, Sierra Leone. (*Ceropegia*, *Euphorbia*)

Deilanthe Gr. 'deile', evening; and Gr. 'anthos', flower; because the flowers of the type species open in the evening. (*Aizoaceae*)

deilanthoides Gr. '-oides', resembling; and for the genus *Deilanthe* (*Aizoaceae*). (*Delosperma*)

deinacanthus Gr. 'deinos', dreadful, terrible; and Gr. 'akantha', thorn, spine; for the spination. (*Melocactus*)

dejagerae For Ina de Jager (fl. 1929–1930), without further data. (*Drosanthemum*, *Ruschia*)

dejectus Lat., low, fallen; perhaps for the low growth. (*Crassula*, *Opuntia*)

dekenahii For Albert Jacob ("Japie") Dekenah (*1907), professional photographer and plant enthusiast in Riversdale, RSA. (*Haworthia magnifica* var.) – (**2**) For Ivor Dekenah (*1904), South African magistrate and plant enthusiast, collecting plants mainly in the area of Fraserberg, Northern Cape. (*Antimima*)

dekindtii For Eugene Dekindt, German who collected plants in Angola 1899–1902. (*Euphorbia*)

dekosterianus For a Mr. De Koster (fl. 1864), without further data. (*Beschorneria yuccoides* ssp.)

delaetianus For Frans (Frantz) de Laet (1866–1928 [1929?]), Belgian coffee importer and succulent expert and horticulturist in Contich (Kontich). (*Coryphantha*, *Dracophilus*, *Hoodia officinalis* ssp., *Opuntia*)

delaetii As above. (*Argyroderma*, *Echinocereus longisetus* ssp., *Gymnocalycium*)

delagoensis For the occurrence at Delagoa Bay, Moçambique, where the type was collected. (*Aspidoglossum*, *Kalanchoe*, *Urginea*)

delamateri For Rick DeLamater (1954–1989), student of botany at Arizona State University (USA). (*Agave*)

deleeuwiae For Mrs. de Leeuw (fl. 1966), without further data. (*Delosperma*)

delgadilloanus For Dr. José Delgadillo (fl. 2001), botanist at the Universidad Autónoma de Baja California in Enseñada, Mexico. (*Cylindropuntia californica* var.)

delicatulus Dim. of Lat. 'delicatus', delicate; for the growth. (*Drosanthemum*)

delicatus Lat., delicate; (**1**) for the growth. (*Brachystelma*, *Coryphantha*, *Sedum minimum* ssp.) – (**2**) for the more slender flower tubes. (*Jasminocereus thouarsii* var.)

deliciosus Lat., delicious; for the edible fruits. (*Carpobrotus*)

delicus Lat., weaned; application obscure. (*Sedum eriocarpum* ssp.)

Delonix Gr. 'delos', evident, conspicuous; and Gr. 'onyx', claw; for the long claw of the petals. (*Fabaceae*)

Delosperma Gr. 'delos', evident, conspicuous; and Gr. 'sperma', seed; because the seeds are visible in the open fruit. (*Aizoaceae*)

delphinensis Lat., of the dolphin (Fr. 'dauphin'); for the occurrence near Fort Dauphin, Madagascar. (*Aloe*, *Euphorbia*)

delphinoides Gr. 'delphinos', dolphin; and Gr. '-oides', resembling; for the resemblance of the leaf shape to a dolphin's snout. (*Cheiridopsis*)

deltoideodontus Gr. 'deltoides', delta-shaped; and Gr. 'odous, odontos', tooth; for the leaf marginal teeth. (*Aloe*)

deltoides Lat., deltoid, triangular; for the leaf shape. (*Oscularia*)

deltoideus Lat., deltoid, triangular; for the leaf shape. (*Cissus microdonta* fa., *Crassula*)

delus Gr. 'delos', open, evident; because the yellow staminodes are visible almost as soon as the petals show. (*Phyllobolus*)

deminuens Lat., diminishing; for the smaller overall size. (*Euphorbia strangulata* ssp.)

deminutus Lat., small, reduced, diminutive; (**1**) for the small size of the plants. (*Crassu-*

la setulosa var., *Echeveria setosa* var., *Frailea pumila* ssp., *Rebutia, Ruschia*) – (**2**) for the small leaves. (*Acrodon*)
demissus Lat., low, humble, drooping; for the decumbent habit. (*Aspidoglossum, Euphorbia*)
denboefii For Mr. J. L. den Boef (fl. 1983), Dutch visitor to Kenya who first collected the taxon during a random stop on a group tour. (*Orbea*)
dendriticus Lat., branched like a tree; for the leaf markings. (*Lithops pseudotruncatella* ssp.)
Dendrocereus Gr. 'dendron', tree; and for *Cereus*, a genus of columnar cacti; for the large size of adult plants. (*Cactaceae*)
dendroides Lat., tree-like (from Gr. 'dendron', tree); for the growth form. (*Euphorbia*)
dendroideus Lat., tree-like (from Gr. 'dendron', tree); for the growth form. (*Sedum*)
Dendroportulaca Gr. 'dendron', tree; and for the genus *Portulaca* ("Purslane"; *Portulacaceae*); for the similar fruits but differing in the shrubby growth. (*Portulacaceae*)
Dendrosicyos Gr. 'dendron', tree; and Gr. 'sicyos', cucumber; for the large size of the plants and their family classification. (*Cucurbitaceae*)
denegrii For Mr. Denegri (fl. 1923), Mexican Minister of Agriculture. (*Obregonia*)
denisianus For Dr. Marcel Denis (1897–1929), French botanist and expert on Madagascan *Euphorbia* species. (*Euphorbia*)
Denmoza Anagram of Mendoza, name of the Argentinian province, which is part of the geographic range of the genus. (*Cactaceae*)
densiareolatus Lat. 'densus', dense; and Lat. 'areolatus', with areoles. (*Pilosocereus*)
densiflorus Lat. 'densus', dense; and '-florus', -flowered. (*Bulbine, Cistanthe, Dudleya, Kalanchoe, Pachypodium, Ruschia, Sedum wrightii* ssp.)
densifolius Lat. 'densus', dense; and '-folius', -leaved. (*Lampranthus*)
densipetalus Lat. 'densus', dense; and Lat. 'petalum', petal. (*Lampranthus*)
densipunctus Lat. 'densus', dense; and Lat. 'punctum', dot, spot; for the densely spotted leaves. (*Conophytum quaesitum* ssp.)

densirosulatus Lat. 'densus', dense; and 'rosulatus', rosulate; for the crowded rosette leaves. (*Sinocrassula*)
densispinus Lat. 'densus', dense; and '-spinus', -spined. (*Echinopsis, Frailea buenekeri* ssp., *Mammillaria*)
densus Lat., dense; (**1**) for the compact growth habit. (*Trichodiadema*) – (**2**) presumably for the growth habit. (*Psilocaulon*)
dentatus Lat., toothed; (**1**) for the prominently tubercular ribs. (*Euphorbia heptagona* var.) – (**2**) for the leaf margins. (*Crassula, Haworthia floribunda* var.) – (**3**) for the leaf tip. (*Othonna*)
denticulatus Lat., minutely toothed; (**1**) for the leaves. (*Ceropegia, Cheiridopsis, Crassula lanceolata* ssp.) – (**2**) for the petal tips. (*Delosperma*)
denticulifer Lat. 'denticulus', small tooth; and Lat. '-fer, -fera, -ferum', -carrying; for leaf-margins. (*Haworthia chloracantha* var.)
dentiens From Lat. 'dentire', becoming teeth; for the small leaf margin teeth. (*Agave cerulata* ssp.)
dentonii For William C. Denton (1886 or 1887–1953), English succulent plant hobbyist who created the hybrid. (*Euphorbia*)
denudatus Lat., denuded, stripped, worn off; for the few appressed spines. (*Gymnocalycium*)
depauperatus Lat., depauperate; for the small 'starved' habit. (*Opuntia, Sedum radiatum* ssp.)
dependens Lat., suspended, hanging down; (**1**) for the growth form. (*Chasmanthera, Cleistocactus, Crassula, Quaqua*) – (**2**) for the branches. (*Lampranthus*)
depressus Lat., depressed, flattened; (**1**) for the low shrubby growth. (*Opuntia, Ruschia*) – (**2**) for the leaves appressed to the ground. (*Glottiphyllum*) – (**3**) for the flattened fused leaf pairs. (*Conophytum*) – (**4**) perhaps for the flattish inflorescence, or for the flatly spreading leaves. (*Massonia*) – (**5**) application obscure. (*Aloe brevifolia* var., *Crassula*)
derenbergianus For Dr. Julius Derenberg (1873–1928), German physician and succulent plant collector in Hamburg, with a spe-

cial interest in Mesembs, and a friend of K. Dinter and G. Schwantes. (*Cheiridopsis, Ebracteola*)

derenbergii As above. (*Echeveria*)

derustensis For the occurrence near De Rust, Western Cape, RSA. (*Haworthia blackburniae* var.)

descampsii For G. Descamps, Belgian plant collector in Zaïre 1890–1896. (*Monadenium*)

descoingsii For Dr. Bernard Descoings (*1931), French botanist and specialist on Madagascar. (*Aloe, Cynanchum, Senecio*)

deserti Lat., of the desert. (*Agave, Aloe, Escobaria*)

deserticola Lat. 'desertus', deserted, desert; and Lat. '-cola', inhabiting; for the occurrence in very arid regions. (*Echinopsis, Juttadinteria*)

desertorum Lat., of the deserts. (*Pelargonium*)

desertus Lat., abandoned, forsaken, solitary; for the occurrence in a deserted landscape. (*Cylindropuntia*)

desmetianus For Louis De Smet (1813–1887), Belgian horticulturist and nursery owner. (*Agave*)

desmondii For R. Desmond Meikle (*1923), British botanist at RBG Kew. (*Euphorbia*)

despainii For Mr. Kim Despain (fl. 1978), US-American student of the flora of Utah. (*Pediocactus*)

despoliatus Lat., deprived, robbed; for the leafless branches. (*Euphorbia*)

devius Lat., off the way, out of the way, deviating; for having 5 instead of 6 calyx segments like related taxa. (*Conophytum*)

devosianus For A. Devos (fl. 1855), (Belgian?) plant collector in Brazil for the Verschaffelt Nurseries. (*Codonanthe*)

dewetii For J. F. de Wet (fl. 1937), Headmaster of Vryheid Junior School, RSA. (*Aloe*)

deweyanus For Lyster H. Dewey (1865–1944), fibre specialist at the US Dept. of Agriculture. (*Agave vivipara* var.)

dewinteri For Dr. Bernard de Winter (*1924), botanist at the Botanical Research Institute, Pretoria, RSA. (*Aloe*)

dhofarensis For the occurrence in the Dhofar [Dhofar] Prov., Oman. (*Cibirhiza, Echidnopsis scutellata* ssp., *Euphorbia*)

dhofaricus As above. (*Jatropha, Portulaca*)

dhufarensis For the occurrence in the Dhufar [Dhofar] Prov., Oman. (*Aloe*)

diabolicus Lat., diabolical; application unknown. (*Adromischus*)

dianthiflorus For the genus *Dianthus* ("Pinks"; *Caryophyllaceae*); and Lat. '-florus', -flowered; for the fringed petals. (*Alsobia*)

diaz-romeroanus For Dr. Belisario Díaz Romero (fl. 1920), Bolivian naturalist. (*Pereskia*)

diazlunanus For Carlos L. Díaz Luna (fl. 1993), Mexican biologist and founder of the herbarium in Guadalajara. (*Pedilanthus*)

dicapuae For Mrs. Ernesta Di Capua (fl. 1904), Italian botanist. (*Caralluma*)

dichondrifolius For *Dichondra micrantha* (*Convolvulaceae*); and Lat. '-folius', -leaved; for the leaves, which are similar in shape, texture and indumentum. (*Pelargonium*)

dichotomus Lat., dichotomous, division in pairs; (**1**) for the branching of the stems. (*Aloe*) – (**2**) for the occasional apparently dichotomous branching. (*Ceropegia, Crassula*) – (**3**) for the branching of the inflorescence. (*Eberlanzia*)

dichrous Gr. 'dichroos', bi-coloured; (**1**) for the branches striped in two shades of green. (*Euphorbia*) – (**2**) for the two-coloured petals. (*Ruschia*)

dickisoniae For Mrs. Shirley Dickison (fl. 1982), US-American cactus collector from Texas. (*Turbinicarpus schmiedickeanus* ssp.)

Dicrocaulon Gr. 'dikros', fork; and Gr. 'kaulos', stem; because the long shoots are overtopped by lateral shoots after flowering, giving the appearance of a fork. (*Aizoaceae*)

dictyanthus Gr. 'diktyon', net; and Gr. 'anthos', flower; for the pattern of the corolla. (*Matelea*)

Didelta Gr. 'di', double; and Gr. 'delta', the letter D or a triangle; perhaps for the double involucral triangular scales (Jackson 1990). (*Asteraceae*)

Didierea For Col. Alfred Grandidier (1836–1921), French pioneer explorer and chronicler of Madagascar. (*Didiereaceae*)

didiereoides Gr. '-oides', resembling; and for the genus *Didierea* (*Didiereaceae*). (*Euphorbia*)

Didymaotus Gr. 'didymos', twice, double; and Gr. 'aotos', flower; for the laterally appearing flowers, one on each side of the plant. (*Aizoaceae*)

didymocalyx Gr. 'didymos', twice, double; and Gr. 'calyx', calyx; for the double whorl of sepals. (*Sedum*)

diegoi For Carlos M. Diego Legrand (*1901), Uruguayan botanist and specialist of the genus *Portulaca*. (*Portulaca*)

dielsianus For Prof. Dr. Friedrich L. E. Diels (1874–1945), German botanist in Berlin and at some time director of the botanical garden Berlin. (*Haworthia cooperi* var., *Ruschia*)

dielsii As above. (*Sedum*)

diersianus Nach Prof. Lothar Diers (fl. 1981), German cactus specialist in Köln. (*Pilosocereus*)

dieterlea For Jennie van Ackeren Dieterle (*1909), US-American botanist at the University of Michigan Herbarium and *Cucurbitaceae* specialist. (*Parasicyos*)

difficilis Lat., difficult; (**1**) probably for the difficulty in classifying the taxon. (*Aspidoglossum*) – (**2**) for the difficulty to cultivate the plants. (*Coryphantha*)

difformis Lat., unevenly or differently formed, unlike what is usual; (**1**) for the dimorphic leaves. (*Rabiea, Stomatium*) – (**2**) for the appearance. (*Agave, Glottiphyllum*)

diffractens Lat., breaking into pieces, shattering; for the easily detached bracts. (*Echeveria*)

diffusus Lat., diffuse, widely spreading, scattered; (**1**) for the growth form. (*Caralluma, Praecereus euchlorus* ssp., *Sedum, Trianthema*) – (**2**) for the low, flat, ill-defined tubercles of the plant bodies. (*Lophophora*) – (**3**) for the scattered flowers. (*Villadia*)

digitatus Lat., digitate; (**1**) for the digitate leaves. (*Adansonia, Adenia, Cucurbita*) – (**2**) for the position of the two different leaves of a pair in a "finger and thumb" manner. (*Phyllobolus*)

Digitostigma Lat. 'digitus', finger; and Gr. 'stigma', stigma, spot, dot; for the long and narrow tubercles with a white-dotted epidermis. (*Cactaceae*)

diguetii For Léon Diguet (1859–1926), French chemist and explorer in Mexico. (*Ferocactus, Fouquieria, Pereskiopsis*)

digynus Gr. 'di-', two; and Gr. 'gyne', woman; for the often double central flowers. (*Portulaca*)

dilatatus Lat., broadened, expanded; (**1**) for the broad leaves. (*Bijlia*) – (**2**) for the broad sepals. (*Ruschia*)

dillenii For Prof. John J. Dillen [Dillenius] (1684–1747), German botanist from Darmstadt, later in Oxford. (*Opuntia*)

dilunguensis For the occurrence in the Dilungu area of the Biano Plateau, Shaba Prov., Zaïre. (*Monadenium*)

dilutus Lat., diluted; probably for the very pale pink petals. (*Lampranthus*)

diminutus Lat., diminutive; for the small size of the plants. (*Sedum*)

dimorphanthus Gr. 'dimorphus', in two forms; and Gr. 'anthos', flower; for the presence of both bisexual and male flowers on the same plant. (*Tribulocarpus*)

dimorphophyllus Gr. 'dimorphus', in two forms; and Gr. '-phyllus', -leaved. (*Sedum*)

dimorphus Gr., in two forms; (**1**) for the dissimilar vegetative and flowering stems. (*Ceropegia*) – (**2**) because cultivated plants are much larger. (*Haworthia marumiana* var.) – (**3**) for the differing leaves on young and old branches. (*Psilocaulon*) – (**4**) for the stellate or cage-like flowers. (*Brachystelma*)

dinae For Dina Buining (fl. 1973), wife of the Dutch cactus enthusiast Albert Buining. (*Arrojadoa*)

dinklagei For W. Dinklage (fl. 1985), horticulturist at the Botanical Garden Heidelberg, Germany. (*Kalanchoe*)

Dinteranthus For Prof. Kurt M. Dinter (1868–1945), German botanist famous for his explorations in Namibia; and Gr. 'anthos', flower. (*Aizoaceae*)

dinteri For Prof. Kurt M. Dinter (1868–1945),

German botanist famous for his explorations in Namibia. (*Aizoanthemum*, *Aloe*, *Avonia*, *Brachystelma*, *Ceropegia*, *Commiphora*, *Lithops*, *Plectranthus*, *Psilocaulon*)

dioicus Lat., dioecious. (*Jatropha*, *Mammillaria*, *Phytolacca*)

diolii For Dr. Maurizio Dioli (fl. 1995), Italian veterinary officer resident in Kenya, later in Ethiopia. (*Aloe*)

Dioscorea For Pedanios Dioscorides, most influential Greek physician and herbalist of the first century C.E. (*Dioscoreaceae*)

dioscoridis From the old name 'Dioscoris' or 'Dioscorida' for Socotra; for the occurrence there. (*Duvaliandra*)

dipageae For Mrs. Di Page (fl. 1993), South African naturalist and specialist in the Swartkops Valley Bushveld vegetation. (*Drosanthemum*)

Dipcadi Perhaps the ancient oriental name for some species today classified as *Muscari* ("Grape Hyacinth"). (*Hyacinthaceae*)

dipetalus Gr. 'di-', double, two; and Gr. 'petalon', petal; for the 2-petalled flowers. (*Pelargonium*)

diphyllus Gr., two-leaved. (*Bulbine*)

diplocyclus Gr. 'diplo-', double; and Gr. 'cyclus', circle; application obscure. (*Aeonium*)

diploglossus Gr. 'diplo-', double; and Gr. 'glossa', tongue; for the inner appendix of the corona segments. (*Aspidonepsis*)

Diplosoma Gr. 'diplo-', double; and Gr. 'soma', body; for the connate leaves of a pair, or for the 2 different leaf pairs formed during the year. (*Aizoaceae*)

dipterus Gr. 'di-' double; and Gr. 'pteron', wing; (**1**) for the two lateral wings on the fruit. (*Tetragonia*) − (**2**) application obscure. (*Hoya*)

Dischidia Gr. 'dischides', cleft in two; for the apically bifid staminal corona. (*Asclepiadaceae*)

Dischidiopsis Gr. '-opsis', similar to; and for the genus *Dischidia* (*Asclepiadaceae*). (*Asclepiadaceae*)

disciformis Lat. 'discus', disc; and Lat. '-formis', -shaped; for the shape of the plant body. (*Strombocactus*)

Discocactus Lat. 'discus', disc; and Lat. 'cactus', cactus; for the depressed-globose to flat disc-like plant bodies. (*Cactaceae*)

discoideus Lat., disc-shaped; (**1**) for the all-disc florets. (*Othonna carnosa* var.) − (**2**) for the rounded shape of the bract cup. (*Monadenium*) − (**3**) for the shape of the corona. (*Brachystelma*)

discolor Lat., many-coloured, with different colours; (**1**) for the spination. (*Mammillaria*) − (**2**) for the colour markings below each areole. (*Opuntia*) − (**3**) because of the differently coloured leaf tips. (*Rhodiola*) − (**4**) application obscure. (*Pachycormus*)

discrepans Lat., differing; for the differences from the closely related *E. tetracanthoides*. (*Euphorbia*)

disepalus Gr. 'di-', two-; and Gr. 'sepalon', sepal. (*Lewisia*)

disgregus Lat. 'dis-', not, separate; and perhaps from Lat. 'grex, gregis', flock; perhaps in the sense of "separate from the flock", i.e. disagree, since the plant was first known under a wrongly applied name. (*Lampranthus*)

Disocactus Shortened compound form of Gr. 'dis-', twice, and Gr. 'isos', the same; and Lat. 'cactus', cactus; for the leaf-like flattened ("with two identical sides") stems. (*Cactaceae*)

dispar Lat., different, unequal; for the different shape of the leaves within a pair. (*Gibbaeum*)

dispermus Gr. 'di-', two; and Gr. 'sperma', seed; for the 2-seeded fruits. (*Parakeelya*, *Sedum*)

dispersus Lat., scattered (from Lat. 'dispergere', to scatter), for the scattered occurrence in contrast to the crowded populations of related taxa. (*Euphorbia*)

Disphyma Gr. 'dis-', double; and Gr. 'phyma', tumour, swelling; for the two large closing bodies of the fruit capsules of most species. (*Aizoaceae*)

dissectus Lat., dissected; for the deeply lobed leaves. (*Jatropha*, *Kedrostis*)

dissimilis Lat., dissimilar; (**1**) for the variable flowers of the taxon. (*Aspidoglossum*) − (**2**) because the taxon has no close affinities with any other. (*Bulbine*, *Lampranthus*, *Rhipsalis*)

dissitispinus Lat. 'dissitus', well-spaced, remote; and Lat. '-spinus', -spined. (*Euphorbia*)
dissitus Lat., well-spaced, remote; for the long internodes. (*Mitrophyllum*)
distans Lat., distant, standing apart; for the long internodes (i.e. with distant nodes). (*Aloe, Antimima, Erepsia*)
distichus Lat., distichous, two-ranked; for the leaf arrangement. (*Boophane, Gasteria*)
distinctissimus Superl. of Lat. 'distinctus', distinct, different. (*Euphorbia*)
distinctus Lat., distinct, different. (*Ceropegia, Huernia, Orbea*)
distortus Lat., distorted; for the branches. (*Tetragonia*)
distylus Gr. 'di', double; and Gr. 'stylos', style. (*Sesuvium*)
diutinus Lat., long-lasting; because the flowers remain open until late in the day. (*Lampranthus*)
divaricatus Lat., spreading, divaricate; (**1**) for the growth. (*Aichryson, Harrisia, Nolana, Stapelia*) – (**2**) for the branching of the inflorescences. (*Aloe, Ruschia*)
divergens Lat., divergent; (**1**) for the spreading leaves. (*Lithops*) – (**2**) for the widely spreading follicles. (*Sedum*)
diversifolius Lat. 'diversus', different; and Lat. '-folius', -leaved. (*Drosanthemum, Haworthia nigra* var., *Hoya, Ruschia, Sinocrassula, Stathmostelma, Trochomeriopsis*)
diversiphyllum Lat., 'diversus', different; and Gr. 'phyllon', leaf; for the unequal size of the leaves on a branchlet. (*Cephalophyllum*)
divisus Lat., divided; application obscure. (*Gunniopsis*)
dixanthocentron Gr. 'di-', two; Gr. 'xanthos', yellow; and Gr. 'kentron', centre; for the often 2 yellow central spines. (*Mammillaria*)
dodii For Anthony H. Wolley-Dod (1861–1948), British soldier and botanist collecting in S Africa 1900–1901. (*Crassula*)
dodomaensis For the occurrence near Dodoma, Central Prov., Tanzania. (*Portulaca*)
dodrantalis Lat., three quarters long, i.e. the distance between thumb and little finger when extended (± 24 cm); for the inflorescence size. (*Aeonium*)

dodsonianus For Dr. John ('Jay') W. Dodson (1901–1999), US-American accountant, succulent plant enthusiast and founder of the International Succulent Institute (now International Succulent Introductions, ISI). (*Pseudolithos*)
dodsonii As above. (*Mammillaria deherdtiana* ssp.)
doei For Brian Doe (fl. 1965), Director of Antiquities in Aden, Yemen. (*Aloe*)
doellii For Wilhelm Döll (fl. 1853), German surveyor in Santiago de Chile who accompanied R. A. Philippi on his trip to the Atacama Desert. (*Peperomia*)
doelzianus For Bruno Dölz (1906–1945), German lawyer and cactus hobbyist, 1941–1945 president of the Deutsche Kakteen-Gesellschaft. (*Oreocereus*)
doinetianus For M. Doinet (fl. 1956), horticulturist in Glain-lez-Liège, Belgium. (*Euphorbia*)
dolabellus Dim. of Lat. 'dolabra', axe, hatchet; for the leaves, which are shaped like small axe blades. (*Peperomia*)
dolabriformis Lat. 'dolabra', axe, hatchet; and Lat. '-formis', -shaped; for the leaf-shape. (*Peperomia, Rhombophyllum*)
doldii For A. C. "Tony" Dold (fl. 2002), botanist at the Schoenland Herbarium, Rhodes University, Grahamstown, RSA. (*Haworthia cooperi* var., *Orbea*)
dolichanthus Gr. 'dolichos', long; and Gr. 'anthos', flower. (*Agave, Dischidia*)
dolichocarpus Gr. 'dolichos', long; and Gr. 'karpos', fruit. (*Caralluma*)
dolichoceras Gr. 'dolichos', long; and Gr. 'keras', horn; for the long horns of the nectary glands. (*Euphorbia*)
dolichocnemus Gr. 'dolichos', long; and Gr. 'knemis', splint; for the long pedicels. (*Socotrella*)
dolichophyllus Gr. 'dolichos', long; and Gr. 'phyllon', leaf. (*Ceropegia*)
dolichopodus Gr. 'dolichos', long; and Gr. 'pous, podos', foot; for the long leaf petiole. (*Plectranthus*)
dolichopus Gr. 'dolichos', long; and Gr. 'pous, podos', foot; for the elongated pachycaul stem. (*Cyphostemma humile* ssp.)

Dolichos Gr. / Lat., long; and also the ancient name of a cultivated legume with long pods; for the long fruits (*Fabaceae*)

dolichosiphon Gr. 'dolichos', long; and Gr. 'siphon', tube; for the long tubular flowers. (*Adenia*)

dolichospermaticus Gr. 'dolichos', long; and Lat. 'spermaticus', -seeded. (*Micranthocereus*)

dolomiticus For the occurrence on dolomite outcrops. (*Antimima, Delosperma, Gasteria batesiana* var., *Pelargonium, Plectranthus, Sempervivum*)

dolosus Lat., deceptive; perhaps because the taxon was previously misidentified. (*Sedum*)

domeykoensis For the occurrence near Domeyko in the Chilean Atacama Desert; Domeyko is named for Prof. Domeyko (fl. 1853), Santiago, Chile. (*Maihueniopsis*)

dominella Dim. of Lat. 'domina', Mistress; application unknown. (*Aloe*)

domingensis For the occurrence in Santo Domingo, Dominican Republic. (*Melocactus intortus* ssp., *Sarcopilea, Talinum*)

dongzhiensis For the occurrence in the Dongzhi District, Anhui Prov., China. (*Sedum*)

donkelaarii For André van Donkelaar (1783–1858), Belgian botanist in Gent. (*Selenicereus*)

dooneri For H. B. Dooner (fl. 1915), collector of plants in Kenya, without further data. (*Sansevieria*)

doranus For Dora Frey (fl. 2002), partner for life of Hansjörg Jucker, intrepid Swiss cactus enthusiast who widely travels in South America. (*Sulcorebutia*)

dorisiae For Doris Amerhauser (*1943), wife of the Austrian *Gymnocalycium* specialist Helmut Amerhauser. (*Gymnocalycium pflanzii* ssp.)

dorotheae For Miss Dorothy Westhead (fl. 1908), London, without further data. (*Aloe*) – (2) For Dr. Dorothea van Huyssteen (fl. 1935), daughter of Dr. D. P. van Huyssteen, a succulent plant collector in Bellville near Cape Town, RSA, who had cultivated the type specimen. (*Lithops*)

Dorotheanthus For Dorothea Schwantes (fl. 1927), mother of the German Mesemb specialist Prof. G. Schwantes; and Gr. 'anthos', flower. (*Aizoaceae*)

Dorstenia For Theodor Dorsten [Dorstenius] (1492–1552), German professor of medicine. (*Moraceae*)

Doryanthes Gr. 'dory', wood, trunk, lance, spear; and Gr. 'anthos', flower. (*Doryanthaceae*)

douglasii For Mr. David Douglas (1798–1834), Scottish plant collector for the Royal Horticultural Society in London, gored to death by a trapped bull when he fell into a wild-cattle pit in Hawaii. (*Sinningia*)

downsii For Philip E. Downs (*1938), British dentist and succulent plant enthusiast, 1967–1987 resident in RSA (*Sansevieria*)

drabii For Igor Dráb (fl. 2002), Czech cactus collector. (*Ariocarpus*)

Dracaena Lat. 'drago, draconis', female dragon (from Gr. 'drakon', dragon); from the vernacular name of *D. draco*, "Dragon's Blood Tree", which is based on the red exudate of the bruised stems. (*Dracaenaceae*)

draco Lat., dragon; see *Dracaena* for an explanation of the etymology. (*Dracaena*)

Dracophilus Gr. 'drakon', dragon, snake; and Gr. 'philos', friend; for the occurrence of the type species on the Drachenberg (Germ. 'Drachen', dragon), SW Namibia. (*Aizoaceae*)

dregeanus For Jean François Drège (1794–1881), German plant collector in S Africa 1826–1833. (*Crassula obovata* var., *Euphorbia, Galenia, Lampranthus*)

dregei As above. (*Begonia, Hoodia*) – (2) For Isaac L. Drège (1853–1921), apothecary in Port Elizabeth, RSA, son of C. F. Drège, collected plants in the Port Elizabeth area. (*Euphorbia ledienii* var.)

drepanophyllus Gr. 'drepane', sickle; and Gr. 'phyllon', leaf; for the leaf shape. (*Esterhuysenia*)

Drimia Gr. 'drimys', sharp, cutting; for the pointed capsules (Genaust 1983). (*Hyacinthaceae*)

Drosanthemum Gr. 'drosos', dew; and Gr. 'anthemon', flowering plant, flower; for the glittering papillae of the leaves. (*Aizoaceae*)

drouhardii For Eugène J. Drouhard (1874–1945), French agronomist and forester, father-in-law of the French botanist Perrier de la Bâthie, collected plants in Madagascar and Moçambique. (*Moringa*)

drummondii Probably for Thomas Drummond (1780–1835), Scottish collector and botanist, working in N America 1831–1835. (*Crassula*)

drupifer Lat. 'drupa', drupe; and Lat. '-fer, -fera, -ferum', -bearing. (*Elaeophorbia*)

drymarioides Gr. '-oides', resembling; for the similarity to *Drymaria cordata* (*Caryophyllaceae*). (*Sedum*)

drymophilus Gr. 'drymo', forest, and Gr. 'philos', loving; for the preferred habitat. (*Gynura*)

dualis Lat., of two; for the two-leaved branchlets. (*Antimima*)

dubitans Lat., doubting; for the uncertain generic placement of the taxon. (*Lampranthus*)

dubius Lat., dubious, doubtful; (**1**) for the uncertain taxonomic status. (*Jordaaniella*) – (**2**) because it is a rare (and possibly now extinct), hence dubious, species known from only one locality. (*Erepsia*)

dubniorum For Milos and Tomás Duben (fl. 2002), Czech cactus hobbyists. (*Ariocarpus*)

duchii For Joaquin Duch (fl. 1976), Venezuelan who discovered the taxon. (*Frailea cataphracta* ssp.)

duckeri For H. C. Ducker (fl. 1940), then in charge of the Cotton Experiment Station, Malawi. (*Aloe*)

Dudleya For Prof. William R. Dudley (1849–1911), US-American botanist at the Stanford University. (*Crassulaceae*)

dugueyi For "Monsieur l'Adjudant-Chef" Duguey (fl. 1916), a friend of the French botanist and physician Raymond Hamet. (*Sedum*)

dulcinomen Latinization of the name of the type locality near Dulces Nombres, Nuevo León, Mexico. (*Sedum*)

dulcis Lat., sweet, pleasant; (**1**) probably in the sense of 'attractive', for the flowers. (*Lampranthus*) – (**2**) perhaps for the pale ("sweet") pink flower colour. (*Astridia*)

dumeticola Lat. 'dumetum', thicket; and Lat. '-cola', -dwelling. (*Euphorbia*)

dumetorum Gen. Pl. of Lat. 'dumetum', thicket, hedge; for the occurrence under trees and bushes. (*Grusonia*, *Mammillaria schiedeana* ssp.)

dummeri For Richard A. Dummer [also written as Dümmer] (1887–1922), South African horticulturist, trained at Kew and since 1914 employed in Uganda where he was killed in a motorcycle accident. (*Orbea*)

dumortieri For Count Barthélemy C. J. Dumortier (1797–1878), Belgian politician and botanist. (*Isolatocereus*)

dumosus Lat., covered with (thorn) bushes (from Lat. 'dumetum', thicket); (**1**) for the growth form and the spination. (*Alluaudia*) – (**2**) perhaps for the habitat. (*Aichryson*, *Rhodiola*)

dumoulinii For Jan Dumoulin (fl. 1970s), Belgium-born curator of the Hester Malan Nature Reserve, RSA. (*Aloe krapohliana* var.)

duncanii For ex-Capt. Frank Duncan (fl. 1937–1945), on whose mining claims the taxon was first discovered. (*Escobaria*)

dunensis Lat., of the dune; for the preferred habitat. (*Delosperma*, *Erepsia*)

dunsdonii For Mr. L. Dunsdon (fl. 1924), without further data. (*Disphyma*)

duoformis Lat. 'duo', two; and Lat. '-formis', -shaped; for the existence of plants with straight central spines and plants with hooked central spines. (*Mammillaria*)

duplessiae For Miss Rosalie du Plessis (later Mrs. C. Gill) (fl. 1932–1955), staff member of the Bolus Herbarium, Cape Town, RSA. (*Drosanthemum*)

duplicatus Lat., double, duplicate; (**1**) because the taxon was first described under an illegitimate homonym name. (*Brachystelma*) – (**2**) because a new name was necessary when the basionym was transferred to the present genus. (*Agave*)

durangensis For the occurrence in the Mexican state of Durango. (*Agave*, *Coryphantha*, *Dasylirion wheeleri* var., *Opuntia*)

duranii For M. Duran, who collected a living plant of the taxon in 1951. (*Euphorbia*)

duripulpa Lat. 'durus', hard; and Lat. 'pulpa', fruit pulp, flesh; for the tough tissue of the plant body. (*Eriosyce napina* ssp.)

durispinus Lat. 'durus', hard; and Lat. '-spinus', -spined; for the rigid spines. (*Mammillaria polythele* ssp.)

duseimatus Gr. 'duseimatos', clad in rags; for the untidy appearance. (*Euphorbia*)

dussianus For Father Antoine Duss (1840–1924), Swiss clergyman and botanist in the West Indies. (*Agave*)

duthiae For Dr. Augusta V. Duthie (1881–1963), South African botanist at Stellenbosch University. (*Ruschia, Stomatium*)

duthiei For John F. Duthie (1845–1922), British botanist in India. (*Sedum*)

Duvalia For Henri Auguste Duval (1777–1814), French physician and botanist in Alençon, Normandy. (*Asclepiadaceae*)

Duvaliandra For the similarity to the genus *Duvalia* (*Asclepiadaceae*). (*Asclepiadaceae*)

dwequensis For the occurrence at the Dwequa River, Ceres Karoo, RSA. (*Tridentea*)

dybowskii Most probably for Jean Dybowski (1855–1928), topographical engineer in Bahia, Brazil. (*Espostoopsis*)

dyckii For Fürst Joseph Salm-Reifferscheid-Dyck (1773–1861), German (Prussian) botanist, botanical artist, horticulturist and succulent plant collector. (*Lampranthus*)

dyeri For Sir William T. Thiselton-Dyer (1843–1928), British botanist, Director of Kew 1885–1905. (*Aloe, Kalanchoe*) – (**2**) For Dr. Robert A. Dyer (1900–1987), South African botanist with a strong interest in succulents, 1944–1963 director of the Botanical Research Institute Pretoria. (*Delosperma, Raphionacme, Rhombophyllum*)

dyvrandae For Agathe Dyvranda (fl. 1914), without further data. (*Villadia*)

dzhavachischvilii For A. Dzhavachischvili (fl. 1969), Georgian plant collector. (*Sempervivum*)

E

eastwoodiae For Alice Eastwood (1859–1953), US-American botanist in California and long-time curator of the herbarium at the California Academy of Sciences. (*Ferocactus*, *Sedum laxum* ssp.)

eastwoodianus As above. (*Sesuvium*)

Eberlanzia For Friedrich G. Eberlanz (1879–1966), teacher and founder of the Museum at Lüderitzbucht, Namibia. (*Aizoaceae*)

eberlanzii As above. (*Lithops karasmontana* ssp.)

ebracteatus Lat. 'e, ex', without; and Lat. 'bracteatus', bracteate; (**1**) for the absence of bracts. (*Cephalophyllum, Eberlanzia, Lampranthus*) – (**2**) for the inconspicuous and thus easily overlooked bracts on the inflorescences. (*Sansevieria, Sedum*)

Ebracteola Lat. 'e, ex', without; and Lat. 'bracteola', bracteole, small bract; for the lack of bracts in some taxa. (*Aizoaceae*)

eburneus Lat., ivory-white; for the flower colour. (*Drosanthemum, Pachypodium rosulatum* var.)

ecalcaratus Lat. 'e-, ex', without; and Lat. 'calcaratus', spurred; because the leaves are spurless. (*Sedum*)

Echeveria For Atanasio Echeverría (fl. 1787), Mexican botanical artist of Basque origin who made drawings (never published) for Sessé & al., Flora Mexicana. (*Crassulaceae*)

echidne Gr. 'echidna', snake, adder; application obscure. (*Ferocactus*)

echidnopsioides Gr. '-oides', resembling; and for the genus *Echidnopsis* (*Asclepiadaceae*). (*Huernia*)

Echidnopsis Gr. 'echidna', snake, adder; and Gr. '-opsis', looking like; for the often creeping stems. (*Asclepiadaceae*)

echinarius Lat., having spines, spiny; for the well-formed central spines. (*Mammillaria elongata* ssp.)

echinatus Lat., prickly; (**1**) probably for the foliage. (*Lepidium*) – (**2**) for the prickly leaves. (*Delosperma*) – (**3**) for the persistent spine-like stipules. (*Pelargonium*) – (**4**) for the spiny fruits. (*Tetragonia*)

echinellus Dim. of Lat. 'echinus', hedge-hog; for the spiny inflorescence. (*Anthorrhiza*)

Echinocactus Lat. 'echinus', hedgehog; and Lat. 'cactus', cactus; for the globose plant bodies and the spination. (*Cactaceae*)

echinocarpus Gr. 'echinos', hedgehog; and Gr. 'karpos', fruit; (**1**) for the spiny fruits. (*Cylindropuntia*) – (**2**) for the spiny surface of the seeds. (*Cyphostemma*)

Echinocereus Lat. 'echinus', hedgehog; and *Cereus*, a genus of columnar cacti; for the spiny columnar plant bodies. (*Cactaceae*)

echinoides Gr. 'echinos', hedgehog; and Gr. '-oides', like, similar to; for the spiny habit of the plants. (*Copiapoa*)

echinoideus Gr. 'echinos', hedgehog; and Gr. '-oides', like, similar to; for the spiny habit of the plants. (*Coryphantha*)

Echinomastus Gr. 'echinos', hedgehog; and Gr. 'mastos', breast; for the spiny tubercles of the plant bodies. (*Cactaceae*)

Echinopsis Gr. 'echinos', hedgehog; and Gr. '-opsis', similar to; for the spiny globose plant bodies. (*Cactaceae*)

echinospermus Gr. 'echinos', hedgehog; and Gr. 'sperma', seed; for the long spines on the seed. (*Portulaca*)

echinulatus Lat., echinulate, with very small prickles; for the small prickles on leaves and inflorescences. (*Monadenium*)

echinus Lat., hedgehog, sea-urchin; for the spiny nature of the plants. (*Coryphantha, Eriosyce taltalensis* ssp., *Euphorbia*)

echios Lat. / Gr. ('echion'), "Viper's Bugloss", generally a plant against snakebite; perhaps for the similarity of the dense fine spination to the rough bristly hairiness of Viper's Bugloss. (*Opuntia*)

ecirrhosus Lat. 'e, ex', without; and Lat. 'cirrhosus', with tendrils; (**1**) for the absence of tendrils. (*Citrullus*) – (**2**) erroneously applied since the taxon does have tendrils. (*Cephalopentandra*)

ecklonianus For Christian Frederick Ecklon (1795–1868), Danish chemist and botanical explorer settling at the Cape. (*Tylecodon wallichii* ssp.)

ecklonii As above. (*Euphorbia, Plectranthus*)
ecklonis As above. (*Aloe, Delosperma, Galenia*)
Ectotropis Gr. 'ektos', out of, on the outside; and Gr. 'tropis', keel, ridge; most probably for the radial expanding keels of the fruit capsules. (*Aizoaceae*)
ectypus Lat. / Gr., embossed, with a relief; for the appearance of the fused leaf pairs. (*Conophytum*)
edentatus Lat. 'e, ex', without; and Lat. 'dentatus', toothed; for the unarmed leaf margins. (*Aloe*)
edentulus Lat. 'e-, ex-', without; and Lat. 'dentulus', small teeth; for the absence of small teeth on the leaf margins. (*Ruschia*)
edithae For Miss Edith Cole (1859–1940), Englishwoman who collected plants during a botanical expedition led by E. Lort-Phillips into N Somalia 1894–1895. (*Caralluma*)
Edithcolea As above. (*Asclepiadaceae*)
edmonstonei For Mr. T. Edmonstone (fl. 1835), English (?) botanical collector accompanying Darwin on his famous trip on HMS Beagle. (*Sesuvium*)
eduardoi For Dr. Eduardo J. S. M. Mendes (*1924), Portuguese botanist, collecting 1955–1956 in Angola, later director of the Lisbon herbarium. (*Euphorbia*)
edulis Lat., edible; (**1**) because the stems are edible. (*Caralluma*) – (**2**) because the tubers are edible. (*Brachystelma, Fockea*) – (**3**) because young inflorescences and leaves where eaten by the local indigenous people. (*Dudleya*) – (**4**) for the edible fruits. (*Carpobrotus*)
edwardii For Edward Taylor (1848–1928), British grower of succulent plants, esp. Mesembs. (*Conophytum piluliforme* ssp.)
edwardsiae For Miss Gwendoline Edwards (1888–1960), South African school teacher and plant collector. (*Drosanthemum, Lampranthus*) – (**2**) For Miss Sue Edwards (fl. 1977), British botanist in Ethiopia. (*Caralluma*)
edwardsii For James L. Edwards (1895–1972), US-American naturalist, explorer and engineer. (*Sedum*)

eendornensis For the occurrence at Eendorn, Warmbad District, Namibia. (*Antimima*)
egregius Lat., excellent. (*Mammillaria lasiacantha* ssp.)
ehrenbergii For Prof. Christian G. Ehrenberg (1795–1876), German biologist and professor at the University of Berlin. (*Sansevieria*)
eichlamii For Friedrich (Federico) Eichlam († 1911) from Hildburghausen, Germany, emigrating to Guatemala in 1892, cactus amateur and collector of Guatemalan cacti. (*Disocactus, Mammillaria voburnensis* ssp., *Myrtillocactus, Opuntia, Stenocereus*)
eilensis For the occurrence near Eil, Somalia. (*Duvalia, Euphorbia, Sansevieria*)
einsteinii For Prof. Albert Einstein (1879–1955), German / Swiss / US-American world-renowned physicist. (*Rebutia*)
eitapensis For the occurrence at Eitape in NE Papua New Guinea. (*Hoya*)
eitenii Either for George E. Eiten (*1923) or for Liene T. Eiten (1925–1979), who jointly collected the type of this taxon. (*Portulaca*)
ekmanii For Dr. Erik L. Ekman (1883–1931), Swedish botanist and explorer in Argentina, Brazil, Cuba and Hispaniola. (*Leptocereus, Mammillaria, Opuntia*)
elachistemmoides Gr. '-oides', resembling; and for the unpublished genus *Elachistostemma* Choux (*Asclepiadaceae*). (*Sarcostemma*)
Elaeophorbia Gr. 'elaia', olive; and for the genus *Euphorbia*; for the olive-like fruits and the relationship. (*Euphorbiaceae*)
elatinoides Gr. '-oides', resembling; and for the similarity to the genus *Elatine* ("Waterwort"; *Elatinaceae*). (*Crassula, Sedum*)
elatior Lat., taller (Comp. of Lat. 'elatus', tall); for the stature of the plant. (*Opuntia, Portulaca*)
elatus Lat., tall; (**1**) for the tall stems. (*Aloe, Opuntia, Yucca*) – (**2**) for the height of the plants. (*Mestoklema*)
elburzensis For the occurrence on Mt. Elburz, Iran. (*Sedum*)
elegans Lat., elegant; for the elegant appearance. (*Aloe, Callisia, Crassula, Duvalia, Echeveria, Jatropha gossypiifolia* var., *Lampranthus, Lampranthus, Mammillaria haageana* ssp., *Monadenium*)

elegantissimus Superl. of Lat. 'elegans', elegant; for the appearance. (*Euphorbia*)
elegantulus Dim. of Lat. 'elegans', elegant, selected, i.e. small and elegant. (*Brachystelma*)
elephantidens Lat. 'elephantus', elephant; and Lat. 'dens', tooth; for the large tubercles. (*Coryphantha*)
elephantipes Lat. 'elephantus', elephant; and Lat. 'pes', foot; for the caudex like an elephant's foot. (*Dioscorea, Yucca*)
elephantopus Lat. 'elephantus', elephant; and Gr. 'pous', foot; for the large caudex. (*Cyphostemma*)
elevatus Lat., elevated; for the position of the current year's fruit above the fruits of the last year. (*Antimima*)
elgonicus For the occurrence on Mt. Elgon on the Kenya-Uganda border. (*Aloe*)
elineatus Lat. 'e-, ex-', without; and Lat. 'lineatus', with lines; because the leaf sheaths are without a visible line of fusion. (*Ruschia*)
elinguis Lat. 'e-, ex', without; and Lat. 'lingua', tongue; because the corona segments have no appendages. (*Schizoglossum*)
eliseae For Mrs. Elise Bodley van Wyk (1922–1997), South African botanical illustrator who painted all known *Tylecodon* species. (*Cotyledon*)
elizae Unknown. (*Kalanchoe*)
elizondoanus For Jorge L. Elizondo († 1989), Mexican botanist with an interest in cacti. (*Opuntia*)
ellacombianus For Canon H. N. Ellacombe (fl. 1912), who first urged L. R. Praeger to undertake a revision of the cultivated Sedums. (*Phedimus*)
ellaphieae For Ellaphie Ward-Hilhorst (1920–1994), well-known botanical artist in RSA, painted numerous succulents. (*Gasteria, Pelargonium, Tylecodon*)
ellemeetianus For W. C. M. de Jonge van Ellemeet (1811–1888), Dutch plant fancier and friend of the German botanist G. Jacobi. (*Agave*)
ellenbeckianus For Dr. H. Ellenbeck, German physician who collected material for Berlin on Baron von Erlanger's expedition to E Africa in 1900–1901. (*Dorstenia, Dracaena*)

ellenbeckii As above. (*Adenia, Aloe, Euphorbia, Jatropha, Monadenium*)
elliotii For George F. Scott-Elliot (1862–1934), Indian-born British botanist. (*Aspidoglossum, Euphorbia*)
elliottii For C. F. Elliott, English botanical collector in Africa before 1900. (*Pterodiscus*)
ellipticus Lat., elliptic; (**1**) for the shape of the stem segments. (*Rhipsalis*) – (**2**) for leaf shape. (*Hoya*) – (**3**) for the shape of the leaflets. (*Pseudobombax*)
ellisianus For Prof. J. Coswell Ellis (fl. 1910), who first collected the taxon. (*Opuntia*)
elongatus Lat., elongate; (**1**) for the elongate stems. (*Mammillaria*) – (**2**) for the leaf shape. (*Brachystelma, Conicosia*) – (**3**) for the narrower leaves with attenuate base. (*Euphorbia perrieri* var.) – (**4**) application obscure. (*Orbea gerstneri* ssp., *Portulaca*)
elquiensis For the occurrence in the valley of the Río Elqui, C-N Chile. (*Eriosyce senilis* ssp.)
elsanus For Mrs. Elsa Pooley (fl. 1969–1985), South African artist and amateur botanist of Rennieshaw, KwaZulu-Natal, who discovered the taxon. (*Raphionacme*)
elsieae For Mrs. Elsie E. Esterhuysen (*1912), botanist at the Bolus Herbarium, Cape Town, RSA. (*Crassula*)
elymaiticus Probably for the grass genus *Elymus*, some species of which have similarly blue-green leaves. (*Rosularia*)
emarcescens Lat., becoming withered; for the rapidly withering leaves. (*Antimima*)
emarcidus Lat., withered; presumably for the persistent dead leaves. (*Sceletium*)
emarginatoides Gr. '-oides', resembling; and for the similarity to *Lampranthus emarginatus* (Aizoaceae). (*Lampranthus*)
emarginatus Lat., emarginate; (**1**) for the emarginate leaf tips. (*Sedum, Sedum oligospermum* var.) – (**2**) for the emarginate petal tips. (*Lampranthus, Trichodiadema*)
emelyae For Mrs. Emily [as 'Emely'] Ferguson (fl. 1928–1933), plant collector in RSA. (*Haworthia*)
emetocatharticus Lat. 'emeticus', emetic; and Lat. 'catharticus', purgative; for the medicinal use. (*Doyerea*)

eminens Lat., standing out; for the conspicuousness in nature. (*Aloe, Stenostelma*)

emoryi For Major William H. Emory (1811–1887), US-American soldier and in charge of the Mexican boundary survey 1850–1854. (*Bergerocactus, Ferocactus, Grusonia*)

emskoetterianus For Robert Emskötter (fl. 1910), German horticulturist in Magdeburg, Germany. (*Escobaria*)

Enarganthe Gr. 'enarges', brilliant, visible, manifest; and Gr. 'anthos', flower. (*Aizoaceae*)

Endadenium Gr. 'endo-', inside, and Gr. 'aden, adenos', gland; for the nectaries that are situated on the inside of the involucre. (*Euphorbiaceae*)

endlicherianus For Prof. Dr. Stephan L. Endlicher (1804–1849), Austrian botanist, sinologist and physician. (*Pelargonium*)

endlichianus For Dr. R. Endlich, collector of plants in Mexico c. 1906. (*Yucca*)

engelmannii For Dr. George Engelmann (1809–1884), German-born US-American physician and botanist in St. Louis, Missouri. (*Echinocereus, Opuntia*)

engleri For Prof. Heinrich Gustav Adolf Engler (1844–1930), German botanist in Berlin and director of the Berlin Botanical Garden and Museum. (*Sedum*) – (**2**) For Mr. E. Engler (fl. 1959), English cactus hobbyist (?) who supported the travels of F. Ritter. (*Eriosyce*)

englerianus For Prof. Heinrich Gustav Adolf Engler (1844–1930), German botanist in Berlin and director of the Berlin Botanical Garden and Museum. (*Hoya, Malephora, Stapelia*)

englishiae For Mrs. N. English (fl. 1917), who collected the type specimen, without further data. (*Prenia*)

enneacanthus Gr. 'ennea', nine; and Gr. 'akantha', thorn, spine; for the number of spines on the original plants. (*Echinocereus*)

enoplus Gr. 'enoplos', armed; for the strong spination. (*Euphorbia*)

enormis Lat., abnormal, irregular; for the distinctive structure of the spination when compared with *Euphorbia clavigera*. (*Euphorbia*)

enotatus Lat., unmarked; for the unspotted leaves. (*Aloe*)

ensifer Lat. 'ensis', sword; and Lat. '-fer, -fera, -ferum', -carrying; for the leaves. (*Agave*)

ensifolius Lat. 'ensis', sword; and Lat. '-folius', -leaved. (*Aloe menyharthii* ssp., *Ceropegia*)

Entandrophragma Gr. 'entos', within; Gr. 'andro-', male-; and Gr. 'phragma', screen, fence, partition'; for the united filaments, which produce an urceolate tube. (*Meliaceae*)

enterophorus Gr. 'enteron', entrails, intestines; and Gr. '-phoros', carrying; perhaps for the appearance of the branches. (*Euphorbia*)

ephedroides Gr. '-oides', resembling; and for the genus *Ephedra* (*Ephedraceae*). (*Euphorbia*)

ephemerus Lat., ephemeral; because the plants are short-lived. (*Heliophila*)

epidendrum Gr. 'epidendrios', living on trees (from Gr. 'epi', on top of; and Gr. 'dendron', tree); for the epiphytic habit. (*Sedum*)

epigeus Gr. 'epigeios', on the ground (from Gr. 'epi', on top of; and Gr. 'ge, gaia', earth); for the above-ground storage organ. (*Adenia, Corallocarpus, Urginea*)

epiphylloides Gr. '-oides', resembling; and for the genus *Epiphyllum* (*Cactaceae*); for the similar branches. (*Euphorbia, Hatiora*)

Epiphyllum Gr. 'epi', on top of; and Gr. 'phyllon', leaf; for the flowers that appear on the seemingly leaf-like flattened stems. (*Cactaceae*)

epiphyticus Lat., epiphytic; for the growth form of the type. (*Sarcorrhiza*)

epiroticus For the occurrence in the region known as Epiros during antique times on the W coast of Greece; also from Gr. 'epeiros', land, mainland. (*Sedum eriocarpum* ssp.)

Epithelantha Gr. 'epi', on top of; Gr. 'thele', nipple, tubercle; and Gr. 'anthos', flower; for the position of the flower. (*Cactaceae*)

eranthes Gr. 'e(r)-', without; and Gr. 'anthos',

flower; because the vegetative stems appear separately from the flowering stems. (*Euphorbia*)

erectiflorus Lat. 'erectus', erect; and Lat. '-florus', -flowered. (*Stapelia*)

erectilobus Lat. 'erectus', erect; and Lat. 'lobus', lobe, for the corolla lobes. (*Huernia*)

erectocentrus Lat. 'erectus', erect; and Lat. 'centrum', centre; for the erect central spines. (*Echinomastus*)

erectocladus Lat. 'erectus', erect; and Gr. 'klados', branch; for the disposition of the stem segments. (*Tunilla*)

erectus Lat., erect; (**1**) for the erect stems. (*Adenia, Corryocactus, Coryphantha, Ruschia, Sesuvium, Tetragonia*) – (**2**) for the erect stems in comparison to related taxa with twining stems. (*Stephania*) – (**3**) for the erect leaves. (*Delosperma*) – (**4**) for the erect inflorescences. (*Sedum rupestre* ssp., *Umbilicus*) – (**5**) for the erect corona lobes. (*Glossostelma*)

eremaeus Gr. 'eremos', solitary, deserted; for the solitary flowers. (*Tetragonia*)

eremastrum Gr. 'eremos', solitary, deserted; and Gr. 'aster, astron', star, a plant; probably for the outlying distribution. (*Orbea wissmannii* var.)

eremophilus Gr. 'eremos', solitary, deserted; and Gr. 'philos', friend; for the habitat preference in desert. (*Aloe*)

erensii For Jan Erens (1911–1982), Dutch horticulturist and collector, emigrated to RSA in 1914, also collecting in East Africa. (*Aloe*)

Erepsia Gr. 'erepsis', cover; for the filamentous staminodes that cover the flower centre. (*Aizoaceae*)

erergotanus For the occurrence near Erer Gota, Harerge Region, Ethiopia. (*Ceropegia*)

eriacanthus Gr. 'erion', wool; and Gr. 'akanthos', spine, thorn; for the pubescent spines. (*Mammillaria*)

ericetorum Gen. Pl. of Lat. 'ericetum', heath, moor; for the habitat preference. (*Aloe*)

erici-magnusii For Eric Magnus (fl. 1942), without further data. (*Sedum*)

ericiflorus For the genus *Erica* ("Heath", *Ericaceae*); and Lat. '-florus', -flowered; because of the resemblance of the flowers to those of Ericas. (*Echidnopsis*)

ericoides Gr. '-oides', resembling; and for the genus *Erica* (*Ericaceae*); for the similar leaves. (*Aeollanthus subacaulis* var., *Crassula*)

erigavensis For the occurrence near Erigavo, Somalia. (*Euphorbia*)

erigeriflorus Lat. '-florus', -flowered; and for the genus *Erigeron* ("Fleabane", *Asteraceae*); for the superficially similar flowers. (*Drosanthemum*)

erinaceus Lat., hedgehog-; (**1**) for the spiny tuber. (*Myrmecodia*) – (**2**) for the prickly spination. (*Gymnocalycium, Opuntia polyacantha* var., *Parodia*) – (**3**) for the prickly appearance of the leaf rosettes. (*Aloe melanacantha* var.) – (**4**) for the papillate corolla. (*Huernia*)

eriocarpus Gr. 'erion', wool; and Gr. 'karpos', fruit; perhaps for the tuberculate fruits. (*Sedum*)

eriocaulis Gr. 'erion', wool; and Lat. 'caulis', stem; for the woolly cephalia. (*Arrojadoa dinae* ssp.)

eriophorus Gr. 'erion', wool; and Gr. '-phoros', -carrying; for the long hairs on the perianth tube. (*Harrisia*)

eriophyllus Gr. 'erion', wool; and Gr. 'phyllon', leaf; for the felt-covered leaves. (*Kalanchoe*)

Eriospermum Gr. 'erion', wool; and Gr. 'sperma', seed; for the hairy seeds. (*Eriospermaceae*)

Eriosyce Gr. 'erion', wool; and Gr. 'syke', fig tree, fig; for the wool-covered fruits. (*Cactaceae*)

erlangeri For Carl Baron von Erlanger, who conducted an expedition to E Africa in 1899–1901. (*Euphorbia*)

ermanicus For the occurrence near Ermani, Georgia. (*Sempervivum*)

ermininus Lat., like an ermine; for the toothed leaf margins, comparing the gaping leaves of a pair with the mouth of an ermine. (*Stomatium*)

ernesti-ruschii For Ernst Rusch jr. (1867–

1957), German farmer in Namibia and discoverer of the taxon. (*Gasteria pillansii* var.)

ernestii For Dr. Ernest E. Galpin (1858–1941), South African banker and amateur botanist. (*Euphorbia, Lampranthus*) – (**2**) For Ernst H. G. Ule (1854–1915), German botanist and botanical explorer of Brazil. (*Melocactus*) – (**3**) For Dr. Ernesto J. Fittkau (fl. 1971), Mexican biologist and brother of Hans W. Fittkau. (*Mammillaria backebergiana* ssp.)

ernianus For Mr. Frantz Erni (1878–1952), farmer and plant collector in Namibia. (*Conophytum taylorianum* ssp.)

ernstii For Ernst van Jaarsveld (*1953), botanist and horticulturist at the Kirstenbosch Botanical Gardens, RSA. (*Conophytum, Plectranthus*)

erosulus Dim. of Lat. 'erosus', erose, gnawed, jagged; for the appearance of the taxon in the field. (*Crassula subacaulis* ssp.)

erosus Lat., erose, gnawed, jagged; (**1**) for the petal tips. (*Antimima*) – (**2**) perhaps for the dentate leaves. (*Pachyrhizus*)

erratus Lat., erroneous; application obscure, perhaps because the name is based on the illegimate homonym *Mesembryanthemum virens*. (*Lampranthus*)

erubescens Lat., becoming red; (**1**) for the spine colour. (*Parodia*) – (**2**) for the bright red colour of mature inflorescences. (*Monadenium*) – (**3**) for the corona colour. (*Aspidoglossum*)

eruca Perhaps a Latinization of Spanish 'oruga', caterpillar; because of the caterpillar-like stems. (*Portulaca, Stenocereus*)

erythracanthus Gr. 'erythros', red; and Gr. 'akanthos', thorn, spine; for the spine colour. (*Parodia mammulosa* ssp.)

erythraeae For the occurrence in Eritrea. (*Sansevieria*)

erythraeus Gr. 'erythros', red; for the leaf colour. (*Sempervivum marmoreum* ssp.)

Erythrina Gr. 'erythros', red; for the mostly red flowers. (*Fabaceae*)

erythrocarpus Gr. 'erythros', red; and Gr. 'karpos', Frucht. (*Rhipsalis baccifera* ssp.)

erythrophaeus Gr. 'erythros', red; and Gr. 'phaios', dark; probably for the coloured calyx. (*Columnea*)

erythrophyllus Gr. 'erythros', red; and Gr. 'phyllon', leaf; for the leaf colour. (*Aloe*)

Erythrophysa Gr. 'erythros', red; and Gr. 'physa', bladder; for the fruits. (*Sapindaceae*)

erythropodus Gr. 'erythros', red; and Gr. 'pous, podos', foot; for the red sap of the root tuber. (*Jatropha*)

erythrospermus Gr. 'erythros', red; and Gr. 'sperma', seed. (*Mammillaria, Sedum*)

erythrostemma Gr. 'erythros', red; and Gr. 'stemma', garland, crown; for the red corona. (*Hoya*)

erythrostictus Gr. 'erythros', red; and Gr. 'stiktos', dotted, colourful; for the flower colour. (*Hylotelephium*)

escayachensis For the occurrence near Escayache, Prov. Mendez, Dept. Tarija, Bolivia. (*Echinopsis*)

eschauzieri For Dr. Louis Eschauzier (also mis-spelled 'Eschanzier' or 'Eschaus[s]ier') († 1906?), owner of a ranch in San Luis Potosí, Mexico. (*Mammillaria bocasana* ssp.)

Escobaria For the brothers Rómulo and Numa Escobar of Mexico City and Juárez, honouring their work, without further details. (*Cactaceae*)

Escontria For Don Blas Escontría († 1906), Mexican "Ministro de Fomento" and with a great interest in all subjects relating to the scientific development of his country. (*Cactaceae*)

escuintlensis For the occurrence near Escuintla, Guatemala. (*Hylocereus*)

esculentus Lat., edible; (**1**) for the potato-like edible roots. (*Plectranthus*) – (**2**) for the astringent tubers. (*Tylosema*) – (**3**) for the use as cattle-feed. (*Euphorbia*) – (**4**) because there are reports that the flowers are edible. (*Aloe*)

esmeraldanus For the occurrence near La Esmeralda, Antofagasta Prov., N Chile. (*Copiapoa, Eriosyce*)

esperanzae For Sra. Esperanza Benavides de Valázquez (fl. 1996), a local community leader and mother of the Mexican singing poet Guillermo Valázquez; also from Span.

'esperanza', hope, in the hope that the taxon can be saved from unscrupulous collecting. (*Strombocactus disciformis* ssp.)

espinosae For Prof. Marcial R. Espinosa B. (1874–1959), Chilean botanist (cryptogamist) and founder of the section of cryptogamic botany at the Chilean National Natural History Museum in Santiago. (*Tetragonia*)

espinosus Lat. 'e-, ex', lacking, without; and Lat. 'spinosus', thorny, spiny; for the spineless stems. (*Euphorbia*)

Espostoa For Nicolas E. Esposto (fl. 1920), Peruvian botanist at the Escuela Nacional de Agricultura at Lima. (*Cactaceae*)

Espostoopsis Gr. '-opsis', similar to; and for the genus *Espostoa* (*Cactaceae*). (*Cactaceae*)

estebanensis For the occurrence on Isla San Esteban, Baja California, Mexico. (*Mammillaria*)

Esterhuysenia For Mrs. Elsie E. Esterhuysen (*1912), botanist at the Bolus Herbarium, Cape Town, RSA. (*Aizoaceae*)

esterhuyseniae As above. (*Bulbine, Delosperma, Erepsia, Gibbaeum, Lampranthus, Ruschia*)

estevesii For Eddie Esteves Pereira (fl. 1989), Brazilian cactus enthusiast and collector in Goiânia, Goiás. (*Euphorbia, Facheiroa cephaliomelana* ssp., *Melocactus, Micranthocereus, Pilosocereus, Tacinga saxatilis* ssp.)

ettyuensis For the occurrence in Prov. Ettyu, Honshu, Japan. (*Hylotelephium sieboldii* var.)

etuberculatus Lat. 'e, ex', without; and Lat. 'tuberculatus', tuberculate; for the smooth stem surface. (*Euphorbia*)

euchlorus Gr. 'eu', truly; and Gr. 'chloros', green; for the beautifully green body colour. (*Praecereus*)

eugeniae For Eugenia van Vliet (fl. 1976), wife of Dirk van Vliet, Dutch (?) cactus collector in Brazil. (*Parodia mammulosa* ssp.)

Eulychnia Gr. 'eu-', good, well; and Gr. 'lychnos', candlestick, torch; for the columnar stems. (*Cactaceae*)

eumassawanus Gr. 'eu', truly; and for *Aloe massawana*, with which the taxon was previously confused, and which despite its name does not come from Massawa. (*Aloe*)

eumorphus Gr. 'eu-', good, well; and Gr. 'morphe', shape; probably for the pretty flowers. (*Sinningia*)

euniceae For Mrs. Eunice E. Burmeister (fl. 1963), a friend of Herkie Horn who collected the taxon. (*Lithops aucampiae* ssp.)

Euphorbia Gr. 'euphorbos', well-fed; also to honour Euphorbos, physician to King Juba of Mauritania after whom the king named the first succulent species he discovered himself in the Atlas Mts. (probably *E. resinifera* A. Berger). (*Euphorbiaceae*)

euphorbioides Gr. '-oides', resembling; and for the resemblance to some species of the genus *Euphorbia* (*Euphorbiaceae*). (*Neobuxbaumia, Othonna*)

Eureiandra Gr. 'eu', truly; '-rei-', unknown; and Gr. 'aner, andros', stamen; application obscure. (*Cucurbitaceae*)

eurychlamys Gr. 'eurys', broad, wide; and Gr. 'chlamys', cloak; for the conspicuous broad bracts. (*Echeveria*)

eurypleurus Gr. 'eurys', broad, wide; and Gr. 'pleuron', rib. (*Gymnocalycium*)

eurystigmatus Gr. 'eurys', broad, wide; and Gr. 'stigma', scar, stigma; and also from the former genus name *Eurystigma*. (*Mesembryanthemum*)

eustacei For Charles Eustace Pillans (1850–1919), civil servant in the Cape Department of Agriculture, RSA, and father of the botanist Neville S. Pillans. (*Euphorbia*)

euxinus Lat., pertaining to the Black Sea; for the occurrence in the Pontic (Euxinian) region. (*Sedum*)

evadens Lat., escaping, going out; perhaps for the insufficient material available to the describing author. (*Agave*)

evansii For Dr. Illtyd B. Pole-Evans (1877–1968), Welsh botanist and plant pathologist, lived in RSA from 1905 and travelled widely. (*Euphorbia*)

evermannianus For Dr. Barton W. Evermann (1853–1932), US-American naturalist and ichthyologist, and director of the Museum of the California Academy of Sciences. (*Mammillaria*)

evolutus Lat., evolved, unrolled, unfolded; application obscure. (*Antimima*)

evrardii For Dr. Francis Évrard (1885–1957), French botanist in Paris, collecting plants in Indo-China 1920–1930. (*Aeschynanthus*)

ewaldianus For Ernst Ewald (1946–2001), German grower of epiphytic cacti in Hamburg. (*Rhipsalis*)

ewersii For Johann P. G. Ewers (fl. 1829), Officer of the Imperial Russia. (*Hylotelephium*)

exalatus Lat. 'ex-', without; and Lat. 'alatus', winged; for the wingless fruits. (*Sceletium*)

exasperatus Lat. 'e, ex', beyond, very; and Lat. 'asperatus', roughened, covered with short hard points; for the papillate interior of the corolla. (*Stapeliopsis*)

excavatus Lat., hollowed, excavate; (**1**) for the upper face of the leaves, which is concave near the leaf tip. (*Ihlenfeldtia*) – (**2**) for the upper leaf face. (*Mesembryanthemum*) – (**3**) for the holes left by the fallen flowers in the rachis of the inflorescence (*Basella*)

excedens Lat., surpassing. (*Antimima, Oscularia*)

excelsus Lat., tall, high; (**1**) for the growth habit. (*Aloe, Euphorbia, Opuntia, Plectranthus*) – (**2**) for the tall inflorescences. (*Doryanthes, Echeveria, Gasteria*)

exhibens Lat., exhibiting; application obscure. (*Pelargonium*)

exiguus Lat., weak, feeble, little; (**1**) for the small growth and the small flowers. (*Ruschia*) – (**2**) perhaps beause the flowers are much shorter than the bracts. (*Galenia*) – (**3**) for the short corolla lobes. (*Ceropegia*)

exilis Lat., small, meagre, slender; (**1**) for the small size of the plants. (*Crassula, Euphorbia*) – (**2**) application obscure. (*Brachystelma*)

exilispinus Lat. 'exilis', small, meagre, slender; and Lat. '-spinus', -spined. (*Euphorbia*)

eximius Lat., out of the ordinary, distinguished; probably for the unusual conspicuous brown margins of the calyx lobes. (*Lampranthus*)

exoticus Lat., exotic; here in the sense of extraordinary for the hybrid nature of the taxon. (*Schlumbergera*)

expansus Lat., expanded, spread out; (**1**) for the position of the leaves. (*Agave americana* var., *Crassula*) – (**2**) for the large broad leaves. (*Sceletium*)

expatriatus Lat., expatriate; because no wild locality is known. (×*Cremneria*)

explanatus Lat., flattened, outspread; for the orientation of the stamens. (*Lampranthus*)

exsertus Lat., exserted, sticking out, protruding; application obscure. (*Crassula*)

exspersus Lat. 'e, ex', beyond, very; and Lat. 'sparsus', untidy, spread out; perhaps for the untidy growth habit. (*Drosanthemum*)

exstipulatus Lat. 'e-, ex', without; and Lat. 'stipulatus', stipulate; for the minute and inconspicuous stipules. (*Pelargonium*)

exsurgens Lat., raisinig up, ascending; for the stigma lobes, which surpass the stamens. (*Antimima*)

extensus Lat., extended, spread out; for the growth habit. (*Ruschia*)

extimus Lat., most remote; because the subspecies grows farthest from the American continent. (*Dudleya virens* ssp.)

extrorsus Lat., opening on the outside, turned outwards; for the outwards-bent fruiting follicles and the widely spreading margins of the sutures of the dehising follicles. (*Crassula*)

eyassianus For the occurrence near Lake Eyassi, Tanzania. (*Euphorbia*)

eylesii For Mr. Frederick Eyles (1864–1937), English journalist, farmer, miner, and government officer, living in Zimbabwe from 1899, 1923–1928 government botanist, who discovered the taxon. (*Aspidoglossum*)

eyriesii For Mr. J. B. Eyriès (fl. 1822, 1830), cactus collector in Le Havre, France, translator of the travelogue of Prince Maximilian Wied-Neuwied ("Reise nach Brasilien 1813 –1817"). (*Echinopsis*)

eytianus For the occurrence near Eyti, Dept. Santa Cruz, Bolivia. (*Gymnocalycium*)

F

fabrisii For Dr. Humberto A. Fabris (1924–1976), Argentinian botanist in La Plata. (*Echinopsis, Rebutia*)

Facheiroa From the Brazilian vernacular name for many columnar cacti; from Port. 'facheiro', being a torch, carrying a torch. (*Cactaceae*)

fadeniorum Lat. Gen. Pl., for Dr. Robert ("Bob") B. Faden (*1942), US-American botanist at the Smithsonian Institution, Washington, specialist on *Commelinaceae* and ferns, and Mrs. Audrey J. Faden (née Evans) (*1941), Kenyan-born naturalist, who collected the type specimen in 1977. (*Kalanchoe*)

fagaroides Gr. '-oides', resembling; and for the genus *Fagara* (syn., = *Zanthoxylum*, *Rutaceae*); for the similarity, probably because both are aromatic. (*Bursera*)

falcatus Lat., falcate, curved like a sickle; (**1**) for the shape of the stem segments. (*Consolea*) – (**2**) for the leaf shape. (*Agave striata* ssp., *Aloe, Lampranthus, Ruschianthus*)

falciformis Lat. 'falx, falcis', sickle; and Lat. '-formis', -shaped; for the leaf shape. (*Oscularia*)

fallax Lat., deceptive; perhaps for the difficulties in identifying this taxon. (*Adromischus, Bulbine, Crassula, Talinum*)

famatamboay From the local vernacular name of the plants in Madagascar. (*Euphorbia*)

famatimensis Intentional Latinization, for the occurrence in the Sierra de Famatina, La Rioja, Argentina. (*Echinopsis*)

familiaris Lat., pertaining to the family; for the clustering growth. (*Pygmaeocereus*)

Fanninia For George Fannin (fl. 1868), owner of the farm in RSA where the taxon was discovered. (*Asclepiadaceae*)

fanshawei For Dennis B. Fanshawe (1915–1993), Forest officer and plant collector in Kitwe, Zambia, better known for his collections in S America. (*Euphorbia*)

fantasticus From Lat. 'phantasma', phantom; for the extraordinary construction of the flowers. (*Ceropegia*)

farinaceus Lat., farinose, mealy; for the leaves. (*Kalanchoe*)

farinifer Lat. 'farina', flour, meal; and Lat. '-fer', carrying; for the glaucous-farinose leaves. (*Graptopetalum saxifragoides* var.)

farinosus Lat., farinose, mealy; for the densely farinose leaves. (*Dudleya, Sedum, Wooleya*)

fartaqensis For the occurrence at Ras Fartaq, SE Yemen. (*Echidnopsis*)

fasciatus Lat., banded (from Lat. 'fascia', bundle); (**1**) for the cross-banded basal sheaths. (*Rhadamanthus*) – (**2**) for the leaf coloration. (*Haworthia, Sansevieria*)

fascicaulis Lat. 'fascis', bundle, cluster; and Lat. 'caulis', stem; for the tufted habit. (*Euphorbia*)

fasciculaceus Lat., clustered, bundled; for the flowers radiating from the inflorescence centre. (*Pelargonium*)

fascicularis Lat., clustered, bundled; for the growth form. (*Aspidoglossum, Crassula, Haageocereus, Portulaca*)

fasciculatus Lat., clustered; (**1**) for the clusters of cane-like branches. (*Echinocereus, Fouquieria, Tylecodon buchholzianus* var.) – (**2**) for the leaves on the stem tips. (*Plectranthus*) – (**3**) for the cyathia clustered in false umbels. (*Euphorbia*) – (**4**) for the flowers. (*Eureiandra*)

fassoglensis For the occurrence at Fazoghli, Sudan. (*Tylosema*)

fastigiatus Lat., fastigiate, with clustered branches. (*Crassula subulata* var., *Dioscorea, Mesembryanthemum, Rhodiola*)

Faucaria Lat. 'faux, fauces', mouth, entrance, gorge; for the toothed leaf margins that make the diverging leaves of a pair resemble open jaws of an animal. (*Aizoaceae*)

faucicola Lat. 'fauces', gorges; and Lat. '-cola', -dwelling. (*Euphorbia*)

faucius From Lat. 'faux, fauces', mouth, entrance, gorge; for the type locality in the Verlatekloof (gorge), Western Cape, RSA. (*Tylecodon*)

faustianus For Carlos Faust († 1952), German

merchant emigrating to Spain, plant enthusiast who 1924 converted an old vineyard into the succulent plant garden "Marimurtra" (also written "Mar y Murtra") in Blanes, Spain. (*Cleistocactus acanthurus* ssp.)

favosus Lat., honeycombed, covered with regular angular depressions; application obscure. (*Bulbine*)

faxonianus For Charles E. Faxon (1846–1926), US-American botanist (?). (*Yucca*)

fechseri For Helmut Fechser (*1918), cactus collector in Los Olivos, Prov. Buenos Aires, Argentina. (*Gymnocalycium andreae* var.)

feddei For Friedrich K. G. Fedde (1873–1942), German botanist, editor and publisher. (*Sedum*)

fedtschenkoanus For Boris A. Fedtschenko [Fedchenko] (1873–1947), Russian botanist, son of Olga Fedtschenko, director of the imperial Botanical Garden at St. Petersburg. (*Pseudosedum*)

fedtschenkoi As above. (*Kalanchoe, Sedum*)

felgeri For Dr. Richard S. Felger (fl. 1972, 2002), US-American botanist. (*Agave*)

felinus Lat., pertaining to cats or pine martens; for the toothed leaf margins likened to cat teeth. (*Faucaria*)

fendleri For Augustus Fendler (1813–1883), Prussian-born botanist and plant collector, emigrating 1836 to the USA. (*Echinocereus*)

Fenestraria Lat. 'fenestra', window; for the translucent patch on the leaf tip. (*Aizoaceae*)

fenestratus Lat., windowed; for the translucent leaf tips. (*Antimima*)

fenzlii For Prof. Eduard Fenzl (1808–1879), Austrian botanist in Vienna. (*Cistanthe*)

fera-rubra Lat. 'ferus', wild; and Lat. 'ruber, rubra, rubrum', red; for the fierce red spination. (*Mammillaria rhodantha* ssp.)

ferganensis For the occurrence in the Ferganskyi Mts., Kirgistan. (*Pseudosedum*)

fergusoniae For Mrs. Emily Ferguson (fl. 1928–1933), plant collector in RSA. (*Antimima, Glottiphyllum, Lampranthus, Pelargonium, Pleiospilos compactus* ssp., *Trichodiadema, Tylecodon*)

fernambucensis Mis-spelled for the occurrence in the state of Pernambuco, Brazil. (*Cereus*)

fernandopoensis For the occurrence on Fernando Po Island (now Bioko Island), Equatorial Guinea. (*Sansevieria longiflora* var.)

fernowii For Prof. Bernhard E. Fernow (1851–1923), German-born US-American forester, chief of the US Bureau of Forestry within the USDA and one of the leaders in the movement to protect forests. (*Harrisia*)

feroacanthus Lat. 'ferus', wild; and Gr. 'akanthos', thorn, spine. (*Opuntia*)

Ferocactus Lat. 'ferus', wild; and Lat. 'cactus', cactus; for the heavy spination of some taxa. (*Cactaceae*)

ferox Lat., fierce; (**1**) for the spines. (*Echinopsis, Euphorbia, Gymnocalycium gibbosum* ssp., *Myrmecodia*) – (**2**) for the prickly leaves. (*Aloe*) – (**3**) for the strong leaf marginal teeth. (*Agave salmiana* var.)

ferrarii For Ing. Omar Ferrari (fl. 1976), Argentinian agronomist and cactus collector. (*Acanthocalycium, Cleistocactus*)

ferreirianus For Enrique Ferreira (fl. 1953), former Mexican consul in San Diego, California, USA. (*Echinocereus*)

ferreophilus Lat. 'ferrum', iron; and Gr. 'philos', friend; for the occurrence on iron-rich soil. (*Melocactus azureus* ssp.)

ferreyrae For Ramón A. Ferreyra (*1912), Peruvian botanist. (*Peperomia*)

ferricola Lat. 'ferrum', iron; and Lat. '-cola', inhabiting; for the occurrence on iron-rich soil. (*Discocactus*)

ferrugineo-pubescens Lat. 'ferrugineus', rusty, orange-brown-red; and Lat. 'pubescens', pubescent; for the pubescence of the plants. (*Cissus rotundifolia* var.)

festivus Lat., festive, gay, bright; presumably for the attractive appearance. (*Ruschia*)

festucifolius For the genus *Festuca* ("Fescue", "Bluegrass"; *Poaceae*); and Lat. '-folius', -leaved; for the grass-like leaves. (*Brachystelma*)

Fevillea For Louis E. Feuillée (1660–1732), French clergyman, astronomer, botanist and explorer, collected in Central America and the West Indies 1707–1712. (*Cucurbitaceae*)

fianarantsoae For the occurrence near Fianarantsoa, Madagascar. (*Euphorbia*)
fibrosus Lat., fibrous; for the presence of fibres in the leaves. (*Aloe*)
ficifolius For the genus *Ficus* ("Fig", Moraceae); and Lat. '-folius', -leaved. (*Neorautanenia, Obetia*)
ficiformis For the genus *Ficus* ("Fig", Moraceae); and Lat. '-formis', -shaped; for the shape of the fused leaf pair. (*Conophytum*)
ficksbergensis For the occurrence near Ficksberg, Orange Free State, RSA. (*Delosperma*)
ficoides Presumably from *Ficoides*, a pre-Linnean name for *Mesembryanthemum* in the broadest sense (from Lat. 'ficus', fig). (*Senecio*)
Ficus Lat. name of the edible fig (*Ficus carica*). (*Moraceae*)
ficus-indicus Lat. 'ficus', name of the edible fig (*Ficus carica*); and Lat. 'indicus', from India; for the edible fruit and the origin from the West Indies (then thought to be India). (*Opuntia*)
fidaianus For Mr. H. F. Fida (fl. 1934), German cactus hobbyist in Mannheim and editor of "Der Kakteenfreund". (*Weingartia*)
fiebrigii For Dr. Carl F. Fiebrig (1869–1951), botanist and zoologist, founder and director of the botanical garden and museum in Asunción, Paraguay. (*Rebutia*)
fiedlerianus For Rudolf Fiedler sen. (fl. 1903), German master cabinet maker and cactus hobbyist in Berlin. (*Copiapoa*)
fieldianus For Captain Marshall Field (fl. 1922), US-American philanthropist and patron of science, funded 1922 a botanical expedition to South America. (*Cleistocactus*)
fievetii For Gerard Fievet (fl. 1965), French wine-grower and succulent plant enthusiast in Fianarantsoa, Madagascar. (*Aloe*)
fiherenensis For the occurrence in the Fiherenana River valley, Madagascar. (*Alluaudiopsis, Euphorbia*)
filamentosus Lat., with thread-like hairs (Lat. 'filum', thread); (**1**) for the axillary hairs. (*Anacampseros*) – (**2**) for the leaf margins. (*Yucca*)

filicaulis Lat. 'filum', thread; and Lat. 'caulis', stem. (*Adromischus, Crassula expansa* ssp., *Lampranthus, Othonna*)
filicifolius Lat. 'filix, filicis', fern; and Lat. '-folius', -leaved; for the resemblance of the leaves to a small fern frond. (*Bursera*)
filifer Lat. 'filum', thread; and Lat. '-fer, -fera, -ferum', carrying; (**1**) for the thread-like bristle at the leaf tip. (*Graptopetalum*) – (**2**) for the threads along the leaf margins. (*Agave, Yucca*)
filiflorus Lat. 'filum', thread; and Lat. '-florus', -flowering; for the hair-like processes on the nectary glands. (*Euphorbia*)
filifolius Lat. 'filum', thread; and Lat. '-folius', -leaved; for the thread-like leaves. (*Brachystelma, Bulbine, Nolana*)
filiformis Lat. 'filum', thread; and Lat. '-formis', -shaped; (**1**) for the thin stems. (*Crassula, Drosanthemum*) – (**2**) for the linear leaves. (*Monadenium*) – (**3**) for the corolla lobes. (*Ceropegia*) – (**4**) application obscure. (*Galenia*)
filipendulus Lat. 'filum', thread; and Lat. 'pendulus', pendulous, hanging down; application obscure. (*Ceropegia*)
filipes Lat. 'filum', thread; and Lat. 'pes', foot; for the thin pedicels. (*Echidnopsis chrysantha* ssp., *Sedum*)
filipetalus Lat. 'filum', thread; and Lat. 'petalum', petal; for the filiform petals. (*Ruschia*)
filsonii For Rex B. Filson (fl. 1966), the discoverer of the taxon. (*Portulaca*)
fimbrialis Lat., fimbriate; for the leaf margins. (*Aloe*)
fimbriatus Lat., fimbriate; (**1**) for the margins of young leaves. (*Echeveria*) – (**2**) for the leaf tip appendage. (*Orostachys*) – (**3**) for the margins of the involucral lobes. (*Euphorbia*) – (**4**) for the margins of the roof formed by the corolla lobes. (*Ceropegia*) – (**5**) for the corona. (*Cynanchum*) – (**6**) for the scales of the pericarpel. (*Stenocereus*) – (**7**) for the basal part of the stigmas (*Hereroa*) – (**8**) application obscure. (*Caralluma adscendens* var.)
fimbriifer Lat. 'fimbria', thread, fringe; and Lat. '-fer, -fera, -ferum', -carrying; for the hairs on the corolla lobes. (*Ceropegia*)

finckii For the Mr. Finck (fl. 1862), who collected the type (probably Hugo Finck († 1895), who collected in Mexico and sent specimens to Kew). (*Pedilanthus*)

finlaysonii Perhaps for George Finlayson (1790–1823), British (?) botanist. (*Hoya*)

firingalavensis For the occurrence at Firingalava, Madagascar. (*Adenia*)

firmus Lat., firm, stable; for the rigid stems. (*Ruschia*)

fischeri For Alexander A. Fischer von Waldheim (1839–1920), Russian botanist. (*Sedum*) – (**2**) For Dr. G. A. Fischer (1848–1886), medical doctor and naturalist, explorer in Kenya and Tanzania in 1882 and 1885 who died of blackwater fever. (*Sansevieria*) – (**3**) For Walter Fischer (fl. 1914), German horticulturalist at the Botanical Garden Göttingen, who discovered the taxon in Argentina. (*Pterocactus*) – (**4**) For Ladislav Fischer (fl. 2002), Czech cactus collector. (*Gymnocalycium*)

fischerianus For Slávek Fischer (fl. 2002), Czech cactus collector in Horice. (*Rebutia*)

fissifolius Lat. 'fissus', cleft, split; and Lat. '-folius', -leaved; for the pinnate leaves. (*Pelargonium*)

fissispinus Lat. 'fissus', cleft, split; and Lat. '-spinus', -spined; for the paired spines, which are united basally and appear split apically. (*Euphorbia*)

fissoides Gr. '-oides', resembling; and for the similarity to *Mesembryanthemum fissum* (now *Argyroderma fissum*). (*Antegibbaeum*)

fissuratus Lat., fissured, grooved; for the appearance of the tubercles. (*Ariocarpus*)

fissus Lat., split; (**1**) for the gaping leaves of a pair. (*Argyroderma*) – (**2**) probably for an anomalously split corolla. (*Nematanthus*)

fitchii For William R. Fitch (fl. 1913), who accompanied J. N. Rose on collecting trips to the West Indies and W Texas in 1913. (*Echinocereus reichenbachii* ssp.)

fittkaui For Father Hans W. Fittkau (1913–2002), German (Prussian) priest and succulent plant collector, 1960 emigrating to Mexico, 1993 returning to Germany. (*Mammillaria*, *Pachyphytum*)

flabelliformis Lat. 'flabellum', fan; and Lat. '-formis', -shaped; probably for the shape of the leaflets. (*Erythrina*)

flaccidus Lat., flaccid, not able to hold one's own weigth; (**1**) for the soft and weak stems. (*Saphesia*) – (**2**) for the soft and weak leaves. (*Haworthia herbacea* var., *Yucca*) – (**3**) for the drooping inflorescences. (*Sedum*)

flagelliformis Lat. 'flagellum', (small) whip; and Lat. '-formis', -shaped; for the long slender stems. (*Disocactus*)

flammeus From Lat. 'flamma', flame, blaze; for the flower colour. (*Drosanthemum*)

flammosus From Lat. 'flamma', flame, blaze; for the flower colour. (*Cyrtanthus*)

flanaganii For Henry G. Flanagan (1861–1919), South African citrus farmer interested in botany. (*Aspidoglossum*, *Cotyledon orbiculata* var., *Crassula*, *Euphorbia*, *Raphionacme*)

flavescens Lat., yellowish, pale yellow; for the flower colour. (*Monsonia*)

flavicentrus Lat. 'flavus', yellow; and Lat. 'centrum', centre; for the yellow central spines. (*Mammillaria*)

flavidispinus Lat. 'flavidus', yellow; and Lat. '-spinus', -spined. (*Thelocactus bicolor* ssp.)

flavidus Lat., yellow; for the flower colour. (*Brachystelma pygmaeum* ssp., *Sedum laxum* ssp.)

flaviflorus Lat. 'flavus', yellow; and Lat. '-florus', -flowered. (*Micranthocereus*, *Turbinicarpus schmiedickeanus* ssp.)

flavipulvinatus Lat. 'flavus', yellow; and Lat. 'pulvinatus', cushion-shaped, strongly convex; for the colour and shape of the areoles. (*Pilosocereus*)

flavisetus Lat. 'flavus', yellow; and Lat. '-setus', -bristled. (*Coleocephalocereus buxbaumianus* ssp.)

flavispinus Lat. 'flavus', yellow; and Lat. '-spinus', -spined. (*Opuntia engelmannii* var., *Uebelmannia pectinifera* ssp.)

flavistylus Lat. 'flavus', yellow; and Lat. 'stylus', style. (*Rebutia*)

flavocroceus Lat. 'flavus', yellow; and Lat. 'croceus', saffron-yellow; for the flower colour. (*Malephora*)

flavopurpureus Lat. 'flavus', yellow; and Lat.

'purpureus', purple, dark red; for the flowers. (*Stapelia*)

flavovirens Lat. 'flavus', yellow; and Lat. 'virens', becoming green; (**1**) for the body colour. (*Ferocactus, Mammillaria gigantea* ssp.) – (**2**) for the flower colour. (*Caralluma*)

flavus Lat. 'flavus', yellow, golden-yellow, pale yellow; for the flower colour. (*Aspidonepsis, Caralluma, Conophytum, Crassula, Drosanthemum, Hoodia, Orbea huillensis* ssp., *Schizoglossum, Schizoglossum stenoglossum* ssp.)

fleckii For Dr. E. Fleck, German geologist, travelled across the Kalahari Desert to Lake Ngami in 1888. (*Euphorbia*)

fleurentiniorum Lat. Gen. Pl., for Jacky and Martine Fleurentin (fl. 1977), French medical technician and wife, resident in Yemen. (*Aloe*)

fleuretteanus For Mrs. Fleurette Andriantsjlavo (fl. 2000), Head of the Direction de la Planification des Eaux et Forêts, Madagascar. (*Aloe*)

flexibilispinus Lat. 'flexibilis', bendable, tending to be flexible; and Lat. '-spinus', -spined. (*Pilosocereus*)

flexicaulis Lat. 'flexus', bent; and Lat. 'caulis', stem; for the flexuose inflorescence. (*Bulbine, Peperomia*)

flexiflorus Lat. 'flexus', bent; and Lat. '-florus', -flowered; for the downwards bent flowers. (*Agave parviflora* ssp.)

flexifolius Lat. 'flexus', bent; and Lat. '-folius', -leaved. (*Lampranthus*)

flexilifolius Lat. 'flexilis', flexible; and Lat. '-folius', -leaved. (*Aloe*)

flexilis Lat., flexible; for the flexible stems. (*Lampranthus*)

flexispinus Lat. 'flexus', bent; and Lat. '-spinus', -spined; (**1**) for the spination. (*Opuntia engelmannii* var.) – (**2**) for the tortuous end spine at the leaf tip. (*Agave*)

flexuosus Lat., full of bends; for the inflorescence. (*Bulbine*)

floccosus Lat., floccose, with tufts of soft hairs; (**1**) for the bristly-hairy glochids. (*Austrocylindropuntia*) – (**2**) for the woolly stem apex. (*Eriosyce taltalensis* var.) – (**3**) for the tufts of hairs produced by flowering areoles. (*Pilosocereus, Rhipsalis*)

floresianus For Robert Flores (fl. 1958), nurseryman in Salinas, California, USA. (*Echeveria semivestita* var.)

floresii For R. Flores (fl. 1949), US-American (?) cactus collector in Spreckels, California, assistant on F. Schwarz's collecting trips. (*Echinocereus sciurus* ssp.)

floribundus Lat., profusely flowering. (*Brachystelma, Ceropegia, Crassula multicava* ssp., *Delonix, Delosperma, Drosanthemum, Haworthia, Ruschia*)

florifer Lat. 'flos, floris', flower; and Lat. '-fer, -fera, -ferum', -carrying; for the free-flowering nature of the plants. (*Cynanchum, Phedimus*)

fluminensis For the occurrence in the area of Rio de Janeiro (Lat. 'Flumen Januarii'). (*Coleocephalocereus, Nematanthus, Tradescantia*)

fluminis Lat., of the river; for the occurrence near rivers. (*Euphorbia*)

fluvialis Lat., of the river; for the occurrence near rivers. (*Euphorbia subsalsa* ssp., *Portulaca*)

fobeanus For Friedrich Fobe (1864–1941), German cactus horticulturist and head of the estate nursery of G. Hempel in Ohorn, Sachsen, Germany. (*Echinocereus chisoensis* var.)

Fockea For Charles Focke (1802–1856), Dutch botanist, collecting especially in Surinam. (*Asclepiadaceae*)

foetens Lat., stinking, foetid; for the foetid smell of the latex. (*Euphorbia mauritanica* var.)

foetidissimus Superl. of Lat. 'foetidus', stinking, evil-smelling; for the foetid odour of crushed plant parts. (*Cucurbita*)

foetidus Lat., stinking, evil-smelling; (**1**) for the scent of the crushed leaves. (*Furcraea, Plectranthus*) – (**2**) for the scent of the latex. (*Dorstenia*) – (**3**) for the flower odour. (*Brachystelma, Caralluma, Orbea sprengeri* ssp., *Piaranthus geminatus* var.)

foleyi For Mr. W. J. Foley, member of the South African Museum Herbarium from 1916 to 1918. (*Bulbine*)

foliolosus Lat., with small leaves. (*Astroloba*)

foliosus Lat., leafy, many-leaved. (*Ceropegia, Lampranthus, Pilea, Portulaca, Psilocaulon, Ruschia*)

folotsioides Gr. '-oides', resembling; and for the genus *Folotsia* (*Asclepiadaceae*); for the similar corona. (*Cynanchum*)

fontinalis Lat., growing by a spring; for the type locality Matjesfontein, RSA. (*Stapelia pillansii* var.)

forbesii For Dr. Henry O. Forbes (1851–1932), Scottish naturalist and collector. (*Aloe*)

fordii For Lyman M. Ford (fl. 1922) in San Diego, California, USA. (*Ferocactus*)

forficatus Lat., scissors-shaped, forked; application obscure. (*Erepsia*)

formicarum Gen. Pl. of Lat. 'formica', ant. (*Hydnophytum*)

formosanus For the occurrence on Formosa (Taiwan). (*Dischidia, Sedum*)

formosus Lat., handsome; (**1**) for the appearance of the plants. (*Dudleya, Echinopsis, Eureiandra, Fouquieria, Huernia, Lampranthus, Mammillaria, Parodia*) – (**2**) for the beautiful white inflorescences. (*Monadenium heteropodum* var.) – (**3**) for the beautiful large flowers. (*Matucana*)

fornicatus Lat., arched, vaulted, with small arched scale-like appendages; for the outer corona segments. (*Stathmostelma*)

forolensis For the occurrence on Forole [Furrole] Mountain at the Kenya / Ethiopia border. (*Euphorbia*)

forreri For Mr. A. Forrer (fl. 1887), who collected the type. (*Sedum*)

forrestii For George Forrest (1873–1932), British traveller and plant collector in China (1904–1932). (*Rhodiola yunnanensis* ssp., *Sedum, Sinocrassula indica* var.)

forskaolianus For Pehr Forsskål (1732–1763), Finnish botanist of Swedish parents, botanical traveller in Egypt and Arabia (1761–1763), died of malaria in Yemen. (*Sansevieria, Sarcostemma*)

forsterianus For Edward Forster (1765–1849), English botanist. (*Sedum*)

fortiflorus Lat. 'fortis', strong, powerful; and Lat. '-florus', -flowered. (*Agave*)

fortissimus Superl. of Lat. 'fortis', strong, powerful; for the growth habit. (*Euphorbia*)

fortuitus Lat., fortuitous, lucky; for the chance discovery of the taxon. (*Ceropegia, Euphorbia*)

fosbergii For Dr. Francis R. Fosberg (1908–1993), US-American botanist. (*Cylindropuntia*)

fosteri For Cyril Foster (fl. 1933), Krugersdorp, RSA. (*Aloe*)

fosterianus For Mulford B. Foster (1888–1978), US-American plantsmen and Bromeliad specialist. (*Peniocereus*)

Fouquieria For Pierre-Eloi (some sources have Pierre Edouard) Fouquier (1776–1850), French physician and physician in ordinary of King Louis-Philippe. (*Fouquieriaceae*)

fourcadei For Dr. Henri G. Fourcade (1855–1948), French forester and land surveyor, emigrated to RSA in 1881. (*Drosanthemum, Ruschia, Trichodiadema*)

fourcroydes Gr. '-oides', resembling; and for the genus '*Fourcroya*' (*Furcraea*) (*Agavaceae*). (*Agave*)

fouriei For Stephanus P. Fourie (fl. 1987) of the Transvaal Nature Conservation Division, RSA. (*Aloe*)

fractiflexus Lat., zig-zag; for the zig-zag undulations of the branch angles. (*Euphorbia*)

fragilis Lat., fragile, brittle; (**1**) for the fragile branches. (*Conophytum wettsteinii* ssp., *Crassula expansa* ssp., *Portulaca, Tylecodon*) – (**2**) for the easily broken rosettes. (*Aloe*) – (**3**) because the plants are uniquely fragile. (*Bulbine, Opuntia*)

fragosus From Lat. 'fragum', strawberry; probably for the similarly red fruit. (*Tinospora*)

fragrans Lat., scented, fragrant; for the flowers. (*Callisia, Harrisia, Sedum*)

Frailea For Manuel Fraile (fl. 1922), US-American horticulturist who maintained the cactus collection at the US Department of Agriculture in Washington D.C. (*Cactaceae*)

fraileanus As above. (*Mammillaria*)

frailensis For the occurrence near Punta Frailes, Baja California, Mexico. (*Agave sobria* ssp.)

framesii For Percival ("Percy") Ross Frames (1863–1947), South African solicitor, and collector and grower of succulents. (*Aloe, Argyroderma, Cephalophyllum, Delosperma, Drosanthemum, Lampranthus, Malephora, Piaranthus, Quaqua, Ruschia*)

francescae For Françoise M-L. Williamson (née Clerc) (*1935), Swiss teacher and plant collector in Namibia and RSA, wife of the Zimbabwean / South African dental surgeon and succulent plant collector Dr. Graham Williamson. (*Bulbine*)

francesiae For Miss Frances M. Leighton (later Mrs. Isaac) (*1909), botanist at the Bolus Herbarium, University of Cape Town, RSA. (*Lampranthus*)

franchetii For Adrien R. Franchet (1834–1900), French botanist in Paris. (*Sedum*)

francisci For Frans (Frantz) de Laet (1866–1928), Belgian succulent plant expert and horticulturist in Contich (Kontich). (*Lithops*) – (**2**) For Frank J. Stayner (1907–1981), horticulturist and 1959–1969 curator of the Karoo Botanic Garden Worcester. (*Lampranthus*)

francisii For Francis (Frank) K. Horwood (1924–1987), eminent English-born cultivator of succulents, emigrating to the USA in the 70ies. (*Sansevieria*)

franckianus For Harry Franck (fl. 1907), Frankfurt, owner of a large collection of succulents. (*Euphorbia*)

francoiseae For Françoise M-L. Williamson (née Clerc) (*1935), Swiss teacher and plant collector in Namibia and RSA, wife of the Zimbabwean / South African dental surgeon and succulent plant collector Dr. Graham Williamson. (*Conophytum wettsteinii* ssp.)

francoisii For E. François (fl. 1946), owner of the farm near Fort Dauphin, Madagascar, where the taxon was discovered. (*Euphorbia*)

francombei For Colin Francombe (fl. 1994), Ranch Manager in Kenya. (*Aloe*)

franksiae For Ms. Millicent Franks (later Mrs. Flanders) (1886–1961), botanical artist and assistant to Prof. Medley Wood at the Natal Herbarium, RSA. (*Brachystelma, Euphorbia*)

franzosinii For Francesco Franzosini (fl. 1892), Italian nobleman and owner of a fine garden at the Lago Maggiore. (*Agave*)

fraternus Lat., brotherly, closely allied; (**1**) for the close affinity with *Conophytum minutum*. (*Conophytum*) – (**2**) for the relationship to another taxon. (*Brownanthus*)

frederici For Frederick A. Rogers (1876–1944), British missionary and amateur botanist, lived in RSA from 1904. (*Ruschia*) – (**2**) For Frederik T. Herselman (fl. 1968), without further data. (*Lithops dinteri* ssp.)

fredericii For Frederick Huntly Holland (1873–1955), South African businessmen and naturalist. (*Delosperma*)

frequens Lat., frequent; implying widespread, because the taxon is found frequently when travelling. (*Sansevieria*)

frerei For Sir Henry B. Frere (fl. 1865), English diplomat and Governor of Bombay, later in RSA. (*Caralluma*)

freudenbergeri For Gerhard Freudenberger (fl. 1981), German cactus collector in Bad Rappenau. (*Echinocereus*)

fricii For Alberto V. Fric (1882–1944), Czech horticulturist and for 12 years intrepid explorer and cactus collector in the Americas. (*Cereus, Stenocereus*)

friedrichiae For Margarete Friedrich (fl. 1914), teacher in Warmbad, Namibia. (*Conophytum, Euphorbia*)

friedrichii For Adolfo M. Friedrich (1897–1987), German photographer emigrating 1925 to Brazil and settling 1930 in Paraguay, war photographer and cactus collector during the Chaco war between Paraguay and Bolivia. (*Frailea*) – (**2**) For Dr. Heimo Friedrich (1911–1987), Austrian botanist and specialist in the cultivation of pharmacological plants, interested in the classification of cacti. (*Echinopsis*)

friesianus For Robert E. Fries (1876–1966), Swedish botanist and son of Theodor M. Fries. (*Portulaca*)

friesii As above. (*Monadenium*)

frigidus Lat., cold; for the high-Andean occurrence. (*Cistanthe, Cumulopuntia*)

friisii For Ib Friis (*1945), Danish botanist at the University of Copenhagen. (*Aloe*)

Frithia For Frank Frith (1872–1954), horticulturist with the RSA Railways, and succulent plant collector. (*Aizoaceae*)

frithii As above. (*Peersia*)

fritschii For K. Fritsch (1864–1934), Austrian botanist who referred this taxon erroneously to *N. fluminensis*. (*Nematanthus*)

frohningiorum For Hans and Uta Frohning (fl. 2001), German cactus hobbyists. (*Weberocereus*)

frommii For the German botanical collector Fromm (fl. 1913). (*Stathmostelma spectabile* ssp.)

frutescens Lat., becoming shrubby (from Lat. 'frutex', shrub). (*Bulbine, Conophytum, Delosperma, Espostoa, Grahamia, Leipoldtia, Sedum, Stoeberia*)

fruticosus Lat., shrubby (from Lat. 'frutex', shrub). (*Adenia, Euphorbia, Galenia, Graptopetalum, Matucana, Plectranthus, Tetradenia, Tetragonia*)

fruticulosus Dim. of Lat. 'fruticosus'; i.e. small shrubby. (*Ceraria, Dischidia*)

fucosus Lat., (reddish) coloured. (*Orbea verrucosa* var.)

fugitans Lat., fleeing, departing; (**1**) for the short-lived flowers. (*Lampranthus*) – (**2**) application obscure. (*Ruschia*)

fui For Shu Hsia Fu (*1916), Chinese botanist. (*Sedum*)

fulgens Lat., shining, bright-coloured; (**1**) for the leaves. (*Echeveria*) – (**2**) for the flowers. (*Portulaca, Senecio*)

fulgidus Lat., shining, brightly coloured; (**1**) for the spination. (*Cylindropuntia*) – (**2**) for the brilliant flowers. (*Pelargonium*)

fuliginosus Lat., sooty, full of soot; for the dark appearance of the plants in the wild. (*Opuntia*)

fulleri For Ernest R. Fuller (fl. 1920–1928), postmaster at Kakamas, Kenhardt and Pofadder, Northern Cape, RSA, and active field collector of succulent plants. (*Cephalophyllum, Conophytum, Drosanthemum, Ebracteola, Lithops julii* ssp., *Stomatium*) – (**2**) For Major Andrew B. I. Fuller (fl. 1967), plant collector in SW Arabia. (*Aloe, Rhytidocaulon*)

fulviceps Lat. 'fulvus', tawny, yellowish-brown; and Lat. '-ceps', -headed; (**1**) for the cephalium colour. (*Pachycereus*) – (**2**) for the leaf colour. (*Lithops*)

fulvicomus Lat. 'fulvus', tawny, yellowish-brown; and Lat. 'coma', hair tuft, mane; for the spine colour. (*Cumulopuntia*)

fulvilanatus Lat. 'fulvus', tawny, yellowish-brown; and Lat. 'lanatus', woolly; for the hairs produced by the flower-bearing areoles. (*Pilosocereus*)

fulvisetus Lat. 'fulvus', tawny, yellowish-brown; and Lat. 'seta', bristle; (**1**) for the spination. (*Rebutia*) – (**2**) for the bristles on the flower tube. (*Frailea pygmaea* ssp.)

fulvus Lat., tawny, yellowish-brown; for the spine colour. (*Eriosyce odieri* ssp., *Lasiocereus*)

funalis Lat. 'funis', rope; for the narrowly cylindrical stems resembling a thick rope. (*Tacinga*)

funifer Lat. 'funis', rope; and Lat. '-fer, -fera, -ferum', -carrying; for the leaf margins with strong fibres. (*Hesperaloe*)

furcatus Lat., forked; (**1**) for the predominant branching mode of the stems. (*Othonna*) – (**2**) for the spines spreading from a common base. (*Euphorbia*) – (**3**) for the forked apex of the outer corona lobes. (*Brachystelma*)

Furcraea For Antoine F. de Fourcroy (1755–1809), French politician and chemist, 1784 director at the Jardin des Plantes in Paris. (*Agavaceae*)

furfuraceus Lat., scurfy, covered in bran-like scales; for the scurfy leaves. (*Sedum*)

furseorum Lat. Gen. Pl., for Admiral J. P. Furse (fl. 1969) and some family member, who collected the type of the taxon (*Sempervivum*)

furtus Lat., theft, robbery; the type material was stolen from the Living Collections at Kew. (*Caralluma*)

furusei For Miyoshi Furuse (fl. 1954), Japanese plant collector. (*Hylotelephium*)

furvus Lat., dark; for the dark purple stigma. (*Lampranthus*)

fuscomarginatus Lat. 'fuscus', sombre brown; and Lat. 'marginatus', margined; for the darker leaf margins. (*Hoya*)

fuscus Lat., sombre brown; (**1**) for the reddish-brown leaves. (*Crassula*) – (**2**) for the colour of the hairs enveloping the flower base. (*Parodia*) – (**3**) for the flower colour. (*Agave, Ceropegia*) – (**4**) for the colour of the cyathial glands. (*Euphorbia*) – (**5**) application unclear. (*Polyachyrus, Sedum*)

fusiformis Lat. 'fusus', spindle; and Lat. '-formis', -shaped; (**1**) for the tuberous root. (*Euphorbia*) – (**2**) for the shape of the leaves. (*Sedum*) – (**3**) for the shape of the fruits. (*Ibervillea*)

fwambense For the occurrence at Fwambo, Northern Prov., Zambia. (*Monadenium*)

G

gabbii For William M. Gabb (1839–1888), US-American paleontologist and geologist collecting plants in Mexico. (*Mammillaria brandegeei* ssp.)
gaertneri For Karl F. Gaertner († 1850), physician of German descent in Blumenau, Brazil. (*Hatiora*)
gagei For Prof. Dr. Gage (fl. 1910), then director of the Botanical Garden in Calcutta, India. (*Sedum*)
galapageius For the occurrence on the Galapagos Islands. (*Opuntia*)
galapagensis As above. (*Nolana*)
galapagosus As above. (*Talinum*)
galeanensis For the occurrence near Galeana, Nuevo León, Mexico. (*Acharagma roseanum* ssp.)
galeatus Lat., provided with a helmet; for the corolla. (*Ceropegia*)
Galenia For Galen of Pergamum (129–c. 216), Greek physician, philosopher and natural scientist. (*Aizoaceae*)
galenioides Gr. '-oides', resembling; and for the genus *Galenia* (*Aizoaceae*). (*Aizoanthemum*, *Tetragonia*)
galeottianus For Henri G. Galeotti (1814–1858), French-born Belgian geologist, botanist and explorer of Mexico. (*Bursera*)
galerasensis For the occurrence at Galeras, Dept. Ayacucho, Peru. (*Cumulopuntia*)
galericulatus Lat., with a little cap or hat, with false hair, with a periwig; for the operculum of the fruits, which has the outline of a lawyer's wig. (*Zaleya*)
galgalanus For the occurrence at Galgalo in the Bosaso-Region, Somalia. (*Euphorbia*)
galgallensis For the occurrence at Galgallo in the Bosaso-Region, Somalia. (*Duvalia*)
galioides Gr. '-oides', resembling; and for the genus *Galium* ("Woodruff", "Bedstraw"; *Rubiaceae*); for the whorled leaves. (*Peperomia*)
galpiniae For Marie E. Galpin (née de Jongh) († 1933), wife of the South African amateur botanist and banker Ernest E. Galpin. (*Lampranthus*)
galpinii For Dr. Ernest E. Galpin (1858–1941), South African banker and amateur botanist. (*Delosperma*, *Raphionacme*, *Schizoglossum bidens* ssp., *Senecio*)
gamkensis For the occurrence near the Gamka River, Western Cape, RSA. (*Euphorbia*)
gamoepensis For the occurrence at Gamoep, Bushmanland, Northern Cape, RSA. (*Cheiridopsis*)
gamugofana For the occurrence in the Gamu Gofa Region in Ethiopia. (*Euphorbia septentrionalis* ssp.)
ganderi For Frank F. Gander (fl. 1938), without further data. (*Cylindropuntia*)
gaponii For Victor Gapon (fl. 2001), Russian cactus hobbyist in Moscow and one of the initiators of the Russian "Gymnophil" group. (*Gymnocalycium*)
garambiensis For the occurrence near Garambi, Taiwan. (*Kalanchoe*)
garaventae For Augustin Garaventa (fl. 1959), Chile; without further data. (*Eriosyce*)
garciae For Juan A. García Luna (fl. 1997), Mexican horticulturist at the CANTE Botanical Garden. (*Pachyphytum*, *Thelocactus*)
garciae-mendozae For Dr. Abisaí García Mendoza (fl. 2002), Mexican botanist and specialist on Mexican *Agavaceae*. (*Agave*)
garibinus From the Khoi name 'Gariep' for the Orange or Oranje River, meaning 'large, huge'; for the distribution. (*Crassula*)
gariepensis From the Khoi name 'Gariep' for the Orange or Oranje River, meaning 'large, huge'; for the distribution. (*Aloe*, *Bowiea*, *Phyllobolus*, *Stapelia*)
gariepinus From the Khoi name 'Gariep' for the Orange or Oranje River, meaning 'large, huge'; for the distribution. (*Ceraria*, *Euphorbia*)
gariusanus Unknown; application obscure. (*Mesembryanthemum*)
gasserianus For Jakob Gasser (1870–1932), Swiss cactus nurserymen in Zürich. (*Mammillaria*)
Gasteria Gr. 'gaster', stomach; for the stomach-shaped basally inflated perianth. (*Aloaceae*)

gastonis-bonnieri For Prof. Dr. Gaston Bonnier (1853–1922), French botanist in Paris. (*Kalanchoe*)

gatbergensis For the occurrence near Gat Berg, Eastern Cape, RSA. (*Euphorbia*)

gatesii For Howard E. Gates (1895–1957), US-American explorer of Baja California and owner of a large succulent plant nursery in Norco, California. (*Dudleya, Ferocactus gracilis* ssp., *Pachycereus, Pereskiopsis*)

gattefossei For Jean Gattefossé (1899–1960), French botanical collector in Morocco. (*Sedum*)

gaumeri For Dr. F. Gaumer (fl. 1920), who collected plants in Mexico. (*Jatropha, Mammillaria heyderi* ssp., *Pachycereus*)

gautengensis For the occurrence in Gauteng Prov., RSA. (*Delosperma*)

gautii For J. H. Gaut (fl. 1905), without further data. (*Echinomastus*)

gayanus For Claude Gay (1800–1873), French botanist and traveller, esp. in Chile. (*Nolana*)

geayi For Martin François Geay (1859–1910), French pharmacist, natural history collector and traveller. (*Pachypodium*)

gebseri For Walter Gebser (fl. 1960), who collected the type in Namibia, without further data. (*Lithops schwantesii* ssp.)

geldorensis For the occurrence on the Geldora Pass, Somalia. (*Euphorbia*)

gelidus Lat., very cold, icy; for the occurrence in Siberia. (*Rhodiola*)

geminatus Lat., double; (**1**) for the paired leaves. (*Braunsia*) – (**2**) for the occasionally occuring two flowers. (*Piaranthus*)

geminiflorus Lat. 'geminus', double, or Lat. 'gemini', twins; and Lat. '-florus', -flowered. (*Agave, Ruschia*)

geminispinus Lat. 'geminus', double, or Lat. 'gemini', twins; and Lat. '-spinus', -spined; for the usually two central spines. (*Mammillaria*)

geminus Lat., double; (**1**) for the two leaves per branch. (*Cerochlamys*) – (**2**) presumably for the leaf pair. (*Gibbaeum*)

gemmeus Lat., made from gems; for the ruby-red nectary glands. (*Euphorbia*)

gemmifer Lat. 'gemma', bud, eye, pearl; and Lat. '-fer, -fera, -ferum', -carrying; for the occasional production of bulbils. (*Crassula*)

gemmiflorus Lat. 'gemma', bud, eye, pearl; and Lat. '-florus', -flowered; for the rugulose glabrous corolla. (*Tridentea*)

gemugofana For the occurrence in the Gemu [Gamu] Gofa Region in Ethiopia. (*Orbea*)

geniculatus Lat., with a knee, with a knot; (**1**) for the knee-like bend in the inflorescence. (*Ornithogalum*) – (**2**) for the knee-like bend in the pedicel. (*Caralluma adscendens* var.) – (**3**) for the knee-like bend in the corolla lobes. (*Ceropegia fimbriata* ssp.)

geniculiflorus Lat. 'geniculum', knee, knot; and Lat. '-florus', -flowered; application obscure. (*Aptenia*)

genoudianus For Mlle. J. Genoud (fl. 1955), botanical assistant at the Scientific Research Institute, Madagascar. (*Euphorbia*)

gentilis Lat., related; for the close relationships to several other taxa. (*Euphorbia*)

gentryi For Dr. Howard S. Gentry (1903–1993), US-American botanist, explorer and *Agave* specialist. (*Agave, Echinocereus scheeri* ssp.)

geoffreyi For Geoffrey James (fl. 1931), without further data. (*Stomatium*) – (**2**) For Geoffrey Hinton (fl. 2003), son of the Mexican farmer and botanical collector George Sebastián Hinton (*1949). (*Coryphantha hintoniorum* ssp.)

Geohintonia For George Sebastián Hinton (*1949), Mexican farmer and plant collector in Nuevo León, grandson of George B. Hinton (senior). (*Cactaceae*)

geometricus MLat., geometrical; for the polygonal pattern formed by the tubercles of the stem segments. (*Tephrocactus*)

geometrizans MLat., geometrical; for the arcuate pattern between the annual growths of the stems. (*Myrtillocactus*)

georgii For Mr. E. Georgi (fl. 1931), cactus collector in Saltillo, Nuevo León, Mexico. (*Coryphantha*)

germanae Gen. of Lat. 'germana', sister, brother; application obscure. (*Kalanchoe*)

geroldii For Raymond Gerold (fl. 1994), plant trader in Madagascar. (*Euphorbia*)

Gerrardanthus Gr. 'anthos', flower; and for William T. Gerrard († 1866), British naturalist and traveller in RSA and Madagascar, collected in RSA ± 1860–1865. (*Cucurbitaceae*)

gerrardii For William T. Gerrard († 1866), British naturalist and traveller in RSA and Madagascar, collected in RSA ± 1860–1865. (*Brachystelma*, *Cynanchum*)

gerstneri For Father Jacob Gerstner (1888–1948), Bavarian missionary and botanist, emigrating to Africa 1924 and living in KwaZulu-Natal, RSA. (*Aloe*, *Delosperma*, *Nananthus*, *Orbea*, *Stomatium*)

gesinae For Mrs. Gesina de Boer-Weyer (fl. 1955), wife of the Dutch *Lithops* specialist H. W. de Boer. (*Lithops*)

gessertianus For a Mr. Gessert (fl. 1923), without further data. (*Psilocaulon*)

gettliffei For George F. R. Gettliffe (1873–1948), South African farmer formerly working for the Irrigation Department. (*Stapelia*)

geyeri For Dr. Albertus L. Geyer (1894–1969), South African journalist, diplomat and plant collector with a special interest in succulent plants. (*Lithops*)

giajae For Dr. Jean Giaja (fl. 1910), "Maître de Conférences" at Belgrad University. (*Sedum*)

gibarensis For the occurrence near the town of Gibara, Cuba. (*Consolea nashii* ssp.)

Gibbaeum Lat. 'gibba', swelling, gibbosity; for the hump or chin of the lower surface of the longer leaf of each pair. (*Aizoaceae*)

gibbiflorus Lat. 'gibba', swelling, gibbosity; and Lat. '-florus', -flowering; for the flower shape. (*Echeveria*)

gibbosus Lat., gibbous, asymmetrically swollen; (**1**) for the irregular swellings of stems and branches. (*Adansonia*, *Pelargonium*) – (**2**) for the flatly tuberculate plant body. (*Gymnocalycium*) – (**3**) for the hump or chin of the lower surface of the longer leaf of each pair. (*Gibbaeum*)

gibbsiae For Miss Lilian S. Gibbs (1870–1925), British pioneer plant collector in Southern Rhodesia (now Zimbabwe). (*Euphorbia trichadenia* var.)

gielsdorfianus For Karl Gielsdorf (1888–1973), German horticulturist and for many years responsible for the cactus collection at the Botanical Garden Berlin-Dahlem. (*Turbinicarpus*)

giessii For Heinrich J. W. Giess (1910–2000), farmer and botanist of German origin at the Windhoek herbarium, Namibia. (*Aizoon*, *Commiphora*, *Crassula ausensis* ssp., *Euphorbia*)

giffenii For Prof. Malcolm Hutchinson Giffen (*1902), South African botanist at the University of Fort Hare. (*Delosperma*, *Drosanthemum*)

giffordianus For Prof. E. M. Gifford (fl. 1978), University of California, Davis. (*Jatropha*)

gigantensis For the occurrence in the Sierra de la Giganta in S Baja California, Mexico. (*Agave*, *Cylindropuntia alcahes* var.)

giganteus Lat., gigantic, enormous; (**1**) for the size of the plants. (*Mammillaria*, *Opuntia echios* var., *Oxalis*) – (**2**) for the large rosettes. (*Echeveria*, *Haworthia limifolia* var., *Prometheum serpentinicum* var.) – (**3**) for the flower size. (*Stapelia*)

gigantiflorus Lat. 'giganteus', gigantic, enormous; and Lat. '-florus', -flowered. (*Stathmostelma*)

gigas Gr., giant; for the size of the plants. (*Dorstenia*, *Pseudolithos*, *Ruschianthemum*)

gikyi For the Madagascan nature conservationist Monsieur Giky, whose tomb is near the type locality of the taxon, which has the local vernacular name "felangiky". (*Ceropegia*)

gilbertii For John Gilbert Baker (1834–1920), British botanist at Kew. (*Agave*) – (**2**) For Mike G. Gilbert (*1943), English botanist resident in Ethiopia and Kenya 1968–1982, with a strong interest in succulent plants, especially Aloes, Asclepiads and Euphorbias. (*Aloe*)

gilensis For the occurrence in Gila County, Arizona, USA. (*Marah*)

gilgianus For Dr. Ernst F. Gilg (1867–1933), German botanist in Berlin. (*Ceropegia*)

gillespieae For Lynn J. Gillespie (fl. 1989), US-American botanist. (*Marsdenia*)

gillettii For Jan B. Gillett (1911–1995), English botanist at Kew, resident in Kenya 1963–1984, specialist in *Indigofera* and *Commiphora*. (*Aloe, Euphorbia, Monadenium, Plectranthus*)

gillianii For Miss Gillian Meadows (fl. 1969), who produced drawings to illustrate a paper on *Sempervivum*. (*Sempervivum*)

gilliesii For John Gillies (1792–1834), Scottish physician and botanist, working and collecting in Argentina 1820–1828. (*Portulaca*)

gilvus Lat., yellowish; for the leaf colour. (*Echeveria*)

giselae For Prof. Gisela Gallegos Hernández (fl. 1997), wife of the Mexican botanist J. G. Martínez-Avalos. (*Mammillaria schiedeana* ssp.)

githagineus Lat., with red or purple streaks on a green ground, as the calyx of *Agrostemma githago* ("Corn Cockle"; *Caryophyllaceae*). (*Pelargonium*)

glabellus Lat., glabrous; for the glabrous corolla. (*Aspidoglossum*)

glaber Lat., naked, glabrous; (**1**) for the completely or almost glabrous plants. (*Gunniopsis, Jatropha pelargoniifolia* var., *Kalanchoe deficiens* var., *Sedum*) – (**2**) for the stems. (*Brachystelma*) – (**3**) for the leaves. (*Columnea, Crassula garibina* ssp., *Gasteria carinata* var., *Hymenogyne, Rosularia*) – (**4**) for the often completely absent spination. (*Weberocereus*)

glabratus Lat., devoid of hairs, glabrous; (**1**) for the almost entirely glabrous plants. (*Synadenium*) – (**2**) for the lack of (hair-like) marginal spines. (*Haworthia*)

glabrescens Lat., becoming glabrous; (**1**) for the weak and sometimes absent spination. (*Eriosyce odieri* ssp.) – (**2**) for the inconspicuous papillae that cover the herbaceous parts and that are much smaller than in related species. (*Drosanthemum*) – (**3**) for the perianth surface. (*Aloe*) – (**4**) application obscure. (*Aspidoglossum*)

glabricaulis Lat., 'glaber, glabra, glabrum', glabrous; and Lat. 'caulis', stem. (*Stapelia*)

glabriflorus Lat. 'glaber, glabra, glabrum', glabrous; and Lat. '-florus', -flowered. (*Brachystelma*)

glabrifolius Lat. 'glaber, glabra, glabrum', glabrous; and Lat. '-folius', -leaved. (*Crassula tomentosa* var., *Gynura drymophila* var., *Hoya calycina* ssp., *Sempervivum*)

glabriphyllus Lat. 'glaber, glabra, glabrum', glabrous; and Gr. 'phyllon', leaf; for the glabrous upper face of the leaves. (*Pelargonium*)

gladiatus Lat., sword-like; for the elongated shape of the leaves. (*Monadenium*)

glaebosus Perhaps from Lat. 'gl[a]eba', clump; for the tufted growth. (*Sedum*)

glaetzleanus For Dr. Wolfgang Glätzle (*1951), Austrian chemist and cactus hobbyist. (*Echinopsis calochlora* ssp.)

glandularis Lat., with glands; for the large stipular glands. (*Euphorbia*)

glandulifer Lat. 'glandula', gland; and Lat. '-fer, -fera, -ferum', -carrying; (**1**) for the glandular stems and leaves. (*Galenia*) – (**2**) for the glandular leaves. (*Adenia hastata* var., *Aspidoglossum, Jatropha, Sedum dasyphyllum* var.)

glanduliflorus Lat. 'glandula', gland; and Lat. '-florus', -flowered; for the hairy corolla. (*Stapelia*)

glanduliger Lat. 'glandula', gland; and Lat. '-ger, -gera, -gerum', -carrying; for the spines transformed into nectar glands. (*Coryphantha*)

glandulosus Lat., glandular-hairy; (**1**) for the dense glandular-pubescent indumentum. (*Aeonium*) – (**2**) for the indumentum of stems and branches. (*Trianthema*) – (**3**) for the indumentum of pedicels and sepals. (*Tripogandra*)

glareicola Lat. 'glarea', gravel; and Lat. '-cola', -dwelling. (*Brownanthus*)

glareosus Lat., pebbly, full of pebbles; for the preferred habitat. (*Mammillaria brandegeei* ssp.)

glassianus For Charles E. Glass (1934–1998), US-American horticulturist and succulent plant specialist, editor of the US-American cactus journal for 26 years, living in Mexico from 1991 onwards. (*Calibanus*)

glassii As above. (*Coryphantha, Graptopetalum, Mammillaria*)

glaucescens Lat. 'glaucus', blue-green, and

Lat. '-escens', becoming; (**1**) for the colour of the plants. (*Ferocactus, Kalanchoe, Melocactus*) − (**2**) for the colour of older leaves. (*Synadenium*) − (**3**) for the slightly glaucous leaves. (*Carpobrotus*)

glaucifolius Lat. 'glaucus', blue-green, and Lat. '-folius', -leaved. (*Trianthema*)

glaucochrous Gr. 'glaucos', bluish, blue-green; and Gr. 'chroa', body colour; for the colour of the stems. (*Pilosocereus*)

glaucophyllus Gr. 'glaucos', bluish, blue-green; and Gr. 'phyllon', leaf. (*Dasylirion, Haworthia limifolia* var., *Rosularia sempervivum* ssp., *Sedum*)

glaucus Lat., glaucous, grey-green, blue-grey; (**1**) for the body colour. (*Acanthocalycium, Echinopsis, Sclerocactus*) − (**2**) for the colour of the caudex. (*Adenia*) − (**3**) for the stems. (*Ruschia*) − (**4**) for the stems and leaves. (*Tetragonia*) − (**5**) for the leaf colour. (*Aloe, Bulbine, Gasteria, Haworthia, Heliophila, Jatropha, Lampranthus, Monsonia, Sansevieria raffillii* var., *Yucca*)

glaziovii For Auguste F. M. Glaziou (1828–1906), French botanical traveller collecting 1861–1895 in Brazil. (*Arthrocereus*)

glechomoides Gr. '-oides', resembling; and for the genus *Glechoma* ("Ground Ivy"; *Lamiaceae*); for the similar leaves. (*Pelargonium*)

glenensis For the occurrence near Glen, Free State, RSA. (*Brachystelma, Hereroa*)

glinoides Gr. '-oides', resembling; and for the genus *Glinus* (*Aizoaceae*). (*Aizoon*)

globifer Lat. 'globus', globe, sphere; and Lat. '-fer, -fera, -ferum', -carrying; (**1**) for the globose offsets. (*Sempervivum*) − (**2**) for the globose clusters and rosettes. (*Haworthia pulchella* var.)

globosiflorus Lat. 'globosus', globose; and Lat. '-florus', -flowered. (*Haworthia nortieri* var.)

globosus Lat., globose; (**1**) for the globose caudex. (*Adenia*) − (**2**) for the globose branches. (*Euphorbia*) − (**3**) for the globose leaves. (*Drosanthemum*) − (**4**) for the globose fused leaf pairs. (*Conophytum, Meyerophytum*) − (**5**) for the globose inflorescences. (*Monadenium, Raphionacme*) − (**6**) for the globose corolla. (*Echidnopsis, Huernia keniensis* var.) − (**7**) for the globose receptacle of the flowers. (*Lampranthus*)

globulariifolius For the genus *Globularia* (*Globulariaceae*); and Lat. '-folius', -leaved; for the similar leaf shape. (*Rosularia*)

globularioides Gr. '-oides', similar to; and for the genus *Globularia* (*Globulariaceae*); for the similarly capitate inflorescences. (*Crassula*)

globulicaulis Lat. 'globulus', globule, little ball; and Lat. 'caulis', stem. (*Euphorbia*)

globulifer Lat. 'globulus', globule, little ball; and Lat. '-fer, -fera, -ferum', -carrying; for the capitate inflorescences. (*Kalanchoe*)

globuliflorus Lat. 'globulus', globule, little ball; and Lat. '-florus', -flowered; for the flower shape. (*Echeveria, Sedum*)

globuligemma Lat. 'globulus', globule, little ball; and Lat. 'gemma', bud; for the globular flower buds. (*Aloe*)

globulosus Lat., like a small globe; (**1**) for the globular rosettes. (*Echeveria*) − (**2**) for the globular inflorescences. (*Hoya*)

glochidiatus Lat., barbed; (**1**) for the forked tips of the spines. (*Euphorbia*) − (**2**) for the hooked central spines. (*Mammillaria*)

glomeratus Lat., glomerate, collected closely together in heads; (**1**) for the growth form. (*Cheiridopsis, Gasteria, Maihueniopsis, Parodia rudibuenekeri* ssp.) − (**2**) for the inflorescence. (*Crassula, Lampranthus*)

glomerifolius Lat. 'glomerus', small heap, head; and Lat. '-folius', -leaved; for the globose clustered leaves. (*Sedum*)

glomeruliflorus Lat. 'glomerulus', glomerule, small heap; and Lat. '-florus', -flowered; (**1**) for the clustered flowers. (*Agave*) − (**2**) for the clustered male flowers. (*Corallocarpus*)

gloriosus Lat., glorious. (×*Gasteraloe, Yucca*)

glossistigma Gr. 'glossa', tongue; and Gr. 'stigma', stigma; for the appearance of the stigma, slightly shifted to one side and returned. (*Trianthema*)

Glossostelma Gr. 'glossa, glotta', tongue; and Gr. 'stelma', crown, garland, wreath; for the structure of the corona. (*Asclepiadaceae*)

Glottiphyllum Gr. 'glossa, glotta', tongue;

and Gr. 'phyllon', leaf; for the leaf shape in some species. (*Aizoaceae*)

glutinicaulis Lat. 'glutinus', glutinous, sticky; and Lat. 'caulis', stem; for the sticky stem surface. (*Pachyphytum*)

glutinosus Lat., glutinous, sticky; for the sticky leaves. (*Aeonium*)

gneissicola German 'gneiss', gneiss rock; and Lat. '-cola', inhabiting. (*Aloe capitata* var.)

gnomus MLat., dwarf fabled being; for the small size of the plants and their parts. (*Dudleya*)

godingianus For Dr. F. W. Goding (fl. 1918), US Consulate-General in Guayaquil, Ecuador, who assisted Dr. J. N. Rose in his botanical explorations of Ecuador. (*Armatocereus*)

godmaniae For Dame Alice Godman (fl. 1919). (*Drosanthemum, Lampranthus*)

goebelianus For Prof. Dr. Karl I. E. Goebel (1855–1932), German botanist and director of the Botanical Garden München. (*Coleocephalocereus, Rhipsalis*)

goetzei For Walter Goetze († 1899), German naturalist and explorer in Tanzania 1898–1899, who died of blackwater fever in Tanzania. (*Adenia, Dorstenia, Euphorbia, Monadenium*)

goianus For the occurrence in the state of Goiás, Brazil. (*Pilosocereus*)

goiasensis As above. (*Cereus jamacaru* ssp.)

goldianus For Dudley Gold (1897–1990), US-American succulent plant collector resident in Mexico. (*Echeveria halbingeri* var.)

goldii As above. (*Mammillaria saboae* ssp.)

goldmanianus For Edward A. Goldman (1873 –1946), US-American botanist. (*Agave shawii* ssp.)

goldmanii As above. (*Beaucarnea, Echeveria, Sedum*)

gomerensis For the occurrence on Gomera, Canary Islands. (*Aeonium*)

gonjianii For Barkev Gonjian (fl. 1960, 1973), Argentinian cactus collector in Buenos Aires and for many years president of the Círculo de Coleccionistas de Cactus y Crasas de la República Argentina. (*Pterocactus, Rebutia*)

gonzalezii For M. en C. Francisco Gonzalez Medrano (*1939), Mexican botanist who first found the taxon. (*Echinocereus parkeri* ssp.)

goochiae For Mrs. Gooch (fl. 1840), maiden name of the mother of the English naturalist and geologist Philip. B. Webb (1793–1854). (*Aeonium*)

gooddingii For Leslie N. Goodding (1880–1967), US-American botanist. (*Talinum*)

goodiae For Mrs. R. Good (fl. 1928), without further data. (*Ruschia*)

goodii For Mr. E. A. Good (fl. 1923), without further data. (*Cephalophyllum*)

goodridgii For John Goodridge (fl. 1846), English natural history traveller, participant of the voyage of the HMS Herald. (*Mammillaria*)

gordonianus For Gordon King (fl. 1937), son of Mrs. Isabella King, in Port Elizabeth, Eastern Cape, RSA. (*Haworthia cooperi* var.)

gordonii For Col. Robert J. Gordon (1743-1795), son of a Scottish father and a Dutch mother, soldier, explorer, naturalist, artist and illustrator who explored and named the Orange River (RSA) in 1777 and 1779, committed suicide in Cape Town. (*Hoodia*)

gorgoneus Lat. adjective referring to the Gorgades (present-day Cape Verde Islands); for the occurrence. (*Aeonium*)

gorgonis Lat., of the Gorgon; for the medusoid growth form. (*Euphorbia*)

gosselinianus For Robert Roland-Gosselin (1854–1925), French botanist. (*Opuntia*)

gossweileri For John Gossweiler (1873–1952), Swiss botanist in Portugal and plant collector in Angola around 1900. (*Aloe, Endadenium*)

gossypiifolius For the genus *Gossypium* ("Cotton"; *Malvaceae*); and Lat. '-folius', -leaved. (*Jatropha*)

gossypinus Lat., cotton-like; for the cotton-like filiform bracteoles emerging from the cyathia. (*Euphorbia*)

gottlebei For Gunter Gottlebe (fl. 1992), German resident of Tamatave, Madagascar. (*Euphorbia*)

goudotii For the (French ?) collector Goudot (fl. 1887), without further data. (*Tetradenia*)

gounellei For Edmond Gounelle (fl. 1892), French entomologist, 1892–1893 in Pernambuco, Brazil. (*Pilosocereus*)

govindia Unknown, presumably relating to the distribution in India. (*Zaleya*)

gracilicaulis Lat. 'gracilis', slender, delicate; and Lat. 'caulis', stem. (*Aloe, Euphorbia*)

gracilidelineatus Lat. 'gracilis', slender, delicate; and Lat. 'delineatus', drawn, marked; for the fine leaf markings. (*Lithops*)

gracilifrondosus Lat. 'gracilis', slender, delicate; and Lat. 'frondosus', leavy, full of leaves; for the narrow leaflets. (*Commiphora*)

gracilipes Lat. 'gracilis', slender, delicate; and Lat. 'pes', foot; (**1**) for the slender inflorescences. (*Agave, Caralluma*) – (**2**) for the slender pedicels. (*Kalanchoe, Lampranthus, Ruschia*)

gracilirameus Lat. 'gracilis', slender, delicate; and Lat. 'rameus', belonging to a branch. (*Euphorbia*)

gracilis Lat., slender, delicate; (**1**) for the rather small plant bodies. (*Coryphantha*) – (**2**) for the growth habit. (*Brachystelma, Caralluma adscendens* var., *Codonanthe, Corpuscularia, Delosperma, Duvalia, Echeveria, Eriosyce marksiana* var., *Ferocactus, Hammeria, Haworthia, Kedrostis, Mammillaria vetula* ssp., *Parakeelya, Parodia oxycostata* ssp., *Pedilanthus, Sedum, Seyrigia, Trichodiadema, Turbinicarpus schmiedickeanus* ssp.) – (**3**) for the slender stems. (*Aloe, Harrisia*) – (**4**) for the slender ultimate stem parts. (*Kleinia*) – (**5**) for the leaves. (*Aspidoglossum, Beaucarnea, Erepsia, Hereroa, Monadenium, Ruschia, Sansevieria, Scopelogena*) – (**6**) for the slender spines. (*Discocactus bahiensis* ssp.) – (**7**) for the narrow leaflets. (*Delonix leucantha* ssp.) – (**8**) for the delicate flowers. (*Quaqua*) – (**9**) application obscure. (*Schizoglossum bidens* ssp.)

gracilispinus Lat. 'gracilis', thin, slender; and Lat. '-spinus', -spined. (*Myrmecodia*)

gracilistylus Lat. 'gracilis', thin, slender; and Lat. 'stylus', style. (*Conophytum bilobum* ssp.)

gracilius Comp. of Lat. 'gracilis', slender; for the growth habit. (*Pachypodium rosulatum* var.)

gracillimus Superl. of Lat. 'gracilis', slender, delicate; (**1**) in comparison with the related *Brachystelma gracile*. (*Brachystelma*) – (**2**) for the habit of the plants. (*Agave, Antimima, Frailea*) – (**3**) for the very slender stems. (*Drosanthemum*) – (**4**) for the small and delicate leaves. (*Peperomia*)

gradyi For Prof. Grady L. Webster (*1927), US-American botanist and renowned *Euphorbiaceae* specialist. (*Euphorbia*)

graeseri For Alfred Gräser (1895–1973), German horticulturist and owner of a well-known cactus nursery in Nürnberg. (*Hatiora*)

graessneri For Richard Grässner (1875–1942), German cactus hobbyist and nurseryman in Perleberg near Berlin. (*Parodia haselbergii* ssp.)

Grahamia For Prof. Robert Graham (1786–1845), Scottish physician and botanist. (*Portulacaceae*)

grahamii For James D. Graham (fl. 1856), colonel in the Corps of Engineers, US Army, and chief of the scientific corps of the US & Mexican Boundary Survey. (*Grusonia, Mammillaria*) – (**2**) For the botanist M. D. Graham (fl. 1993), without further data. (*Cyphostemma*)

grahlianus For Paul Grahl (fl. 1899), German cactus hobbyist in Erfurt. (*Frailea*)

grallatus Lat., stilt-like; for the hairs on the leaves. (*Plectranthus*)

gramineus Lat., grass-like; (**1**) for the slender terminal branches. (*Delosperma*) – (**2**) for the linear leaves. (*Jatropha dioica* var.)

graminicola Lat. 'gramen, graminis', grass; and Lat. '-cola', -dwelling; for the habitat preference. (*Aloe lateritia* var.)

graminifolius Lat. 'gramen, graminis', grass; and Lat. '-folius', -leaved; for the leaf shape. (*Agave duplicata* ssp., *Dasylirion, Haworthia blackburniae* var., *Ischnolepis*)

grammanthoides Gr. '-oides', similar to, and for the genus *Grammanthes* (now a synonym of *Crassula*). (*Crassula*)

grammontensis For the occurrence at Morne Grammont, Haiti. (*Agave antillarum* var.)

grammophyllus Gr. 'gramma', line; and Gr. 'phyllon', leaf; for the linear leaves. (*Sedum*)

grandialatus Lat. 'grandis', large; and Lat. 'alatus', winged; for the broad wings of the branches. (*Euphorbia*)

grandicalcaratus Lat. 'grandis', large; and Lat. 'calcaratus', spurred; for the large nectar tube. (*Pelargonium*)

grandicornis Lat. 'grandis', large; and Lat. '-cornis', -horned; for the exceptionally long spines. (*Euphorbia*)

grandicuspis Lat. 'grandis', large; and Lat. 'cuspis', cusp; for the pointed leaf tips. (*Sansevieria*)

grandidens Lat. 'grandis', large; and Lat. 'dens', tooth; (**1**) for the prominent teeth of the branch angles. (*Euphorbia*) – (**2**) for the large teeth of the leaf margins. (*Stomatium*)

grandidentatus Lat. 'grandis', large; and Lat. 'dentatus', toothed; for the large teeth of the leaf margins. (*Aloe, Plectranthus*)

grandidieri For Alfred Grandidier (1836–1921), French naturalist active in Madagascar. (*Adansonia, Cynanchum, Kalanchoe, Uncarina*)

grandiflorus Lat. 'grandis', large; and Lat. '-florus', -flowering. (*Aspidoglossum, Brachystelma burchellii* var., *Cistanthe, Copiapoa, Dicrocaulon, Ferocactus chrysacanthus* ssp., *Glottiphyllum, Huernia keniensis* var., *Kalanchoe, Maihueniopsis, Portulaca, Pseudobombax, Raphionacme, Rhipsalis, Sedum moranense* ssp., *Selenicereus, Sempervivum, Stapelia, Tripogandra, Tylecodon, Yucca*)

grandifolius Lat. 'grandis', large; and Lat. '-folius', -leaved. (*Elaeophorbia, Leipoldtia weigangiana* ssp., *Sedum ebracteatum* ssp.)

grandilobus Lat. 'grandis', large; and Lat. 'lobus', lobe; for the massively lobed stems. (*Epiphyllum*)

grandipetalus Lat. 'grandis', large; and Lat. 'petalum', petal. (*Sedum*)

grandis Lat., large; (**1**) for the large growth. (*Echinocereus, Graptopetalum, Pachycereus*) – (**2**) for the large leaves. (*Mitrophyllum*) – (**3**) for the large flowers. (*Edithcolea*)

grandisepalus Lat. 'grandis', large; and Lat. 'sepalum', sepal. (*Sedum*)

grandyi For Abraham Osman Grandy (fl. 1913), a "good friend" of the french botanist and physician Raymond Hamet. (*Sedum*)

graniformis Lat. 'granum', seed, kernel; and Lat. '-formis', -shaped; for the leaf shape. (*Rhinephyllum*)

graniticola Ital. 'granito', granite; and Lat. '-cola', inhabiting; for the preferred habitat. (*Euphorbia*)

graniticus Ital. 'granito', granite; and Lat. '-icus', pertaining to; for the preferred habitat. (*Antimima*)

grantiae Possibly for Dr. Adèle Grant (née Lewis) (1881–1967), US-American botanist teaching in RSA ± 1925–1930. (*Delosperma*)

grantianus For Major Chapman Grant (fl. 1932) of the US Army. (*Leptocereus*)

grantii For Capt. (later Col.) James Augustus Grant (1827–1892), Scottish explorer and plant collector who travelled with J. H. Speke to discover the source of the Nile. (*Senecio sempervivus* ssp., *Synadenium*)

granulatus Lat., granulated; for the leaf surface. (*Haworthia venosa* ssp., *Hereroa*)

granulicaulis Lat. 'granula', small grain; and Lat. 'caulis', stem; for the papillate and thus slightly rough stems. (*Psilocaulon*)

granvikii Most probably for Hugo Granvik (1889-?), Swedish ornithologist, also collecting plants in Africa. (*Crassula*)

Graptopetalum Gr. 'graptos', marked, inscribed; and Gr. 'petalon', petal; for the usually spotted or blotched petals. (*Crassulaceae*)

gratiae For Grace V. Britten (*1904), botanical assistant at the Albany Museum Herbarium, Grahamstown, RSA, and a keen cultivator esp. of succulents. (*Delosperma, Faucaria*)

gratus Lat., pleasing, agreeable; for the appearance of the plant. (*Aloe, Coryphantha, Plectranthus*)

graveolens Lat., strong-smelling; for the strong scent of the crushed leaves. (*Peperomia*)

gravidus Lat., heavy, weighty; for the appearance of the plants. (*Eberlanzia*)
greatheadii For Dr. J. B. Greathead (fl. 1903), without further data. (*Aloe*)
greenei For Prof. Edward L. Greene (1843–1915), US-American botanist. (*Dudleya*)
greenii For C. G. or G. H. Green (fl. 1880), without further data. (*Aloe, Haworthia coarctata* fa.)
greenwayi For Dr. Percy ("Peter") J. Greenway (1897–1980), British botanist in Tanzania from 1928 and in Kenya from 1957. (*Euphorbia, Portulaca*)
greenwoodii For Edward Greenwood (fl. 1970), US-American engineer collecting cacti in Mexico. (*Coryphantha elephantidens* ssp.)
gregarius Lat., belonging to the herd, growing in company; (1) for the similarity to other species. (*Nematanthus*) – (2) because the taxon grows in large populations. (*Euphorbia*)
greggii For Josiah Gregg (1806–1850), US-American cactus collector. (*Epithelantha micromeris* ssp., *Peniocereus, Sedum*)
gregorii For Augustus C. Gregory (1819–1905) who discovered the taxon during the North Australian Exploring Expedition 1855–1856. (*Adansonia*)
grenvilleae For Lord and Lady W. W. Grenville who introduced the taxon to England in 1810. (*Pelargonium*)
grevei For Mr. Grevé († 1895), French colonist and naturalist, settled in Madagascar as farmer and collector of natural history specimens for the French naturalist A. Grandidier. (*Talinella*)
griffithsii For Prof. Dr. David Griffiths (1867–1935), British-born US-American horticulturist and mycologist, and student of prickly pears at the US Department of Agriculture. (*Sedum*)
grijalvensis For the occurrence in the valley of the Río Grijalva, Chiapas, Mexico. (*Agave*)
griquensis For the occurrence near Griquatown, Northern Cape, RSA. (*Ruschia*)
grisebachii For August H. R. Grisebach (1814–1879), German botanist and director of the Botanical Garden Göttingen. (*Sedum*)

griseolus Lat., greyish; for the grey confluent spine shields. (*Euphorbia*)
griseus Lat., grey; (1) for the grey colour of the plants. (*Crassula, Ruschia*) – (2) for the grey-brown flaking bark. (*Sedum*) – (3) for the grey-felted leaves. (*Begonia*) – (4) for the leaf colour. (*Agave, Aloe, Echeveria, Hallianthus*) – (5) for the spine colour. (*Stenocereus*)
groendrayensis For the occurrence on Groendraai Farm, Namibia. (*Lithops pseudotruncatella* ssp.)
groenewaldii For B. H. Groenewald (fl. 1938), South African amateur specialist on *Aloe*. (*Euphorbia*)
grossei For Hermann Grosse who collected plants in Paraguay in 1898 and 1903. (*Cleistocactus*)
grossidentatus Lat. 'grossus', thick, coarse; and Lat. 'dentatus', toothed; for the leaf margins. (*Jatropha*)
grossus Lat., thick, coarse. (*Phyllobolus, Stapelia montana* var.)
Grusonia For Hermann Gruson (1821–1895), German merchant in Magdeburg and amateur grower of succulents. (*Cactaceae*)
grusonii As above. (*Echinocactus, Mammillaria*)
guadalajaranus For the occurrence near the city of Guadalajara, Jalisco, Mexico. (*Agave, Sedum*)
guadalupensis For the occurrence on Guadalupe Island W of Baja California. (*Baeriopsis, Cistanthe, Dudleya*) – (2) For the occurrence in the Guadalupe Mts., Texas, USA. (*Escobaria*)
guadarramensis For the occurrence in the Sierra de Guadarrama, Segovia, Spain. (*Sempervivum cantabricum* ssp.)
guamacho From the local vernacular name of the plants used in Colombia and Venezuela. (*Pereskia*)
guanajuatensis For the occurrence in the state of Guanajuato, Mexico. (*Portulaca*)
guanchezii For the (Venezuelan ?) botanical collector Guanchez (fl. 1986). (*Marsdenia*)
guantanamanus For the occurrence in Guantánamo Prov., Cuba. (*Consolea moniliformis* ssp.)
guatemalensis For the occurrence in Guatem-

ala. (*Beaucarnea, Echeveria, Epiphyllum, Furcraea, Opuntia, Sedum, Villadia*)
gueinzii For Wilhelm Gueinzius (±1814–1874), German apothecary and naturalist, emigrated to RSA in 1838. (*Euphorbia*)
guelzowianus For Robert Gülzow (fl. 1928), German cactus nurserymen in Berlin. (*Mammillaria*)
guentheri For a (botanical?) collector Mr. Guenther (fl. 1909). (*Monadenium*) – (**2**) For Ernesto Günther (fl. 1931), botanical philanthropist in Valparaiso, Chile, for significantly contributing to the cost of the botanical travels of the German geographer and botanist C. Troll in Bolivia. (*Espostoa*)
guerichianus For Prof. Georg Gürich (1859–1938), German geologist at Hamburg University, prospecting for gold in Namibia. (*Euphorbia, Mesembryanthemum*)
guerichii As above. (*Sesamothamnus*)
guerkeanus For Prof. Dr. Robert L. A. M. (Max) Gürke (1854–1911), German botanist in Berlin, co-author with K. Schumann, and president of the Deutsche Kakteen-Gesellschaft. (*Coryphantha*)
guerrae For Dr. Guilherme Guerra (fl. 1960), Director of Agriculture and Forests, Angola. (*Aloe*)
guerrerensis For the occurrence in the state of Guerrero, Mexico. (*Agave, Furcraea*)
guerreronis As above. (*Mammillaria*)
guianensis For the occurrence in Guiana. (*Mesembryanthemum*)
guiengola For the occurrence on Guiengola Limestone, a limestone formation in Oaxaca, Mexico. (*Agave*) – (**2**) For the occurrence at Cerro Guiengola, Oaxaca, Mexico. (*Euphorbia*)
guilanchi From the local vernacular name of the plants in Zacatecas, Mexico. (*Opuntia*)
guillaumetii For Dr. Jean L. Guillaumet (*1934), French plant ecologist. (*Aloe*)
guillauminianus For Prof. André Guillaumin (1885–1974), French botanist and Director of the Paris Natural History Museum. (*Euphorbia, Mammillaria*)
guillemetii Probably a spelling error for Dr. Jean L. Guillaumet (*1934), French plant ecologist. (*Euphorbia beharensis* var.)

gumaroi For Señor Gumaro Manzo (fl. 2000), Mexican cactus enthusiast. (*Euphorbia*)
gummifer Lat. 'gumma', rubber, sticky material; and Lat. '-fer, -fera, -ferum', -carrying; (**1**) for the sap, which dries into a rubber-like substance. (*Adenia, Euphorbia, Mammillaria heyderi* ssp.) – (**2**) for the body cavities filled with gum-like mucilage. (*Uebelmannia*)
gummosus Lat., full of gum; for the stems. (*Stenocereus*)
Gunniopsis Gr. '-opsis', similar to; and for the genus *Gunnia* (now a synonym of *Gunniopsis*, Aizoaceae). (*Aizoaceae*)
gurgenidzeae For Mrs. M. Z. Gurgenidze (fl. 1965–1991), Georgian botanist in Tbilisi, and *Sempervivum* specialist. (*Sempervivum*)
guthriae For Miss Louise Guthrie (1879–1966), botanical assistant and artist at the Bolus Herbarium, Cape Town, RSA. (*Oscularia*)
guthriei For Prof. Frances Guthrie (1831–1899), British lawyer, mathematician and botanist, lived in RSA from 1861. (*Delosperma*)
gutierrezii For Antonio Gutierrez (fl. 1985), Brazilian cactus collector in Rio Grande do Sul. (*Parodia mueller-melchersii* ssp.)
guttatus Lat., sprinkled with dots or drops; (**1**) for the blotched leaves. (*Agave, Lenophyllum*) – (**2**) for the blotched corolla. (*Huernia*)
gydouwensis For the occurrence at Gydouw, Ceres Distr., Western Cape, RSA. (*Lampranthus*)
gymnocalycioides Gr. '-oides', resembling; and for the genus *Gymnocalycium* (Cactaceae). (*Euphorbia*)
Gymnocalycium Gr. 'gymnos', naked; and Gr. 'kalyx', calyx; for the perianth tube devoid of spines, bristles or hairs. (*Cactaceae*)
gymnocladus Gr. 'gymnos', naked; and Gr. 'klados', branch; for the leafless twigs. (*Euphorbia*)
gymnopodus Gr. 'gymnos', naked; and Gr. 'podos', foot; perhaps because the type was found with an exposed tuber. (*Brachystelma*)
Gynostemma Gr. 'gyne', woman, female or-

gans; and Gr. 'stemma', garland, crown; application obscure. (*Cucurbitaceae*)

Gynura Gr. 'gyne', woman, female organ, and Gr. 'oura', tail; the stigmas have a hairy tip. (*Asteraceae*)

gypsicola Lat. 'gypsum', gypsum; and Lat. '-cola', -dwelling. (*Sedum*)

gypsophilus Gr. 'gypsos', gypsum; and Gr. 'philos', friend; for the occurrence on gypsum soils. (*Agave*, *Dorstenia*, *Sedum*)

H

haageanus For Friedrich Adolph Haage (1796–1866), German cactus horticulturist in Erfurt, founder of the renowned Haage Nursery. (*Mammillaria*) – (**2**) For Walther Haage (1899–1992), German horticulturist, son of Ferdinand Haage and after his death (1930) owner of the renowned Haage cactus nursery in Erfurt. (*Cereus*)

haagei For Walther Haage (1899–1992), German horticulturist, son of Ferdinand Haage and after his death (1930) owner of the renowned Haage cactus nursery in Erfurt. (*Gibbaeum*)

Haageocereus For Walther Haage (1899–1992), German horticulturist, son of Ferdinand Haage and after his death (1930) owner of the renowned Haage cactus nursery in Erfurt; and *Cereus*, a genus of columnar cacti. (*Cactaceae*)

Haagespostoa Combination derived from the names of the parent genera. (*Cactaceae*)

haagnerae For Mrs. C. H. "Peggy" Haagner (fl. 1986), South African naturalist. (*Lavrania*)

habdomadis Contraction from Gr. 'hepta', seven, and Gr. 'hebdomades', week; for the occurrence at Sevenweekspoort, Western Cape, RSA. (*Haworthia mucronata* var.)

hadhramauticus For the occurrence in the Had[h]ramaut region in S Yemen. (*Huernia*)

hadiensis For the occurrence at Hadiyah, Yemen. (*Plectranthus*)

hadramauticus For the occurrence in the Had[h]ramaut region in S Yemen. (*Euphorbia*)

hadrostachyus Gr. 'hadros', stout; and Gr. 'stachys', spike; for the inflorescences. (*Peperomia*)

haeckelianus For Prof. Dr. Ernst H. P. A. Haeckel (1834–1919), German evolutionary biologist in Jena. (*Aptenia*)

haemanthifolius For the genus *Haemanthus* (*Amaryllidaceae*); and Lat. '-folius', -leaved. (*Aloe*)

haemanthoides Gr. '-oides', resembling; and for the genus *Haemanthus* (*Amaryllidaceae*). (*Boophane*)

Haemanthus Gr. 'haima, haimatos', blood; and Gr. 'anthos', flower; for the dark red flowers of some species. (*Amaryllidaceae*)

haematacanthus Gr. 'haima, haimatos', blood; and Gr. 'akanthos', spine, thorn; for the dark red spination. (*Ferocactus*)

haematanthus Gr. 'haima, haimatos', blood; and Gr. 'anthos', flower; for the dark red flowers. (*Echinopsis*)

haeneliae For Miss Christine Hänel (fl. 1996), who collected the type. (*Raphionacme*)

hahnianus For Adolf Hahn († 1954), German cactus collector and nurseryman in Berlin-Lichterfelde. (*Mammillaria*)

haitiensis For the occurrence in Haiti. (*Mammillaria prolifera* ssp.)

hakonensis For the occurrence in the Hakone Distr., Honshu, Japan. (*Sedum*)

halbingeri For Christian Halbinger (fl. 1931) of Mexico City. (*Echeveria*, *Mammillaria*)

halei For J. P. Hale (fl. 1889), landowner in Baja California who assisted T. S. Brandegee on his 1889 expedition. (*Mammillaria*)

halenbergensis For the occurrence at Halenberg [Haalenberg], Namibia. (*Conophytum*)

halimoides Gr. '-oides', resembling; and for the genus *Halimus* (a synonym of *Sesuvium*, *Aizoaceae*) (from Lat. 'halimos', salty); for the similar growth habit. (*Portulaca*, *Tetragonia*)

halipedicola Lat. 'halipedum', coastal plain; and Lat. '-cola', inhabiting; for the preferred habitat. (*Euphorbia*, *Orbea*, *Synadenium*)

Hallianthus Gr. 'anthos', flower; and for Harry Hall (1906–1986), English-born horticulturist, curator of the Darrah Collection Manchester before emigrating to RSA in 1947 and taking charge of the succulent plant collection at Kirstenbosch National Botanic Garden. (*Aizoaceae*)

hallianus For Harry Hall (1906–1986), English-born horticulturist, curator of the Darrah Collection Manchester before emigrating to RSA in 1947 and taking charge of the succulent plant collection at Kirstenbosch National Botanic Garden. (*Senecio*)

hallii As above. (*Adromischus marianiae* var., *Antimima*, *Argyroderma framesii* ssp., *Astridia*, *Bulbine*, *Cephalophyllum*, *Cylindrophyllum*, *Drosanthemum*, *Erepsia*, *Euphorbia*, *Hartmanthus*, *Huernia*, *Jacobsenia*, *Lampranthus*, *Lithops*, *Othonna*, *Sansevieria*, *Schlechteranthus*, *Trichodiadema*, *Tylecodon*)

Halosicyos Gr. 'hals, halos', salt; and Gr. 'sicyos', cucumber; for the preferred occurrence on saline soils. (*Cucurbitaceae*)

hamaderohensis For the occurrence at Hamaderoh, Socotra. (*Cissus*)

hamatacanthus Lat. 'hamatus', hooked; and Gr. 'akanthos', spine, thorn; for the hooked central spines. (*Ferocactus*)

hamatilis Lat., having hooks; for the hooked leaf tips. (*Antimima*)

hamatus Lat., hooked; (**1**) for the recurved stem tubercles. (*Euphorbia*, *Selenicereus*) – (**2**) for the recurved leaf tips. (*Ruschia*) – (**3**) for the hooked appendages of the corona segments. (*Schizoglossum*)

hametiorum Gen. Pl. of 'Hamet'; for the ancestors of the French physician and botanist Raymond Hamet. (*Kalanchoe*)

hammeri For Steven A. Hammer (*1951), US-American pianist, horticulturist and Mesemb specialist, and monographer of the genus *Conophytum*. (*Conophytum*)

Hammeria As above. (*Aizoaceae*)

hammerschmidii For Prof. Dr. Lorenzo Justinian Hammerschmid (1914–1970), German Franciscan Father and from 1947 onwards missionary in San Ignacio de Velasco, Bolivia. (*Echinopsis*)

hancockii For Allan Hancock (fl. 1949), US-American captain of the marine laboratory ship Velero IV. (*Echinocereus maritimus* ssp.)

handelii For Freiherr Heinrich von Handel-Mazzetti (1882–1940), Austrian botanist and explorer. (*Rhodiola*)

handiensis For the occurrence on the Jandia pensinsula, Fuerteventura, Canary Islands. (*Euphorbia*)

hangzhouensis For the occurrence in the Hangzhou Distr., Zhejiang, China. (*Sedum*)

hantamensis For the occurrence in the Hantam Mts., Northern Cape, RSA. (*Antimima*)

harardheranus For the occurrence near the village of Harardhere, Mudug Region, Somalia. (*Pseudolithos*)

harazianus For the occurrence on Mt. Haraz, Yemen. (*Delosperma*)

hardyi For David S. Hardy (1931–1998), horticulturist and former curator at the Botanical Garden Pretoria, RSA. (*Aloe*, *Cynanchum*, *Cyphostemma*, *Euphorbia platyclada* var., *Orbeanthus*, *Stapelianthus*)

harlanus For the occurrence near Harla, Harar Prov., Ethiopia. (*Aloe*)

harlowii For Capt. Charles H. Harlow (fl. 1909), US-American naval officer and commandant of the Guantánamo Naval Station. (*Melocactus*)

harmonianus For the occurrence near the village Harmonia, Rio Grande do Sul, Brazil. (*Frailea castanea* ssp.)

harmsii For Hermann A. T. Harms (1870–1942), German botanist. (*Echeveria*)

Harrisia For William Harris, collecting 1907–1911 on Jamaica, and Superintendent of the Public Gardens and Plantations of Jamaica. (*Cactaceae*)

harrisii As above. (*Agave*)

Hartmanthus For Dr. Heidrun E. K. Hartmann (*1942), German botanist at Hamburg University and *Aizoaceae* specialist; and Gr. 'anthos', flower. (*Aizoaceae*)

haselbergii For Dr. von Haselberg (fl. 1885), cactus hobbyist in Stralsund, Germany. (*Parodia*)

haseltonianus For Scott E. Haselton (1895–1991), US-American horticultural journalist, cactus enthusiast and founder of the journal of the Cactus and Succulent Society of America. (*Copiapoa*)

hassei For Herman E. Hasse (1836–1915), US-American physician and amateur botanist. (*Dudleya virens* ssp.)

hastatus Lat., hastate, lance-shaped; (**1**) for the leaf shape. (*Adenia*) – (**2**) for the robust central spine. (*Stenocactus*)

hastifer Lat. 'hasta', spear, lance; and Lat. '-fer, -fera, -ferum', -carrying; for the spination. (*Matucana aurantiaca* ssp., *Thelocactus*)

Hatiora Anagram of the invalid generic name

Hariota, for Thomas Hariot [Harriot] (1560–1621), English mathematician, astronomer and natural history traveller. (*Cactaceae*)

hatschbachii For Gert Hatschbach (fl. 1960, 1990), botanist and director of the Municipal Botanical Museum, Curitiba, Brazil. (*Apodanthera, Portulaca*)

haudeanus For Michael Haude (1940–1994), German horticultural engineer and head of a cactus nursery. (*Mammillaria saboae* ssp.)

hauniensis For Kobenhaven (Lat. Havnia [Haunia]), where the plant was found growing in the Botanical Garden. (*Agave*)

haussknechtii For Heinrich K. Haussknecht (1838–1903), German botanist in Weimar and explorer of the botany of the Orient. (*Rosularia*)

hausteinianus For Prof. Dr. Erik Haustein (*1910), German botanist at the University Erlangen, 1949–1968 editor of the German periodical "Kakteen und andere Sukkulenten". (*Parodia*)

havardianus For Valéry Havard (1846–1927), French-born US-American physician and botanist. (*Agave*)

havasupaiensis For the occurrence in the Havasupai Canyon, Arizona, USA. (*Sclerocactus parviflorus* ssp.)

havlasae For Elsie Havlasa (fl. 1921), wife of a friend of the Czech botanist Karel Domin. (*Pelargonium*)

Haworthia For Adrian H. Haworth (1768–1833), English zoologist and botanist and succulent plant specialist. (*Aloaceae*)

haworthii As above. (*Aeonium, Lampranthus, Ruschia, Senecio, Tetragonia*)

haworthioides Gr. '-oides', similar to; and for the genus *Haworthia* (*Aloaceae*). (*Aloe, Bulbine, Drimia*)

haygarthii For Walter J. Haygarth (1862–1950), Durban, RSA, collecting in RSA and contributing some drawings to Wood's "Natal Plants". (*Ceropegia*)

haynei For Friedrich G. Hayne (1763–1832), German botanist and pharmacist in Berlin. (*Matucana*)

hazelianus For Mrs. Hazel O. Munch (née Elske) (1912–2001), South African moving to Zimbabwe with her parents when a child, wife of Raymond C. Munch, farmer and succulent plant grower near Rusape, Zimbabwe. (*Aloe*)

heathii For Dr. F. H. Rodier Heath (fl. 1910–1937), English grower of succulent plants. (*Gibbaeum*)

hebdingii For René Hebding (1930–2002), French head gardener at 'Les Cèdres', the private botanical garden of the Marnier-Lapostolle family. (*Senecio*)

hecatandrus Gr. 'hecat-, hecto-', hundred-; and Gr. 'aner, andros', man, [botany] stamen; for the numerous stamens per flower. (*Trianthema*)

heckelii For Dr. Edouard M. Heckel (1843–1916), French landscape architect and creator of the Marseille Botanical Garden. (*Sedum*)

heckneri For J. H. Heckner (fl. 1930), Australia-born Government Surveyor in Oregon, USA (*Lewisia cotyledon* var., *Sedum laxum* ssp.)

hedbergii For K. Olov Hedberg (*1923), Swedish botanist in Uppsala, and specialist on East African vegetation. (*Crassula*)

hederifolius For the genus *Hedera* ("Ivy"; *Araliaceae*); and Lat. '-folius', -leaved. (*Othonna*)

hedigerianus For A. Hediger (fl. 1990), director of the primary school at Lubumbashi, Shaba Prov., Zaïre. (*Monadenium*)

hedyotoides Gr. '-oides', resembling; and for the genus *Hedyotis* (*Rubiaceae*); application obscure. (*Euphorbia*)

hedysaroides Gr. '-oides', resembling; and for the genus *Hedysarum* (*Fabaceae*). (*Oxalis*)

hegnaueri For Prof. Dr. Robert Hegnauer (*1919), Swiss-born phytochemist at Leiden University, Netherlands. (*Sedum luteolum* nssp.)

heidelbergensis For the occurrence at Heidelberg, Western Cape, RSA. (*Haworthia*)

heidiae For Heidi Krähenbühl, wife of the Swiss cactus hobbyist Felix Krähenbühl (1917–2001). (*Mammillaria*) – (**2**) For Heidi Neuhuber (*1965), wife of the Austrian cactus collector Gert Neuhuber. (*Gymnocalycium*)

heimenii For Gerhard Heimen (fl. 1999), German cactus hobbyist who first found the taxon in Brazil. (*Arrojadoa*)

heinrichianus For W. Heinrich (fl. 1942), German cactus hobbyist and around 1942 organizer of the illustrations archive of the German Cactus Society. (*Eriosyce*)

helenae For Mrs. Helen Decary (fl. 1929), wife of Raymond Decary, French financial administrator and botanist in Madagascar. (*Aloe*) – (**2**) For Mrs. Helena van Heerde (fl. 1937), wife of Pieter van Heerde, Springbok, Northern Cape, RSA. (*Conophytum tantillum* ssp.)

helianthoides Gr. '-oides', resembling; and for the genus *Helianthus* ("Sunflower"; *Asteraceae*). (*Apatesia*)

helideranus For the occurrence near Helidera, Bosaso Region, Somalia. (*Aloe*)

Heliophila Gr. 'helios', sun; and Gr. 'phile', friend. (*Brassicaceae*)

heliosus Gr. 'helios', sun; and Lat. '-osus', full of; for the numerous closely set areoles with stellate spination. (*Rebutia*)

helleri For Edmund Heller (1875–1939), US-American zoologist, traveller and collector. (*Opuntia*)

helmiae For Mrs. M. Helm (fl. 1932–1937), *Haworthia* enthusiast. (*Gibbaeum*)

helmsii Probably for Richard Helms (1842–1914) of the Department of Agriculture in Sydney, Australia. (*Crassula*)

helmutianus For Helmut Regnat (fl. 2002), German collector of succulents, esp. *Crassulaceae*. (*Echeveria*)

helmutii For Helmut E. Meyer (*1908), horticulturist at Stellenbosch University Botanic Garden, RSA, son of the succulent plant enthusiast Louis G. Meyer. (*Conophytum stephanii* ssp., *Lithops*)

hemicryptus Gr. 'hemi-', half; and Gr. 'kryptos', hidden; (**1**) for the half-underground caudex. (*Dioscorea*) – (**2**) for the half-buried rosettes. (*Haworthia variegata* var.)

hemisphaericus Lat., hemispherical; (**1**) for the shape of the plant bodies. (*Mammillaria heyderi* ssp.) – (**2**) for the rosettes, which are semiglobose in section. (*Crassula*) – (**3**) probably for the leaf form. (*Adromischus, Galenia*)

hemmingii For C. F. Hemming (fl. 1964), of the Desert Locust Survey. (*Aloe*)

hempelianus For Georg Hempel (1847–1904), German cactus hobbyist, businessmen and estate owner in Ohorn, Sachsen, Germany. (*Oreocereus*)

hempelii As above. (*Echinocereus fendleri* ssp.)

hemsleyanus For William B. Hemsley (1843–1924), British botanist at Kew. (*Matelea, Sedum*)

hendrickxii For Fred L. Hendrickx (fl. 1955), Belgian agronomist in Central Africa. (*Aloe*)

hengduanensis For the occurrence in the Hengduan Mts., Sichuan, China. (*Sedum*)

henrici-robertii For Monsieur Henri-Robert (fl. 1913), "Bâtonnier de l'Ordre des Avocas", a friend of the French botanist and physician Raymond Hamet. (*Sedum*)

henricii For Dr. M. Henrici (1892–1971), South African botanist. (*Lampranthus*)

henricksonii For Dr. James S. Henrickson (*1940), US-American botanist. (*Escobaria chihuahuensis* ssp.)

heptacanthus Gr. 'hepta', seven; and Gr. 'akantha', spine, thorn; for the spine number. (*Discocactus*)

heptagonus Gr. 'hepta', seven; and Gr. 'gonia', corner, angle; for the number of branch ribs. (*Euphorbia*)

herbaceus Lat., herbaceous, green and slightly fleshy; (**1**) for the growth habit. (*Monadenium, Tetragonia*) – (**2**) for the leaves. (*Haworthia*)

herberti For Dr. Herbert Maughan Brown (fl. late 1920s, 1930s), physician and plant collector in RSA. (*Phyllobolus*)

herbeus Lat., herbaceous, herb-like; for the annual herbaceous shoots. (*Delosperma*)

herbsthoferianus For Ing. Gunther Herbsthofer (*1958), Austrian businessman in Linz and supporter of the *Gymnocalycium* specialist G. Neuhuber. (*Gymnocalycium ochoterenae* ssp.)

Hereroa For the occurrence in the region inhabited by the Herero tribe, Namibia. (*Aizoaceae*)

hereroensis As above. (*Aloe, Cyphostemma, Portulaca, Trianthema*)

herman-schwartzii For Dr. Herman Schwartz (fl. 2002), US-American physician, succulent plant enthusiast and owner of Strawberry Press. (*Euphorbia*)

hermannii For Hermann Petignat († 2000), Swiss-born hotel-owner, artist and horticulturist in Madagascar. (*Ceropegia*) – (**2**) For August Hermann (senior) (fl. 1953), US-American cactus collector in Phoenix, Arizona. (*Pediocactus*)

hermansdorpensis For the occurrence in the Hermansdorp District, Western Cape, RSA. (*Pelargonium*)

hermarius For Hermias (Mias) C. Kennedy (fl. 1966–1968), succulent plant collector in Bellville, RSA; and from the original provisional name *C. rimarium*, from Lat. 'rimarium', of the clefts, for the preferred habitat. (*Conophytum smorenskaduense* ssp.)

hermeticus MLat. 'hermeticus', a term used in alchemy for confidential recipes; for the occurrence in the "hermetically closed" Protected Diamond Area ("Sperrgebiet") in Namibia. (*Lithops*)

herminiae For Prof. Herminia Castellanos (fl. 1941), wife of the Argentinian botanist Alberto Castellanos. (*Hatiora*)

hernandezii For Jorge Hernandez Camacho (fl. 1960), Colombian botanist and plant collector. (*Melocactus andinus* ssp.) – (**2**) For Eulalio Hernandez (fl. 1983), nephew and field companion of the Mexican cactus collector Felipe Otero. (*Mammillaria*)

herniariifolius Lat. '-folius', -leaved; for the similarity of the leaves to those of *Herniaria* ("Herniary", "Rupturewort"; *Caryophyllaceae*). (*Galenia*)

herreanthus For Adolar G. J. "Hans" Herre (1895–1979), German horticulturist and collector in RSA, 1925–1960 curator of Stellenbosch University botanical garden; and Gr. 'anthos', flower; for the previous classification of the taxon in the genus *Herreanthus*. (*Conophytum*)

herreanus For Adolar G. J. "Hans" Herre (1895–1979), German horticulturist and collector in RSA, 1925–1960 curator of Stellenbosch University botanical garden. (*Avonia*, *Senecio*)

herrei As above. (*Antimima*, *Astridia*, *Astroloba*, *Brunsvigia*, *Cephalophyllum*, *Cheiridopsis*, *Cleretum*, *Crassula nudicaulis* var., *Cyrtanthus*, *Euphorbia*, *Haworthia glauca* var., *Hereroa*, *Lithops*, *Malephora*, *Monsonia*, *Othonna*, *Schwantesia*, *Tromotriche*)

herrerae For Alfonso L. Herrera (1870–1942), Mexican biologist and naturalist, founder of the zoological garden at Chapultepec in 1923. (*Ferocactus*, *Mammillaria*) – (**2**) For Dr. Fortunato L. Herrera y Garmendia (1873–1945), Peruvian botanist and professor at the University of Cuzco. (*Oxalis*)

herteri For Dr. Wilhelm (Guillermo) Herter (1884–1958), Swiss botanist and physician, 1907 emigrating to Uruguay and later director of the Montevideo Botanical Garden, later returning to Germany (Hamburg). (*Parodia*)

hertlingianus For Mr. Hertling, German merchant and trader of drugs for Bayer in Peru, accompanied C. Backeberg on his 1931-trip in Peru. (*Browningia*)

hertrichianus For William Hertrich (1878–1966), curator of the Huntington Botanical Gardens, California, USA. (*Echinopsis*, *Mammillaria*)

herzogianus For Prof. Dr. Theodor C. J. Herzog (1880–1961), German bryologist and botanist in München, travelling 1907-1912 in Bolivia. (*Echinopsis tarijensis* ssp., *Matucana haynei* ssp., *Neoraimondia*)

Hesperaloe Gr. 'hespera', evening; for the occurrence in North America (i.e. in the West, where the sun disappears in the evening); and for the superficial similarity to *Aloe* (*Aloaceae*). (*Agavaceae*)

hesperanthus Gr. 'hespera', evening; and Gr. 'anthos', flower; for the opening time of the flowers. (*Hereroa*)

Hesperoyucca Gr. 'hespera', evening; for the occurrence in W North America (i.e. in the West, where the sun disappears in the evening); and for the similarity to *Yucca* (*Agavaceae*). (*Agavaceae*)

hesseae For the (botanical?) collector Mrs. Hesse (fl. 1954), without further data. (*Cistanthe parryi* var.)

hesteri For J. Pinckney Hester (fl. 1945), US-American cactus enthusiast in Fredonia, Arizona, and one of the leading explorers of the Big Bend area of Texas, USA. (*Escobaria*)

hestermalensis For the occurrence in the Hester Malan Nature Reserve, Northern Cape, RSA. (*Dorotheanthus bellidiformis* ssp.)

heterocarpus Gr. 'heteros', different; and Gr. 'karpos', fruit; for the variable fruits. (*Pteronia*)

heterochromus Gr. 'heteros', different; and Gr. 'chromos', colour; (**1**) for the branches variegated with green and yellow-green. (*Euphorbia*) – (**2**) for the spine coloration. (*Thelocactus*)

heterodontus Gr. 'heteros', different; and Gr. 'odous, odontos', tooth; for the variously toothed leaf margins. (*Rhodiola*)

heterodoxus Gr. 'heteros', different; and Gr. 'doxa', splendour, glory; perhaps for the noticeably different inflorescences. (*Euphorbia*)

heteropetalus Gr. 'heteros', different; and Gr. 'petalon', petal; for the unequal size of the petals. (*Erepsia, Ruschia*)

heterophyllus Gr. 'heteros', different; and Gr. 'phyllos', leaf. (*Aspidoglossum, Pelargonium*)

heteropodus Gr. 'heteros', different; and Gr. 'podion', foot; for the differently shaped tubercles at base and tip of the stems, which were thought to be diagnostic. (*Monadenium*)

heterosepalus Gr. 'heteros', different; and Gr. 'sepalon', sepal; for the unequal size of the sepals. (*Echeveria*)

heterospinus Gr. 'heteros', different; and Lat. '-spinus', -spined; for the differing spine size at the base and tip of mature plants. (*Euphorbia*)

heterotrichus Gr. 'heteros', different; and Gr. 'thrix, trichos', hair; application obscure. (*Crassula perfoliata* var.)

heuffelii For Johann A. Heuffel (1800–1857), Hungarian physician and botanist. (*Sempervivum*)

hexaedrophorus Gr. 'hexa-', six; Gr. 'hedra', plane, seat; and Gr. '-phoros', carrying; for the 6-angled tubercles of the plant body. (*Thelocactus*)

hexagonus Gr., with six angles; (**1**) for the six-ribbed stems. (*Cereus*) – (**2**) for the usually six-ribbed stems. (*Caralluma*)

hexamerus Gr. 'hexa-', six; and Gr. 'meros', part; for the 6 sepals and 6 fruit locules. (*Ruschia*)

hexapetalus Gr. 'hexa-', six; and Gr. 'petalon', petal. (*Furcraea*)

heybensis For the occurrence on Buur Heybe, an inselberg in S Somalia. (*Aloe*)

heyderi For Privy Councellor Eduard Heyder (1808–1884), German cactus hobbyist in Berlin. (*Mammillaria*)

heynei For Benjamin Heyne (1770–1819), East India Company botanist stationed in Madras, India. (*Jatropha*)

hians Lat., gaping, widely opening; for the gaping fissure between the lobes of the fused leaf pair. (*Conophytum*)

hickenii For Dr. Cristóbal M. Hicken (1875–1933), Argentinian botanist and founder of the Instituto Botánico Darwinion, San Isidro, Argentina. (*Pterocactus*)

hiemalis Lat., pertaining to winter; for the flowering time. (*Lampranthus*)

hiemiflorus Lat. 'hiems', winter; and Lat. '-florus', -flowering. (*Agave*)

hiernii For William P. Hiern (1839–1925), British botanist who catalogued Welwitsch's plant collections from Angola. (*Elaeophorbia*)

hierrensis For the occurrence on Hierro, Canary Islands. (*Aeonium*)

hijazensis For the occurrence in Hijaz Prov., Saudi Arabia. (*Aloe*)

hildebrandtii For Dr. Johann M. Hildebrandt (1847–1881), German naturalist and widely-travelled collector in Africa and Madagascar. (*Aloe, Dorstenia, Kalanchoe, Moringa, Senecio, Senecio nyikensis* var.)

hildegardiae For Hildegard Winter (1893–1975), sister of the German cactus specialist F. Ritter, who sold the seeds collected by him. (*Cleistocactus*)

hildmannianus For Heinrich Hildmann (fl. 1870–1895), German cactus horticulturist and owner of a cactus nursery in Birkenwerder near Berlin. (*Cereus*)

hileiabaianus For the occurrence in the perhumid Atlantic forest ('Hileia Baiana') in Bahia, Brazil. (*Rhipsalis baccifera* ssp.)

hilliardiae For Dr. Olive Mary Hilliard (née Hillary) (*1926); South African botanist and *Streptocarpus* specialist. (*Schizoglossum*)

hillii For Leslie J. Hill (1908–2003), South African chartered accountant, businessman and philanthropist supporting education and conservation in the Western Cape, and ethusiastic field collector and grower of plants, esp. succulents. (*Astridia*)

hilmarii For Hilmar Lückhoff († 1994), without further data. (*Deilanthe, Tanquana*)

himalensis For the occurrence in the Himalayas. (*Rhodiola*)

himanthocladus Gr. 'himas, himantos', strap; and Gr. 'klados', branch; for the shape of the stem segments. (*Pseudorhipsalis*)

hindii For Dr. David J. N. Hind (*1957), English botanist at the Kew herbarium. (*Apodanthera*)

hindsianus For Richard B. Hinds (1812–1847), British naval surgeon, attached as surgeon naturalist to HMS Sulphur. (*Bursera, Euphorbia californica* var.)

hintermannii For Dr. Hintermann (fl. 1940), veterinary surgeon in Casablanca and president of Le Club des Plantes Grasses de Casablanca, who raised this hybrid. (*Senecio*)

hintonii For George Boole Hinton (senior) (1882–1942), English-born metallurgist, farmer and plant collector in Mexico, grandfather of George S. Hinton. (*Euphorbia, Sedum*) – (**2**) For George Sebastián Hinton (*1949), Mexican farmer and botanical collector in Nuevo León, grandson of George B. Hinton (senior). (*Ariocarpus bravoanus* ssp., *Aztekium, Thelocactus rinconensis* ssp.)

hintoniorum For George Sebastián Hinton jr. (*1949) and his family, Mexican farmer and botanical collector in Nuevo León, Mexico. (*Coryphantha, Sedum, Turbinicarpus beguinii* ssp.)

hiriartiae For Patricia Hiriart Valencia (fl. 1992), Mexican botanist and professor. (*Beaucarnea*)

hirschii For Dr. G. Hirsch (fl. 1954), German botanist in Heidelberg who accompanied W. Rauh 1954 on his expedition to Peru. (*Austrocylindropuntia*)

hirschtianus For Karl Hirscht († 1925), German cactus hobbyist in Berlin. (*Peniocereus*)

hirsutissimus Lat., hairiest (Superl. of Lat. 'hirsutus', hairy); for the very hairy leaves. (*Operculicarya, Portulaca*)

hirsutus Lat., roughly hairy; (**1**) because the whole plant is distinctly hairy. (*Ceropegia, Monadenium, Sedum*) – (**2**) for the stems. (*Stapelia*) – (**3**) for the hairy stems and leaves. (*Tetragonia, Trichodiadema*) – (**4**) for the leaves. (*Crassula, Cucumis, Dischidia, Raphionacme*)

hirtellus Lat., shortly bristly (Dim. of Lat. 'hirtus', roughly hairy); (**1**) for the stems and leaves. (*Brachystelma, Kedrostis*) – (**2**) for the stems and pedicels. (*Drosanthemum*)

hirticrassus Lat. 'hirtus', hairy; and Lat. 'crassus', thick; for the overall pubescence. (*Senecio*)

hirtifolius Lat. 'hirtus', hairy; and Lat. '-folius', -leaved. (*Tylecodon*)

hirtipes Lat. 'hirtus', hairy; and Lat. 'pes', foot; for the hairy leaves. (*Crassula*)

hirtipetalus Lat. 'hirtus', hairy; and Lat. 'petalum', petal. (*Pelargonium*)

hirtulus Lat., shortly hairy; for the shorter hairs in comparison with the typical variety. (*Jatropha erythropoda* var.)

hirtus Lat., (roughly) hairy; (**1**) for the hairy young shoots. (*Delosperma*) – (**2**) for the hairiness of the plant. (*Tradescantia*) – (**3**) for the hairs present on most plant parts. (*Sempervivum globiferum* ssp.) – (**4**) for the hairy leaves. (*Pelargonium*) – (**5**) for the hairy corolla lobes. (*Schizoglossum bidens* ssp.)

hirundo Lat., swallow; for the shape of the corpuscle of the pollinaria. (*Aspidoglossum*)

hislopii For Mr. A. Hislop (fl. 1920–1922), botanical collector in Southern Rhodesia (now Zimbabwe). (*Euphorbia milii* var., *Huernia*)

hispanicus Lat. 'Hispania', Spain; for the occurrence there. (*Aizoanthemum*, *Pistorinia*, *Sedum*)

hispidissimus Superl. of Lat. 'hispidus', hispid, covered with coarse erect hairs like an unshaved beard five days old; for the stem surface. (*Galenia*)

hispidus Lat., hispid, covered with coarse erect hairs like an unshaved beard five days old; for the hairy leaves. (*Crassula mesembryanthoides* ssp., *Crassula subulata* var.)

hispifolius Lat., hispid, covered with coarse erect hairs like an unshaved beard five days old; and Lat. '-folius', -leaved. (*Drosanthemum*)

histrix Lat./Gr., porcupine, hedgehog; for the spiny plant bodies. (*Ferocactus*)

hitchcockii For Albert S. Hitchcock (1865–1935), US-American botanist (*Opuntia*)

hlangapies For the occurrence at Hlangapies (Zulu place name, "Langgewacht") in Mpumalanga, RSA. (*Aloe*)

hobsonii For H. E. Hobson (fl. 1913), without further data. (*Rhodiola*)

hoehnei For Frederico C. Hoehne (1882–1959), Brazilian botanist in São Paulo. (*Portulaca*)

hoelleri For Werner Höller (fl. 1995), head gardener at the University Botanical Garden Bonn, Germany. (*Rhipsalis*)

hoerleinianus For a Mr. Hoerlein [Hörlein ?] (fl. 1923), without further data. (*Lampranthus*)

hoferi For Anton ("Toni") Hofer (fl. 2002), Swiss cactus collector. (*Turbinicarpus*)

hoffmannii For Ralph Hoffmann (fl. 1995–2002), Swiss horticulturist and succulent plant enthusiast near Zürich. (*Aloe*)

hofstaetteri For Siegfried Hofstätter (fl. 1992), German plant importer. (*Ceropegia*, *Euphorbia*)

hohenauensis For the occurrence at Hohenau, Paraguay. (*Rhipsalis floccosa* ssp.)

hojnyi For L. Hojný (fl. 1998), Czech succulent plant collector. (*Pachypodium*)

holei For R. S. Hole (fl. 1913), then Director at the Forest Research Institute, Dehra Dun, India. (*Sedum*)

holensis For the occurrence near the Hol River, Vanrhynsdorp Distr., Western Cape, RSA. (*Lampranthus*, *Ruschia*)

hollandii For Frederick Huntly Holland (1873–1955), South African businessmen and naturalist. (*Delosperma*, *Lampranthus*)

hollianus Unknown. (*Pachycereus*)

holmesiae For Mrs. Susan Carter Holmes (*1933), English botanist at RBG Kew, and specialist on *Euphorbia* and *Aloe* in Tropical Africa. (*Euphorbia*)

holochlorinus Gr. 'holo', entire; and Lat. 'chlorinus', yellow-green; for the entirely green branches. (*Euphorbia*)

holochrysus Gr. 'holo', entire; and Gr. 'chrysos', golden yellow; for the flower colour. (*Aeonium arboreum* var.)

holopetalus Gr. 'holo', entire; and Gr. 'petalon', petal; application obscure. (*Sedum*)

hommelsii For Mr. Cees H. Hommels (fl. 1987), Dutch botanist and cytologist. (*Sedum lorenzoi* nssp.)

hondala From the local vernacular name of the plants in Sri Lanka. (*Adenia*)

hondurensis For the occurrence in Honduras. (*Disocactus nelsonii* var., *Opuntia*, *Selenicereus*)

Hoodia For a Mr. Hood (fl. 1830), British surgeon in London, and succulent plant grower. (*Asclepiadaceae*)

hookeri For Sir William J. Hooker (1785–1865), British botanist, first director of the Royal Botanic Gardens Kew 1841–1865. (*Agave*, *Calibanus*, *Ceropegia*, *Epiphyllum*) – (**2**) For Sir Joseph D. Hooker (1817–1911), British botanist and explorer, pioneer plant geographer, director of the Royal Botanic Gardens Kew 1865–1885, son of William J. Hooker. (*Lithops*, *Pachyphytum*, *Rhodiola*)

hopetownensis For the occurrence near Hopetown, Northern Cape, RSA. (*Euphorbia*)

horakii For Bohuslav Horák (1877–1942), Czech botanist. (*Sedum grisebachii* var.)

horichii For Clarence K. Horich (1930–1994), German horticulturist, resident in Costa Rica from 1957, and well-known plant collector. (*Pseudorhipsalis*)

horizontalis Lat., horizontal; for the horizontally spreading flowers. (*Umbilicus*)

horizonthalonius Lat. 'horizontalis', horizontal; and Gr. 'halonion', small place (for the areoles); for the horizontally oriented areoles. (*Echinocactus*)

horombensis For the occurrence in the Horombe Mts., Madagascar. (*Euphorbia, Pachypodium*)

horridispinus Lat. 'horridus', bristly, prickly, rough; and Lat. '-spinus', spined. (*Gymnocalycium monvillei* ssp.)

horridus Lat., bristly, prickly, rough; (**1**) for the rigid spines. (*Myrmecodia*) – (**2**) for the numerous strong spines. (*Acanthocereus, Euphorbia, Melocactus, Parodia microsperma* ssp., *Pereskia, Rhipsalis baccifera* ssp., *Uebelmannia pectinifera* ssp.) – (**3**) for the strong leaf marginal teeth. (*Agave*)

horripilus Lat. 'horrere', be seized with horror, project; and Lat. 'pilus', hair; for the spination. (*Turbinicarpus*)

horrispinus Lat. 'horrere', be seized with horror, project; and Lat. '-spinus', spined. (*Cereus*)

horstii For Leopoldo Horst (1918–1987), German-born Brazilian cactus collector and exporter in Rio Grande do Sul. (*Cleistocactus baumannii* ssp., *Discocactus, Frailea gracillima* ssp., *Gymnocalycium, Parodia, Parodia ottonis* ssp.)

hortenseae For Hortense Muir (fl. 1927), daughter of the Scottish physician and naturalist Dr. John Muir. (*Muiria*)

horwoodii For Francis ("Frank") K. Horwood (1924–1987), English horticulturist and succulent plant propagator, later emigrating to California. (*Euphorbia, Pseudolithos*)

hossei For Prof. Dr. Carl C. Hosseus (1878–1950), German botanist, collected in Thailand, settling in Argentina 1912 and 1916–1946 professor of botany in Córdoba. (*Gymnocalycium*)

hottentotorum For the Hottentots, a Khoi tribe living in the area where the taxon is native. (*Quaqua*)

hottentottus As above. (*Crassula sericea* var., *Euphorbia*)

houlletianus For Mr. Houllet, "sous-chef" of the glasshouse section of the "Muséum Impérial d'Histoire Naturelle" in Paris (fl. 1857), travelled 1838 with Guillaumin in Brazil. (*Lepismium*)

howardii For Dr. Thaddeus M. Howard (*1929), US-American botanist. (*Agave*)

howei For D. F. Howe (fl. 1974), the collector of the type material. (*Echinocereus engelmannii* var.)

howellii For Thomas J. Howell (1842–1912), US-American botanist, pioneer of the Oregon and NW American flora. (*Lewisia cotyledon* var.) – (**2**) For John T. Howell (*1903), US-American botanist at the California Academy of Sciences and specialist for the Galápagos flora. (*Portulaca*)

howeyi For Mr. Howey (fl. 1925), without further data. (*Opuntia*)

howmanii For H. Roger G. Howman (*1909), Zimbabwean Native Commissioner in several areas of the former Southern Rhodesia. (*Aloe*)

Hoya For Thomas Hoy (†1821), gardener at Syon House, England. (*Asclepiadaceae*)

huachucensis For the occurrence in the Huachuca Mountains, Arizona, USA. (*Agave parryi* var.)

huagalensis For the occurrence at Hacienda Huagal, Dept. Cajamarca, Peru. (*Matucana*)

huajuapensis For the occurrence near Huajuapan de León, Oaxaca, Mexico. (*Opuntia*)

huanucoensis For the occurrence near Huánuco, Dept. Huánuco, Peru. (*Espostoa*)

huascensis For the occurrence near Huasco, Chile. (*Eriosyce crispa* var.)

huascha From the local vernacular name of the plants in Argentina with the meaning of "orphan". (*Echinopsis*)

huasiensis For the occurrence at Inca Huasi, Dept. Chuquisaca, Bolivia. (*Rebutia*)

huastecensis For the occurrence in the Cañón de Huasteca, Nuevo León, Mexico. (*Echinocereus viereckii* ssp.)

hubertii For Prof. Dr. Hubert Winkler (1875–1941), German botanist widely travelling in Asia and Africa. (*Euphorbia*)

Huernia For Justus Heurnius [van Heurne] (1587–1652), Dutch physician and missionary and first European to collect plants at the Cape of Good Hope, RSA, retaining the original typographical error in the protologue. (*Asclepiadaceae*)

huernioides Gr. '-oides', resembling; and for the genus *Huernia* (*Asclepiadaceae*). (*Orbea*)

Huerniopsis For the genus *Huernia* (*Asclepiadaceae*); and Gr. '-opsis', like. (*Asclepiadaceae*)

hugo-schlechteri For Hugo Schlechter (fl. 1926), German lithographer, father of the botanist Rudolf Schlechter. (*Titanopsis*)

huillensis For the occurrence at Huil[l]a, Angola. (*Adenia*, *Orbea*)

huilunchu From the local vernacular name of the plants in Prov. Ayopaya, Dept. La Paz, Bolivia. (*Cereus*)

huincoensis For the occurrence near Huinco in the Santa Eulalia valley, Dept. Lima, Peru. (*Corryocactus*)

huitcholensis For the occurrence in the Sierra de los Huitcholes, Jalisco, Mexico. (*Echinocereus polyacanthus* ssp.)

huitzilopochtli For Huitzilopochtli, deity of sun and war in the Aztec religion. (*Mammillaria*)

hultenii For Eric O. G. Hultén (1894–1981), Swedish botanist and explorer. (*Sedum*)

humbert-capuronii For Prof. Henri [Jean-Henri] Humbert (1887–1967), and René P. R. Capuron (1921–1971), French botanists in Madagascar. (*Cynanchum*)

humbertianus For Prof. Henri [Jean-Henri] Humbert (1887–1967), French botanist in Madagascar, and specialist of the Madagascan flora. (*Cissus*)

humbertii As above. (*Alluaudia*, *Aloe*, *Ceropegia*, *Crassula*, *Peperomia*, *Seyrigia*)

humblotianus For Léon Humblot (fl. 1848), French plant collector. (*Dorstenia cuspidata* var.)

humboldtianus For Friedrich H. A. von Humboldt (1769–1859), German scientist who travelled widely in S America and Mexico with Aimé Bonpland. (*Jatropha*)

humboldtii As above. (*Mammillaria*)

humifusus Lat., spread out on the ground and mat-forming. (*Acrosanthes*, *Cypselea*, *Nolana*, *Opuntia*, *Sedum*, *Tylosema*)

humilis Lat., low, modest, low-growing. (*Adromischus*, *Aloe*, *Armatocereus*, *Cistanthe*, *Copiapoa*, *Cyphostemma*, *Dicrocaulon*, *Echeveria*, *Frithia*, *Huernia*, *Kalanchoe*, *Rhodiola*, *Talinum*)

hunua For the occurrence near Hunua, a locality near Auckland, New Zealand. (*Crassula*)

huotii For Mr. Huot (fl. 1853), cactus horticulturist working in the Monville collection. (*Echinopsis*)

hurlingii For Mr. J. Hurling (fl. 1928–1933), dairy farmer and nurseryman (together with Neil) in Bonnievale, Western Cape, RSA. (*Haworthia reticulata* var., *Lampranthus*)

hurstii For Capt. H. E. Hurst (fl. 1941) of Puerto Plata, Dominican Republic, who first collected this taxon. (*Harrisia*)

hurteri For Mr. Don G. Hurter (fl. 1915), Quezaltenango, Guatemala, who photographed Guatemalan Agaves for W. Trelease. (*Agave*)

hutchisonianus For Ted Hutchison (fl. 1934), US-American cactus collector in Azusa, California. (*Mammillaria*)

hutchisonii For Paul C. Hutchison (1924–1997), US-American botanist at Berkeley, California, and later owner of a nursery. (*Lewisia kelloggii* ssp., *Peperomia*)

huttoniae For Mrs. Caroline Hutton (née Atherstone) (1826–1908), South African wife of the English civil servant Henry Hutton, who emigrated 1844 to RSA. (*Euphorbia inermis* var.)

huttonii For Henry Hutton (1825–1896), English-born civil servant, emigrated 1844 to RSA. (*Brachystelma*)

hyacinthoides Gr. '-oides', resembling; and for the genus *Hyacinthus* ("Hyacinth"; *Hyacinthaceae*). (*Dipcadi*, *Sansevieria*)

hyalacanthus Gr. 'hyalos', glass, crystal; and Gr. 'akantha', thorn, spine; for the spination. (*Cleistocactus*)

hyalinus Lat., glassy, crystal-like; for the spination. (*Mammillaria lasiacantha* ssp.)

hybopleurus Gr. 'hybos', tubercle, hump; and Gr. 'pleuron', rib; for the tuberculate ribs. (*Gymnocalycium*)

hybridus Lat., hybrid; (**1**) because the taxon is of hybrid origin. (*Disocactus*, *Mesembryan-*

themum) – (**2**) because the taxon was (erroneously) thought to be of hybrid origin. (*Phedimus*)

Hydnophytum Gr. 'hydnon', truffle, edible mushroom; and Gr. 'phyton', plant; for the truffle-like tubers. (*Rubiaceae*)

hydrocotylifolius Lat. '-folius', -leaved; for the similarity of the leaves to those of *Hydrocotyle* ("Marsh Pennywort"; *Apiaceae*). (*Tetilla*)

Hydrophylax Gr. 'hydro', water; and Gr. 'phylax', guardian, custodian; perhaps for the occurrence near the coast. (*Rubiaceae*)

hylaeus Gr. 'hylaios', pertaining to the forest; for the occurrence in forested areas. (*Espostoa*)

Hylocereus Gr. 'hyle', forest; and for *Cereus*, a genus of columnar cacti; for the habitat in forests. (*Cactaceae*)

Hylotelephium Gr. 'hyle', forest; and Gr. 'telephion', antique Gr. name for *Hylotelephium telephium* and other taxa; for the frequent occurrence in open forests. (*Crassulaceae*)

Hymenogyne Gr. 'hymen', membrane; and Gr. 'gyne', woman, [Bot.] female organs; for the seemingly winged seeds. (*Aizoaceae*)

Hypagophytum Gr. 'hypagein', to deceive, to mislead; and Gr. 'phyton', plant; for the former erroneous classification as *Sempervivum*. (*Crassulaceae*)

hypertrophicus Gr. 'hyper', beyond, over, above, very; and Gr. 'trophe', food, nurishment; because the plants are extraordinarily fat-looking and succulent, i.e. 'over-fed'. (*Mesembryanthemum*)

hyphaenoides Gr. '-oides', resembling, and for the palm genus *Hyphaene*; application obscure. (*Operculicarya*)

hypogaeus Gr., underground; (**1**) for the semi-underground plant bodies. (*Copiapoa*) – (**2**) for the underground stem and main branches. (*Euphorbia*)

hypolasia Gr. 'hypo', beneath; and Lat. 'Asia', Asia; for the occurrence in Papua New Guinea 'beneath' (south of) Asia. (*Hoya*)

hypoleucus Gr. 'hypo', beneath; and Gr. 'leukos', white; for the grey-tomentose lower face of the leaves. (*Cyphostemma*)

hyptiacanthus Gr. 'hyptios', recurved; and Gr. 'akantha', spine, thorn. (*Gymnocalycium, Opuntia*)

hystrichoides Gr. 'hystrix', porcupine, hedgehog; and Gr. '-oides', resembling; for the spiny plant bodies. (*Echinopsis*)

hystricinus Lat., like a hedgehog; for the spination. (*Opuntia polyacantha* var.)

hystrix Lat./Gr., porcupine, hedgehog; (**1**) for the dense spination. (*Cleistocactus, Cumulopuntia, Matucana haynei* ssp.) – (**2**) for the spiny persistent stipules. (*Pelargonium*) – (**3**) for the bristly-papillate corolla. (*Huernia*)

I

ianthinanthus Gr. 'ianthinos', violet; and Gr. 'anthos', flower. (*Tunilla*)

ianthothele Gr. 'ianthos', violet; and Gr. 'thele', tubercle; for the colour of the tuberculate ribs of the stems. (*Lepismium*)

ibicuatensis For the occurrence near Ibicuati, Prov. Cordillera, Dept. Santa Cruz, Bolivia. (*Echinopsis*)

ibityensis For the occurrence on Mt. Ibity, C Madagascar. (*Aloe*)

icensis For the occurrence near Ica, Dept. Ica, Peru. (*Haageocereus*)

icosagonoides Gr. '-oides', resembling; for the similarity to *Cleistocactus icosagonus*. (*Haageocereus*)

icosagonus Gr. 'eikosi', twenty; and Gr. 'gonia', corner, angle; for the number of ribs originally observed. (*Cleistocactus*)

ictericus Lat., yellowish as suffering from jaundice; for the pale yellowish-green colour of the plant. (*Monanthes*)

ignavus Lat., lazy, idle, sluggish; for the slowness with which the old sheathing leaves are broken up. (*Conophytum ectypum* ssp.)

ignescens From Lat. 'ignescere', to catch fire, set fire; for the flower colour. (*Cumulopuntia*)

ignoratus Lat., ignored, unnoticed; because the taxon went unnoticed for a long time. (*Monsonia*)

iharanae For the occurrence near Iharana (formerly Vohémar), N Madagascar. (*Euphorbia*)

Ihlenfeldtia For Prof. Dr. Hans-Dieter Ihlenfeldt (*1932), German botanist at Hamburg University, specializing in *Aizoaceae* and *Pedaliaceae*. (*Aizoaceae*)

illegitimus Lat., illegitimate, not right; for the unusual flower colour. (*Ceropegia*)

illepidus Lat., impolite, rude, disagreeable; application obscure, perhaps for the general appearance of the plants. (*Mestoklema*)

illichianus For a Mr. Illich (fl. 1907), without further data. (*Crassula globularioides* ssp.)

imalotensis For the occurrence in the Imaloto valley, Madagascar. (*Aloe*)

imbricans Lat., becoming imbricate, like tiles on a roof; for the leaf arrangement. (*Lampranthus*)

imbricatus Lat., imbricate, overlapping like tiles on a roof; (**1**) for the leaf arrangement. (*Ceropegia, Delosperma, Dischidia, Hoya, Rhodiola, Villadia*) – (**2**) for the tuberculate stem segments. (*Cylindropuntia*) – (**3**) for the arrangement of the bracteoles on the persistent peduncles and pedicels. (*Ruschia*)

imerina For the occurrence in the region of the Imerina tribe, C Madagascar. (*Euphorbia*)

imerinensis As above. (*Aloe, Cynanchum compactum* var.)

imitans Lat., imitating; (**1**) for the similarity of the stems to those of *Epiphyllum anguliger*. (*Weberocereus*) – (**2**) for the similarity to species of *Cephalophyllum* (*Aizoaceae*). (*Cheiridopsis*) – (**3**) for the similarity of the leaves and fruit capsules to those of some species of the genus *Delosperma* (*Aizoaceae*). (*Trichodiadema*)

imitatus Lat., imitated; for the similarity to *Euphorbia brevis*. (*Euphorbia*)

immaculatus Lat., unspotted; (**1**) for the (usually) unspotted leaves. (*Adromischus marianiae* var., *Aloe*) – (**2**) for the uniformly coloured corolla. (*Duvalia*)

immelmaniae For Mrs. Immelman of Piquetberg, RSA, who collected the plant around 1927. (*Lampranthus, Stapelia*)

immersus Lat., immersed; for the growth form with the stem below ground-level. (*Euphorbia*)

imminutus Lat., diminished, reduced in size; for the smaller habit. (*Euphorbia ephedroides* var.)

imparispinus Lat. 'impar', unequal; and Lat. '-spinus', -spined; for the irregular spine length. (*Euphorbia*)

Impatiens Lat., impatient, sensitive, touchy; for the explosive fruits. (*Balsaminaceae*)

imperatae For Imperato Ferrante (1550–1625), Italian apothecary and naturalist. (*Euphorbia milii* var.)

imperialis Lat., imperial, stately, grand; for the large beautiful flowers. (*Echinopsis*)

implexicoma Lat. 'implexus', tangled, interwoven; and Lat. 'coma', tuft of hair, mane; most probably for the intricately intertwined filiform stems. (*Tetragonia*)

implexus Lat., tangled, interwoven; for the growth habit. (*Senecio*)

implicatus Lat., entangled, interwoven; for the growth form. (*Cynanchum*)

impressus Lat., impressed, printed; (**1**) for the impressions left on the leaves by the central bud cone. (*Agave*) – (**2**) for the longitudinal impressed line on the leaves. (*Ruschia*)

inachabensis For the occurrence at Inachab, Namibia. (*Mesembryanthemum*)

inaequalis Lat., unequal, different; (**1**) for the unequal leaves. (*Rhinephyllum*) – (**2**) for the unequal length of the first-formed leaves. (*Cephalophyllum*) – (**3**) for the unequal length of the sepals. (*Delosperma, Lampranthus*)

inaequidens Lat. 'in-', not; Lat. 'aequus', equal; and Lat. 'dens', tooth; for the unequal leaf marginal teeth. (*Agave*)

inaequilateralis Lat. 'in-', not; Lat. 'aequus', equal; and Lat. 'latus, lateris', [lateral] margin; for the irregularly shaped stem segments. (*Opuntia*)

inaequispinus Lat. 'in-', not; Lat. 'aequus', equal; and Lat. '-spinus', -spined; for the variation in the spination on each branch. (*Euphorbia*)

inaguensis For the occurrence on Little Inagua Island, Bahamas. (*Agave*)

inamarus Lat. 'in-', not; and Lat. 'amarus', bitter; because the leaves do not taste bitter. (*Aloe*)

inamoenus Lat., not beautiful; (**1**) for the stem segments that are dull-coloured and wrinkled in the dry season. (*Tacinga*) – (**2**) perhaps for the unspotted leaves. (*Adromischus*)

inandensis For the occurrence at Inanda, KwaZulu-Natal, RSA. (*Crassula*)

inanis Lat., empty, useless, vain; perhaps for the small plant size. (*Crassula*)

inapertus Lat. 'in-', not; and Lat. 'apertus', open; for the tubular and hardly opening flowers. (*Opuntia*)

inarticulatus Lat. 'in-', not; and Lat. 'articulatus', jointed; for the stems. (*Euphorbia*)

incachacanus For the occurrence at Incachaca, Prov. Cochabamba, Bolivia. (*Lepismium*)

incanus Lat., hoary, white; for the grey-hairy stems and leaves. (*Brachystelma, Peperomia*)

incarnatus Lat., flesh-coloured; for the flower colour. (*Quaqua, Stathmostelma*)

incarus For the occurrence in the Land of the Inca, Peru. (*Villadia*)

ince Turkish 'ince', elegant; for the delicate slender stems and the general appearance of the plants. (*Sedum*)

inclaudens Lat. 'in-', not; Lat. 'claudens', closing; because the flowers last several days and do not close for the night. (*Erepsia, Esterhuysenia*)

inclusus Lat., included; because pedicel and flower base are enclosed by large bracts. (*Octopoma, Ruschia*)

incomptus Lat., unadorned, unkempt, untidy; perhaps for the growth. (*Delosperma*)

inconfluens Lat., not flowing together; because the lines on the leaves are hardly connected. (*Haworthia mucronata* var.)

inconspicuus Lat., inconspicuous; (**1**) for the size of the plants. (*Aloe, Hoya, Raphionacme, Sedum*) – (**2**) for the small flowers. (*Delosperma, Lampranthus*) – (**3**) for the small flowers and the few petals. (*Ruschia*)

inconstantius Lat., inconstant, fickle; for the variable habit. (*Euphorbia*)

incrassatus Lat., thickened; for the basal tuber. (*Pelargonium*)

incultus Lat. 'in-', not; and Lat. 'cultus', cultivated; for the occurrence at a distance from local areas of cultivation. (*Euphorbia*)

incumbens Lat., incumbent, folded inwards and lying upon; for the position of the staminodes. (*Ruschia*)

incurvatus Lat., curved inwards; for the placement of the staminodes. (*Ruschia*)

incurvulus Dim. of Lat. 'incurvus', curved inwards; for the orientation of the leaf tips. (*Haworthia cymbiformis* var.)

incurvus Lat., curved inwards; for the leaves. (*Hereroa, Lampranthus*)

indagatorum Gen. Pl. of Lat. 'indagator', explorer; according to the protologue commemorating the explorers who first sighted Watling Island in the Bahamas, and Dr. Britton and his associates for their botanical collections. (*Agave*)

indecorus Lat. 'in-', not; Lat. 'decorus', graceful, noble; for the undistinguished appearance. (*Euphorbia, Ruschia*)

indensis For the occurrrence near Indé, Durango, Mexico. (*Coryphantha*)

indicus For the occurrence in India. (*Caralluma, Dorstenia, Sinocrassula*)

induratus Lat., hardened; for the pungent leaf tips and the hardened fruit remains, making the plant spiny. (*Ruschia*)

indurescens Lat., becoming hard; for the branches, which become woody with age. (*Euphorbia*)

inermis Lat., unarmed, without spines or prickles; (**1**) for the entire leaf margin. (*Aloe, Beaucarnea*) – (**2**) for the membranous stipules. (*Pelargonium antidysentericum* ssp.) – (**4**) Lat., unarmed, without spines or prickles. (*Adenia, Euphorbia, Monsonia, Opuntia echios* var., *Selenicereus*)

inexpectatus Lat., unexpected; (**1**) for the unexpected flower colour, contrasting with that of the type subspecies. (*Conophytum tantillum* ssp.) – (**2**) because the taxon had been confused with another species and its distinct status was unexpected. (*Dinteranthus*)

infestus Lat., unsafe, hostile; for the long spines. (*Opuntia*)

inflatus Lat., inflated; (**1**) for the inflated calyx. (*Nolana*) – (**2**) for the inflated corolla tube. (*Ceropegia*) – (**3**) for the inflated corolla. (*Madangia*) – (**4**) for the inflated fruits. (*Bulbine*)

ingens Lat., huge; (**1**) for the plant size. (*Euphorbia*) – (**2**) for the plant size in comparison with other taxa of the genus. (*Dudleya*)

ingenticapsa Lat. 'ingens', huge; and Lat. 'capsa', capsule; for the exceptionally large fruits. (*Euphorbia*)

ingezalahianus For Mr. Ingezalaha (fl. 1955), District Chef in Fianarantsoa, Madagascar. (*Euphorbia*)

ingomensis For the occurrence at Ingoma, KwaZulu-Natal, RSA. (*Schizoglossum*)

ingwersenii For Walter E. T. Ingwersen (1885–1960), botanist collecting in the Caucasus. (*Sempervivum*)

innesii For Clive F. Innes (1909–1999), English horticulturist and succulent plant enthusiast. (*Selenicereus*)

inopinatus Lat., unexpected; because the discovery of a new taxon in the small genus was unexpected. (*Pachypodium rosulatum* var.)

inornatus Lat., unadorned; (**1**) for the unexceptional appearance. (*Euphorbia*) – (**2**) for the lack of bristles on the leaf tips (in comparison with species of *Trichodiadema*, where the taxon was first placed). (*Drosanthemum*) – (**3**) for the inconspicuously coloured flowers. (*Ceropegia*) – (**4**) because the plants never flowered in cultivation. (*Conophytum*)

inquisivensis For the occurrence near Inquisivi, Dept. La Paz, Bolivia. (*Yungasocereus*)

insigniflorus Lat. 'insignis', distinguished, remarkable; and Lat. '-florus', -flowered. (*Huernia*)

insignis Lat., distinguished, remarkable; (**1**) for the appearance. (*Brighamia, Ceiba, Ceropegia, Cynanchum, Erepsia, Portulaca, Stapelianthus*) – (**2**) for the reddish colour of the rosettes. (*Sempervivum armenum* var.)

insolitus Lat., unusual, uncommon, strange; for the differences in comparison with related taxa. (*Drosanthemum*)

insularis Lat., pertaining to an island; (**1**) for the occurrence on the island of Socotra. (*Echidnopsis*) – (**2**) for the occurrence on the islands of the Fernando de Noronha Archipelago off the coast of Brazil. (*Cereus*) – (**3**) for the occurrence on some of the islands of the Galápagos archipelago. (*Opuntia*) – (**4**) for the occurrence on an island. (*Dudleya blochmaniae* ssp., *Dudleya virens* ssp., *Mammillaria*)

integer Lat. 'integer, integra, integrum', entire; for the leaf margin. (*Aloe, Furcraea cabuya* var., *Stomatium*)

integrifolius Lat. 'integer, integra, integrum', entire; and Lat. '-folius', -leaved. (*Crassula sarmentosa* var., *Kalanchoe*, *Rhodiola*)

interjectus Lat., intercalated, thrown in between; for the systematic position intermediate between related taxa. (*Opuntia vitelliniflora* ssp.)

intermedius Lat., intermediate; for the resemblance to several other taxa. (*Crassula*, *Drosanthemum*, *Eriosyce heinrichiana* ssp., *Gunniopsis*, *Haworthia maculata* var., *Huernia brevirostris* ssp., *Othonna*, *Ruschia*, *Sclerocactus parviflorus* ssp., *Umbilicus horizontalis* var., *Yucca baileyi* var.)

intermixtus Lat., mixed, intermingled; because herbarium specimens were mixed with those of another taxon. (*Agave*)

interratus Lat., burried, deposited in the ground; for the underground rhizomes. (*Nolina*)

interruptus Lat., interrupted; application obscure. (*Aspidoglossum*)

intertextus Lat. 'inter', mixed, intermingled; and Lat. 'textus', woven (from Lat. 'texere', to weave); for the interwoven spination. (*Echinomastus*, *Gymnocalycium bodenbenderianum* ssp., *Matucana*)

intervallaris Lat., with intervals, spaced; (**1**) for the long internodes. (*Antimima*, *Lampranthus*) – (**2**) for the often regular intervals between the branchlets on the elongated branches. (*Mossia*)

intisy From the local vernacular name of the plants in Madagascar. (*Euphorbia*)

intonsus Lat., unshaved, i.e. bristly; (**1**) for the rough stem surface. (*Delosperma*) – (**2**) for the tuft of bristles on the leaf tip. (*Trichodiadema*)

intortus Lat., contorted; application obscure. (*Melocactus*)

intricatus Lat., intricate, entangled; for the inflorescence. (*Ruschia*, *Schizobasis*)

introrsus Lat., towards the inside; but here a printer's error for 'intonsus', but retained as epithet as the plant is distinct from *Mesembryanthemum* (*Trichodiadema*) *intonsum*. (*Trichodiadema*)

intrusus Lat., intruding; for the enclosure of the flower base by the uppermost leaves. (*Ruschia*)

inundaticola Lat. 'inundatus', flooded; and Lat. '-cola', -dwelling; for the occurrence on a flood plain. (*Euphorbia*)

invaginatus Lat., sheathed, covered; for the floral bracts covering the cyathia. (*Euphorbia*)

invalidus Lat., invalid; because the taxon was originally illustrated under the name of a different species. (*Delosperma*)

invenustus Lat. 'in-', not; and Lat. 'venustus', graceful; for the ungraceful appearance. (*Monadenium*)

inversus Lat., inverted; application obscure. (*Quaqua*)

invictus Lat., invincible; for the strong and fierce spination. (*Grusonia*)

involucratus Lat., wrapped up, having an involucre; application obscure. (*Trachyandra*)

involutus Lat., involute; for the inrolled leaves. (*Agave*)

inyangensis For the occurrence on Mt. Inyanga, Zimbabwe. (*Aloe*)

ionanthus Gr. 'ion', violet; and Gr. 'anthos', flower; for the flower colour. (*Tripogandra*)

Ipomoea Gr. 'ips, ipos', a worm, bindweed; and Gr. 'homoios', similar; for the twining stems ("twisting like a worm") or the similarity to *Convolvulus arvensis* ("Common Bindweed"). (*Convolvulaceae*)

iquiquensis For the occurrence near the city of Iquique, Chile. (*Eriosyce recondita* ssp., *Eulychnia*)

iranicus For the occurrence in Iran. (*Sempervivum*)

irmae For Mrs. Irma Burger (fl. 1997), wife of Willem Burger who owns Aggeneys Farm, RSA. (*Conophytum*)

irwinii For Dr. Howard S. Irwin (*1928), US-American botanist at the New York Botanical Garden. (*Portulaca*)

isabellae For Mrs. Isabella King (fl. 1938), housewife in Port Elizabeth, Eastern Cape, RSA. (*Haworthia gracilis* var.)

isacanthus Gr. 'iso', equal; and Gr. 'akantha', spine, thorn; for the equally long spines and stipular spines. (*Euphorbia*)

isaloensis For the occurrence in the Isalo mountains, Madagascar. (*Aloe*)

Ischnolepis Gr. 'ischnos', dry; and Gr. 'lepis',

scale; perhaps for the filiform slender corona segments. (*Asclepiadaceae*)

ishidae For Bunzaburo Ishida (fl. 1921), without further data. (*Rhodiola*)

islayensis For the occurrence in Prov. Islay, Dept. Arequipa, Peru. (*Eriosyce*)

Isolatocereus MLat. 'isolatus', detached, isolated (from Ital. 'isola', island; and for *Cereus*, a genus of columnar cacti, because the type species occurs as large isolated individuals. (*Cactaceae*)

ispartae For the occurrence at Isparta, SW Turkey. (*Sempervivum*)

isthmensis For the occurrence on the Isthmus of Tehuántepec, Mexico. (*Agave*)

itampolensis For the occurrence near Itampolo, Madagascar. (*Euphorbia famatamboay* ssp., *Euphorbia neobosseri* var.)

itremensis For the occurrence in the Itremo mountains, Madagascar. (*Aloe*, *Euphorbia*)

ivohibensis For the occurrence on the Pic d'Ivohibe, Madagascar. (*Kalanchoe jongmansii* ssp.)

ivorii For Ivor Dekenah (*1904), South African magistrate and plant enthusiast, collecting plants mainly in the area of Fraserberg, Northern Cape. (*Antimima*)

iwarenge From the Japanese vernacular name Iwa-renge for the taxon. (*Orostachys malacophylla* var.)

J

Jacaratia Probably from the vernacular Brazilian name for one of the species. (*Caricaceae*)

jaccardianus For Henri Jaccard (1844–1922), Swiss botanist and high school teacher. (*Sedum*)

jacksonii For T. H. E. Jackson (fl. 1955), Acting Senior Civil Affairs Officer, Ethiopia. (*Aloe*, *Echidnopsis*)

Jacobsenia For Dr. Hermann J. H. Jacobsen (1898–1978), German horticulturist, longtime director of the Botanical Garden of Kiel University, and author of important succulent plant literature. (*Aizoaceae*)

jacobsenii As above. (*Senecio*)

jaegerianus For Edmund C. Jaeger (1887–1983), US-American botanist. (*Yucca brevifolia* var.)

jaenensis For the occurrence near Jaén, Dept. Cajamarca, Peru. (*Praecereus euchlorus* ssp.)

jahandiezii For Émile Jahandiez (1876–1938), French botanist and horticulturist at the Jardin d'Acclimatation de Carqueiranne, France, active in N Africa etc. (*Sedum*)

jaiboli From the local Warihio Indian vernacular name of the taxon in Sonora, Mexico. (*Agave*)

jainii For Sudhanshu K. Jain (*1926), Indian botanist. (*Ceropegia*)

jaliscanus For the occurrence in the Mexican state of Jalisco. (*Agave*, *Mammillaria*, *Opuntia*, *Sedum*)

jaliscensis As above. (*Yucca*)

jalpanensis For the occurrence near Jalpan, Querétaro, Mexico. (*Coryphantha*)

jamacaru From the local vernacular name "Jamacaru" or "Mandacaru" used widely for the plants in E Brazil. (*Cereus*)

jamaicensis For the occurrence on Jamaica. (*Opuntia*, *Pedilanthus tithymaloides* ssp.)

jamesii For Mr. H. W. James (fl. 1931), of Cradock, Eastern Cape, RSA, field collector of succulents. (*Drosanthemum*, *Rabiea*, *Stomatium*)

jansei For Mr. Anthonie J. T. Janse (1877–1970), Dutch entomologist, emigrated to RSA in 1899. (*Delosperma*)

jansenvillensis For the occurrence near Jansenville, Eastern Cape, RSA. (*Euphorbia*)

japonicus For the occurrence in Japan. (*Orostachys*, *Sedum uniflorum* ssp.)

jarmilae For Jarmila Haldová (fl. 2002), wife of the Czech botanist J. J. Halda. (*Begonia*, *Conophytum*)

jarucoensis For the occurrence at Escaleras de Jaruco, Cuba. (*Agave*)

Jasminocereus For the genus *Jasminum* ("Jasmine", *Oleaceae*); and *Cereus*, a genus of columnar cacti; for scented flowers of these columnar cacti. (*Cactaceae*)

Jatropha Gr. 'iatros', physician; and Gr. 'trophe', food; for the medicinal use of the seeds of some taxa, and for the edible root tubers of *Manihot esculenta* ("Cassava"), formerly also placed in this genus. (*Euphorbiaceae*)

jauernigii For Johann Jauernig (*1937), Austrian mechanic and cactus hobbyist. (*Turbinicarpus*)

jenkinsii For T. J. Jenkins (fl. 1929), assistant at the Transvaal Museum around 1930. (*Crassula setulosa* var.)

Jensenobotrya For Emil Jensen (1889–1963), German amateur botanist emigrating to Namibia in 1936; and Gr. 'botrys', grape, for the thick succulent grape-like leaves. (*Aizoaceae*)

jiaodongensis For the occurrence in the Jiaodong Prov. (today Shandong Prov.), China. (*Sedum*)

jiguu From the local vernacular name of the plants in Kenya. (*Cyphostemma*)

jinianus Unknown. (*Sedum*)

jiuhuashanensis For the occurrence on Jiuhua-Shan (Mt.), Anhui, China. (*Sedum*)

jiulungshanensis For the occurrence on Jiulung-Shan (Mt.), Zhejiang, China. (*Sedum*)

jobiensis For the type locality Jobi in Irian Jaya. (*Myrmecodia*)

joconostle From the local Nahuatl vernacular names Xoconochtli / Joconostle for the sour-tasting fruits of the plants in Mexico. (*Opuntia*)

johannis For John J. Lavranos (*1926), Greek-born insurance broker and botanist, and intrepid collector of succulents throughout S and E Africa. (*Euphorbia*)

johnsonii For Joseph E. Johnson (1817–1882), US-American amateur botanist of St. George, Utah. (*Echinomastus*) – (**2**) For William H. Johnson, Director of the Department of Agriculture of The Moçambique Company from 1906–1910. (*Euphorbia knuthii* ssp.) – (**3**) Perhaps for William H. Johnson, Director of the Department of Agriculture of The Moçambique Company from 1906–1910. (*Ceropegia*) – (**4**) For Harry Johnson (1894–1987), US-American nurseryman in California. (*Echeveria*, *Weberbauerocereus*)

johnstonianus For Dr. Ivan M. Johnston (1898–1960), US-American botanist at the Arnold Arboretum and the Harvard University. (*Ferocactus*)

johnstonii Perhaps for Henry H. Johnston (1856–1939), Scottish physician travelling in S Africa 1899–1902. (*Brachystelma*) – (**2**) For Dr. Ivan M. Johnston (1898–1960), US-American botanist at the Arnold Arboretum and the Harvard University. (*Mammillaria*, *Peniocereus*) – (**3**) For Peter Johnston (fl. 1996), English succulent plant enthusiast of Guernsey and sponsor of the expedition on which the taxon was found. (*Gibbaeum*) – (**4**) For Marshall C. Johnston (*1930), US-American botanist. (*Portulaca*)

jongmansii For Dr. Willem J. Jongmans (1878–1957), Dutch palaeobotanist and curator of the Musée Royal de Botanique in Leiden, Netherlands. (*Kalanchoe*)

Jordaaniella For Prof. Dr. Pieter G. Jordaan (1913–1987), South African botanist at Stellenbosch University. (*Aizoaceae*)

joubertii For Adriaan J. Joubert (*1901), South African science teacher and expert on the flora of the Little Karoo, RSA. (*Conophytum*, *Hereroa*)

juarezensis For the occurrence in the Sierra de Juárez, Oaxaca, Mexico. (*Echeveria*)

jubaephyllus Combined from the specific epithets of the parents *E. regis-jubae* and *E. aphylla*. (*Euphorbia*)

jubatus Lat., with a mane, crested; for the crest-like appearance of the inflorescence. (*Euphorbia*)

jucundus Lat., pleasant, nice; for the attractive appearance. (*Aloe*, *Conophytum*, *Tridentea*)

judaicus For the occurrence in Judaea, Israel. (*Caralluma europaea* var.)

juengeri For Ernst Jünger (1895–1998), German writer, on the occasion of his 100. birthday in March 1995. (*Rhipsalis*)

juglans For the genus *Juglans* ("Walnut"; *Juglandaceae*); for the similarity of the branch tips to walnut shells. (*Euphorbia*)

jujuyensis For the occurrence in Prov. Jujuy, Argentina. (*Sedum*)

juliani-marnieri For Julien Marnier-Lapostolle (1902–1976), French connoisseur and collector of succulents, owner of the Grand Marnier company and founder of the private botanical garden "Les Cèdres". (*Cynanchum*)

julii For Dr. Julius Derenberg (1873–1928), German physician and succulent plant collector in Hamburg, with a special interest in Mesembs, and friend of K. Dinter and G. Schwantes. (*Lithops*)

jumellei For Henri Jumelle (1866–1935), French botanist and ultimately director of the Musée Colonial. (*Cynanchum*)

junceus For the genus *Juncus* ("Rush"; *Juncaceae*), i.e. rush-like; (**1**) for the slender green stems. (*Ceropegia*, *Psilocaulon*, *Senecio*) – (**2**) for the leaves. (*Senecio*)

juncifolius For the genus *Juncus* ("Rush", *Juncaceae*); and Lat. '-folius', -leaved. (*Albuca*, *Ornithogalum*)

junciformis For the genus *Juncus* ("Rush", *Juncaceae*); and Lat. '-formis', -shaped; for the branches. (*Cynanchum*)

junggaricus For the occurrence in the Dzungaria Basin, NW China. (*Rhodiola*)

jurgensenii For Mr. Jürgensen (fl. 1840), German botanical collector, collected for H. Galeotti in Mexico after the latter's return to Europe in 1840. (*Sedum*)

jussiaeicarpus For the genus *Jussiaea* (*Onagraceae*); and Gr. 'karpos', fruit; i.e. with fruits like *Jussiaea*. (*Begonia*)

justi-corderoyi For Justus Corderoy (1832–1911), English miller and succulent plant cultivator at Blewbury near Didcot, Berkshire (now Oxfordshire). (*Crassula*)

Juttadinteria For Mrs. Jutta Dinter (née Helena Jutta Schilde) (fl. 1906–1935), wife of the German botanist Prof. Kurt M. Dinter. (*Aizoaceae*)

juttae As above. (*Adromischus schuldtianus* ssp., *Cyphostemma*, *Euphorbia*, *Hoodia*, *Synaptophyllum*)

juvanii For Franc Juvan (1875–1960), gardener at the Ljubljana Botanical Gardens, Slovenia. (*Sempervivum wulfenii* ssp.)

juvenna Pseudo-Lat., from English 'juvenile', misread on the original label of a cultivated plant, labelled as a possible juvenile form. (*Aloe*)

K

kaessneri For Theodor Kässner [later anglicized to Kassner] (fl. 1901–1908), botanist at Kew and the British Natural History Museum, who collected in Kenya and Tanzania in 1901–1902, and in Zaïre in 1908. (*Monadenium*)

kaffirdriftensis For the occurrence at Kaffirdrift on the Fish River, Eastern Cape, RSA. (*Haworthia reinwardtii* fa.)

kaibabensis For the occurrence on the Kaibab Plateau, N Arizona, USA. (*Agave utahensis* ssp.)

kalahariensis For the occurrence in the Kalahari Desert. (*Cucumis*)

Kalanchoe Phonetic transcription from the Chinese "Kalan Chauhuy" (for *K. spathulata* ?), with the meaning "that which falls and grows", perhaps for the bulbils (although no bulbilliferous taxa are native to China); or (Genaust 1996) from ancient Indian 'kalanka-', spot, rust; and 'chaya', gloss; perhaps for the glossy and perhaps sometimes reddish leaves of the Indian *K. laciniata*. (*Crassulaceae*)

kalisana Lat. from Kiswahili 'kali sana', very sharp, fierce; for the spination. (*Euphorbia*)

kamelinii For R. V. Kamelin (*1938), Russian botanist. (*Pseudosedum*)

kamerunicus For the occurrence in Cameroon. (*Euphorbia*)

kamponii For Kampon Tansacha (fl. 1995), landscape architect and owner of a private botanical garden in Bangkok, Thailand. (*Euphorbia*)

kamtschaticus For the occurrence in the Kamtschatka Region, Siberia, Russia. (*Phedimus*)

kanabensis For the occurrence at Kanab, Utah, USA. (*Yucca angustissima* var.)

kanalensis For the occurrence (cultivated) at Kanala, New Caledonia. (*Euphorbia*)

kaokoensis For the occurrence in Kaokoland, Namibia. (*Euphorbia*, *Orbea maculata* ssp.)

karasbergensis For the occurrence on the Great Karasberg, RSA. (*Aloe striata* ssp.)

karasmontanus For the occurrence in the Karas Mts., Namibia (Lat. 'montanus', montane). (*Anacampseros*, *Lithops*)

karatavicus For the occurrence in the Kara Tau Range, Kazakhstan. (*Pseudosedum*)

karatto From the local vernacular name "Karat" of the taxon on St. Kitts (West Indies). (*Agave*)

karibaensis For the occurrence in the Kariba Distr., Zimbabwe. (*Adenia*)

karooicus For the occurrence in the Karoo (original Khoi name for the region) in RSA. (*Aizoon*, *Pelargonium*, *Plinthus*, *Rhadamanthus*)

karrachabensis For the occurrence at Karrachab, Richtersveld, Northern Cape, RSA. (*Ruschia*)

karroensis For the occurrence in the Karoo (original Khoi name for the region) in RSA. (*Euphorbia*)

karroicus As above. (*Delosperma*)

karroideus As above. (*Antimima*)

karrooensis As above. (*Drosanthemum*)

karrooicus As above. (*Ruschia*)

karwinskianus For Baron W. F. Karwinsky von Karwin (1780–1855), German mining engineer collecting botanical specimens in Mexico. (*Mammillaria*, *Opuntia*)

karwinskii As above. (*Agave*)

kasamanus For the occurrence near Kasama, N Zambia. (*Euphorbia perplexa* var.)

kaschgaricus For the occurrence in the region Kashgaria, Xinjiang, China. (*Rhodiola*)

katangensis For the occurrence in the Katanga region, E Zaïre. (*Stathmostelma*)

katbergensis For the occurrence on the Katberg mountain, Stutterheim Distr., Eastern Cape, RSA. (*Delosperma*)

kaurabassanus For the occurrence at Kaurabassa, Moçambique. (*Pyrenacantha*)

kautskyi For Roberto Kautsky (fl. 1982), Brazilian plant collector. (*Schlumbergera*)

kawaguchii For the collector E. Kawaguchi (fl. 1914). (*Rhodiola alsia* ssp.)

keayi For Dr. Ronald W. J. Keay (1920–1998), British botanist and forestry officer in Nigeria. (*Aloe*, *Raphionacme*)

kedongensis For the occurrence in the Kedong Valley, Kenya. (*Aloe*)

Kedrostis Gr. 'kedrostis', White Bryony, i.e. the ancient name of a scrambling cucurbit, used as generic name for a different group of plants by Medikus. (*Cucurbitaceae*)

kefaensis For the occurrence in the Kef[f]a Region, Ethiopia. (*Aloe*)

keithii For Capt. D. R. Keith (fl. 1935), retired Indian army officer who farmed at Palata Farm, Isiteki, KwaZulu-Natal, RSA. (*Aloe*, *Euphorbia*)

kellerianus For Prof. Dr. A. Keller, Swiss botanist visiting Ethiopia in 1891 as a member of Count Ruspoli's expedition. (*Pterodiscus*)

kellermanii For Prof. William A. Kellerman (1850–1908), US-American botanist and mycologist, died during a botanical expedition in Guatemala. (*Pereskiopsis*)

kelloggii For Dr. Albert Kellogg (1813–1887), US-American physician and botanist, collecting esp. in California. (*Lewisia*)

kelvinensis For the occurrence near the city of Kelvin, Arizona, USA. (*Cylindropuntia*)

kenhardtensis For the occurrence at Kenhardt, Northern Cape, RSA. (*Ruschia*)

keniensis For the occurrence in Kenya. (*Brachystelma*, *Ceropegia*) – (**2**) For the occurrence at the base of Mt. Kenya. (*Huernia*)

kennedyanus For Hermias (Mias) C. Kennedy (fl. 1966–1968), succulent plant collector in Bellville, Western Cape, RSA. (*Huernia*)

kennedyi As above. (*Lithops villetii* ssp.)

kentaniensis For the occurrence near Kentani, Eastern Cape, RSA. (*Streptocarpus*)

keramanthus Gr. 'keramion', pot; and Gr. 'anthos', flower; for the broadly tubular pot-like flowers. (*Adenia*)

keraudreniae For Monique Keraudren-Aymonin (1928–1981), French botanist and Madagascar specialist. (*Cynanchum lineare* ssp., *Stapelianthus*)

kerberi For Edmund Kerber (fl. 1895), collected cacti in Mexico and sent them to Berlin. (*Stenocereus*)

kerchovei For Comte Oswald C. E. M. G. de Kerchove de Denterghem (1844–1906), Belgian horticulturist, botanist, politician and administrator. (*Agave*)

kermesinus Lat., crimson; for the flower colour. (*Portulaca*)

kerrii For Arthur F. G. Kerr (1877–1942), English botanist at the Calcutta Botanic Garden, India. (*Euphorbia*, *Hoya*)

kerzneri For Mr. Sol Kerzner (fl. 1995), owner of the holiday resort in RSA where the type was collected. (*Brachystelma*)

ketabrowniorum Lat. Gen. Pl. of 'Brown'; and Lat. 'et', and; for Ken D. F. Brown (*1957), artist, and his wife Anne E. (née Powys) (*1964), natural history consultant; Kenyan explorers and field collectors. (*Aloe*)

kewensis For the Royal Botanic Gardens Kew, where the taxon was found growing. (*Cyanotis*, *Kalanchoe*)

Khadia From the Tswana / Sotho name "Khadi" for a beer brewed traditionally using the fleshy roots of a variety of taxa, including Mesembs and perhaps also species of this genus. (*Aizoaceae*)

khalidbinsultanii For Prince Khalid bin Sultan bin Abdul Aziz Al Saud (*1949), Saudi Arabian nobleman and sponsor of nature conservation. (*Huernia*)

khamiesbergensis For the occurrence at Khamiesberg, Northern Cape, RSA. (*Conophytum*, *Stapeliopsis*)

khamiesensis For the occurrence near Khamieskroon, Northern Cape, RSA. (*Aloe*)

khandallensis For the occurrence near Khandalla, India. (*Euphorbia*)

kibwezensis For the occurrence at Kibwezi, Kenya. (*Euphorbia bussei* var.)

kieslingii For Dr. Roberto Kiesling (*1941), Argentinian botanist and cactus specialist in San Isidro, Buenos Aires. (*Gymnocalycium*)

kilifiensis For the occurrence near Kilifi, Kenya. (*Aloe*)

kimberleyanus For Michael ("Mike") J. Kimberley (*1934), Zimbabwean legal practicioner, editor of "Excelsa" and "Ingens", and his wife Rosemary ("Rose") C. Kimberley (*1937), South African teacher, both succulent plant enthusiasts in Zimbabwe. (*Aloe inyangensis* var., *Monadenium*)

kimberleyi For the occurrence in the Kimberley region, Australia. (*Trianthema*)

kimnachii For Myron Kimnach (*1922), US-American botanist, specialist on *Crassula*-

ceae and epiphytic cacti, and former longstanding director of the Huntington Botanical Gardens, California. (*Disocactus, Echeveria, Epiphyllum crenatum* var., *Pachyphytum, Sedum*)

kingdonii For Frank Kingdon-Ward (1885–1958), English horticulturist, explorer and plant collector. (*Sedum*)

kingianus For Mrs. E. B. King (fl. 1937), *Haworthia* collector. (*Haworthia*)

kirilowii For Iwan P. Kirilow (1821–1842), Russian botanist. (*Rhodiola*)

kirkii For Thomas Kirk (1828–1898), English-born botanist in New Zealand. (*Crassula*) – (2) For Sir John Kirk (1832–1922), surgeon, Consul-General and British political agent in Zanzibar, collected plants in East Africa. (*Adenia, Sansevieria, Synadenium*) – (3) For Lt.-Col. John W. C. Kirk (1878–1962), son of Sir John Kirk. (*Huernia*)

kiska-loro From the vernacular name of the plants in Argentina; from Quechua 'kiska' (= 'quisca'), spine, spiny plant, cactus; and 'loro', parrot; perhaps because the fruits are eaten by parrots. (*Opuntia anacantha* var.)

kituloensis For the occurrence on the Kitulo Plateau, S Tanzania. (*Ceropegia*)

kladiwaianus For Dr. Leo Kladiwa (1920–1987), Austrian physician in Vienna and cactus specialist. (*Echinopsis*)

klapperi For Ingo Klapper (fl. 1998), German cactus hobbyist. (*Echinocereus*)

klaverensis For the occurrence near Klaver, Vanrhynsdorp Division, Northern Cape, RSA. (*Antimima, Leipoldtia*)

kleinia For the genus *Kleinia* (now a synonym of *Senecio, Asteraceae*), named for Jacob T. Klein (1685–1759) who first flowered this species and wrote on it in 1730. (*Senecio*)

kleiniae For the similarity to *Senecio kleinia* (*Asteraceae*). (*Cylindropuntia*)

kleiniiformis For the genus *Kleinia* (now a synonym of *Senecio, Asteraceae*); and Lat. '-formis'; shaped. (*Senecio*)

klimpelianus For Georg Klimpel († 1959), German horticulturist and owner of a cactus nursery in Berlin-Kleinmachnow. (*Acanthocalycium*)

klinghardtensis For the occurrence in the Klinghardt Mts., Namibia. (*Conophytum, Pelargonium*)

klinghardtianus For the occurrence in the Klinghardt Mountains, Namibia. (*Delosperma*)

klinglerianus For Father Elmar Klingler (1905–1995), Austrian Franciscan Father working for most of his life in Santa Teresita, Bolivia, plant enthusiast, collected Bolivian cacti for European nurseries to finance his missionary work. (*Echinopsis*)

klinkerianus For Christian Klinker (*±1868), German cactus collector and horticulturist in Schleswig. (*Turbinicarpus schmiedickeanus* ssp.)

klipbergensis For the occurrence at Klipberg, Darling Distr., Western Cape, RSA. (*Ruschia*)

klissingianus For C. L. Klissing, German horticulturist and nursery owner, financed Hugo Baum's travels 1925 in Mexico in exchange for the plants collected. (*Mammillaria*)

klossii For the botanical collector Kloss (fl. 1951). (*Dischidia acutifolia* ssp.)

kniphofioides Gr. '-oides', resembling; and for the genus *Kniphofia* (*Asphodelaceae*). (*Aloe*)

knippelianus For Carl Knippel (fl. 1895), German cactus horticulturist in Halberstadt. (*Echinocereus, Frailea, Mammillaria*)

knizei For Karel Knize (fl. 1969, 2002), contemporaneous Czech cactus collector in Lima, Peru. (*Cintia*)

knobelii For Johann C. Knobel (1879–?), South African missionary and trader and keen naturalist. (*Orbea*) – (2) For Jurgens C. J. Knobel (1881–?), director of Prisons, Pretoria, and interested in the development of the public gardens in Pretoria. (*Euphorbia*)

knowltonii For Fred G. Knowlton († 1958), US-American of Bayfield, Colorado, who discovered the taxon. (*Pediocactus*)

knox-daviesii For C. N. Knox-Davies (fl. 1966), without further data. (*Delosperma*)

knuthianus For Count Frederic M. Knuth von Knuthenborg (1904–1970), Danish botanist, plant collector and cactus hobbyist. (*Echinopsis*, *Turbinicarpus*)

knuthii For Prof. Paul E. O. W. Knuth (1854–1899), German botanist at Kiel and specialist in flower biology. (*Euphorbia*)

knysnanus For the occurrence at Knysna, Western Cape, RSA. (*Ruschia*)

kochii For M. Koch (fl. 1898), who collected the type. (*Gunniopsis*)

koekenaapensis For the occurrence near Koekenaap, RSA. (*Antimima*)

koelmaniorum For Arthur ("At") Koelman (1915–1994), South African schoolteacher and horticulturalist, a founder of the Succulent Society of South Africa, pioneer in *Aloe* hybridizing, and his wife Maria ("Ria") M. J. Koelman (1917–1993). (*Haworthia*)

koelzii For the botanical collector Koelz (fl. 1970), without further data. (*Pseudosedum*)

koenigii For the Koenig Family, on whose ranch in New Mexico the taxon was first discovered. (*Escobaria orcuttii* var.)

kofleri For C. Kofler (fl. 1966), without further data. (*Delosperma*)

kolarensis For the occurrence in the Kolar Distr., India. (*Brachystelma*)

kolbei For Friedrich C. Kolbe (1854–1936), South African priest, philosopher and amateur botanist. (*Jacobsenia*)

komaggasensis For the occurrence at Komaggas, Northern Cape, RSA. (*Aloe striata* ssp.)

komarovii For Vladimir L. Komarov (1869–1945), Russian botanist in St. Petersburg and editor of the 30-volume 'Flora URSS'. (*Orostachys aliciae* var.)

komkansicus For the occurrence at Komkans, Northern Cape, RSA. (*Antimima*)

konasita Kiswahili 'kona', angle; and Kiswahili 'sita', six; for the 6-angled stems. (*Ceropegia*)

kondoi For Prof. Kondo (fl. 1989), Japanese *Euphorbia* enthusiast. (*Euphorbia*)

kongboensis For the occurrence in Konbgo Prov., Tibet. (*Rhodiola primuloides* ssp.)

koolwijkianus For Antonius J. Koolwijk (1836–1913), Dutch missionary, 1871–1886 on Aruba etc. where he collected plants. (*Melocactus curvispinus* ssp.)

korethroides Gr. 'korethron', broom, brush; and Gr. '-oides', resembling; for the brushlike spination of the plant bodies. (*Echinopsis*)

korneliuslemsii For Kornelius Lems (1931–1968), Dutch-American botanist. (*Aeonium*)

kosaninii For Prof. Nedelyko Kosanin (1874–1934), Serbian botanist in Belgrad. (*Sempervivum*)

kotschoubeyanus For Prince Vasily von Kotschoubey (1812–1850), Russian numismatist and son of the Royal Minister at St. Petersburg Prince Victor Kotschoubey (1768–1834), supporter of the expeditions 1840–1843 of Baron von Karwinsky in Mexico (*Ariocarpus*)

kotschyanus For Carl G. T. Kotschy (1813–1866), Austrian botanical explorer of the Orient, later assistant and curator of the herbarium of the Natural History Museum in Vienna. (*Sedum*)

koubergensis For the occurrence at Kouberg, Northern Cape, RSA. (*Conophytum lithopsoides* ssp.)

kougabergensis For the occurrence on the Kougaberg, Eastern Cape, RSA. (*Stapelia*)

kozelskyanus For Mr. Kozelsky (fl. 1966), Czech cactus hobbyist. (*Gymnocalycium riojense* ssp.)

kracikii For Karel Kracík (fl. 2002), Czech cactus collector. (*Coryphantha*)

kraehenbuehlii For Felix Krähenbühl (1917–2001), Swiss cactus hobbyist in Arlesheim near Basel. (*Mammillaria*)

kraeuselianus For Prof. Richard Kräusel (1890–1966), German palaeobotanist visiting Namibia 1953–1954. (*Commiphora*)

krahnii For Wolfgang Krahn (fl. 1960, 2003), German cactus collector, travelled in Peru the 1960s together with Paul C. Hutchison. (*Matucana*)

krainzianus For Hans Krainz (1906–1980), Swiss horticulturist, 1931–1971 director of the Municipal Succulent Plant Collection Zürich. (*Copiapoa*, *Turbinicarpus pseudomacrochele* ssp.)

krainzii As above. (*Ceropegia dichotoma* ssp.)

krapohlianus For H. J. C. Krapohl (fl. 1908), Land Surveyor in RSA. (*Aloe*)

krausii For Peter Kraus (fl. 1955), Toledo, Chile, who discovered the taxon. (*Eriosyce*)

kraussii For Dr. Ferdinand F. von Krauss (1812–1890), German scientist, director of the Stuttgart Natural History Museum, traveller and collector in RSA 1838–1840. (*Aloe*)

kritzingeri For Mr. Kobus Kritzinger (*1953), Cape Department of Nature and Environmental Conservation (RSA). (*Tylecodon*)

kroenleinii For Marcel Kroenlein (1928–1994), long-time director of the Jardin Exotique in Monaco. (*Cereus*, *Gymnocalycium*)

krugerae For Mrs. Anna Maria Kruger (fl. 1957), Bolivian botany student of M. Cárdenas. (*Sulcorebutia*)

kuboosanus For the occurrence at Kuboos, Richtersveld, Northern Cape, RSA. (*Ruschia*)

kubusensis For its occurrence at Kubus, Richtersveld, Northern Cape, RSA. (*Adromischus marianiae* var.)

kulalensis For the occurrence on Mt. Kulal, N Kenya. (*Aloe*)

kundelunguensis For the occurrence on the Kundelungu Plateau, Shaba Prov., Zaïre. (*Ceropegia*, *Monadenium*)

kuntzei For Dr. Carl Ernst Otto Kuntze (1843–1907), German traveller and botanist. (*Brownanthus*)

kunzei For Prof. Dr. Gustav Kunze (1793–1851), German physician and botanist, from 1835 professor of botany at Leipzig University. (*Eriosyce*) – (**2**) For Dr. Richard Ernest Kunze (1838–1919), collected cacti for the German Haage nursery about 1900, mainly around Phoenix, Arizona. (*Grusonia*)

kupperianus For Prof. Dr. Walter Kupper (1874–1953), Swiss-born botanist in München, Germany. (*Rebutia deminuta* ssp.)

kurdicus For the occurrence in Kurdistan, a region in SE Turkey and adjacent Iran and Iraq. (*Rosularia sempervivum* ssp.)

kuriensis For the occurrence on the island Abd-El-Kuri off the coast of Socotra. (*Portulaca*)

kurtzii For Dr. Fritz (Federico) Kurtz (1854–1920), German botanist in Córdoba, Argentina. (*Grahamia*)

kutubuensis For the type locality Lake Kutubu in Papua New Guinea. (*Myrmecodia*)

kwebensis For the occurrence in the Kwebe Hills in Ngamiland in present-day Botswana. (*Stapelia*)

L

labatii For Jean-Noël Labat (*1959), French botanist in Paris. (*Euphorbia*)

labiatus Lat., lipped; for the two-lipped flowers. (*Cyrtanthus*)

labworanus For the occurrence in the Labwor Hills, Uganda. (*Aloe*)

lacandonicus For the occurrence in the Selva Lacandona, Chiapas, Mexico. (*Yucca*)

lacei For John H. Lace (1857–1918), who made one of the first large botanical collections in Myanmar (Burma). (*Euphorbia*)

lacer Lat. 'lacer, lacera, lacerum', torn, mangled, cut up; for the roughly dentate keel of the leaves. (*Erepsia*)

lacertosus Lat., bulky, powerful, strong; for the bulky appearance. (*Euphorbia magnicapsula* var.)

Lachenalia For Werner de [von] Lachenal (1736–1800), Swiss botanist in Basel. (*Hyacinthaceae*)

laciniatus Lat., laciniate; for the leaf shape. (*Kalanchoe*)

laconicus Lat., pertaining to Laconia; for the occurrence in Laconia, Peloponnisos, S Greece. (*Sedum*)

lacteus Lat., milk-white; (**1**) for the whitish markings on the branches. (*Euphorbia*) – (**2**) for the flowers. (*Crassula*)

lactifluus Lat. 'lac, lactis', milk; and Lat. 'fluere', to flow; for the abundant milky latex. (*Euphorbia*)

lacunosus Lat., covered with depressions; for the upper leaf face. (*Hoya*)

ladismithensis For the occurrence near Ladismith in the Little Karoo, Western Cape, RSA. (*Cotyledon tomentosa* ssp.)

laetivirens Lat. 'laetus', bright; and Lat. 'virens', green; for the leaf colour. (*Kalanchoe*)

laetus Lat., bright; (**1**) for the bright crimson flowers. (*Aloe*) – (**2**) application obscure. (*Armatocereus, Lampranthus*)

laevigatus Lat., smooth, smoothened; application obscure. (*Brachystelma, Stenocereus*)

laevis Lat., smooth, flat; (**1**) for the comparatively short or at times absent spination. (*Opuntia*) – (**2**) for the smooth branches, peduncles and bracts. (*Monadenium*) – (**3**) for the glabrous corolla. (*Huernia*) – (**4**) application obscure. (*Amphibolia, Miraglossum*)

lagarinthoides Gr. '-oides', resembling; and for the genus *Lagarinthus* (Asclepiadaceae). (*Jatropha*)

lagascae For Mariano Lagasca y Segura (1776–1839), Spanish botanist at the Madrid Botanical Garden. (*Sedum*)

lageniformis MLat. 'lagena', flask, bottle; and Lat. '-formis', -shaped; for the shape of the stems. (*Echinopsis*)

lagopus Gr. / Lat., hare foot; (**1**) for the stem segments covered with dense hair. (*Austrocylindropuntia*) – (**2**) for the shape of the fleshy roots. (*Bulbine*)

lagunae For the occurrence at the Laguna de Amatitlán, Guatemala. (*Agave*) – (**2**) For the occurrence in the Sierra de la Laguna, Baja California, Mexico. (*Opuntia*)

laikipiensis For the occurrence in the Laikipia Distr., Kenya. (*Euphorbia, Orbea*)

lakhonensis For the occurrence in the former Presidency of Lakhon, Laos. (*Portulaca pilosa* ssp.)

lamarckii For Prof. Jean B. P. A. de Monet de Lamarck (1744–1829), celebrated French naturalist and Professor of Natural History at the Jardin des Plantes, Paris. (*Euphorbia*)

lambii For Edgar Lamb (1905–1980), English succulent plant nurseryman, father of Brian Lamb. (*Euphorbia*)

lamellatus Lat., with lamellae; for the segments of the staminal corona. (*Aspidoglossum*)

lamerei For Monsieur Lamère (fl. 1899), French customs official at Fort Dauphin, Madagascar, who collected the type. (*Pachypodium*)

lamii For Prof. Dr. Herman J. Lam (1892–1977), Dutch botanist in Leiden, 1919–1933 in Bogor (Indonesia). (*Myrmecodia*)

Lampranthus Gr. 'lampros', glossy, shiny; and Gr. 'anthos', flower; for the bright shiny flowers of most species. (*Aizoaceae*)

lamprochlorus Gr. 'lampros', glossy, shiny; and Gr. 'chloros', yellowish-green, pale green; for the body colour. (*Echinopsis*)

lamprophyllus Gr. 'lampros', glossy, shiny; and Gr. 'phyllon', leaf. (*Bulbine*)

lamprospermus Gr. 'lampros', glossy, shiny; and Gr. 'sperma', seed. (*Cereus, Crassula colligata* ssp.)

lampusae For the occurrence near Lampusa (now Lapithos), Cyprus. (*Sedum*)

lanatus Lat., woolly; (**1**) for the woolly hairs intermixed with the spines, and the woolly cephalium. (*Espostoa*) – (**2**) for the texture of the upper leaf surface. (*Monadenium pudibundum* var.) – (**3**) for the hairy corona. (*Aspidoglossum*)

lancasteri For Alan Percy-Lancaster (1944–1995), South African amateur botanist and succulent plant enthusiast. (*Brachystelma*)

lanceolatus Lat., spear-shaped, lanceolate (widest below the middle); for the leaf shape. (*Adenia, Anacampseros, Crassula, Dischidia, Dudleya, Hoya, Kalanchoe alternans* var.*, Kalanchoe, Pelargonium, Portulaca umbraticola* ssp.*, Raphionacme, Sedum, Tetragonia*)

lancerottensis For the occurrence on Lanzarote, Canary Islands. (*Aeonium, Sedum*)

lanceus From Lat. 'lancea', lance; for the leaf shape. (*Prenia pallens* ssp.)

lancifolius Lat. 'lancea', lance; and Lat. '-folius', -leaved. (*Aptenia*)

langebaanensis For the occurrence at Langebaan, Western Cape, RSA. (*Ruschia*)

langsdorfii For Georg H. von Langsdorff (1774–1852), German explorer and surgeon, later Russian consul in Brazil and plant collector. (*Parodia*)

lanianuliger Dim. of Lat. 'lanius', butcher; and Lat. '-ger, -gera, -gerum', carrying, bearing; perhaps for the somewhat strong central spines. (*Espostoa*)

laniceps Lat. 'lana', wool; and Lat. '-ceps', headed; for the strongly woolly areoles. (*Cleistocactus*)

laniflorus Lat. 'lana', wool; and Lat. '-florus', -flowered; for the wool on the pericarpel and perianth tube. (*Cipocereus*)

laniger Lat. 'lana', wool; and Lat. '-ger, -gera, -gerum', -carrying, -bearing; for the dense pubescence of the stems. (*Cyphostemma*)

lankanus For the occurrence in Sri Lanka. (*Brachystelma*)

lankesteri For Charles Lankester (1879–1969), British naturalist in Costa Rica. (*Pseudorhipsalis*)

lanosus Lat., woolly; for the long wool of the areoles. (*Cereus*)

lanssensianus For Etienne Lanssens (fl. 1986), Belgian cactus enthusiast. (*Melocactus*)

lanuginosus Lat., woolly, downy; (**1**) for the leaves. (*Crassula*) – (**2**) for the tufts of wool formed by the floriferous areoles of the stems. (*Pilosocereus*)

lanugispinus Lat. 'lanugo', woolly covering (from Lat. 'lana', wool); and Lat. '-spinus', -spined; for the minute hairs covering the spines. (*Haageocereus*)

lapiazicola Fr. 'lapiaz', karst limestone; and Lat. '-cola', -dwelling; for the preferred habitat. (*Adenia*)

Lapidaria Lat. 'lapis, lapidis', pebble, stone; for the resemblance of the plants to groups of stones. (*Aizoaceae*)

lapidicola Lat. 'lapis, lapidis', pebble, stone; and Lat. '-cola', -dwelling; for the preferred habitat. (*Ruschia*)

lapidiformis Lat. 'lapis, lapidis', pebble, stone; and Lat. '-formis', -shaped; for the appearance of the single compact leaf pair. (*Didymaotus*)

Laportea For Francis-Louis Laporte de Castelnau (1812–1880), French zoologist and entomologist. (*Urticaceae*)

laredoi For Mathias Laredo (fl. 1978), Mexican gardener working for the cactus enthusiast Ing. Gustavo Aguirre Benavides in Parras de la Fuente, Coahuila, Mexico. (*Escobaria*)

laricus Lat., for the occurrence in the region of Laristan in present-day Iran. (*Euphorbia*)

larreyi For a Mr. Larrey (fl. 1898), without further data. (*Opuntia*)

Larryleachia For Leslie (Larry) C. Leach (1909–1996), English-born electrical engineer and self-taught botanist in Zimbabwe and later in RSA, specialist on succulent Asclepiads and Euphorbias. (*Asclepiadaceae*)

lasiacanthus Gr. 'lasios', hairy, densely woolly; and Gr. 'akanthos', spine, thorn; for the fine spination. (*Mammillaria, Opuntia*)

lasianthus Gr. 'lasios', hairy, densely woolly; and Gr. 'anthos', flower; (**1**) for the hair-like papillae on the floral bracts. (*Tetragonia*) – (**2**) perhaps for the hairy sepals. (*Crassula*)

Lasiocereus Gr. 'lasios', hairy, densely woolly; and *Cereus*, a genus of columnar cacti; for the densely wool-covered flower tubes. (*Cactaceae*)

lateganiae For Mrs. L. Lategan (fl. 1937), farmer's wife in VanWyksvlei, Oudtshoorn Distr., Western Cape, RSA. (*Haworthia scabra* var.)

latentibulbosus Lat. 'latentus', hidden, concealed; and Lat. 'bulbosus', bulbous; for the bulbils on the sterile stems. (*Sedum*)

lateriflorus Lat. 'latus, lateris', side; and Lat. '-florus', -flowered; because on the type specimen a branch at the base of the terminal inflorescence appeared to continue the growth of the stem, giving the erroneous impression of a lateral inflorescence. (*Euphorbia*)

lateritius Lat., (dark) brick red; for the flower colour. (*Aloe, Echinopsis, Kalanchoe, Portulaca foliosa* var.)

latibracteatus Lat. 'latus', broad, wide; and Lat. 'bracteatus', bracteate. (*Crassula*)

laticephalus Lat. 'latus', broad, wide; and Gr. 'kephale', head; for the broad compact inflorescence. (*Crassula congesta* ssp.)

laticipes Lat. 'latex, laticis', latex, rubber; and Lat. 'pes', foot; for the pliable stems. (*Senecio*)

laticoronus Lat. 'latus', broad, wide; and Lat. 'corona', corona. (*Orbea*)

latifilamentus Lat. 'latus', broad, wide; and Lat. 'filamentum', filament. (*Sedum*)

latifolius Lat. 'latus', broad, wide; and Lat. '-folius', -leaved. (*Bulbine, Dischidia, ×Gasteraloe lapaixii* nvar., *Hoya, Schizoglossum stenoglossum* ssp., *Sedum laxum* ssp., *Solenostemon, Stomatium*)

latipetalus Lat. 'latus', broad, wide; and Lat. 'petalum', petal. (*Drosanthemum, Hereroa, Malephora, Phyllobolus*)

latisepalus Lat. 'latus', broad, wide; and Lat. 'sepalum', sepal. (*Kalanchoe*)

latispinus Lat. 'latus', broad, wide; and Lat. '-spinus', -spined; for the broad flattened central spines. (*Ferocactus*)

latus Lat., broad, wide; (**1**) for the broader leaves in comparison with a related taxon. (*Agave virginica* ssp., *Lenophyllum*) – (**2**) for the fused leaf pairs that are broader than tall. (*Conophytum maughanii* ssp.)

lauchsii For Gerhard Lauchs (fl. 2003), German journalist, succulent plant hobbyist, and editor of the journal "Kakteen und andere Sukkulenten". (*Kleinia*)

laui For Alfred B. Lau (fl. 1970-), German plant-explorer and self-named missionary in Mexico, and later in Belize. (*Copiapoa, Coryphantha pseudoechinus* ssp., *Echeveria, Echinocereus, Epiphyllum, Eriosyce, Mammillaria, Neobuxbaumia, Turbinicarpus*)

laurentii For Prof. Émile Laurent (1861–1904), French agronomist and botanical collector in tropical Africa, died on a collecting trip. (*Sansevieria trifasciata* var.)

lausseri For Alfons Lausser (fl. 1986), German cactus collector. (*Thelocactus, Turbinicarpus pseudomacrochele* ssp.)

lautneri For Mr. Jürgen Lautner (fl. 2003), German horticulturist, technical curator of the Old Botanical Garden in Göttingen, and plant collector. (*Selenicereus grandiflorus* ssp.)

lavisiae For Mary G. Lavis (later Mrs. O'Connor-Fenton) (*1903), South African herbarium assistant at the Bolus Herbarium, Cape Town, RSA. (*Delosperma*)

lavisii For Sidney Warren Lavis (fl. 1927–1928), South African bishop, father of Mary G. Lavis. (*Drosanthemum, Lampranthus, Ruschia*)

lavrani For John J. Lavranos (*1926), Greek-born insurance broker and botanist, and intrepid collector of succulents throughout S and E Africa. (*Bulbine, Caralluma, Euphorbia, Haworthia sordida* var., *Huernia*)

Lavrania As above. (*Asclepiadaceae*)

lavranosii As above. (*Aloe doei* var., *Aloe*)

lawsonii For George M. Lawson (1865–1945), British missionary in RSA with an interest in succulent plants. (*Antimima*)

laxiflorus Lat. 'laxus', lax; and Lat. '-florus', -flowered; for the lax inflorescence. (*Kalanchoe, Monanthes, Plectranthus, Ruschia*)

laxifolius Lat. 'laxus', lax; and Lat. '-folius', -leaved. (*Lampranthus*)

laxipetalus Lat. 'laxus', lax; and Lat. 'petalum', petal; for the arrangement of the petals. (*Delosperma, Ruschia*)

laxus Lat., lax; (**1**) for the branching pattern. (*Leipoldtia, Nolana, Ruschia*) – (**2**) for the open inflorescence. (*Aichryson, Pelargonium, Sedum, Villadia*) – (**3**) for the arrangement of the petals. (*Drosanthemum*)

laza From the local vernacular name of the plants in Madagascar. (*Cyphostemma*)

lazaro-cardenasii For General Lázaro Cárdenas (1895–1970), Mexican civil servant from the state of Michoacán, 1934–1940 president of Mexico. (*Peniocereus*)

leachii For Leslie (Larry) C. Leach (1909–1996), English-born electrical engineer and self-taught botanist in Zimbabwe and later in RSA, specialist on succulent Asclepiads and Euphorbias. (*Aloe, Crassula, Echidnopsis, Huernia*)

lealii For Fernando da Costa Leal (fl. 1859–1860), Portuguese army officer and administrator of Huila Prov., Angola, who assisted F. Welwitsch during his travels in Angola. (*Pachypodium*)

leandrianus For Jacques D. Leandri (1903–1982), French botanist in Madagascar. (*Basella, Euphorbia*)

leandrii As above. (*Aloe, Senecio, Uncarina*)

leblancae For Mlle. Alice Leblanc (fl. 1910, 1913), an intimate acquaintance of the French physician and botanist Raymond Hamet. (*Kalanchoe, Sedum*)

lebomboensis For the occurrence near Lebombo, KwaZulu-Natal, RSA. (*Delosperma*)

lechuguilla Dim. of Span. 'lechuga', salad; perhaps from the local vernacular name of the plants in Mexico. (*Agave*)

lecomtei For Dr. (Paul) Henri Lecomte (1856–1934), French botanist and from 1906 director of the Laboratoire de Phanérogamie in Paris. (*Cynanchum*)

lecontei For John L. LeConte (1825–1883), US-American botanist. (*Ferocactus cylindraceus* ssp.)

Ledebouria For Prof. Dr. Carl F. von Ledebour (1785–1851), German botanist widely travelling in Russia. (*Hyacinthaceae*)

ledermannii For Carl L. Ledermann (1875–1958), German (but born in Switzerland) botanical explorer and plant collector in W Africa and New Guinea. (*Ceropegia*)

ledienii For Fr. Ledien (fl. 1890), head gardener at the Dresden Botanical Garden, Germany. (*Euphorbia*)

ledingii For Mr. A. M. Leding (fl. 1936), US-American cactus lover at State College, New Mexico. (*Echinocereus*)

leeanus For Lambert W. Lee who made botanical collections during a geological survey in 1876 in Oregon, USA. (*Lewisia*) – (**2**) For John Lee (fl. 1845), nurseryman in Hammersmith, London. (*Gymnocalycium*)

leedalii For G. Philip Leedal (1927–1982), British geologist and priest, 1950–1953 working for the Geological Survey, Tanzania, from 1961 as missionary in S Tanzania, active amateur field botanist and author of handbooks on mountain plants, killed in a motorcycle accident. (*Aloe*)

leedyi For John L. Leedy (fl. 1947), without further data. (*Rhodiola integrifolia* ssp.)

leei For W. T. Lee (fl. 1924), who first collected the taxon. (*Escobaria sneedii* ssp.)

leendertziae For Reino Leendertz (later Mrs. Potts) (1869–1965), Dutch botanist at the Pretoria Museum (formerly called Transvaal Museum), RSA. (*Delosperma, Stapelia*)

lehmannii For Prof. Johann G. C. Lehmann (1792–1860), professor of botany in Hamburg (Germany). (*Corpuscularia, Parakeelya*)

leibergii For John B. Leiberg (1853–1913), Swedish-born US-American forester, bryologist and plant collector. (*Sedum*)

leightoniae For Miss Frances M. Leighton (later Mrs. Isaac) (*1909), botanist at the Bolus Herbarium, University of Cape Town, RSA. (*Delosperma, Lampranthus*)

leightonii For James Leighton (1855–1930), Scottish-born horticulturist in RSA. (*Haworthia cooperi* var.)

leiocarpus Gr. 'leios', smooth; and Gr. 'karpos', fruit. (*Sedella*)

leiophyllus Gr. 'leios', smooth; and Gr. 'phyllon', leaf. (*Dasylirion*)

Leipoldtia For Dr. Christian F. L. Leipoldt (1880–1947), famous author, poet, medical practitioner and plant collector in RSA. (*Aizoaceae*)

leipoldtii As above. (*Antimima, Conophytum minusculum* ssp., *Drosanthemum, Lampranthus, Pelargonium*)

leistneri For Dr. Otto A. Leistner (*1931), German-born botanist, living in Africa from 1937 (Tanzania, later RSA), 1955–1996 on staff at the National Botanical Institute, Pretoria, RSA. (*Euphorbia*)

lemaireanus For [Antoine] Charles Lemaire (1800–1871), French philologist, self-taught botanist, botanical editor and specialist on cacti, working in Belgium for a long time. (*Euphorbia*)

lemairei As above. (*Melocactus*)

lembckei For Hans Lembcke (1918–1985), German horticulturist and garden designer in Hamburg, 1953–1971 resident in Chile, and specialist on Chilean cacti. (*Eriosyce napina* ssp.)

lenewtonii For Prof. Dr. Leonard ("Len") E. Newton (*1936), English botanist at Kumasi University, Ghana, and later at the Kenyatta University, Kenya. (*Cynanchum, Euphorbia, Huernia*)

leninghausii For Guillermo Leninghaus (fl. 1894), cactus collector of German descent in Porto Alegre, Rio Grande do Sul, Brazil. (*Parodia*)

lenkorianus For the occurrence in the Talysh Mts. near Lenkoran, Azerbaidzhan. (*Sedum*)

lenophylloides Gr. '-oides', resembling; and for the genus *Lenophyllum* (*Crassulaceae*). (*Sedum*)

Lenophyllum Gr. 'lenos', trough, tank, bath; and Gr. 'phyllon', leaf; for the often channelled leaves. (*Crassulaceae*)

lensayuensis For the occurrence at the Lensayu Rocks, Kenya. (*Aloe*)

lentus Lat., pliant, flexible, sluggish, viscous; application obscure, perhaps for the slow growth or the somewhat flexible spination. (*Mammillaria*)

Leocereus For Antonio Pacheco Leão (1872–1931), director of the Botanical Garden of Rio de Janeiro, Brazil (Port. 'leão' = Lat. 'leo', lion); and *Cereus*, a genus of columnar cacti. (*Cactaceae*)

leonensis For the occurrence in Sierra Leone. (*Euphorbia*) – (**2**) For the occurrence in the valley of the Río León, Prov. Azuay, Ecuador. (*Cleistocactus*) – (**3**) For the occurrence in the state of Nuevo León, Mexico. (*Echinocereus pentalophus* ssp.)

leonii For Dr. Frère León [Sauget y Barbis, Joseph S.; Hermano Léon] (1871–1955), French-born botanist settling in Cuba in 1905, author of the Flora de Cuba. (*Leptocereus*)

leonilae For Dr. Leonila Vasquez (fl. 1961), Mexican entomologist at the Instituto de Biologia at UNAM, Mexico. (*Fouquieria*)

leontopodus Gr. 'leon, leontos', lion, and Gr. 'pous, podos', foot; for the type locality Gaan Libah, which translated from the Somali language means 'Lion's Foot'. (*Euphorbia*)

lepidanthus Gr. 'lepidos', scale; and Gr. 'anthos', flower; for the peculiar scaly flowers. (*Pachycereus*)

Lepidium Lat. name for the Garden Cress, from the Dim. of Gr. 'lepis', scale; for the small scale-like fruits. (*Brassicaceae*)

lepidocarpus Gr. 'lepis, lepidos', scale; and Gr. 'karpos', fruit; for the scale-covered fruits. (*Epiphyllum*)

lepidocaulis Gr. 'lepis, lepidos', scale; and Gr. 'kaulos' (Lat. 'caulis'), stem; for the scaly stem. (*Othonna*)

lepidus Lat., nicely dwarf; for the decorative nature of the plants. (*Aloe*)

Lepismium From Gr. 'lepisma', skin, husk, scale; for the way how the flowers of some species burst through the epidermis (*Cactaceae*)

leptacanthus Gr. 'leptos', fine or delicate; and Gr. 'akanthos', spine, thorn; for the delicate spination. (*Mammillaria rekoi* ssp.)

leptaleon Gr. 'leptaleos', fine, delicate; for the slender branches. (*Lampranthus*)

leptanthus Gr. 'leptos', fine or delicate; and Gr. 'anthos', flower; for the slender flowers. (*Gymnocalycium*)

leptarthros Gr. 'leptos', fine or delicate; and Gr. 'arthron', joint; for the slender stem segments. (*Psilocaulon*)

leptocalyx Gr. 'leptos', fine or delicate; and Gr. 'kalyx', calyx. (*Ruschia*)

leptocarpus Gr. 'leptos', fine or delicate; and Gr. 'karpos', fruit. (*Uncarina*)

leptocaulis Gr. 'leptos', fine or delicate; and Gr. 'kaulon', stem; for the slender stem segments. (*Cylindropuntia*)

Leptocereus Gr. 'leptos', fine or delicate; and *Cereus*, a genus of columnar cacti; for the relatively slender stems. (*Cactaceae*)

leptopetalus Gr. 'leptos', fine or delicate; and Gr. 'petalon', leaf; for the foliage. (*Lepidium*)

leptophyllus Gr. 'leptos', fine or delicate; and Gr. 'phyllon', leaf. (*Ceropegia, Nolana, Sedum, Senecio mweroensis* ssp.)

leptosepalus Gr. 'leptos', fine or delicate; and Gr. 'sepalon', sepal. (*Lampranthus*)

leptosiphon Gr. 'leptos', fine or delicate; and Gr. 'siphon', tube; for the narrow perianth tube. (*Aloe*)

leptus Gr., fine or delicate; for the stems. (*Drosanthemum, Sedum triactina* ssp.)

lerouxiae For Mrs. Olive le Roux (fl. 1923), without further data. (*Ruschia*)

leroyi For Monsieur I. Leroy (fl. 1964), who first collected the taxon. (*Ceropegia*)

lesliei For Dr. Thomas N. Leslie (1858–1942), English-born builder, emigrated to RSA in 1881, active plant collector and photographer. (*Lithops, Rabiea, Stomatium*)

letestuanus For Georges M. P. C. Le Testu (1877–1967), French botanist in Paris and specialist in the W African flora. (*Euphorbia, Monadenium*)

letestui As above. (*Brachystelma*)

letonae For the occurrence (cultivated) at the farm Sucesión Letona in El Salvador. (*Agave vivipara* var.)

lettyae For Miss Cythna L. Letty (later Mrs. Forssman) (1895–1985), botanical artist for the Botanical Research Institute, Pretoria, RSA, and field collector. (*Aloe*)

leubnitziae For Elsbeth von Leubnitz (fl. 1884), wife of Dr. Eduard Pechuel-Loesche, who collected in Namibia in 1884. (*Arthraerva*)

leucacanthus Gr. 'leukos', white; and Gr. 'akantha', spine, thorn. (*Thelocactus*)

leucanthemus Gr. 'leukos', white; and Gr. 'anthemon', flowering plant, flower. (*Rebutia*)

leucantherus Gr. 'leukos', white; and Gr. 'anthera', anther. (*Antimima*)

leucanthus Gr. 'leukos', white; and Gr. 'anthos', flower. (*Crassula multiflora* ssp., *Delonix, Echinocereus, Echinopsis, Sempervivum*)

leucoblepharus Gr. 'leukos', white; and Gr. 'blepharis', eye-lash; for the white ciliate leaf margins. (*Aeonium*)

leucocarpus Gr. 'leukos', white; and Gr. 'karpos', fruit. (*Sedum*)

leucocentrus Gr. 'leukos', white; and Gr. 'kentron', centre; for the central spines. (*Mammillaria geminispina* ssp.)

leucocephalus Gr. 'leukos', white; and Gr. 'kephale', head; for the white, woolly flower-bearing zones of the stems. (*Pilosocereus*)

leucodendron Gr. 'leukos', white; and Gr. 'dendron', tree; for the waxy pale-grey colour of older branches. (*Euphorbia*)

leuconeurus Gr. 'leukos', white; and Gr. 'neuron', vein, nerve; for the whitish colour of the lateral veins of the lower leaf face. (*Euphorbia*)

leucophyllus Gr. 'leukos', white; and Gr. 'phyllon', leaf. (*Adromischus*)

leucospermus Gr. 'leukos', white; and Gr. 'sperma', seed. (*Ruschia*)

leucostele Gr. 'leukos', white; and Gr. 'stele', pillar, column; for the appearance of the plants. (*Stephanocereus*)

leucothrix Gr. 'leukos', white, glossy; and Gr. 'thrix, trichos', hair; for the white hairs on the leaves. (*Tylecodon*)

leucotrichus Gr. 'leukothrix, leukotrichos', white-hairy; (**1**) for the white hairs covering the plants. (*Echeveria, Oreocereus, Sinningia*) – (**2**) for the whitish flexible spination. (*Opuntia*)

levis Lat., flat, smooth, polished; for the leaf surface. (*Stomatium*)

levitestatus Lat. 'levis', flat, smooth, polished; and Lat. 'testatus', with a testa; for the almost smooth seed coat (testa). (*Melocactus*)

levyi For B. Levy (1896–?), US-America-born pharmaceutical chemist resident in (then) Rhodesia. (*Huernia*)

Lewisia For Meriwether Lewis (1774–1809), US-American army officer, explorer and plant collector in the US-American Northwest, and 1808–1809 Governor of Alabama. (*Portulacaceae*)

lewisiae For Dr. Gwendoline J. Lewis (1909–1967), South African botanist at the South African Museum, Cape Town, and later at the Kirstenbosch Botanical Garden, specialist on *Iridaceae*. (*Lampranthus*)

lewisianus For Berkeley R. Lewis (fl. 1955), Colonel of the US-American Army, who accompanied G. E. Lindsay on a long botanical trip to Baja California. (*Mammillaria brandegeei* ssp.)

libanoticus For the occurrence in the Lebanon. (*Rosularia sempervivum* ssp.)

liberalis Lat., liberal, freely; for the "liberal supply of petals and staminodes". (*Lampranthus*)

libericus For the occurrence in Liberia. (*Sansevieria*)

liciae For Madame Alice Rasse (fl. 1909), an acquaintance of the French botanist Raymond Hamet. (*Rhodiola chrysanthemifolia* ssp.)

liclicensis For the occurrence near Liclic, Prov. San Marcos, Dept. Cajamarca, Peru. (*Peperomia*)

liebenbergii For Louis C. C. Liebenberg (1900–?), agriculturist at the RSA Department of Agriculture, and botanical collector. (*Adromischus, Delosperma*)

liebmannianus For Frederik M. Liebmann (1813–1856), Danish botanist, travelling in Cuba and Mexico 1840–1843. (*Sedum*)

lievenii Unknown. (*Pseudosedum*)

lignescens Lat., becoming woody; for the nature of the branches. (*Brownanthus, Crassula tetragona* ssp., *Phyllobolus*)

lignosus Lat., woody; for the woody branches. (*Drosanthemum, Euphorbia, Euphorbia mauritanica* var.)

lilacinus Lat., lilac-coloured; for the leaf colour. (*Echeveria*)

lilae For Mrs. Lila Trujillo (1913–1987), Venezuelan teacher and promoter of human rights at Campo Elias and later Caracas, aunt of the Venezuelan botanist Baltasar Trujillo. (*Opuntia*)

liliputanus Lat., small enough to inhabit Lilliput from Gulliver's travels; (**1**) for the minute plant bodies. (*Blossfeldia*) – (**2**) for the small leaves. (*Gasteria bicolor* var.)

limariensis For the occurrence in the valley of the Río Limarí, Prov. Limarí, C Chile. (*Eriosyce*)

limbatus Lat., bordered; application obscure. (*Antimima*)

limifolius Lat. 'lima', file, ratchet; and Lat. '-folius', -leaved; for the rough leaf surface. (*Haworthia*)

limitatus Lat., limited; for the very limited geographical range. (*Opuntia*)

limonensis For the occurrence at El Limón, Jalisco, Mexico. (*Mammillaria*)

limoniacus Perhaps from MLat. 'limones', lemons; application obscure. (*Hoya*)

limpidus Lat., clear, transparent, pure; for the translucent window area at the leaf tips. (*Conophytum*)

limpopoanus For the occurrence in the Limpopo River valley, S Africa. (*Euphorbia*)

linaresensis For the occurrence near Linares, Nuevo León, Mexico. (*Mammillaria melanocentra* ssp.)

lindbergianus For Gustav A. Lindberg (1832–1900), Swedish botanist and city treasurer at Stockholm, travelling 1854–1855 in Brazil, specialist of *Rhipsalis*. (*Rhipsalis*)

lindenianus For Dr. Seymour Linden (*1921), US-American chemist and succulent plant enthusiast. (*Conophytum tantillum* ssp.)

lindenii As above. (*Aloe, Ceropegia, Monadenium*)

lindheimeri For Ferdinand J. Lindheimer

(1801–1879), German botanist from Frankfurt / Main, later living in Texas. (*Opuntia engelmannii* var.)

lindiensis For the occurrence in the Lindi area, S Tanzania. (*Adenia*)

lindleyi For Prof. Dr. John Lindley (1799–1865), English botanist and professor of botany at the University College, London. (*Aeonium, Sinningia*)

lindmanii For Carl A. M. Lindman (1856–1928), Swedish botanist and traveller, from 1905 at the Natural History Museum in Stockholm. (*Kalanchoe*)

lindsayi For Dr. George E. Lindsay (1917–2002), US-American botanist and specialist for the Baja California flora, director of the San Diego Natural History Museum 1957–1963, director of the California Academy of Sciences 1963–1982. (*Cylindropuntia, Echinocereus ferreirianus* ssp., *Ferocactus, Mammillaria,* ×*Myrtgerocactus*)

linearifolius Lat. 'linearis', like a line, linear; and Lat. '-folius', -leaved; for the narrow leaves. (*Aloe, Kalanchoe, Nolana, Othonna retrorsa* var., *Rosularia modesta* var., *Yucca*)

linearis Lat., like a line, linear; (**1**) for the leaf shape. (*Aeollanthus subacaulis* var., *Brachystelma, Ceropegia, Columnea, Cynanchum, Delosperma, Dudleya, Hoya, Sedum, Talinum*) – (**2**) for the shape of the corolla segments. (*Quaqua*)

lineatus Lat., striped; (**1**) for the dark red striations of stems, peduncles and petioles. (*Sinningia*) – (**2**) for the longitudinal markings on the leaves. (*Aloe*) – (**3**) for the venation of the petals. (*Rosularia*)

lineolatus Lat., provided with small lines; for the lines on the leaf sheaths. (*Ruschia*)

linguifolius Lat. 'lingua', tongue; and Lat. '-folius', -leaved; for the leaf shape. (*Cremnophila*)

linguiformis Lat. 'lingua', tongue; and Lat. '-formis', -shaped; (**1**) for the shape of the stem segments. (*Opuntia engelmannii* var.) – (**2**) for the leaf shape. (*Glottiphyllum*)

lingulatus Lat., tongue-shaped; for the leaves. (*Cistanthe*)

liniflorus For the genus *Linum* ("Flax"; *Linaceae*); and Lat. '-florus', -flowered. (*Parakeelya*)

linkii For Prof. Dr. Johann H. F. Link (1767–1851), German botanist, and successor of Willdenow in Berlin. (*Parodia*)

linophyllus For the genus *Linum* ("Flax"; *Linaceae*); and Gr. 'phyllon', leaf; for the flax-like narrow leaves. (*Ceropegia*)

lioutchenngoi For Liou Tchen-ngo (1897–1975), Chinese botanist. (*Orostachys malacophylla* ssp.)

lisa From the local vernacular name "maguey lisa" (Span. 'lisa', smooth) of the plants in Mexico. (*Agave mapisaga* var.)

lisabeliae For Lisabel I. Hall (*1919), South African botanist and teacher, wife of Harry Hall. (*Ruschia*)

lisianthoides Gr. '-oides', resembling; and for the genus *Lisianthus* (*Gentianaceae*). (*Glossostelma*)

lissocarpus Gr. 'lissos', smooth; and Gr. 'karpos', fruit. (*Eriosyce marksiana* var.)

Litanthus Gr. 'lithos', stone; and Gr. 'anthos', flower; because flowers are produced directly from the bulb, which looks like a stone. (*Hyacinthaceae*)

lithophilos Gr. 'lithos', stone; and Gr. 'philos', friend; for the habitat preference. (*Hylotelephium verticillatum* var.)

Lithops Gr. 'lithos', stone; and Gr. 'ops', eye, face; because the plants resemble the stones amongst which they grow. (*Aizoaceae*)

lithopsoides Gr. '-oides', resembling; and for the genus *Lithops* (*Aizoaceae*). (*Conophytum*)

litoralis Lat., littoral, coastal; for the coastal occurrence. (*Delosperma, Echinopsis, Eriosyce subgibbosa* var., *Phedimus*)

litoreus Lat., pertaining to the sea shore; for the preferred habitat. (*Sedum*)

littlewoodii For Roy Charles Littlewood (1924–1967), British horticulturist, emigrated to RSA in 1957 and worked at the Karoo Garden, Worcester, Western Cape. (*Lampranthus, Leipoldtia weigangiana* ssp., *Phyllobolus digitatus* ssp., *Ruschia, Trichodiadema*)

littoralis Lat., littoral, coastal; for the coastal occurrence. (*Agave, Aloe, Coccinia rehman-*

nii var., *Ferocactus viridescens* var., *Opuntia*, *Stenocereus thurberi* ssp., *Trianthema*)

littoreus Lat., pertaining to the sea shore; for the preferred habitat. (*Momordica*)

litvinovii For Dimitri I. Litvinov [Litwinow] (1854–1929), Russian botanist and explorer of Central Asia. (*Rhodiola*)

lividiflorus Lat. 'lividus', leaden purple; and Lat. '-florus', -flowered. (*Euphorbia*)

lividus Lat., leaden purple; (**1**) for the leaf colour. (*Haworthia pubescens* var.) – (**2**) for the flower colour. (*Dischidia*)

llanuraensis For the occurrence in the 'Llanura Costera del Pacifico' in Sonora, Mexico. (*Echinocereus nicholii* ssp.)

lloydii For Francis E. Lloyd (1868–1947), English-born cytologist, working in Canada and USA. (*Escobaria*, *Mammillaria*, *Thelocactus hexaedrophorus* ssp.)

loandae For the occurrence near Luanda, capital of Angola. (*Raphionacme*)

lobatus Lat., lobed; for the leaf shape. (*Gerrardanthus*, *Kalanchoe*, *Pelargonium*)

lobulatus Lat., weakly lobed (Dim. of Lat. 'lobatus', lobed); (**1**) for the leaf margins. (*Rhodiola*) – (**2**) perhaps for the two-lobed corona. (*Raphionacme*)

localis Lat., local; perhaps for the presumed local occurrence. (*Lithops*)

lockwoodii For S. Lockwood-Hill (fl. 1940), magistrate in Laingsburg, Western Cape, RSA. (*Haworthia*)

lodarensis For the occurrence at Lodar, Saudi Arabia. (*Huernia*)

lodewykii For Lodewyk van Heerde (fl. 1935–1947), without further data. (*Antimima*)

loeschianus For Alfred Lösch (1865–1946), German succulent plant enthusiast. (*Conophytum*, *Schwantesia*)

loesenerianus For Dr. Ludwig E. T. Loesener (1865–1941), German botanist in Berlin and friend of the German botanist R. Schlechter. (*Huernia*)

loganii For Mr. J. D. Logan (fl. 1933), without further data. (*Aloinopsis*, *Antimima*, *Stomatium*)

lokenbergensis For the occurrence near Lokenberg, Calvinia Distr., Northern Cape, RSA. (*Antimima*)

lolwensis For the occurrence in association with Lake Victoria, locally called Lolwe in the Luo language. (*Aloe*)

lomatophylloides Gr. '-oides', resembling; and for the former genus *Lomatophyllum* (Aloaceae). (*Aloe*)

lomi Combination of the first syllables of the parental species names *E. lophogona* and *E. milii*. (*Euphorbia*)

longaevus Lat. 'longus', long; and Lat. 'aevum', life time, life span; for the assumed longevity of the plants. (*Furcraea*)

longianus For Frank R. Long (1884–1961), English-born horticulturalist, working on rubber in Malaysia and then emigrating to RSA in 1920 and becoming Superintendent of Parks in Port Elizabeth, Eastern Cape. (*Haworthia*)

longiareolatus Lat. 'longus', long; and Lat. 'areolatus', with areoles. (*Opuntia basilaris* var.)

longibracteatus Lat. 'longus', long; and Lat. 'bracteatus', bracteate. (*Agave*, *Conophytum*, *Haworthia turgida* var., *Ornithogalum*, *Senecio melastomifolius* ssp.)

longicarpus Lat. 'longus', long; and Gr. 'karpos', fruit. (*Melocactus ernestii* ssp.)

longicaulis Lat. 'longus', long; and Lat. 'caulis', stem. (*Aeschynanthus*)

longiciliatus Lat. 'longus', long; and Lat. 'ciliatus', ciliate; for the long hairs along the leaf margins. (*Crassula setulosa* var.)

longicomus Lat. 'longus', long; and Lat. 'coma', hair, mane; for the white hair-like spination. (*Weberbauerocereus*)

longicornis Lat. 'longus', long; and Lat. 'cornu', horn; for the long curved spines. (*Coryphantha*)

longicoronae Lat. 'longus', long; and Lat. 'corona', crown, corona. (*Cynanchum luteifluens* var.)

longidens Lat. 'longus', long; and Lat. 'dens', tooth; for the long tubercles of the stems. (*Orbea*)

longidentatus Lat. 'longus', long; and Lat. 'dentatus', toothed, dentate; application obscure. (*Pseudosedum*)

longiflorus Lat. 'longus', long; and Lat. '-florus', -flowered; (**1**) for the overall flower size. (*Agave, Aloe trichosantha* ssp., *Caralluma, Echeveria, Kalanchoe, Mammillaria, Sansevieria, Senecio, Thorncroftia*) – (**2**) for the long and narrow petals. (*Pelargonium*)

longifolius Lat. 'longus', long; and Lat. '-folius', -leaved. (*Agave ocahui* var., *Astridia, Brachystelma, Bulbine, Ceropegia, Euphorbia, Euphorbia milii* var., *Hoya, Kalanchoe, Myrmecodia, Nolina, Pachyphytum, Pelargonium, Psammophora, Raphionacme, Sesuvium*)

longifuniculatus Lat. 'longus', long; and Lat. 'funiculatus', with a funiculus. (*Sedum*)

longii For Frank R. Long (1884–1961), English-born horticulturalist, emigrated to RSA and became Superintendent of Parks in Port Elizabeth, Eastern Cape. (*Huernia, Tromotriche*)

longimammus Lat. 'longus', long; and Lat. 'mamma', nipple, tubercle; for the long body tubercles. (*Mammillaria*)

longipedicellatus Lat. 'longus', long; and Lat. 'pedicellatus', pedicelled, stalked; for the long flower stalk. (*Stapelia*)

longipedunculatus Lat. 'longus', long; and Lat. 'pedunculatus', pedunculate. (*Ceraria, Dorstenia buchananii* var.)

longipes Lat. 'longus', long; and Lat. 'pes', foot; (**1**) for the long petioles of the leaves. (*Tylecodon*) – (**2**) for the long peduncles. (*Agave, Talinum*) – (**3**) for the long pedicels. (*Antimima, Crassula, Drosanthemum, Pectinaria, Sedum, Tromotriche*)

longipetalus Lat. 'longus', long; and Lat. 'petalum', petal. (*Lewisia*)

longiscapus Lat. 'longus', long; and Lat. 'scapus', scape; for the long inflorescence. (*Bulbine, Cistanthe*)

longisepalus Lat. 'longus', long; and Lat. 'sepalum', sepal. (*Bergeranthus, Echeveria pringlei* var., *Sedum fui* var.)

longiserpens Lat. 'longus', long; and Lat. 'serpens', creeping; for the growth form. (*Cleistocactus*)

longisetus Lat. 'longus', long; and Lat. '-setus', with bristles; for the long bristle-like spines. (*Echinocereus*)

longispinus Lat. 'longus', long; and Lat. '-spinus', -spined. (*Euphorbia, Gymnocalycium andreae* var., *Hoodia*)

longissimus Lat., longest (Superl. of Lat. 'longus', long); (**1**) for the long leaves. (*Dasylirion, Myrmecodia*) – (**2**) for the extremely long flowers. (*Echeveria*)

longistamineus Lat. 'longus', long; and Lat. 'stamineus', staminate; for the long filaments. (*Copiapoa, Lampranthus*)

longistylus Lat. 'longus', long; and Lat. 'stylus', style. (*Aloe, Mesembryanthemum, Sinocrassula*)

longituberculosus Lat. 'longus', long; and Lat. 'tuberculosus', tuberculate; for the elongate stem tubercles. (*Euphorbia*)

longitubus Lat. 'longus', long; and Lat. 'tubus', tube; for the flower architecture. (*Huernia, Moringa, Raphionacme, Sansevieria metallica* var., *Sansevieria suffruticosa* var.)

longus Lat. 'longus', long; (**1**) for the long leaves. (*Glottiphyllum*) – (**2**) for the long fused leaf pairs. (*Conophytum*)

longyanensis For the occurrence at Long-yan, Tibet [Xizang, China]. (*Sedum*)

lootsbergensis For the occurrence on the Lootsberg mountain, Eastern Cape, RSA. (*Delosperma*)

lophanthus Gr. 'lophos', crest; and Gr. 'anthos', flower; application obscure. (*Agave*)

lophogonus Gr. 'lophos', crest; and Gr. 'gonia', angle; for the crested stem angles. (*Euphorbia*)

Lophophora Gr. 'lophos', crest; and Gr. '-phoros', -carrying; for the arrangement of the tufted wool of the areoles. (*Cactaceae*)

lophophoroides Gr. '-oides', resembling; and for the genus *Lophophora* (*Cactaceae*). (*Turbinicarpus*)

lophophorus Gr. 'lophos', crest; and Gr. '-phoros', -carrying; for the tuberculate plant body. (*Baynesia*)

loranthiflorus For the genus *Loranthus* (*Loranthaceae*); and Lat. '-florus', -flowered; for the similarly cylindrical corolla. (*Ceropegia*)

lorentzianum For Prof. Paul (Pablo) G. Lorentz (1835–1881), German botanist, 1869–

1874 in Córdoba (Argentina), later high-school teacher in Concepción (Uruguay). (*Lepismium*)

lorenzoi For Lorenzo Dotson-Smith (fl. 1978), without further data. (*Sedum*)

loreus Lat., made from straps, made from thongs; for the elongated branches with long internodes. (*Cephalophyllum*)

loricatus Lat., armoured; for the protection through the persistent peduncles. (*Euphorbia*)

lorifolius Lat. 'lorum', strap; and Lat. '-folius', -leaved. (*Monadenium pseudoracemosum* var.)

loristipula Lat. 'lorum', strap; and Lat. 'stipula', stipule; for the shape of the stipules. (*Jatropha*)

lossowianus For Dr. Otto von Lossow (1888–1947), German physician, alpinist and plant collector, born in Munich, emigrated to Namibia. (*Jensenobotrya*)

louisae For Louisa Hutchison (fl. 1934), mother of the US-American cactus collector Ted Hutchison, who discovered the taxon. (*Mammillaria hutchisoniana* ssp.)

louisianensis For the occurrence in the state of Louisiana, USA. (*Yucca*)

louwii For Dr. Wynand J. Louw (fl. 1975), botanist, herbarium curator and lecturer at the University of Potchefstroom, RSA. (*Euphorbia*) – (**2**) For Mr. Piet Louw (fl. 1980), farmer at Vanrhynsdorp, RSA. (*Bulbine*)

lowei For Richard T. Lowe (1802–1874), British clergyman and botanist, chaplain in Madeira 1832–1854. (*Monanthes*)

lozanoi For Filamon Lozano (fl. 1905), one of its Mexican discoverers. (*Echeveria*)

luapulanus For the occurrence in the Luapula Distr. in Zambia. (*Aloe, Euphorbia*)

lubangensis For the occrrence near Lubango, Angola. (*Kalanchoe*)

lubbersii For George E. K. Lubbers (fl. 1984), Johannesburg, discoverer of the taxon. (*Anacampseros subnuda* ssp.)

lucayanus For the occurrence near Lucaya, Grand Bahama Island. (*Opuntia*)

luchuanicus For the occurrence in the Lu-Chuan Distr., Yunnan, China. (*Sedum*)

luciae For Mlle. Lucy Dufour (fl. 1908), perhaps an acquaintace of the French botanist and physician Raymond Hamet, without further data. (*Kalanchoe*)

lucidus Lat., shining; for the ± shiny leaves. (*Plectranthus, Sedum*)

lucile-allorgeae For Dr. Lucile Allorge (fl. 2002), Madagascar-born French botanist at the National Museum in Paris, with many field trips to Madagascar, daughter of the French botanist Pierre Boiteau. (*Aloe*)

luckhoffii For Dr. James Lückhoff (fl. 1925–1931), physician in Cape Town, RSA, who accompanied Dr. R. Marloth on field trips. (*Aloinopsis, Antimima, Conophytum, Delosperma, Diplosoma, Gibbaeum*)

ludlowii For Frank Ludlow (1886–1972), British explorer of Bhutan and Tibet. (*Rhodiola*)

luederitzii For the occurrence near Lüderitz, Namibia. (*Crassula*)

luethyi For Dr. Jonas M. Lüthy (*1961), Swiss botanist and cactus enthusiast. (*Mammillaria*)

luetzelburgii For Dr. Philipp von Luetzelburg (1880–1948), German botanist and explorer, collecting repeatedly in Brazil. (*Stephanocereus*)

lugardiae For Charlotte E. Lugard (née Howard) (fl. 1897-98), English painter and wife of Major E. J. Lugard. (*Ceropegia, Monadenium*)

lugardii For Major E. J. Lugard (fl. 1897–1898), Englishman who discovered the taxon while representing Kew on an expedition. (*Hoodia currorii* ssp., *Orbea, Sesamothamnus*)

lukoseanus For the occurrence by the Lukose River, C Tanzania. (*Euphorbia*)

lumbricalis Lat., worm-shaped; for the worm-like procumbent branches. (*Euphorbia*)

lumbricoides Lat. 'lumbricus', earthworm; and Gr. '-oides', resembling; for the creeping terete stems. (*Lepismium*)

lumholtzii For Dr. Carl Lumholtz (fl. 1890), US-American archaeologist of Norwegian origin. (*Sedum*)

lunatus Lat., crescent-shaped; for the leaf shape. (*Leipoldtia, Oscularia*)

lungtsuanensis For the occurrence at Lungtsuan, Chekiang Prov., China. (*Sedum*)

luntii For William Lunt (1871–1904), British gardener at Kew, collecting in S Arabia in 1893. (*Aloe, Orbea*)

lunulatus Dim. of Lat. 'lunatus', crescent-shaped; for the leaf scars. (*Lampranthus*)

lupulinus Lat., hop-like; for the similarity of the inflorescence to that of Hop, *Humulus lupulus*. (*Euphorbia*)

lupulus Dim. of Lat. 'lupus', wolf; for the occurrence in the Wolfkloof, Western Cape, RSA. (*Haworthia herbacea* var.)

luribayensis For the occurrence in the valley of Luribay, Prov. Loayza, Dept. La Paz, Bolivia. (*Cleistocactus*)

luridus Lat., drab yellow, dirty brown; (**1**) for the colour of the leaves under water stress. (*Agave*) – (**2**) for the flower colour. (*Pelargonium, Pterodiscus*)

lutatus Lat., muddy, dirty (from Lat. 'lutum', loam'); for the leaf colour. (*Astridia*)

luteifluens Lat. 'luteus', yellow; and Lat. 'fluens', flowing; for the yellow latex. (*Cynanchum*)

luteolus Lat., pale yellow, yellowish; (**1**) for the flower colour. (*Malephora, Pelargonium, Sedum*) – (**2**) for the corona. (*Stapelia praetermissa* var.)

luteoruber Lat. 'luteus', yellow; and Lat. 'ruber, rubra, rubrum', red; perhaps for the petal colour. (*Sinocrassula indica* var.)

luteoviridis Lat. 'luteus', yellow; and Lat. 'viridis', green; (**1**) for the leaf colour. (*Gibbaeum*) – (**2**) for the flower colour. (*Sedum*)

lutescens Lat., becoming yellow; for the gradual change from scarlet buds to yellow open flowers. (*Aloe*)

luteus Lat., yellow; (**1**) for the bract colour. (*Euphorbia atropurpurea* fa.) – (**2**) for the flower colour. (*Aloe madecassa* var., *Crassula namaquensis* ssp., *Echeveria, Malephora, Opuntia, Orbea, Portulaca, Prenia pallens* ssp., *Rhinephyllum, Sedum surculosum* var.) – (**3**) presumably for the flower colour. (*Delosperma*)

lutzii For Monsieur L. Lutz (fl. 1913), French pharmacist and at the time general secretary of the Société Botanique de France. (*Sedum*)

lychnidiflorus For the genus *Lychnis* ("Campion", "Lampflower", *Caryophyllaceae*); and Lat. '-florus', -flowered. (*Pereskia*)

lycioides Gr. '-oides', resembling; and for the genus *Lycium* (*Solanaceae*). (*Nolana*)

lydenburgensis For the occurrence in the Lydenburg Distr., Mpumalanga, RSA. (*Delosperma, Euphorbia, Urginea*)

lydiae For Lydia Triebner (fl. 1948), wife of Wilhelm Triebner, Namibia. (*Conophytum*)

lydius For the occurrence in Lydia in W Turkey. (*Sedum*)

lynchii For Richard Irwin Lynch (1850–1924), English gardener and botanist, 1879–1919 curator of the Cambridge Botanical Garden. (×*Gasteraloe*)

lyratifolius From Lat. 'lyra', lyre; and Lat. '-folius', -leaved; for the leaf shape. (*Cleretum*)

M

maassii For Mr. W. Maass (fl. 1907), German cactus hobbyist at Zehlendorf near Berlin, 1907 secretary of the Deutsche Kakteen-Gesellschaft. (*Parodia*)

macbridei For James F. Macbride (1892–1976), US-American botanist working esp. on the flora of Peru. (*Portulaca*)

macdonaldiae For Mrs. General MacDonald (fl. 1850), who sent material from Honduras to Kew, without further details. (*Selenicereus*)

macdonaldii For J. Andrew McDonald (fl. 2002), US-American botanist at the University of Texas at Austin. (*Sedum*)

macdougalii For Prof. Dr. Daniel Trembly MacDougal (1865–1958), US-American botanist at the New York Botanical Garden. (*Fouquieria*, *Mammillaria heyderi* ssp., *Tumamoca*)

macdougallii For Thomas [Tom] MacDougall (1895–1973), Scottish-born plantsmen and explorer in S Mexico. (*Disocactus*, *Echeveria*, *Furcraea*, *Graptopetalum*, *Ortegocactus*, *Peniocereus*, *Sedum*)

macdowellii For Mr. J. A. MacDowell (fl. 1894), plant collector in Mexico. (*Thelocactus*)

macellus Dim. of Lat. 'macer, macra, macrum', lean; for the slender stems. (*Delosperma*, *Euphorbia*)

macer Lat. 'macer, macra, macrum', lean; for the slender stems. (*Aloe*)

macgillivrayi Probably for Paul H. MacGillivray (1834–1895), Scottish botanist, zoologist and medical practitioner, from 1855 in Australia. (*Hoya*)

Machairophyllum Gr. 'machaira', dagger; and Gr. 'phyllon', leaf; for the leaf shape of some taxa. (*Aizoaceae*)

machrisii For Mr. and Mrs. Maurice A. Machris (fl. 1956), who sponsored an expedition to Brazil in 1956, during which the taxon was discovered. (*Pilosocereus*)

machucae For José Antonio Machuca Núñez (fl. 1999), Mexican agronomist and plant collector, esp. in Jalisco, Mexico. (*Mammillaria*, *Pachyphytum*)

mackieanus For Mr. Mackie (fl. 1837), nurseryman in Lakenham, England. (*Gymnocalycium*)

macleayi For Prof. K. N. G. MacLeay (fl. 1955), botanist at Khartoum University, Sudan. (*Aloe*)

macloughlinii For Alfred G. McLoughlin (1886–1960), officer in the Department of Native Affairs in RSA and keen naturalist. (*Orbea*)

macowanianus For Prof. Peter MacOwan (1830–1909), English-born teacher and botanist in RSA and curator of the Cape Government Herbarium and the Cape Town Botanic Gardens. (*Crassula*)

macowanii As above. (*Euphorbia tuberculata* var., *Ruschia*, *Stapelia*)

macracanthus Gr. 'makros', large; and Gr. 'akanthos', thorn, spine; for the long spines. (*Consolea*, *Melocactus*)

macradenius Gr. 'makros', large; and Gr. 'aden', gland; for the large nectary glands. (*Peersia*)

macrandrus Gr. 'makros', large; and Gr. 'aner, andros', man, [botany] stamen. (*Peperomia*)

macranthus Gr. 'makros', large; and Gr. 'anthos', flower. (*Ceropegia*, *Cynanchum*, *Disocactus*, *Echeveria*, *Jatropha*)

macraxinus Gr. 'makros', large; and Gr. 'axine', ax, wedge, hatchet; for the occurrence in the Big Hatchet Mts., New Mexico, USA. (*Escobaria orcuttii* var.)

macroacanthus Gr. 'makros', large; and Gr. 'akantha', thorn, spine; for the large terminal leaf spine. (*Agave*)

macrocalyx Gr. 'makros', large; and Gr. 'kalyx', calyx. (*Drosanthemum*)

macrocarpus Gr. 'makros', large; and Gr. 'karpos', fruit. (*Agave papyrocarpa* ssp., *Aloe*, *Cyphostemma*, *Huernia*, *Jatropha*, *Lampranthus*, *Marah*, *Opuntia galapageia* var., *Pedilanthus*, *Rhodiola*, *Schreiteria*, *Tetragonia*, *Trochomeria*)

macrocentrus Gr. 'makros', large; and Gr. 'kentron', centre; for the conspicuous red colouring of the flower centre. (*Opuntia*)

macrocephalus Gr. 'makros', large; and Gr. 'kephalos', head; for the large terminal pseudocephalia. (*Neobuxbaumia*)

macrochele Gr. 'makros', large; and Gr. 'chele', claw; for the spination. (*Turbinicarpus schmiedickeanus* ssp.)

macrochlamys Gr. 'makros', large; and Gr. 'chlamys', cloak; for the conspicuous calyx. (*Kalanchoe*)

macrocladus Gr. 'makros', large; and Gr. 'klados', shoot; for the large size of the plants. (*Aloe*)

macrodiscus Gr. 'makros', large; and Gr. 'diskos', disc; for the large flat disc-like plant bodies. (*Ferocactus*)

macroglossus Gr. 'makros', large; and Gr. 'glossa', tongue; for the large ray florets. (*Senecio*)

macrogonus Gr. 'makros', large; and Gr. 'gonia', corner, margin; for the few and broad ribs. (*Echinopsis*)

macrolobus Gr. 'makros', large; and Lat. 'lobus', lobe; (**1**) for the long corolla segments. (*Rhytidocaulon*) − (**2**) possibly for the relatively large corona lobes. (*Cynanchum*)

macromeris Gr. 'makros', large; and Gr. 'meros', part; perhaps for the large perianth segments. (*Coryphantha*)

macropetalus Gr. 'makros', large; and Gr. 'petalon', petal. (*Brachystelma*)

macrophyllus Gr. 'makros', large; and Gr. 'phyllon', leaf. (*Hoya, Jatropha*)

macropodus Gr. 'makros', large; and Gr. 'pous, podos', foot; for the large tubers. (*Sinningia*)

macropterus Gr. 'makros', large; and Gr. 'pteron', wing; for the winged fruits. (*Tetragonia*)

macrorhizus Gr. 'makros', large; and Gr. 'rhiza', root; (**1**) for the swollen roots. (*Jatropha, Opuntia, Portulaca*) − (**2**) for the large tuber. (*Gerrardanthus, Peperomia*)

macrorrhizus Gr. 'makros', large; and Gr. '[r]rhiza', root; for the swollen main root. (*Crassula corallina* ssp.)

macrosepalus Gr. 'makros', large; and Gr. 'sepalon', sepal. (*Lampranthus, Sedum daigremontianum* var., *Sedum susannae* var.)

macrosiphon Gr. 'makros', large; and Gr. 'siphon', tube; for the large flowers. (*Aloe*)

macrospermus Gr. 'makros', large; and Gr. 'sperma', seed. (*Portulaca*)

macrostachya Gr. 'makros', large; and Gr. 'stachys', spike; probably for the dense inflorescence. (*Sinningia*)

macrostigma Gr. 'makros', large; and Gr. 'stigma', stigma, (*Delosperma, Lampranthus*)

maculatus Lat., spotted; (**1**) for the marbled epidermis of the stems. (*Drosanthemum, Peniocereus*) − (**2**) for the spotted leaves. (*Adromischus, Aloe, Haworthia*) − (**3**) for the spotted corolla. (*Brachystelma, Orbea, Ornithogalum*) − (**4**) for the spotted annulus of the flowers. (*Duvalia*) − (**5**) for the spots caused by insect pests on the leaves of the cultivated type material. (*Echeveria paniculata* var.)

maculosus Lat., spotted, mottled; for the mottled leaves. (*Agave, Sinocrassula indica* var.)

macvaughianus For Prof. Dr. Rogers McVaugh (*1909), US-American botanist and specialist in the Mexican flora. (*Matelea*)

madagascariensis For the occurrence in Madagascar. (*Adansonia, Basella, Ceropegia, Cynanchum, Decarya, Didierea, Perrierosedum, Phylohydrax, Plectranthus, Senecio longiflorus* ssp., *Stapelianthus*)

Madangia For the occurrence in Madang Prov., Papua New Guinea. (*Asclepiadaceae*)

madecassus Probably an attempt to latinise a local variant of the name Madagascar. (*Aloe*)

maderensis For the occurrence in the Sierra de la Madera, Coahuila, Mexico. (*Agave asperrima* ssp.)

madiensis For the occurrence in the region inhabited by the ethnic group of the Madi in Uganda. (*Raphionacme*)

madisoniorum For Marshall P. Madison (*1895), US-American lawyer, and his wife Elena Eyre Madison, of San Francisco and Atherton, California, USA, and for the Madison Fund of San Francisco and its support of the Botanical Garden at the University of California at Berkeley. (*Matucana*)

madrensis For the occurrence in the Mexican Sierra Madre Occidental. (*Sedum*, *Yucca*)

maduraiensis For the occurrence at Madurai, Tamil Nadu, India. (*Sansevieria*)

maechlerorum Lat. Gen. Pl.; for Wendelin Mächler (fl. 2002), Swiss cactus hobbyist in Pfungen near Zürich, and his son Wendelin Mächler († c. 1997). (*Eriosyce aspillagae* ssp.)

Maerua Unknown. (*Capparidaceae*)

mafekingensis For the occurrence near Mafeking, capital of the North-West Prov., RSA. (*Brachystelma*)

mafingensis For the occurrence on the Mafinga Hills, Malawi. (*Monadenium*)

magae For "Madame Mag F. (fl. 1914), "en témoignage de très affectueuse tendresse", on the part of the French botanist Raymond Hamet. (*Sedum*)

magallanii For Pedro Magallan (fl. 1945), probably a collector for the Mexican plant trader F. Schmoll. (*Mammillaria*)

magellensis For the occurrence near Mt. Majella in the Abruzzi Mts. between Prov. Pescara and Prov. Chieti, C Italy. (*Sedum*)

magenteum Lat., magenta-red; for the flower colour. (*Pelargonium*)

magnicapsula Lat. 'magnus', large; and Lat. 'capsula', capsule; for the large fruits. (*Euphorbia*)

magnificus Lat., magnificient; (**1**) for the stately appearance of the plants. (*Lampranthus*, *Mammillaria*, *Parodia*, *Pilosocereus*) – (**2**) for the beautiful rosettes. (*Haworthia*) – (**3**) for the magnificient inflorescences. (*Monadenium*) – (**4**) for the large flowers. (*Hoya*) – (**5**) for the bright red flowers. (*Sinningia*)

magniflorus Lat. 'magnus', large; and Lat. '-florus', -flowered. (*Huernia zebrina* ssp., *Sedum*)

magnimammus Lat. 'magnus', large; and Lat. 'mamma', breast, teat; for the very large tubercles of the plant bodies. (*Discocactus heptacanthus* ssp., *Mammillaria*)

magnus Lat., large; (**1**) for the large rosettes. (*Agave shrevei* ssp.) – (**2**) for the large body. (*Arthrocereus melanurus* ssp.)

maguirei For Dr. Bassett Maguire (1904–1991), US-American botanist at the New York Botanical Garden. (*Lewisia*)

mahabalei For Tryambak S. Mahabalé (1909–1983), Indian botanist. (*Ceropegia*)

mahabobokensis For the village of Mahaboboka between Toliara and Fianarantsoa, Madagascar, where the taxon was first found. (*Euphorbia*)

mahafalensis For the occurrence in the Mahafaly region, S Madagascar. (*Cynanchum*, *Euphorbia*, *Jatropha*)

maheshwarii For Prof. Panchanan Maheshwari (1904–1966), Indian botanist. (*Jatropha*)

mahonii For a Mr. J. Mahon (fl. 1900), possibly for John Mahon (1870–1906), Irish gardener at RBG Kew 1891–1897, forester in Malawi 1897–1899, later curator of Entebbe Botanic Garden, Unganda, until 1903, died of sleeping sickness. (*Delosperma*)

mahraensis For the occurrence in Al-Mahra Prov., Yemen. (*Aloe*)

Maihuenia From the local vernacular name "Maihuén" in the Mapuche language of S Chile (*Cactaceae*)

Maihueniopsis Gr. '-opsis', similar to; and for the genus *Maihuenia* (*Cactaceae*). (*Cactaceae*)

mainiae For Mrs. F. M. Main (fl. 1900), who first collected the taxon. (*Mammillaria*)

mainty From the local vernacular name "Famata Mainty" of the plants in Madagascar. (*Euphorbia*)

maireanus For Prof. Dr. René C. J. E. Maire (1878–1949), French botanist, from 1911 professor of botany at the University of Alger, Algeria. (*Sedum*)

mairei For E. E. Maire (fl. 1912), who collected the type. (*Ceropegia*)

maiusculus Lat., a little bit larger. (*Ceropegia*)

maiz-tablasensis For the occurrence between Ciudad Maiz and Las Tablas, San Luis Potosí, Mexico. (*Coryphantha*)

major Comp. of Lat. 'magnus', great, i.e. greater; (**1**) for the large tuber. (*Squamellaria*) – (**2**) for the larger size. (*Aloe boylei* ssp., *Cylindropuntia acanthocarpa* var., *Dischidia*, *Echeveria bella* fa., *Euphorbia horrida* var., *Haworthia emelyae* var., *Hawor-

thia floribunda var., *Sedum, Turbinicarpus viereckii* ssp.)

majus Comp. of Lat. 'magnus', great, i.e. greater; for the larger size. (*Oscularia, Sphyrospermum*)

makallensis For the occurrence in the Makalle Distr., Tigre, Ethiopia. (*Euphorbia*)

makinoi For Tomitaro Makino (1862–1957), Japanese botanist and botanical artist. (*Sedum*)

makurupiniensis For the occurrence near the Makurupini River, Zimbabwe / Moçambique. (*Aloe ballii* var.)

malacophyllus Gr. 'malakos', soft; and Gr. 'phyllon', leaf; (**1**) for the softly fleshy leaves. (*Orostachys*) – (**2**) for the softly hairy leaf-surface. (*Jatropha*)

malangeanus For the occurrence in the Malanje area, Angola. (*Adenia*)

maleolens Lat. 'malus', bad; and Lat. 'olens', smelling; (**1**) for the unpleasant fishy odour of the whole plant. (*Antimima*) – (**2**) for the unpleasant odour of the latex and the cyathial gland secretions. (*Euphorbia*)

Malephora Gr. 'male', armpit; and Gr. '-phoros', carrying; for the leaf sheaths, through which the stem continues. (*Aizoaceae*)

malevolus Lat., envious, malevolent; presumably for the sharp spines. (*Euphorbia*)

malherbei For Mr. M. Malherbe (fl. 1935), without further data. (*Aloinopsis*)

maliterrarum Lat. 'malus', bad; and Gen. Pl. of Lat. 'terra', land; for the occurrence in badland formations. (*Coryphantha*)

mallei Lat. 'malleus', hammer; for Steven A. Hammer (*1951), US-American pianist, horticulturist and Mesemb specialist, and monographer of the genus *Conophytum*. (*Avonia*)

malus Lat., apple; for the apple-shaped urceolate flowers. (*Echidnopsis*)

malvifolius For the genus *Malva* ("Mallow", *Malvaceae*); and Lat. '-folius', -leaved. (*Pyrenacantha*)

malvinus Lat., mauve; for the flower colour. (*Plectranthus*)

mamfwensis For the occurrence near Mamfwe, Shaba Prov., Zaïre. (*Monadenium*)

mamillatus Lat., with tubercles; for the stem segments. (*Cylindropuntia fulgida* var.)

mamillosus Lat., with tubercles; for the plant bodies. (*Echinopsis*)

mammifer Lat. 'mam[m]a', nipple; and Lat. '-fer, -fera, -ferum', -carrying; for the plant body whose ribs are dissolved into tubercles. (*Frailea*)

Mammillaria Lat. 'mam[m]illa', nipple; for the tuberculate plant bodies. (*Cactaceae*)

mammillaris From Lat. 'mam[m]illa', nipple; (**1**) for the stem tubercles. (*Euphorbia, Mammillaria, Quaqua*) – (**2**) for the nipple-like leaves. (*Adromischus*)

Mammilloydia For the assumed relationship of the species, showing characteristics both of *Mammillaria* and *Neolloydia*. (*Cactaceae*)

mammulosus Dim. of Lat. 'mamma', breast, tubercle; i.e. full of small tubercles; for the tuberculate plant body. (*Parodia*)

manaia For the occurrence near Manaia, Egmont Coast, New Zealand. (*Crassula*)

Mandevilla For H. John Mandeville (1773–1861), English diplomat in Argentina. (*Apocynaceae*)

mandragora Lat. 'mandragoras', mandrake (*Mandragora officinalis, Solanaceae*); for the similarly tuberous roots. (*Turbinicarpus*)

mandraliscae For Count Enrico Mandralisca (fl. 1878), Italian nobleman, benefactor and art collector in Sicily. (*Senecio talinoides* ssp.)

mandrarensis For the occurrence in the basin of the Mandraré River, Madagascar. (*Kalanchoe*)

mangelsdorffii For Ralph D. Mangelsdorff (fl. 1998), Frankfurt am Main, Germany. (*Euphorbia*)

manginii For Louis Mangin (fl. 1912), professor at the Natural History Museum of Paris. (*Kalanchoe*)

mangokyensis For the occurrence in the Mangoky Basin, Madagascar. (*Euphorbia*)

mangulensis For the occurrence near Mangula, Northern Prov., Zimbabwe. (*Euphorbia gossypina* ssp.)

maniaensis For the occurrence near the Mania River, Madagascar. (*Aloe laeta* var.)

maninus Perhaps Dim. of Lat. 'manus', hand; perhaps for the hand-like appearance of the clusters of erect leaves. (*Chasmatophyllum*)

mapimiensis For the occurrence in the Bolsón de Mapimi, Coahuila, Mexico. (*Echinocereus*)

mappia From the vernacular name "Mappou" for the plants on Mauritius. (*Cyphostemma*)

Marah Perhaps from Lat. 'amarus', bitter; for the bitter taste of the roots. (*Cucurbitaceae*)

maraisii For W. R. B. Marais (fl. 1930), discoverer of the taxon, without further data. (*Haworthia*)

marcanoi For Prof. E. J. Marcano (fl. 1992), University of Santo Domingo and discoverer of the taxon. (*Pereskia*)

marcescens Lat., marcescent, withering but not falling; for the inflorescences, which dry and persist during summer. (*Dudleya cymosa* ssp.)

marchesii For. Prof. Eduardo Marchesi (fl. 1968), Montevideo, Uruguay, who discovered the taxon. (*Parodia scopa* ssp.)

marcidulus Dim. of Lat. 'marcidus', withered, weak; for the flaccid branches. (*Lampranthus*)

marcosii For Marcos Sierra (*1978?), Mexican cactus grower at CANTE, San Miguel de Allende, Guanajuato, and discoverer of the taxon. (*Mammillaria*)

marenae For Maren B. Parsons (fl. 1910), who first found the taxon. (*Grusonia*)

margaretae For Miss Margaret Johnson (*1946), English botanical cytologist at Kew. (*Euphorbia*) – (**2**) For Margarethe Friedrich (fl. 1919), German teacher in Warmbad, Namibia, and eager collector of succulent plants, often in company with the German botanist Kurt Dinter. (*Lapidaria*)

margarethae For Mrs. Margaretha Wiese (*1923), succulent plant grower and wife of Buys Wiese, owner of the farm Quaggaskop, RSA. (*Bulbine*)

margaritaceus Lat., pearl-like; for the pale pinkish (pearl-coloured) fruits. (*Melocactus violaceus* ssp.)

margaritae For the occurrence on Isla Santa Margarita off the Baja California coast, Mexico. (*Agave*)

margaritifer Lat. 'margarita', pearl; and Lat. '-fer, -fera, -ferum', -carrying; for the leaves covered white warts like pearls. (*Nananthus*)

marginalis Lat., marginal; for the pronounced red leaf margins. (*Crassula pellucida* ssp.)

marginatus Lat., marginate; (**1**) for the dryish leaf margins. (*Talinum*) – (**2**) for the thickish leaf margins. (*Haworthia*, *Rhodiola*) – (**3**) for the white membranous leaf margins. (*Graptopetalum*) – (**4**) for the readdish leaf margins. (*Conophytum*) – (**5**) for the "margin" on the ribs formed by the confluent areoles. (*Copiapoa*, *Pachycereus*) – (**6**) for the deeper colour of the petal margins. (*Lampranthus*)

mariae For Marie E. Galpin (née de Jongh) († 1933), wife of the South African amateur botanist and banker Ernest E. Galpin. (*Delosperma*) – (**2**) For Mary G. Lavis (later Mrs. O'Connor-Fenton) (*1903), South African herbarium assistant at the Bolus Herbarium, Cape Town, RSA. (*Lampranthus*) – (**3**) For Mary Bellerue-Bleck (1933–1999), US-American horticulturist and succulent plant specialist, for some time co-owner of Abbey Garden Nursery, and 1983–1990 curator of the succulent plant collection of the Johannesburg City Botanical Garden. (*Echidnopsis*) – (**4**) For Mrs. M. (Mary?) Villet (fl. 1937), without further data. (*Ruschia*)

marianae For Marian Marloth (fl. 1923), wife of the German botanist and pharmacist Rudolf Marloth. (*Ruschia*)

marianiae For Mrs. Mariane Crossman (fl. 1900), British botanical artist who visited RSA for painting. (*Adromischus*)

marianus For Mary Ann Gentry (fl. 1942), wife of the US-American botanist and *Agave* specialist Howard S. Gentry. (*Peniocereus*)

mariensis For the occurrence at Cap Ste. Marie, S Madagascar. (*Cynanchum*)

marientalensis For the occurrence near Mariental, S Namibia. (*Tridentea*)

marinus Lat., of the sea; for the occurrence near the coast. (*Drosanthemum*)

mariposensis For the occurrence near Mariposa, Brewster County, Texas, USA. (*Echinomastus*)

maritae For Marita Specks (fl. 1996), wife of the German succulent plant collector and nurserymen Ernst Specks. (*Brachystelma, Euphorbia*)

maritimus Lat., maritime; for the preferred habitat. (*Cistanthe, Echinocereus, Hydrophylax, Mammillaria, Quaqua, Sesuvium, Tetragonia*)

markgrafii For Prof. Dr. Friedrich Markgraf (1897–1987), German botanist in Berlin, 1945–1957 at the Botanical Garden München, and 1958 onwards director of the Botanical Garden Zürich. (*Brasilicereus*)

marksianus For Herman Marks (fl. 1960), US-American cactus hobbyist in Salinas, California, who supported the travels of F. Ritter. (*Eriosyce*)

marksianus For Mr. H. Marks (fl. 1946), travel companion in Mexico of the discoverer of the taxon Fritz Schwarz. (*Mammillaria*)

marlothianus For Prof. Dr. Hermann Wilhelm Rudolf Marloth (1855–1931), celebrated German botanist, analytical chemist and pharmacist, living in South Africa from 1883, professor of chemistry at Stellenbosch University 1889–1892. (*Euphorbia*)

marlothii As above. (*Adromischus filicaulis* ssp., *Aloe, Anacampseros, Brownanthus, Conophytum jucundum* ssp., *Gibbaeum, Larryleachia, Monsonia, Odontophorus, Schwantesia, Trichodiadema*)

Marlothistella For Prof. Dr. Hermann Wilhelm Rudolf Marloth (1855–1931), celebrated German botanist, analytical chemist and pharmacist, living in South Africa from 1883, professor of chemistry at Stellenbosch University 1889–1892; and Lat. 'stella', star; for the sar-shaped fruit. (*Aizoaceae*)

marmoratus Lat., marbled; for the leaf markings. (*Agave, Aloe somaliensis* var., *Kalanchoe, Lithops*)

marmoreus Lat., marbled, irregularly striped or veined; application obscure. (*Sempervivum*)

marnieri For Julien Marnier-Lapostolle (1902 –1976), French connoisseur and collector of succulents, owner of the Grand Marnier company and founder of the private botanical garden "Les Cèdres". (*Rosularia alpestris* ssp., *Senecio, Seyrigia*)

marnierianus As above. (*Alluaudiopsis, Conophytum, Crassula rupestris* ssp., *Cynanchum, Huernia, Kalanchoe*)

marrubatus From Lat. 'marrubium', "Horehound" (*Marrubium vulgare, Lamiaceae*); for the similarity of the plants. (*Plectranthus*)

marsabitensis For the occurrence at Marsabit, North-East Distr., Kenya. (*Euphorbia, Monadenium ritchiei* ssp.)

Marsdenia For William Marsden (1754–1836), Irish orientalist and traveller in Sumatra, secretary to the British Admirality. (*Asclepiadaceae*)

marsoneri For Oreste Marsoner (fl. 1932), Argentinian cactus collector. (*Echinopsis, Gymnocalycium, Rebutia*)

martianus For Dr. Carl F. P. von Martius (1794–1868), German botanical traveller, ethnologist and botanist in München, travelling in Brazil 1817–1820, founder of the "Flora Brasiliensis". (*Disocactus*)

martinae For Mrs. Martine Bardot-Vaucoulon (*1948), French school teacher and botanist working in Madagascar. (*Euphorbia*)

martinezii For Prof. Maximo (Maximiliano) Martínez (1888–1964), Mexican botanist, author of books on Mexican plants, and founder of the Mexican botanical society. (*Stenocereus*) – (**2**) For Esteban Martínez Salas (fl. 1984, 2000), Mexican botanical collector, esp. in the Río Balsas area. (*Jatropha*)

martinianus For Dr. William P. Martin (fl. 1940), field companion of the US-American botanist L. D. Benson in Arizona. (*Opuntia*)

martinii For Raymond Martin (fl. 1854), cactus hobbyist in Toulouse, France, and owner of a large cactus collection (*Harrisia*)

martleyi For J. F. Martley (fl. 1932), without further data. (*Lampranthus*)

marumianus For Dr. Martinus van Marum (1750–1837), Dutch botanist. (*Haworthia*)

marylanae For Miss Marylan Coelho (fl.

2003), Brazilian biology student from Bahia. (*Arrojadoa*)
masaicus For the occurrence on the territory of the Mas[s]ai People. (*Aspidoglossum*)
mashonicus For the occurrence in Mashonaland, Zimbabwe. (*Euphorbia griseola* ssp.)
masirahensis For the occurrence on Masirah Island off the coast of Oman. (*Euphorbia*)
masonianus For Maurice L. Mason (1912–1993), British farmer and succulent plant enthusiast near Norfolk in East Anglia. (*Sansevieria*)
masonii For Dr. Herbert L. Mason (1896–1994), US-American botanist at the University of California, Berkeley. (*Portulaca*)
massaicus For the occurrence in the area formerly known as Masailand, Kenya. (*Portulaca*)
massawanus For the presumed occurrence at Massawa [Mits'iwa], Eritrea. (*Aloe*)
Massonia For Francis Masson (1741–1805), British horticulturist collecting esp. in S Africa. (*Hyacinthaceae*)
massonii As above. (*Crassula alpestris* ssp.)
matabelensis For the occurrence in Matabeleland, SW Zimbabwe. (*Euphorbia*)
mataikona For the occurrence at Mataikona, a locality on the E coast of the North Island of New Zealand. (*Crassula*)
matanzanus For the occurrence near Matanzas, Cuba. (*Melocactus*)
matapensis For the occurrence near Matapé, Sonora, Mexico. (*Agave shrevei* ssp., *Nolina*)
mataranensis For the occurrence near Matarani, Prov. Tarata, Dept. Cochabamba, Bolivia. (*Echinopsis*)
mataranus For the occurrence near Matara near San Marcos, Prov. Cajamarca, Peru. (*Armatocereus*)
matehualensis For the occurrence near Matehuala, San Luis Potosí, Mexico. (*Neolloydia*)
Matelea Unknown. (*Asclepiadaceae*)
mathewsii For Mr. Joseph William Mathews (1871–1949), British horticulturist, living in RSA from 1895, curator of the National Botanic Garden Kirstenbosch 1913–1936. (*Drosanthemum*)

mathildae For Mrs. Mathilde Wagner (fl. 1968), wife of Willi Wagner, Cadereyta de Montes, Mexico, who 1951 took over the Finca of his uncle F. Schmoll and developed it into a cactus nursery. (*Mammillaria*)
matoensis For the occurrence in the state of Mato Grosso do Sul, Brazil. (*Gymnocalycium marsoneri* ssp.)
Matucana For the occurrence near the town of Matucana, Prov. Lima, Peru. (*Cactaceae*)
matucanensis As above. (*Armatocereus*)
matudae For Prof. Dr. Eizi Matuda (1894–1978), Mexican botanist of Japanese origin. (*Mammillaria*, *Thelocactus tulensis* ssp.)
maturus Lat., ripe, mature; for the early flowering season in comparison with other taxa. (*Lampranthus*)
matutinus Lat., of the morning; for the morning flowering. (*Lampranthus*)
matznetteri For Josef Matznetter (1870–1956), co-founder of the Austrian Cactus Society and owner of a cactus nursery in Vienna, Austria. (*Gymnocalycium andreae* ssp.)
maughanii For Dr. Herbert Maughan Brown (fl. late 1920s, 1930s), physician and plant collector in RSA. (*Conophytum*, *Dorotheanthus*, *Haworthia truncata* var., *Pectinaria*)
mauritanicus For the presumed occurrence in 'Mauritania' (= NW Africa). (*Euphorbia*)
mauritiensis For the occurrence on Mauritius. (*Portulaca*, *Rhipsalis baccifera* ssp.)
maurus Lat., African, Mauric; for the occurrence in N Africa. (*Caralluma burchardii* ssp., *Sedum*)
mawii For Capt. A. H. Maw (fl. 1940), owner of the property in Malawi where the type was collected. (*Aloe*)
maximilianus For Maximilian ("Max") Schlechter (1874–1960), German trader and plant collector, lived in Namibia and RSA from 1896, brother of the German botanist Rudolf Schlechter. (*Braunsia*, *Schlechteranthus*) – (**2**) For a friend of Geheimrat Heyder named Maximilian and "deceased too early" (before 1846), without further details. (*Echinopsis*)
maximus Lat., the greatest; for the size of the plants. (*Adromischus*, *Haworthia*, *Hylotelephium telephium* ssp., *Ibervillea*, *Ruschia*)

maxonii For Dr. William R. Maxon (1877–1948), US-American botanist. (*Echeveria, Leptocereus*)

maxwellii For Maxwell Bolus (1890–1956), South African farmer, nephew of Dr. H. M. L. Bolus. (*Antimima, Delosperma*)

mayottensis For the occurrence on Mayotte Island, Comoros. (*Aloe*)

Maysara For Harry Mays (fl. 1999), English succulent plant enthusiast and at the time editor of the journal "Haworthiad"; plus the suffix '-ara', indicating plurigeneric hybrids. (*Aloaceae*)

mayuranathanii For P. V. Mayuranathan (fl. 1921–1940), botanist at the Government Museum, Madras, India. (*Euphorbia*)

mazapilensis For the occurrence near Mazapil, Zacatecas, Mexico. (*Echinocereus parkeri* ssp.)

mazatlanensis For the occurrence near Mazatlán, Sinaloa, Mexico. (*Mammillaria*)

mazelianus For Mr. B. Mazel (fl. 1981), Czech engineer and cactus hobbyist. (*Melocactus*)

mbisiensis For the occurrence near Mbisi [Mbizi], Tanzania. (*Glossostelma*)

mccartenii For Niall McCarten (fl. 1972), who first collected the taxon. (*Mammillaria rhodantha* ssp.)

mccoyi For Tom A. McCoy (*1959), US-American consultant and botanical collector, resident in Saudi Arabia. (*Aloe, Huernia, Kleinia, Pseudolithos, Rhytidocaulon*)

mcgillii For a Mr. McGill (fl. 2001), without further data. (*Cylindropuntia alcahes* var.)

mckelveyanus For Susan A. McKelvey (née Delano) (1888–1965), US-American botanist, horticulturist and botanical historian, working with *Agavaceae* and Lilacs. (*Agave*)

mcloughlinii For Major Alfred G. McLoughlin (1886–1960), South African lawyer, collected in NE Africa during military service. (*Aloe*)

mcmurtryi For Douglas McMurtry (fl. 1984), curator of the Municipal Botanic Garden Emmarentia, Johannesburg, RSA. (*Haworthia koelmaniorum* var.)

mcvaughii For Prof. Dr. Rogers McVaugh (*1909), US-American botanist and specialist in the Mexican flora. (*Euphorbia, Jatropha*)

Medinilla For José de Medinilla y Piñeda, Spanish governor of the Marianes in the early 19. century. (*Melastomataceae*)

medishianus For the occurrence at Medishe, Somalia. (*Aloe*)

medius Lat., middle; (1) for the distribution between related taxa. (*Thorncroftia*) – (2) perhaps for the intermediate position between related species. (*Drimia, Khadia*) – (3) perhaps for the medium-sized corolla. (*Ceropegia*)

medley-woodii For Dr. John Medley Wood (1827–1915), British botanist and director of the Botanical Garden in what was then Natal, RSA. (*Senecio*)

meenae For Mrs. Meena Singh (fl. 2002), Indian *Euphorbia* enthusiast and succulent plant collector. (*Euphorbia*)

megacalyx Gr. 'megas, megale', large; and Gr. 'kalyx', calyx; for the prominent calyx. (*Echeveria*)

megacanthus Gr. 'megas, megale', large; and Gr. 'akanthos', spine, thorn. (*Opuntia*)

megalacanthoides Gr. '-oides', resembling; and for the similarity to *Aloe megalacantha* (*Aloaceae*). (*Aloe gilbertii* ssp.)

megalacanthus Gr. 'megas, megale', large; and Gr. 'akantha', thorn, spine; for the large teeth on the leaf margins. (*Aloe*)

megalanthus Gr. 'megas, megale', large; and Gr. 'anthos', flower. (*Marsdenia, Selenicereus*)

megalocarpus Gr. 'megas, megale', large; and Gr. 'karpos', fruit. (*Aloe*)

megalorrhizus Gr. 'megas, megale', large; and Gr. '[r]rhiza', root; for the succulent rhizomes. (*Oxalis*)

megalothelos Gr. 'megas, megale', large; and Gr. 'thele', tubercle; for the large tubercles of the plant bodies. (*Gymnocalycium*)

megapotamicus Gr. 'megas potamos', big river; for the occurrence in the region of the Río de la Plata, Uruguay. (*Opuntia*)

megarrhizus Gr. 'megas, megale', large; and Gr. '[r]rhiza', root; for the large taproot. (*Copiapoa, Opuntia*)

megasepalus Gr. 'megas, megale', large; and Gr. 'sepalon', sepal. (*Brachystelma*)

megaspermus Gr. 'megas, megale', large; and Gr. 'sperma', seed. (*Opuntia, Trianthema*)

megliolii For Dr. Silvio Megliolii (fl. 1971), Argentinian cactus collector and mountain climber. (*Pterocactus*)

meiacanthus Gr. 'meios', smaller; and Gr. 'akantha', thorn, spine; for the smaller spines in comparison with some other taxon. (*Mammillaria heyderi* ssp.)

meintjesii For C. C. C. Meintjes (fl. early 1960s), South African architect interested in succulents, worked in Aden at the time and facilitated the first visit of John Lavranos to the Yemen. (×*Staparesia*, *Stapelia*)

meiringii For the occurrence at Meiringspoort in the Little Karoo, Eastern Cape, RSA. (*Bulbine*) – (**2**) For P. L. Meiring (fl. 1976), property speculator and gardener with the Hurling and Neil Nursery, Bonnievale, Western Cape, RSA. (*Haworthia maraisii* var.)

melaleucus Gr. 'melas, melano-', black; and Gr. 'leukos', white; for the colour of the spination. (*Mammillaria*)

melanacanthus Gr. 'melas, melano-', black; and Gr. 'akantha', spine, prickle. (*Aloe, Myrmecodia*)

melanantherus Gr. 'melas, melano-', black; and Gr. 'anthera', anther. (*Sedum*)

melananthus Gr. 'melas, melano-', black; and Gr. 'anthos', flower. (*Orbea*)

melanocentrus Gr. 'melas, melano-', black; and Gr. 'kentron', centre; for the black central spines. (*Mammillaria*)

melanohydratus Gr. 'melas, melano-', black; and Gr. 'hydros', water; for its occurrence at the locality Swartwater (= Black Water), Northern Cape, RSA. (*Euphorbia*)

melanospermus Gr. 'melas, melano-', black; and Gr. 'sperma', seed. (*Phyllobolus*)

melanotrichus Gr. 'melas, melano-', black; and Gr. 'thrix, trichos', hair; for the black-felted areoles and the black bristles of the flower tube. (*Corryocactus*)

melanovaginatus Gr. 'melas, melano-', black; and Lat. 'vaginatus', provided with a sheath; for the sheathing remains of old leaves. (*Bulbine*)

melanurus Gr. 'melas, melano-', black; and Gr. 'oura', tail; for the black appearance of the original plants. (*Arthrocereus*)

melastomifolius Lat. '-folius', -leaved; and for the similarity of the leaves to those of species of the genus *Melastoma* (*Melastomataceae*). (*Senecio*)

meleagris Lat., guinea-fowl; (**1**) for the dotted leaves. (*Hammeria*) – (**2**) for the purple blotched corolla tube. (*Ceropegia*)

melifluus Lat. 'mel', honey; and from Lat. 'fluere', to flow; for the abundant nectar. (*Hoya*)

melitae For Melita Horst (*1917?), wife of the Brazilian plant collector Leopoldo Horst, who cared for the collected plants when her husband was travelling. (*Frailea cataphracta* ssp.)

mellei For Henry A. Melle (1893–1957), South African agronomist with a farm at Zwartkop, near Pretoria, RSA. (*Carpobrotus*)

melleospinus Lat. 'melleus', sweet like honey; and Lat. '-spinus', -spined; for the spine colour. (*Coryphantha*)

mellitulus Dim. of Lat. 'mellitus', sweet like honey; perhaps for the flower scent ? (*Sedum*)

Melocactus Gr. 'melon', apple, melon; and Gr. 'kaktos', cactus; for the size and shape of the plant body. (*Cactaceae*)

meloformis Lat. 'melo', an apple-like melon (from Gr. 'melon', apple); and Lat. '-formis', -shaped; for the body shape. (*Euphorbia*)

membranaceus Lat., membrane-like; for the thin texture of the corolla lobes. (*Sarcostemma*)

memoralis Lat. 'memorare', remember; in memory of H. Basil Christian (1871–1950), South African, emigrated to Zimbabwe in 1911 and established a large private garden in 1914, which is now the Ewanrigg National Park. (*Euphorbia*)

menachensis For the occurrence at Menacha, Yemen. (*Aloe*)

menarandrensis For the occurrence in the

valley of the Menarandra River, S Madagascar. (*Cynanchum*)

mendelianus For Mr. E. Mendel (fl. 1931), US-American cactophile. (*Mammillaria hahniana* ssp.)

mendesii For Dr. Eduardo J. S. M. Mendes (*1924), Portuguese botanist, collecting 1955–1956 in Angola, later director of the Lisboa herbarium. (*Aloe, Ceropegia, Erythrina*)

mendozae For Mario Mendoza Garcia (fl. 1997), Mexican succulent plant enthusiast and propagator. (*Graptopetalum*)

menniei For Mr. A. Mennie (fl. 1937), who collected the type specimen. (*Antimima*)

mentiens Lat., deceiving, lying; for the similarity to another taxon. (*Phiambolia*)

mentosus Lat. 'mentum', chin; and Lat. '-osus', full of; for the chin-like tubercles of the plant body. (*Sulcorebutia*)

menyharthii For László Menyhárth (1849–1897), Austro-Hungarian missionary and botanist, collected in the Zambesi region. (*Aloe*)

meonacanthus Gr. 'meion', smaller; and Gr. 'akantha', thorn, spine; for the relatively short spines. (*Parodia*)

mercadensis For the occurrence on the Cerro Mercado, C Durango, Mexico. (*Mammillaria*)

meredithii For York Meredith (fl. 1988), Australian plant enthusiast who discovered the taxon, without further data. (*Hoya*)

meridionalis Lat., southern; (**1**) for the distribution. (*Pachypodium rutenbergianum* var.) – (**2**) for the occurrence in S Kenya and adjacent Tanzania S of the equator. (*Euphorbia*)

mesae-verdae For the occurrence near Mesa Verde, Colorado, USA. (*Sclerocactus*)

mesembrianthemopsis Gr. '-opsis', similar to; and for *Mesembryanthemum calcareum* (then spelled *Mesembrianthemum*, now placed in *Titanopsis*; Aizoaceae). (*Crassula*)

mesembryanthemoides Gr. '-oides', resembling; and for the formerly all-embracing genus *Mesembryanthemum* (Aizoaceae); (**1**) for the general similarity. (*Senecio, Sesuvium*) – (**2**) for the short stem segments resembling the leaves of some Aizoaceae. (*Rhipsalis*)

Mesembryanthemum Gr. 'mesembria', noon, south; and Gr. 'anthemon', flower; because the vast majority of the taxa here classified in the course of time are native to South Africa, but perhaps also because most open their flowers around noon; alternatively but unlikely from Gr. 'mesos', middle; Gr. 'embryon', embryo; and Gr. 'anthemon'; for the free axile placentation of the ovules. (*Aizoaceae*)

mesembryanthoides Gr. '-oides', resembling; and for the formerly all-embracing genus *Mesembryanthemum* (Aizoaceae). (*Bulbine, Crassula*)

mesklipensis For the occurrence at Mesklip, Northern Cape, RSA. (*Antimima*)

mesophyticus Gr. 'mesos', middle; and Gr. 'phyton', plant; perhaps for the occurrence at places with an "intermediate" climate. (*Opuntia megasperma* var.)

mesopotamicus For the occurrence in the region of Mesopotamia, i.e. between the Río Paraná and the Río Uruguay, Prov. Entre Rios / Corrientes, Argentina. (*Gymnocalycium*)

messeri For Karl Messer (fl. 1889), teacher and curator in Bremen, Germany. (*Cynanchum*)

Mestoklema Gr. 'mestos', full; and Gr. 'klema', a small branch; for the abundant branching. (*Aizoaceae*)

metallicus Lat., metallic; for the metallic sheen on the leaves. (*Aloe, Sansevieria*)

Meterostachys Gr. 'stachys', spike; for the inflorescences. The first part of the name 'meteros' is probably a fantasy word, as the generic name would be an illegitimate homonym in the form originally published ('Merostachys', from Gr. 'meros', part, place). (*Crassulaceae*)

metriosiphon Gr. 'metrios', medium; and Gr. 'siphon', tube; for the moderately long tubular flowers in comparison to the long ones of *A. dolichosiphon*. (*Adenia*)

meuselii For Prof. Hermann Meusel (*1909), German botanist at Halle University, and specialist in composites. (*Senecio*)

mevei For Dr. Ulrich Meve (*1958), German botanist and *Asclepiadaceae* specialist in Münster and later Bayreuth. (*Cynanchum*)

mexicanus For the occurrence in Mexico. (*Beiselia, Dioscorea, Geohintonia, Peperomia, Portulaca, Sedum, Talinum*)

meyerae For Luise Meyer (née Olpp) (1873–1956), wife of the German-born clergyman and plant collector L. G. Meyer, whom she accompanied on collecting trips. (*Antimima*)

meyeri For Ernst H. F. Meyer (1791–1858), Prussian botanist in Göttingen and Königsberg. (*Ceropegia, Crassula capitella* ssp.) – (**2**) For Louis G. Meyer (1867–1958), German clergyman, went to RSA as a missionary in 1894, and became an active explorer and plant collector. (*Aloe, Cheiridopsis, Conophytum, Lithops, Meyerophytum, Stomatium*) – (**3**) For Rudolph Meyer (fl. 1896–1914), German cactus cultivator in Charlottenburg (Berlin). (*Echinopsis*) – (**4**) For the collector Meyer, active in Argentina before 1945. (*Portulaca*)

meyeri-johannis For Dr. Johannes (Hans) Meyer (1858–1929), German geographer and publisher and the first who climbed Mt. Kilimanjaro. (*Ceropegia, Sedum*)

meyerianus For Ernst H. F. Meyer (1791–1858), Prussian botanist in Göttingen and Königsberg. (*Brachystelma*)

Meyerophytum Gr. 'phyton', plant; and for Louis G. Meyer (1867–1958), German clergyman, went to RSA as a missionary in 1894, and became an active explorer and plant collector. (*Aizoaceae*)

meyranianus For Jorge Meyrán (*1918), Mexican specialist of cacti and *Crassulaceae*. (*Sedum*)

meyranii As above. (*Mammillaria*)

mezcalaensis For the occurrence in the valley of the Río Mezcala, Guerrero, Mexico. (*Neobuxbaumia*)

mezereum Gr. 'mezereon', "Mezereon" (*Daphne mezereum, Thymelaeaceae*); for the similar habit. (*Euphorbia tuckeyana* var.)

mezianus For Prof. Dr. Carl C. Mez (1866–1944), German botanist at Breslau, Halle, Königsberg (1910–1935 director of the Botanical Garden) and finally Freiburg i.Br. (*Cypselea, Galenia*)

micans Lat., gleaming, with slight metallic lustre; (**1**) for the leaf surface. (*Drosanthemum*) – (**2**) application obscure. (*Crassula*)

michelii For the plant collector Cabra-Michel (fl. 1904), or, according to other sources, for the missionary station Cabra-Michel in Zaïre. (*Raphionacme*)

michoacanus For the occurrence in the Mexican state of Michoacán. (*Agave*)

Micholitzia For Wilhelm Micholitz (1854–1932), plant collector for Sander & Co., England. (*Asclepiadaceae*)

micracanthus Gr. 'mikros', small; and Gr. 'akantha', thorn, spine; (**1**) for the spiny stems. (*Euphorbia*) – (**2**) for the small teeth on the leaf margins. (*Aloe*)

Micranthocereus Gr. 'mikros', small; Gr. 'anthos', flower; and *Cereus*, a genus of columnar cacti. (*Cactaceae*)

micranthus Gr. 'mikros', small; and Gr. 'anthos', flower. (*Brachystelma, Callisia, Lepismium, Rhipsalis, Sedum samium* ssp.)

microcarpus Gr. 'mikros', small; and Gr. 'karpos', fruit. (*Osteospermum, Sedum, Tetragonia*)

microceps Gr. 'mikros', small; and Lat. '-ceps', headed; for the small compact rosettes. (*Agave filifera* ssp.)

microdasys Gr. 'mikros', small; and Gr. 'dasys', dense, rough, shaggy; for the fine dense glochids and the absence of spines. (*Opuntia*)

microdontus Gr. 'mikros', small; and Gr. 'odous, odontos', tooth; for the small teeth along the leaf margins. (*Aloe, Cissus*)

microgaster Gr. 'mikros', small; and Gr. 'gaster', stomach; for the small and very inflated corolla tube. (*Ceropegia*)

microhelius Gr. 'mikros', small; and Gr. 'helios', sun; for the radiating prominent radial spines. (*Mammillaria*)

microlepis Gr. 'mikros', small; and Gr. 'lepis', scale; perhaps for the corona. (*Marsdenia*)

micromeris Gr. 'mikros', small; and Gr. 'meros', part; for the numerous small radiating spines. (*Epithelantha*)

micropetalus Gr. 'mikros', small; and Gr. 'petalon', petal; for the small perianth segments. (*Cleistocactus*)

microphyllus Gr. 'mikros', small; and Gr. 'phyllon', leaf. (*Acrosanthes, Antimima, Bursera, Peperomia, Pilea, Sesuvium, Talinella*)

micropterus Gr. 'mikros', small; and Gr. 'pteron', wing; for the hardly winged fruits. (*Tetragonia*)

microsepalus Gr. 'mikros', small; and Gr. 'sepalon', sepal. (*Lampranthus, Sedum*)

microspermus Gr. 'mikros', small; and Gr. 'sperma', seed. (*Browningia, Dinteranthus, Parodia*)

microsphaericus Gr. 'mikros', small; and Gr. 'sphaira', globe, sphere; for the spiny terete stem segments. (*Schlumbergera*)

microstachyus Gr. 'mikros', small; and Gr. 'stachys', spike; for the inflorescences. (*Sedum*)

microstigma Gr. 'mikros', small; and Gr. 'stigma', spot, stigma; (**1**) for the spotted leaves. (*Aloe*) – (**2**) for the shiny green papillae covering the herbaceous parts. (*Dicrocaulon*) – (**3**) for the short stigma lobes. (*Lampranthus*)

microthele Gr. 'mikros', small; and Gr. 'thele', tubercle; for the small tubercles of the plant bodies. (*Mammillaria formosa* ssp.)

middelburgensis For the occurrence at Middelburg, Eastern Cape, RSA. (*Stomatium*)

middendorfianus For Dr. Alexander T. von Middendorff [Middendorf] (1815–1894), Russian biologist and explorer, physician and zoologist. (*Phedimus*)

middlemostii For Alexander J. M. Middlemost (1902–1970), South African horticulturist at the Kirstenbosch National Botanic Garden. (*Lampranthus, Ruschia*)

miegianus For Charles E. Mieg (fl. 1972), US-American cactus collector in Scottsdale, Arizona. (*Mammillaria*)

mieheanus For G. Miehe (fl. 1933), cactus horticulturist at Hannover, Germany. (*Mammillaria*)

migiurtinorum Lat. Gen. Pl., for the occurrence in the coastal region of Migiurtina [Mijertein] in Somalia. (*Cyphostemma, Euphorbia, Kalanchoe*)

migiurtinus For the occurrence in the coastal region of Migiurtina [Mijertein] in Somalia. (*Pseudolithos*)

mihanovichii For Nicolás Mihanovich (1881? –1940), Croatia-born Argentinian engineer and businessman, owner of a shipping company and supporting the travels of the Czech cactus specialist A. Fric in Paraguay. (*Gymnocalycium*)

mijerteinus For the occurrence in Migiurtina [Mijerteina] Prov., Somalia. (*Echidnopsis*)

Mila Anagram of the name of the city of Lima; for the occurrence in Peru. (*Cactaceae*)

milanjianus For the occurrence on Mt. Milanji [Mt. Mulanje], in Nyassaland, Malawi. (*Senecio oxyriifolius* ssp.)

milii For Commander Pierre le Baron Milius (1773–1829), Governor of Mauritius. (*Euphorbia*)

militaris Lat., pertaining to the military or army; for the terminal cephalium likened to a soldiers' cap. (*Pachycereus*)

millaresii For the occurrence near Millares, Dept. Saavedra, Prov. Potosí, Bolivia. (*Gymnocalycium zegarrae* ssp.)

milleri For Anthony G. Miller (*1951), Scotish botanist. (*Echidnopsis*)

millotii For Monsieur Millot (fl. 1912), "professeur de dessin" at the Natural History Museum Paris. (*Kalanchoe*) – (**2**) For Prof. J. Millot (fl. 1949), French zoologist, director of the Institut de Recherche Scientifique, Madagascar, and later director of the Musée de l'Homme, Paris, France. (*Aloe, Euphorbia*)

millspaughii For Prof. Dr. Charles F. Millspaugh (1854–1923), US-American physician and botanist. (*Agave, Consolea, Pedilanthus, Sedum*)

milne-redheadii For Edgar Milne-Redhead (1906–1996), British botanist at Kew and field collector in Africa. (*Aloe*)

minensis For the occurrence in the state of Minas Gerais, Brazil. (*Cipocereus, Portulaca*)

mingjinianus For Ming-Jin Wang (fl. 1965), Chinese botanist. (*Hylotelephium*)

miniatus Lat., vermillion-red, flame-scarlet; for the flower colour (*Kalanchoe*)

minimus Lat., dwarf, very small; (**1**) for the size of the plants. (*Aloe, Antimima, Brachystelma, Bulbine, Conophytum, Echeveria, Escobaria, Haworthia, Sedum*) – (**2**) for the small growth and tiny flowers. (*Monanthes, Pelargonium*) – (**3**) for the smaller flowers in comparison with other taxa. (*Rhytidocaulon macrolobum* ssp.)

minor Lat., smaller, lesser. (*Brachystelma, Cheiridopsis, Crassula perfoliata* var., *Echeveria setosa* var., *Euphorbia mauritanica* var., *Haworthia decipiens* var., *Haworthia heidelbergensis* var., *Haworthia zantneriana* var., *Kalanchoe densiflora* var., *Lewisia rediviva* var., *Opuntia macrocentra* var., *Pediocactus simpsonii* var., *Pleiospilos compactus* ssp., *Senecio stapeliiformis* ssp.)

minus Lat., less, lesser; (**1**) for the small size of the plants (*Crassula natans* var., *Sempervivum*) – (**2**) for the smaller flowers. (*Odontostelma*)

minusculus Lat., somewhat smallish; for the plant size. (*Cistanthe, Conophytum, Rebutia, Tunilla*)

minutiflorus Lat. 'minutus', very small, minute; and Lat. '-florus', -flowered; (**1**) for the small flowers. (*Crassula thunbergiana* ssp., *Thompsonella, Villadia*) – (**2**) for the flowers that are small in relation to those of other species in the genus.. (*Hylocereus*)

minutifolius Lat. 'minutus', very small, minute; and Lat. '-folius', -leaved. (*Antimima, Cyphostemma macrocarpum* var.)

minutissimus Superl. of Lat. 'minutus', very small, minute; for the overall small size. (*Crassula*)

minutus Lat., very small, minute; (**1**) for the plant's size. (*Avonia recurvata* ssp., *Conophytum, Crassula, Crassula vaginata* ssp., *Maihueniopsis, Portulaca, Sempervivum*) – (**2**) for the very small leaves on flowering stems. (*Plectranthus*)

miquelii For Prof. Dr. Friedrich A. W. Miquel (1818–1871), Dutch physician and botanist, director of the botanical gardens of Rotterdam, then Amsterdam, and finally Utrecht. (*Miqueliopuntia*)

Miqueliopuntia For Prof. Dr. Friedrich A. W. Miquel (1818–1871), Dutch physician and botanist, director of the botanical gardens of Rotterdam, then Amsterdam, and finally Utrecht; and for the relationship to the genus *Opuntia* (*Cactaceae*). (*Cactaceae*)

mirabella For the former genus name *Mirabella*, which in turn was named for the occurrence near Mirabela, Minas Gerais, Brazil. (*Cereus*)

mirabilis Lat., wonderful, marvellous, miraculous; (**1**) for the appearance. (×*Cleistocana, Conophytum, Dendroportulaca, Echinopsis, Espostoa, Haworthia, Impatiens, Pelargonium, Trichodiadema*) – (**2**) for the remarkable difference from other species of the genus. (*Phyllanthus*) – (**3**) for the size of the plants. (*Agave atrovirens* var.)

Miraglossum Lat. 'mirus', exceptional, remarkable; and Gr. 'glossa', tongue; for the conspicuous corona segments. (*Asclepiadaceae*)

mirandae For Dr. Faustino Miranda (1905–1964), Mexican botanist. (*Weberocereus glaber* var.)

mireillae For Mireille Laudrin (fl. 2002), wife of the insurance broker and plant explorer John Lavranos and discoverer of the taxon. (*Caralluma*)

miriamiae For Mrs. Miriam Davis (fl. 1918), friend of the Danish botanist Carl Ostenfeld. (*Crassula colorata* var.)

mirkinii For the collector Mirkin (fl. 1939), without further data. (*Ophionella arcuata* ssp.)

mirus Lat., exceptional, remarkable; perhaps for the exceptionally narrow leaves. (*Euphorbia*)

miscellus Lat., mixed; application obscure. (*Orbea*)

miserus Lat., miserable; (**1**) perhaps for the small size of the plants. (*Villadia*) – (**2**) for the frequently leafless state of the plants. (*Euphorbia*) – (**3**) for the flowers without petals. (*Ruschia*)

missionum Gen. Pl. of Lat. 'missio', mission,

i.e. of the missions; application obscure, perhaps for the occurrence near missions. (*Agave*)

missouriensis For the occurrence near the Missouri River, USA. (*Escobaria*)

mistiensis For the occurrence in the vicinity of the Misti volcano, Dept. Arequipa, Peru. (*Cumulopuntia*)

mitejea Anagram of French "Je t'aime", I love you; apparently expressing relationship between the two authors of the name, Mlle. Alice Leblanc and the French botanist and physician Raymond Hamet. (*Kalanchoe*)

mitis Lat., mild, soft, gentle; (**1**) for the smooth tuber. (*Anthorrhiza*) – (**2**) for the soft leaves. (*Agave*) – (**3**) for the absence of spines. (*Neorautanenia*)

mitratus Lat., having a mitre or bishop's cap; for the shape of the leaves. (*Mitrophyllum*)

mitriformis Lat., mitre-shaped; (**1**) for the appearence of the rosette apex. (*Aloe*) – (**2**) for the shape of the cushion formed by the plants. (*Euphorbia*)

Mitrophyllum Gr. 'mitra', mitre, cap; and Gr. 'phyllon', leaf; for the appearance of the leaf pair, which is like the deeply cleft headdress worn by bishops. (*Aizoaceae*)

mitsimbinensis For the Mitsimbina section of the Botanical Garden Tsimbabaza, Antananarivo, Madagascar, where the taxon was found growing. (*Euphorbia*)

mixtecanus For the occurrence in the mountainous region of the Mixteca Alta, Oaxaca / Puebla, Mexico. (*Yucca*) – (**2**) For the occurrence in the region inhabited by the Comunidad Mixtecana, Oaxaca, Mexico. (*Thompsonella*)

mixtus Lat., mixed; because the material was originally mixed up with another species, and also for the illegitimate name first given. (*Euphorbia*)

mlanjeanus For the occurrence on Mt. Mlanje, Malawi. (*Aloe chabaudii* var., *Euphorbia*)

mocinianus For José M. Mociño (1757–1820), first Mexican botanist and member of the "Real Expedición Botánica". (*Sedum*)

modestus Lat., modest; (**1**) for the small size of the plants. (*Aloe*, *Antimima*, *Brachystelma*, *Duvalia*, *Haworthia variegata* var., *Psammophora*, *Rosularia*, *Sedum*) – (**2**) for the smaller plant and bract size. (*Euphorbia atropurpurea* var.) – (**3**) for the small flowers. (*Carpobrotus*)

moelleri For Dr. Heinrich Möller (fl. 1930), physician in Schafhausen, Switzerland, brother of A. Möller who collected cacti in Mexico. (*Grusonia*)

moellerianus As above. (*Mammillaria*)

mojavensis For the occurrence in the Mojave Desert, SW USA. (*Echinocereus*)

molaventi From Lat. 'mola', mill; and Lat. 'ventus', wind; for the windmill-like vertically arranged flowers. (*Brachystelma*)

molederanus For the occurrence at Moledera Hill, Somalia. (*Aloe*)

molestus Lat., molesting, unpleasant; for the spiny nature of the plants. (*Cylindropuntia*, *Echinopsis*)

molinensis For the occurrence near Molinos, Salta, Argentina. (*Tephrocactus*)

mollendorffianus For Mr. H. Mollendorff († 1947), a long-time friend of the Mexican cactus collector Fritz Schwarz. (*Mammillaria rhodantha* ssp.)

mollicomus Lat. 'mollis', soft; and Lat. 'coma', hair tuft, mane; for the pubescence of the leaves. (*Pelargonium*)

molliculus Dim. of Lat. 'mollis', soft; for the softly fleshy plant bodies. (*Copiapoa*)

mollis Lat., soft; (**1**) for the soft and pulpy nature of the bodies formed from the fused leaves. (*Gibbaeum*) – (**2**) for the softly hairy leaves. (*Crassula*, *Kedrostis*, *Nolana*, *Synadenium*) – (**3**) application obscure. (*Malephora*, *Ruschia*)

molokiniensis For the occurrence on Molokini Islet, Hawaii. (*Portulaca*)

molonyae For The Hon. Mrs. Evelyn R. Molony (née Napier) (1902–1952), British, arrived in Kenya 1922, botanist at the Coryndon Memorial Museum, Nairobi (Kenya) 1930–1934 as Miss Napier, after marriage in 1935 working on her husband's farm in Kenya. (*Huernia keniensis* var.)

mombergeri For Peter Momberger (fl. 2002), German cactus collector and trader. (*Turbinicarpus*)

Momordica Probably from Lat. 'mordicus', biting; either for the biting taste of the sap of the ripe fruits of *M. balsamina* (Genaust 1996), or for the chewed appearance of the grooved margins of the seeds. (*Cucurbitaceae*)

monacanthus Gr. 'monos', one, only; and Gr. 'akantha', thorn, spine. (*Euphorbia, Hylocereus, Lepismium*)

monadelphus Gr. 'monos', one, only; and Gr. 'adelphos', brother; for the united filaments. (*Adenia*)

monadenioides Gr. '-oides', resembling; and for the genus *Monadenium* (*Euphorbiaceae*); for the similarly geophytic habit. (*Euphorbia*)

Monadenium Gr. 'monos', one, only; and Gr. 'aden', gland; for the single horse-shoe-shaped nectary gland formed by the fusion of four glands. (*Euphorbiaceae*)

monandrus Gr. 'monos', one, only; and Gr. 'aner, andros', man, [botany] stamen; for the single stamen present in the flowers. (*Cistanthe*)

monanthemus Gr. 'monos', one, only; and Gr. 'anthos', flower; for the solitary flowers. (*Delosperma*)

Monanthes Gr. 'monos', one, only; and Gr. 'anthos', flower; for the few-flowered inflorescences (though only rarely 1-flowered). (*Crassulaceae*)

monanthus Gr. 'monos', one, only; and Gr. 'anthos', flower; for the few-flowered inflorescences. (*Sedum stenopetalum* ssp.)

Monilaria From Lat. 'monile', pearl collar; for the stems with regular constrictions, giving the appearance of a string of beads. (*Aizoaceae*)

moniliformis Lat. 'monile', pearl collar; and Lat. '-formis', -shaped; (**1**) for the chains of tubers formed by the roots. (*Pelargonium*) – (**2**) for the stems with regular constrictions, giving the appearance of a string of beads. (*Monilaria*) – (**3**) probably for the reticulate patterning of the surface of the stem segments. (*Consolea*) – (**4**) for the structure of the inflorescence axis. (*Caralluma*) – (**5**) application obscure. (*Myrmephytum*)

monocarpicus Gr. 'monos', one, only; and Gr. 'karpos', fruit; for the monocarpic nature of the plants, i.e. flowering and fruiting once. (*Sedum creticum* var.)

monophyllus Gr. 'monos', one, only; and Gr. 'phyllon', leaf. (*Bulbine*)

monospermus Gr. 'monos', one, only; and Gr. 'sperma', seed; for the frequently one-seeded fruits. (*Cistanthe*)

monotropus From Gr. 'monotropos', hermit, alone and on its own; for the unique combination of characters. (*Aloe*)

monregalensis From the antique name "Mons Regalis" for the village of Monte Reale in the Abruzzi Mts., Italy. (*Sedum*)

monroi For Claude F. H. Monro (1863–1918), British collector, author of "Indigenous trees of Southern Rhodesia". (*Jatropha*)

Monsonia For Lady Anne Monson, great-grand-daughter of Charles II, visited the Cape around 1775. (*Geraniaceae*)

montagnacii For R. Montagnac (fl. 1942, 1961), French botanist and for many years director of the Services d'Agriculture at Toliara, Madagascar. (*Alluaudia, Cyphostemma, Stapelianthus*)

montaguensis For the occurrence at Montagu, Western Cape, RSA. (*Lampranthus, Ruschia*)

montanus Lat., mountain-; for the habitat. (*Agave, Brachystelma, Copiapoa, Crassula, Cyrtanthus, Echeveria, Echidnopsis, Euphorbia arbuscula* var., *Kalanchoe luciae* ssp., *Monadenium, Plectranthus purpuratus* ssp., *Rhadamanthus, Sedum, Sempervivum, Stapelia, Stenocereus*)

monte-amargensis For the occurrence at Monte Amargo near Caldera, Chile. (*Eriosyce odieri* var.)

monteiroae For Mrs. Monteiro (fl. 1887), English wife of the Portuguese mining engineer, zoologist and naturalist Joachim J. Monteiro, lived in Moçambique after her husband's death in 1878 and collected insects and plants. (*Raphionacme, Stomatostemma*)

monteiroi For Joachim J. Monteiro (1833–1878), Portuguese mining engineer, zoologist and naturalist, resident in Angola from 1858. (*Euphorbia*)

montevideensis For the occurrence in the hills around Montevideo, Uruguay. (*Opuntia*)

monticola Lat. 'mons, montis', mountain; and Lat. '-cola', inhabiting. (*Aloe, Haworthia, Lampranthus, Ottosonderia, Peperomia, Sedum dendroideum* ssp.)

montium-klinghardtii Lat. 'mons, montis', mountain; for the occurrence in the Klinghardt Mts., Namibia. (*Adromischus*)

monvillei For Baron Hippolyte Boissel de Monville (1794–1863), French factory owner and plant collector from near Rouen. (*Gymnocalycium*)

mooneyi For the botanical collector H. F. Mooney (fl. 1958). (*Sedum*)

moramangensis For the occurrence near the city of Moramanga, C Madagascar. (*Cynanchum*)

moranensis For the occurrence near Real de Moran, Hidalgo, Mexico. (*Sedum*)

moranii For Dr. Reid V. Moran (*1916), US-American botanist formerly at the San Diego Natural History Museum, and *Crassulaceae* specialist. (*Agave, Echeveria, Jatropha, Sedum*)

moratii For Prof. Dr. Philippe Morat (*1937), French botanist and Madagascar specialist, from 1986 director of the Laboratoire de Phanérogamie at the Paris Natural History Museum. (*Euphorbia*)

morawetzianus For Victor Morawetz (fl. 1936), US-American philanthropist in New York who financed an expedition of C. Backeberg to South America in the 1930ies. (*Cleistocactus*)

morelensis For the occurrence in the Mexican state of Morelos. (*Senecio praecox* var.)

morganianus For Dr. Meredith W. Morgan (1887–1957), US-American optometrist and succulent plant enthusiast in Richmond, California, USA. (*Mammillaria, Sedum*)

morijensis For the occurrence at Morijo, Kenya. (*Aloe*)

Moringa From the Malabar vernacular name "Moringo" for *Moringa oleifera* (Jackson 1990). (*Moringaceae*)

morricalii For Dale Morrical (1908–1994), US-American electronic technician and cactus amateur. (*Echinocereus viereckii* ssp.)

morrisiae For Mrs. G. Morris (fl. 1937), mother of the South African botanist and succulent plant author Mrs. Doreen Court. (*Haworthia mucronata* var., *Haworthia scabra* var.)

morrisonensis For the occurrence on Mt. Morrison, Taiwan. (*Sedum*)

morrumbalensis For the occurrence on Morrumbala Mt., C Moçambique. (*Crassula*)

mortensenii For Russell H. Mortensen (fl. 1950), who showed the plants to the botanist Léon Croizat. (*Cereus*)

mortimeri For Prof. Keith V. Mortimer (fl. 2002), British dentist and grower of succulents. (*Aloe rigens* var.)

mortolensis For the private botanical garden "La Mortola" on the Italian Riviera, where the German botanist A. Berger was director. (×*Gasteraloe*)

mortonii For Prof. John K. Morton (*1928), English botanist working for many years at the University of Ghana. (*Brachystelma*)

mosaicus Lat., mosaic-like; for the pattern formed by the branch tips at ground level when the body is withdrawn during the dry season. (*Euphorbia*)

moschatus Lat., musk-scented; probably for the flower scent. (*Crassula*)

moseleyanus For H. N. Moseley (1844–1891), British explorer and collector of the type. (*Hydnophytum*)

moserianus For Günther Moser (1919–1994), Austrian cactus hobbyist. (*Frailea grahliana* ssp.)

mossambicensis For the occurrence in Moçambique. (*Adenia*)

mossamedensis For the occurrence in the Moçamedes [Mossamedes] Distr. in Angola. (*Aizoanthemum, Hoodia, Monsonia*)

Mossia For Dr. Charles E. Moss (1870–1930), British botanist emigrating to RSA in 1917. (*Aizoaceae*)

mostii For Carlos Most (fl. 1906), Córdoba, cactus collector in Argentina. (*Gymnocalycium*)

mubendiensis For the occurrence at Mubende, Uganda. (*Aloe*)

mucidus Lat., mouldy, slimy; for the

"mouldy" grey body colour. (*Gymnocalycium*)

mucronatus Lat., mucronate; (**1**) for the leaf tips. (*Antimima, Echeveria, Haworthia, Lampranthus, Portulaca*) – (**2**) for the sepal tips. (*Tumamoca*)

mucronulatus Dim. of Lat. 'mucronatus', mucronate; for the leaf tips. (*Portulaca*)

mudenensis For the occurrence in the Muden Valley, KwaZulu-Natal, RSA. (*Aloe*)

muehlenpfordtii For Dr. Friedrich Mühlenpfordt (fl. 1847), German physician and cactus collector in Hannover. (*Mammillaria*)

mueller-melchersii For Mr. Müller-Melchers (fl. 1929), cactus collector of German descent in Montevideo, Uruguay, supplying the Haage nursery with South American cacti. (*Parodia*)

muelleri For Mr. Müller (fl. 1914), without further data. (*Ruschia*)

Muiria For Dr. John Muir (1874–1947), Scottish physician and naturalist, emigrated to RSA in 1896. (*Aizoaceae*)

muirianus As above. (*Ruschia*)

muirii As above. (*Aloe lineata* var., *Carpobrotus, Conicosia pugioniformis* ssp., *Delosperma, Drosanthemum, Euphorbia, Hereroa, Rhinephyllum, Senecio*)

mulanjensis For the occurrence on Mt. Mulanje, highest mountain in Malawi. (*Sarcostemma*)

muliensis For the occurrence in the Muli Mts., Sichuan, China. (*Ceropegia*)

multangulus Lat., 'multi-', many-; and Lat. 'angulus', angle; for the numerous branch angles. (*Euphorbia proballyana* var.)

multiareolatus Lat., 'multi-', many-; and Lat. 'areolatus', with areoles. (*Neobuxbaumia*)

multibracteatus Lat., 'multi-', many-; and Lat. 'bracteatus', bracteate. (*Pelargonium*)

multicaulis Lat., 'multi-', many-; and Lat. 'caulis', stem. (*Aloe volkensii* ssp., *Crassula, Echeveria, Pseudosedum, Sedum*)

multicavus Lat., 'multi-', many-; and Lat. 'cavus', depression; for the hydathodes on the leaves. (*Crassula*)

multicephalus Lat., 'multi-', many-; and Gr. 'kephale', head; for the clump-forming growth habit. (*Thelocactus rinconensis* ssp.)

multiceps Lat., 'multi-', many-; and Lat. '-ceps', headed; for the much-branched habit. (*Bergeranthus, Crassula, Euphorbia, Sedum*)

multiclava Lat., 'multi-', many-; and Lat. 'clava', club; for the numerous club-shaped branches. (*Euphorbia*)

multicolor Lat. 'multi-', many-; and Lat. 'color', colour; for the multi-coloured perianth. (*Aloe*)

multicostatus Lat., 'multi-', many-; and Lat. 'costatus', ribbed. (*Pilosocereus, Stenocactus*)

multidigitatus Lat., 'multi-', many-; and Lat. 'digitatus', with fingers; for the numerous parallel-erect plant bodies. (*Mammillaria*)

multifidus Lat. 'multi-', many-; and Lat. '-fidus', -divided; (**1**) for the segmented leaves. (*Jatropha, Monsonia, Polyachyrus poeppigii* ssp.) – (**2**) for the branched inflorescences. (*Euphorbia*)

multifilifer Lat., 'multi-', many-; Lat. 'filum', thread; and Lat. '-fer, -fera, -ferum', -carrying; for the leaf margins. (*Agave filifera* ssp.)

multiflorus Lat. 'multi-', many-; and Lat. '-florus', -flowered. (*Callisia, Ceropegia, Crassula, Delosperma, Dioscorea sylvatica* var., *Euphorbia moratii* var., *Fockea, Quaqua, Rauhia, Ruschia, Sedum, Seyrigia, Talinum, Trianthema, Tripogandra*)

multifolius Lat. 'multi-', many-; and Lat. '-folius', -leaved. (*Euphorbia, Haworthia emelyae* var., *Ornithogalum, Urginea*)

multigeniculatus Lat. 'multi-', many-; and Lat. 'geniculatus', with a knee, with a small knot; for the appearance of the stem segments. (*Cylindropuntia*)

multiprolifer Lat. 'multi-', many-; and Lat. 'prolifer, prolifera, proliferum', proliferous; for the numerous offsets usually present. (*Gymnocalycium anisitsii* ssp.)

multipunctatus Lat., 'multi-', many-; and Lat. 'punctatus', dotted; for the pattern of the leaf surface. (*Lithops dinteri* ssp.)

multiradiatus Lat., 'multi-', many-; and Lat. 'radiatus', rayed, provided with rays; (**1**) for the many-flowered inflorescences. (*Pelargonium*) – (**2**) for the numerous petals. (*Lampranthus*)

multiseriatus Lat., 'multi-', many-; and Lat. 'seriatus', in series; for the numerous petals. (*Lampranthus*)

multituberosus Lat., 'multi-', many-; and Lat. 'tuberosus', tuberous; for the numerous lateral root tubers. (*Myrsiphyllum*)

munbyanus For Giles Munby (1813–1876), British botanist working in Algeria for many years. (*Caralluma*)

munchii For Mr. Raymond C. Munch (1901–1985), South African, living in Zimbabwe since 1909, farmer near Rusape, Zimbabwe, with a garden containing a collection of the native flora, esp. Aloes and Cycads. (*Aloe*)

mundtii For J. L. Mundt, German, collected 1815–1850 in the Cape Province of RSA. (*Euphorbia*)

munzii For Prof. Dr. Philip A. Munz (1892–1974), US-American botanist and from 1946 director of the Rancho Santa Ana Botanic Gardens. (*Cylindropuntia, Echinocereus engelmannii* var.)

muralis Lat., wall-; for the occurrence on the walls of buildings at Valverde on Hierro, Canary Islands. (*Monanthes*)

muratdaghensis For the occurrence on the Murat Dagh [Mountain] in W Turkey. (*Prometheum*)

murex Lat., rough, with short hard points; and the name of a genus of spiny shells; for the spiny fruits. (*Sterculia*)

muricatus Lat., muricate, with rough short hard points like the shell of Murex; (**1**) for the rough texture of the branches. (*Euphorbia*) – (**2**) for the central spines. (*Parodia*) – (**3**) for the rough leaves. (*Crassula*) – (**4**) for the leaf margins. (*Ruschia*)

murinus Lat., pertaining to mice; (**1**) for the mouse-grey colour of the plants. (*Dudleya abramsii* ssp.) – (**2**) for the toothed leaf margins, comparing the gaping leaves of a pair to the open mouth of a mouse. (*Stomatium*) – (**3**) for the mouse-grey colour of the inflorescences. (*Aloe*)

murrillii For Dr. William A. Murrill (1869–1957), US-American mycologist at the New York Botanical Garden. (*Selenicereus*)

musapanus For the occurrence at Mt. Musapa, Zimbabwe. (*Aloe*)

muscicola Lat. 'muscus', moss; and Lat. '-cola', inhabiting; for the occurrence in moss carpets. (*Bulbine*)

muscoideus Lat., moss-like; for the habit. (*Sedum*)

muscosus Lat., mossy, moss-like. (*Crassula*)

musculinus From Lat. 'musculus', muscle; for the fleshy ("musculose") leaves; or Lat. 'musculinus', pertaining to small mice (from the Dim. of Lat. 'mus', mouse); for the small teeth on the leaf margins found in some plants. (*Chasmatophyllum*)

mustelinus Lat., of weasels; for the toothed leaf margins, comparing the gaping leaves of a pair with the open mouth of a weasel. (*Stomatium*)

mutabilis Lat., changeable; (**1**) for the variability of the taxon. (*Eriosyce curvispina* var., *Plectranthus*) – (**2**) for the colour change from bud to flower. (*Aloe*)

mutans Lat., changing; (**1**) for the variable number of petals. (*Pelargonium*) – (**2**) for the changing colour of the petals during anthesis. (*Lampranthus*)

mutatus Lat., changed; because the taxon had to be renamed to avoid an illegitimate homonym. (*Lampranthus*)

muticus Lat., blunt, without a point; for the leaf tips. (*Antimima, Haworthia*)

muyaicus For the occurrence in the Muya-Xiang Distr., Sichuan, China. (*Sedum*)

muyurinensis For the occurrence near Muyurina, Prov. Valle Grande, Dept. Santa Cruz, Bolivia. (*Cleistocactus*)

muzinganus For Muzinga Yumba (fl. 1984), technical assistant and florist in Lubumbashi, Zaïre. (*Ceropegia*)

mweroensis For the type locality Mwero, W of Lake Tanganyika, Zambia. (*Senecio*)

mwinilungensis For the occurrence in the Mwinilunga Distr., Zambia. (*Euphorbia*)

myriacanthus Gr. 'myrios', numerous; and Gr. 'akantha', thorn, spine; (**1**) for the spination of the plant bodies. (*Matucana haynei* ssp.) – (**2**) for the numerous tiny prickles on the leaves. (*Aloe*)

myriocladus Gr. 'myrios', numerous; and Gr. 'klados', branch, twig. (*Euphorbia*)

myriostigma Gr. 'myrios', numerous; and Gr. 'stigma', scar, spot; for the numerous minute woolly tufts on the body surface. (*Astrophytum*)

Myrmecodia Gr. 'myrmekodes', full of ants (from Gr. 'myrmex', ant); for the swollen tubers inhabited by ants. (*Rubiaceae*)

Myrmephytum Gr. 'myrmex', ant; and Gr. 'phyton', plant; for the ant-plant association. (*Rubiaceae*)

Myrsiphyllum Gr. 'myrsine', a myrtle branch; and Gr. 'phyllon', leaf; for the leaf-like phyllocladia ('leaves') which resemble those of myrtle (*Myrtus communis*). (*Asparagaceae*)

Myrtillocactus MLat. 'myrtella' / Fr. 'myrtille', blueberry; and Lat. 'cactus', cactus; for the small blueberry-like globose fruits. (*Cactaceae*)

mystax Gr., upper lip, moustache; for the bristles in the axils between the tubercles of the plant body. (*Mammillaria*)

mzimbanus For the occurrence at Mzimba, Malawi. (*Aloe*)

mzimvubuensis For the occurrence in the valley of the Mzimvubu River, Eastern Cape, RSA. (*Adromischus liebenbergii* ssp.)

N

nagasakianus For the occurrence near Nagasaki, Japan. (*Sedum*)
nairobiensis For the occurrence near Nairobi, Kenya. (*Huernia keniensis* var.)
nakurensis For the occurrence at Nakuru, Kenya. (*Delosperma*)
namaensis For the occurrence in Namaland, a region in S Namibia and the Northern Cape of RSA (i.e. the land of the Nama, a Khoi tribe in Namibia). (*Avonia papyracea* ssp., *Bulbine, Commiphora, Galenia*)
Namaquanthus For the occurrence in Namaqualand (S Namibia and Northern Cape of RSA); and Gr. 'anthos', flower. (*Aizoaceae*)
namaquanus For the occurrence in Namaqualand (S Namibia and Northern Cape, RSA). (*Pachypodium*)
namaquensis As above. (*Anacampseros filamentosa* ssp., *Bulbine mesembryanthoides* ssp., *Ceraria, Ceropegia, Cheiridopsis, Crassula, Euphorbia mauritanica* var., *Haworthia arachnoidea* var., *Huernia, Orbea, Pectinaria articulata* ssp., *Prenia pallens* ssp., *Tetragonia*)
namibensis For the occurrence in the Namib Desert in SW Africa. (*Aloe, Brownanthus, Crassula elegans* ssp., *Euphorbia, Rhadamanthus*)
Namibia For the occurrence in Namibia. (*Aizoaceae*)
namibianus As above. (*Raphionacme*)
namibiensis As above. (*Aeollanthus*)
namorokaensis For the occurrence in the Namoroka Natural Reserve, W Madagascar. (*Aloe*)
namuskluftensis For the occurrence at Namuskluft, Namibia. (*Euphorbia*)
namusmontanus For the occurrence on the Namus Mts., Namibia (Lat. 'montanus', mountain-). (*Ruschia*)
Nananthus Gr. 'nanos', small; and Gr. 'anthos', flower; for the small size of the plants. (*Aizoaceae*)

nanchititlensis For the occurrence at Cañada de Nanchititla, México, Mexico. (*Agave*)
nanchuanensis For the occurrence in the Nanchuan Distr., Sichuan, China. (*Sedum*)
nanifolius Lat. 'nanus', dwarf; and Lat. '-folius', -leaved; for the tiny leaves. (*Sedum*)
nanus Lat., dwarf; for the dwarf size. (*Adromischus, Brachystelma, Ceropegia, Espostoa, Euphorbia filiflora* var., *Jatropha, Kedrostis, Odontophorus, Oophytum, Parakeelya, Ruschia, Sedum*)
napifer Lat. 'napus', turnip, beetroot; and Lat. '-fer, -fera, -ferum', -carrying; for the thickened roots. (*Cynanchum, Sedum, Seyrigia*)
napiformis Lat., turnip-shaped; for the tuberous root. (*Delosperma, Talinum*)
napinus From Lat. 'napus', turnip, beetroot; for the tuberous tap roots. (*Eriosyce, Mammillaria*)
narcissifolius For the genus *Narcissus* ("Daffodil", "Narciss"; *Amaryllidaceae*); and Lat. '-folius', -leaved. (*Bulbine*)
nardouwensis For the occurrence at Nardouw Pass, Clanwilliam Distr., Western Cape, RSA. (*Lampranthus*)
narvaecensis For the occurrence near Narvaez, Dept. Tarija, Bolivia. (*Rebutia*)
nashii For George V. Nash (1864–1921), US-American botanist and horticulturist at the New York Botanical Garden, collected extensively in Florida and the West Indies. (*Agave, Consolea, Harrisia*)
natalensis For the occurrence in the former Natal Province of RSA. (*Brachystelma, Bulbine, Crassula, Monsonia, Petopentia*)
natans Lat., swimming; for the habit. (*Crassula*)
naumannii For F. C. Naumann (1841–1902), German explorer. (*Myrmephytum*)
naureeniae For Mrs. Naureen A. Cole (*1935), South African pharmacist and wife of the *Lithops* specialist Prof. D. T. Cole. (*Lithops*)
navae For D. Alonso de Nava Grimón, 6. Marqués de Villanueve del Prado (1757–1832), Tenerife (Canary Islands). (*Euphorbia*)
navajous For the occurrence in the Navajo region, USA. (*Yucca baileyi* var.)

navicula Lat., small boat; for the boat-shaped leaves. (*Ornithogalum*)

navicularifolius Lat. 'navicularis', shaped like a little boat (from Lat. 'navicula'); and Lat. '-folius', -leaved; for the leaf shape. (*Bulbine*)

navicularis Lat., shaped like a little boat (from Lat. 'navicula'); for the leaf shape. (*Callisia*)

nayaritensis For the occurrence in the Mexican state of Nayarit. (*Agave, Echeveria*)

ndotoensis For the occurrence in the Ndoto Mts., Eastern Prov., Kenya; 'ndoto' is a Samburu word meaning "small rocks". (*Kalanchoe*)

nebrownii For Dr. Nicholas E. Brown (1849–1934), English botanist at Kew specializing in African succulents. (*Anacampseros lanceolata* ssp., *Gibbaeum*)

necopinus Lat., unexpected; for the unexpected occurrence. (*Yucca*)

neethlingiae For Dr. Marie Murray Neethling (later Mrs. Vogts) (*1908), South African botanist and agriculturist. (*Delosperma*)

neglectus Lat., neglected; for the prior neglected status. (*Aeollanthus, Agave, Brownanthus, Coryphantha, Kalanchoe*)

negromontanus Lat., from the mountain chain Serra de Montes Negros ('black mountains') in the Moçamedes Distr., Angola. (*Euphorbia*)

neilii For Mr. Neil (fl. 1928–1933), dairy farmer and nurserymen (with J. Hurling) at Bonnievale, Western Cape, RSA. (*Glottiphyllum, Stayneria*)

nejapensis For the occurrence near Nejapa, Oaxaca, Mexico. (*Mammillaria karwinskiana* ssp., *Opuntia*)

Nelia For Prof. Dr. Gert C. Nel (1855–1950), botanist at Stellenbosch University, RSA. (*Aizoaceae*)

nelii As above. (*Chasmatophyllum, Cheiridopsis, Delosperma, Glottiphyllum, Hereroa, Lampranthus, Pleiospilos, Rhombophyllum, Ruschia*)

nelsonii For David Nelson († 1789), English gardener and collector visiting the Cape (RSA) in 1776 and 1788. (*Albuca*) – (**2**) For Edward W. Nelson (1855–1934), US-American naturalist. (*Agave cerulata* ssp., *Disocactus, Villadia*) – (**3**) For Mr. C. Z. Nelson (fl. 1914), enthusiastic cactus grower in Galesburg, Illinois, USA. (*Selenicereus*)

Nematanthus Gr. 'nema, nematos', thread; and Gr. 'anthos', flower; for the thread-like pendent pedicel of the type species. (*Gesneriaceae*)

nematostemma Gr. 'nema, nematos', thread; and Gr. 'stemma', garland; for the filiform tip of the staminal corona segments. (*Cynanchum*)

nemorosus Lat., pertaining to bush and forest; for the habitat. (*Crassula, Faucaria, Pereskia*)

Neoalsomitra Gr. 'neos', new; for the renaming of the type species, which was formerly placed in the genus *Alsomitra* (*Cucurbitaceae*). (*Cucurbitaceae*)

neoarbuscula Gr. 'neos', new (to avoid a homonym); and from Lat. 'arbuscula', small tree; for the similarity to *Cylindropuntia* (*Opuntia*) *arbuscula*. (*Cylindropuntia*)

neoarechavaletae Gr. 'neos', new (to avoid a homonym); and for Dr. José Arechavaleta y Balpardo (1838–1912), Spanish-born pharmacist and botanist in Uruguay, from 1890 onwards director of the Museo Nacional in Montevideo. (*Parodia*)

neobakeri Gr. 'neos', new (to avoid a homonym); and for John G. Baker (1834–1920), British botanist at Kew. (*Senecio*)

neobosseri Gr. 'neos', new (to avoid a homonym); and for Jean M. Bosser (*1922), French botanist and agronomical engineer, and director of ORSTOM in Antananarivo, Madagascar. (*Euphorbia*)

neobuenekeri Gr. 'neos', new (to avoid a homonym); and for Rudolf Heinrich Büneker (fl. 1922), cactus collector of German descent in Rio Grande do Sul, Brazil, brother-in-law of Leopoldo Horst, father of Rudi W. Büneker. (*Parodia scopa* ssp.)

Neobuxbaumia Gr. 'neos', new (to avoid a homonym); and for Prof. Dr. Franz Buxbaum (1900–1979), Austrian botanist, high school teacher and specialist of cactus morphology. (*Cactaceae*)

neochilus Gr. 'neos', new; and Gr. 'cheilos', lip; perhaps for the larger upper lip of the calyx. (*Plectranthus*)

neochrysacanthus Gr. 'neos', new (to avoid a homonym); and for the similarity to the undescribed *Opuntia chrysacantha*. (*Opuntia*)

neocumingii Gr. 'neos', new (to avoid a homonym); and for Hugh Cuming (1791–1865), English collector in South America, since 1822 dealer in natural history objects in Valparaiso, Chile. (*Weingartia*)

Neohenricia Gr. 'neos', new (to avoid a homonym); and for Dr. Marguerite G. A. Henrici (1892–1971), Swiss plant physiologist, living in RSA from 1921. (*Aizoaceae*)

neohorstii Gr. 'neos', new (to avoid a homonym); and for Leopoldo Horst (1918–1987), Brazilian of German origin, cactus collector and exporter in Rio Grande do Sul. (*Parodia*)

neohumbertii Gr. 'neos', new (to avoid a homonym); and for Prof. Jean-Henri Humbert (1887–1967), French botanist in Madagascar. (*Euphorbia*)

Neolloydia Gr. 'neos', new (to avoid a homonym); and for Francis E. Lloyd (1868–1947), US-American botanist. (*Cactaceae*)

neomexicanus For the occurrence in New Mexico, USA. (*Agave*, *Rhodiola integrifolia* ssp., *Yucca harrimaniae* var.)

neonelsonii Gr. 'neos', new (to avoid a homonym); and for Edward W. Nelson (1855–1934), US-American naturalist. (*Agave*)

neopalmeri Gr. 'neos', new (to avoid a homonym); and for the replaced name *Mammillaria palmeri*. (*Mammillaria*)

neopauciflorus Gr. 'neos', new (to avoid a homonym); and for *Jatropha pauciflora*. (*Jatropha*)

neopringlei Gr. 'neos', new (to avoid a homonym); and for Cyrus G. Pringle (1838–1911), US-American plant breeder and explorer-collector. (*Agave*)

Neoraimondia Gr. 'neos', new (to avoid a homonym); and for Prof. Antonio Raimondi (1826–1890), Italian-born botanist, emigrating 1850 to Peru, working as botanist and later State geologist. (*Cactaceae*)

Neorautanenia Gr. 'neos', new (to avoid a homonym); and for Rev. Martti (Martin) Rautanen (1845–1926), Finnish missionary in N Namibia. (*Fabaceae*)

neostayneri Gr. 'neos', new (to avoid a homonym); and for Frank J. Stayner (1907–1981), horticulturist and curator of the Karoo Botanic Garden Worcester 1959–1969. (*Lampranthus*)

neovirens Gr. 'neos', new; and Lat. 'virens', becoming green; because the name is based on the illegitimate name *Mesembryanthemum virens* Haworth. (*Ruschia*)

neovolcanicus For the occurrence in the region known as "Eje Neovolcanico" (Span., "new volcanic axis") in SW Mexico. (*Sedum*)

neovolkensii Gr. 'neos', new (to avoid a homonym); and for Dr. Georg L. A. Volkens (1855–1917), German botanist in Berlin, explorer of the Kilimanjaro 1892–1894. (*Euphorbia nyikae* var.)

Neowerdermannia Gr. 'neos', new (to avoid a homonym); and for Prof. Dr. Erich Werdermann (1892–1959), German botanist in Berlin, specialist on cacti, and director of Botanischer Garten und Museum Berlin. (*Cactaceae*)

nepalensis For the occurrence in Nepal. (*Brachystelma*)

nepalicus As above. (*Rhodiola*)

nephrolobus Gr. 'nephros', kidney; and Lat. 'lobus', lobe; for the shape of the corolla lobes. (*Ceropegia sobolifera* var.)

nephrophyllus Gr. 'nephros', kidney; and Gr. 'phyllon', leaf; for the leaf shape. (*Pelargonium*)

neriifolius For the genus *Nerium* ("Rose-Bay", "Oleander", *Apocynaceae*); and Lat. '-folius', -leaved. (*Euphorbia*)

neronis For Nero, one of the helpers on a botanical expedition of Pillans. (*Stapeliopsis*)

nervosus Lat., veined; for the leaf veins. (*Monadenium*, *Tetradenia*)

neryi For Silvério J. Néry, 1900–1904 president of the Amazonas State, Brazil. (*Melocactus*)

nesemannii For Mr. A. Nesemann (fl. 1934), who collected near Robertson and sent specimens to Grahamstown. (*Euphorbia*)

nesioticus From Gr. 'nesos', island; for the insular type locality. (*Brachycereus, Dudleya, Sedum lanceolatum* ssp.)

netrelianus For a Mr. Netrel (fl. 1853), without further data. (*Gymnocalycium*)

neuhuberi For Gert Neuhuber (*1939), Austrian cactus collector and *Gymnocalycium* specialist, co-founder and former president of the Austrian "Arbeitsgruppe Gymnocalycium". (*Gymnocalycium*)

neumannii For O. Neumann (fl. 1893, 1907), German geographer and / or zoologist (?), travelled in E Africa with Baron Erlanger, (*Kalanchoe petitiana* var.)

nevadensis For the occurrence in Nevada, USA, or in the Sierra Nevada, California, USA. (*Cistanthe parryi* var., *Lewisia*) – (**2**) For the occurrence in the Sierra Nevada, Spain. (*Sedum*)

neves-armondii For Dr. Amaro Ferreira das Neves Armond (fl. 1892), director of the Brazilian National Museum. (*Rhipsalis*)

nevii For Reuben D. Nevius (1827–1913), US-American clergyman and plant collector. (*Sedum*)

ngamicus For the occurrence in the former Ngamiland, RSA. (*Pterodiscus*)

ngomensis For the occurrence near Ngome, KwaZulu-Natal, RSA. (*Brachystelma*)

ngongensis For the occurrence in the Ngong Hills, Kenya. (*Aloe*)

nicholii For Andrew A. Nichol (1895–1961), US-American biologist with a strong interest in cacti. (*Echinocactus horizonthalonius* var., *Echinocereus, Opuntia polyacantha* var.)

nicholsoniae For Baroness Carol Nicholson (fl. 1866), without further data. (*Hoya*)

nickelsiae For Anna B. Nickels (fl. 1893), US-American cactus trader in Texas. (*Coryphantha*)

niduliformis Dim. of Lat. 'nidus', nest; and Lat. '-formis', -shaped; for the interwoven central spines. (*Mammillaria huitzilopochtli* ssp.)

niebuhrianus For Carsten Niebuhr (1733–1815), German-born Danish botanist, explorer in Arabia and elsewhere. (*Aloe*)

nieuwerustensis For the occurrence near Nieuwerust, Vanrhynsdorp Distr., Western Cape, RSA. (*Ruschia*)

niger Lat. 'niger, nigra, nigrum', black; (**1**) for the leaf colour. (*Haworthia mutica* var., *Haworthia*) – (**2**) for the almost black corolla. (*Ceropegia*)

nigerianus For the occurrence in Nigeria. (*Huernia*)

nigrescens Lat., becoming black; because the whole plant turns black when dry. (*Tetragonia*)

nigriareolatus Lat. 'niger, nigra, nigrum', black; and Lat. 'areolatus', with areoles. (*Gymnocalycium*)

nigricans Lat., becoming blackish; (**1**) for the dark brown body colour. (*Rebutia*) – (**2**) for the darker leaf colouration. (*Haworthia arachnoidea* var.) – (**3**) for the blackening of the flowers on drying. (*Trianthema*)

nigricaulis Lat. 'niger, nigra, nigrum', black; and Lat. 'caulis', stem; for the black stem excrescences. (*Tylecodon*)

nigrihorridus Lat. 'niger, nigra, nigrum', black; and Lat. 'horridus', bristly, prickly, rough; for the strong blackish spination. (*Eriosyce subgibbosa* ssp.)

nigrispinoides Gr. '-oides', resembling; and for the similarity of the branches to those of *Euphorbia nigrispina*. (*Euphorbia*)

nigrispinus Lat. 'niger, nigra, nigrum', black; and Lat. '-spinus', -spined. (*Euphorbia, Maihueniopsis, Parodia*)

niloticus Lat., from the Nile River (Lat. 'Nilus', Nile River); for the occurrence there. (*Ceropegia, Sansevieria*)

niquivilensis For the occurrence at Niquivil, Chiapas, Mexico. (*Furcraea*)

nissenii For a Mr. Nissen (fl. ± 1923), without further data. (*Psammophora*)

nitidus Lat., glossy, pretty; (**1**) for the shiny appearance of the leaves. (*Gasteria, Phyllobolus, Sansevieria*) – (**2**) application obscure. (*Schizoglossum*)

nivalis Lat., pertaining to the snow; perhaps for the high-altitude occurrence. (*Peperomia*)

niveus Lat., snowy, snow-white; (**1**) for the white indumentum covering the plant. (*Portulaca*) – (**2**) for the pale leaf colour. (*Agave* vivipara var.) – (**3**) for the flower colour. (*Cephalophyllum, Sedum*)

nivosus Lat., full of snow; (**1**) for the white spination. (*Echinocereus, Parodia*) – (**2**) for the white axillary wool. (*Mammillaria*)

nivulia From the native Brahman name "Nivuli" of the plants in India. (*Euphorbia*)

nizandensis For the occurrence at Nizanda, Oaxaca, Mexico. (*Agave, Cephalocereus*)

nobilis Lat., noble, aristocratic; (**1**) for the large size of the rosettes. (*Aeonium*) – (**2**) for the appearance of the plants. (*Antimima, Rhodiola*)

noctiflorus Lat. 'nox, noctis', night; and Lat. '-florus', -flowered; for the flowers opening at dusk. (*Aridaria*)

nocturnus Lat., nightly, nocturnal; for the nocturnal flowers. (*Hesperaloe, Talinum*)

nodiflorus Lat. 'nodus', node; and Lat. '-florus', -flowered; for the (axillary) flowers at the nodes. (*Mesembryanthemum, Pedilanthus*)

nodosus Lat., knotted, knobby; because the first leaf pair of each season forms a globose to ovoid structure (corpuscle). (*Dicrocaulon*)

nodulosus Lat., with nodules, knobby; (**1**) for the texture of the stems. (*Echeveria*) – (**2**) for the small knob-like flower clusters. (*Crassula capitella* ssp.)

nogalensis For the occurrence at Nogal, Somalia. (*Portulaca, Senecio*)

nokoensis For the occurrence on Noko-San (Mt.), Taiwan. (*Sedum*)

Nolana Either from Lat. 'nola', bell; for the ± campanulate flowers of some taxa; or an alliteration to *Solanum*, since the genera share some similarities. (*Nolanaceae*)

Nolina For P. C. Nolin (fl. 1803), French agriculturalist and horticultural author. (*Nolinaceae*)

nolteei For Frans K. A. Noltee (fl. 2000), Dutch succulent plant hobbyist and nurseryman, now living in Calitzdorp, RSA. (*Tylecodon*)

nonimpressus Lat. 'non', not; and Lat. 'impressus', impressed; for the invisible line of fusion of the leaf sheaths. (*Ruschia*)

noorjahaniae For Mrs. Noorjahan Ansari (fl. 1972), wife of the Indian botanist M. Ansari. (*Ceropegia*)

noorsveldensis For the occurrence in Noorsveld vegetation, RSA. (*Euphorbia horrida* var.)

nordenstamii For Dr. R. Bertil Nordenstam (*1936), Swedish botanist, collected in Namibia and RSA 1962–1964 and 1974. (*Antimima, Drosanthemum*)

norfolkianus For the occurrence on Norfolk Island in the Pacific. (*Euphorbia*)

nortieri For Dr. P. L. Nortier (fl. 1946), physician in the Western Cape, RSA. (*Haworthia*)

notabilis Lat., notable, noteworthy. (*Haworthia maraisii* var.)

Notechidnopsis Gr. 'notos', South, and for the genus *Echidnopsis* (*Asclepiadaceae*); for the similarity to that genus and the more southern distribution. (*Asclepiadaceae*)

nothodugueyi Gr. 'notho-', false, wrong; and for *Sedum dugueyi*, because it resembles that taxon. (*Sedum*)

nothominusculus Gr. 'notho-', false, wrong; to avoid a homonym vs. *Parodia minuscula*. (*Parodia*)

nothorauschii Gr. 'notho-', false, wrong; to avoid a homonym vs. *Parodia rauschii*. (*Parodia*)

nouhuysii For J. J. van Nouhuys (fl. 1930), without further data. (*Huernia*)

novicius Lat., rather new; application obscure, perhaps because the taxon was only recently discovered, or for the fresh green of the new leaves. (*Conophytum flavum* ssp.)

nubicus For the occurrence in 'Nubia' (Sudan). (*Echidnopsis, Euphorbia*)

nubigenus Lat. 'nubes', cloud; and Lat. 'genus', birth, origin; (**1**) for the high-altitude habitat. (*Aloe, Delosperma, Dudleya*) – (**2**) because the collector's camp at the type locality was enveloped in clouds. (*Euphorbia*)

nucifer Lat. 'nux, nucis', hazelnut, nut; and Lat. '-fer, -fera, -ferum', carrying; for the nut-like fruits. (*Brownanthus*)

nuciformis Lat. 'nux, nucis', hazelnut, nut; and Lat. '-formis', -shaped; probably for the appearance of the compact leaf pair. (*Gibbaeum*)

nudicaulis Lat. 'nudus', naked; and Lat. 'caulis', stem; (**1**) for the leafless stem. (*Monadenium*) – (**2**) for the leafless stems, which only branch at the tips. (*Jatropha*) – (**3**) for the leafless peduncles. (*Crassula*)

nudiflorus Lat. 'nudus', naked; and Lat. '-florus', -flowered; for the flowers without hairs. (*Dendrocereus*)

nudus Lat., naked, nude; (**1**) perhaps for the 'naked' stems, because the leaves are congested near the stem tips. (*Sedum*) – (**2**) for the lack of spines. (*Opuntia*) – (**3**) for the lack of bracts on the lower part of the inflorescences of the original plants described. (*Echeveria*)

numaisensis For the occurrence in the Numais Mts., SW Namibia. (*Crassula*)

numeesensis For the occurrence at Numees, Northern Cape, RSA. (*Cephalophyllum*)

nummarioides Gr. '-oides', resembling; and for *Lysimachia nummularia* ("Moneywort", *Primulaceae*); for the similar leaves; or from Lat. 'nummularius', pertaining to coins, for the round leaves. (*Hoya*)

nummularius Lat., pertaining to coins; for the leaf shape. (*Dischidia*)

nunezii For Prof. C. Nuñez (fl. 1921), Mexican botanist and cactophile. (*Mammillaria*)

nussbaumerianus For Ernst Nussbaumer (fl. 1935), head gardener ("inspector") at the Botanical Garden Bremen, Germany. (*Sedum*)

nutans Lat., nodding, pendent; (**1**) for the pendent inflorescences. (*Cremnophila*) – (**2**) perhaps for the flowers. (*Bulbine*)

nuttallianus For Thomas Nuttall (1786–1859), British naturalist and pioneer explorer of the USA 1808–1841. (*Sedum*)

nuttii For W. Harwood Nutt (fl. 1895), missionary in Zambia. (*Aloe*)

nyambensis For the occurrence in the Nyambeni Hills, Meru Distr., Kenya. (*Monadenium ritchiei* ssp.)

nyasae For the occurrence in the former Nyasaland (now Malawi). (*Aspidoglossum*)

nyasicus For its occurrence in the former Nyasaland (now Malawi). (*Sansevieria metallica* var., *Sesuvium*)

nyassae For the occurrence near Lake Nyassa (present-day Lake Malawi). (*Euphorbia*)

nyensis For the occurrence in Nye County, Nevada, USA. (*Sclerocactus*)

nyeriensis For the occurrence at Nyeri, Kenya. (*Aloe*)

nyikae For the occurrence in the 'nyika' or bush-covered coastal plain. (*Euphorbia*, *Kalanchoe*)

nyikensis For the occurrence on the Nyika Plateau, Malawi. (*Glossostelma*, *Senecio*)

nymphaeifolius For the genus *Nymphaea* ("Waterlily", *Nymphaeaceae*); and Lat. '-folius', -leaved; for similarity of the leaves to those of some waterlilies. (*Cissus*)

O

oaxacanus For the occurrence in the Mexican state of Oaxaca. (*Mammillaria albilanata* ssp., *Sedum*)

oaxacensis As above. (*Agave americana* var., *Peniocereus*)

obconellus Lat., obclavate, inversely club-shaped; for the shape of the plant body. (*Mammillaria polythele* ssp.)

obconicus Lat. 'ob-', reversed, inverted; and Lat. 'conicus', conical; (**1**) for the shape of the leaf sheaths. (*Monilaria*) – (**2**) application obscure. (*Lampranthus*)

obcordatus Lat. 'ob-', reversed, inverted; and Lat. 'cordatus', heart-shaped; for the leaf shape. (*Micholitzia, Sedum*)

obcordellus Lat. 'ob-', reversed, inverted; and Dim. of Lat. 'cor, cordis', heart; for the shape of the fused leaf pairs. (*Conophytum*)

obductus Lat., covered over, overspread; perhaps because the plants grow in and under dry leaves etc. and are difficult to see. (*Gymnocalycium, Stapelia*)

obesus Lat., well-fed, plump; for the stem shape. (*Adenium, Euphorbia*)

Obetia Unexplained; perhaps based on Lat. 'obesus', well-fed, plump; either for the thick trunk, or for the plump, lop-sidedly swollen ovary. (*Urticaceae*)

oblanceolatus Lat. 'ob-', reversed, inverted; and Lat. 'lanceolatus', lanceolate, lance-shaped; for the leaf shape. (*Commiphora, Crassula, Sedum*)

obliquus Lat., oblique; (**1**) for the leaf position. (*Rhinephyllum*) – (**2**) for the oblique blunt leaf tip. (*Trichodiadema*)

oblongatus Lat. 'oblongus', oblong; (**1**) for the oblong tubers. (*Pelargonium, Turbina*) – (**2**) for the leaf shape. (*Myrmecodia*)

oblongifolius Lat. 'oblongus', oblong; and Lat. '-folius', -leaved. (*Talinum*)

oblongus Lat., oblong; (**1**) for the shape of the stem segments. (*Rhipsalis*) – (**2**) for the leaf shape. (*Cotyledon orbiculata* var.)

obovatus Lat., obovate, inverted egg-shaped; (**1**) for the leaf shape. (*Crassula, Hoya*) – (**2**) for the shape of the leaf sheaths. (*Monilaria scutata* ssp.)

Obregonia For Álvaro Obregón (1880–1928), Mexican politician and popular president 1920–1924. (*Cactaceae*)

obrepandus Lat. 'ob-', reversed, inverted; and Lat. 'repandus', repand, with a slightly uneven margin; for the obliquely adjacent tubercles of the ribs. (*Echinopsis*)

obscurus Lat., indistinct, obscure; (**1**) for the leaf markings. (*Sansevieria nilotica* var.) – (**2**) because the taxon "seems to have hidden from field botanists until now" [protologue]. (*Amphibolia*) – (**3**) for the undistinguished appearance of the plants. (*Conophytum*) – (**4**) perhaps for the dark red flowers. (*Agave*)

obsubulatus Lat. 'ob-', reversed, inverted; and Lat. 'subulatus', subulate, awl-shaped; for the subulate leaves becoming broader towards the tip. (*Cylindrophyllum*)

obtrullatus Lat. 'ob-', reversed, inverted; and Lat. 'trullatus', trullate, shaped like a bricklayers trowel; for the petal shape. (*Sedum*)

obtusatus Lat. 'obtusus', blunt, obtuse; for the leaf tips. (*Sedum*)

obtusifolius Lat. 'obtusus', blunt, obtuse; and Lat. '-folius', -leaved. (*Crassula muscosa* var., *Echeveria fulgens* var., *Phedimus, Portulaca, Sinocrassula indica* var.)

obtusipetalus Lat. 'obtusus', blunt, obtuse; and Lat. 'petalum', petal. (*Sedum*)

obtusus Lat., blunt, obtuse; (**1**) for the leaf shape. (*Crassula, Delosperma, Haworthia cymbiformis* var., *Kalanchoe, Lenophyllum, Portulaca*) – (**2**) for the obtuse leaf keels. (*Ruschia*)

obvallatus Lat., surrounded with a wall or rampart; for the robust spination of the plant body. (*Stenocactus*)

ocahui From the local vernacular name "Ocahui" or "Ojahui" for the plants in Mexico. (*Agave*)

ocampoi For Enrique Ocampo (fl. 1955), Bolivian student of Prof. M. Cárdenas. (*Parodia*)

ocamponis For Don Melchor Ocampo (fl. 1849, 1866), Mexican scientist and politician. (*Hylocereus*)

occidentalis Lat., western; (**1**) for the occurrence in W Madagascar. (*Aloe*) – (**2**) for the occurrence in the Western Cape, RSA. (*Ceropegia, Trichodiadema*) – (**3**) because this is the W-most taxon of the genus. (*Brachystelma, Graptopetalum*) – (**4**) for the occurrence in W Africa. (*Telfairia*) – (**5**) for the occurrence in W Peru. (*Furcraea, Rhipsalis*) – (**6**) for the occurrence in W Botswana. (*Orbea valida* ssp.) – (**7**) for the occurrence in W Mexico. (*Acanthocereus, Echinocereus stramineus* ssp., *Peniocereus*) – (**8**) for the occurrence in the W USA. (*Opuntia*)

occultans Lat., hiding; (**1**) probably because the plants are small and difficult to locate. (*Tylecodon*) – (**2**) for the hidden stamens. (*Lampranthus*)

occultiflorus Lat. 'occultus', hidden; and Lat. '-florus', -flowered; for the flowers hidden in a pseudocephalium. (*Pilosocereus*)

occultus Lat., hidden; (**1**) because the taxon prefers deep shade, and because it was formerly subsumed under another species. (*Huernia*) – (**2**) for the cryptic appearance of the plants. (*Eriosyce, Gymnocalycium stellatum* ssp.) – (**3**) application obscure. (*Ceropegia*)

ochoterenae For Prof. Isaac Ochoterena (1885–1950), Mexican botanist and from 1915 director of the Biological Institute of the University of Mexico. (*Echinocereus subinermis* ssp., *Fouquieria, Stenocactus*) – (**2**) Perhaps for Prof. Isaac Ochoterena (1885–1950), Mexican botanist, but unlikely since *G. ochoterenae* is a plant from Argentina. (*Gymnocalycium*)

ochraceus Lat., ochre-yellow; for the flower colour. (*Malephora*)

ochroleucus Gr. 'ochros', pale, pale yellowish; and Gr. 'leukos', white; for the flower colour. (*Pelargonium, Sedum*)

octacanthus Gr. 'okto', eight; and Gr. 'akanthos', spine, thorn; for the number of radial spines. (*Coryphantha*)

octojuge Lat. 'octo', eight; and Lat. 'jugum', yoke, ridge, pair; perhaps for the architecture of the fruit capsules with eight locules. (*Octopoma*)

octonarius Lat., in numbers of eight; for the eight locules of the fruit capsules. (*Enarganthe*)

octophyllus Gr. 'okto', eight; and Gr. 'phyllon', leaf. (*Argyroderma*)

octopodes Lat. 'octopus', octopus; for the numerous long stolons. (*Sempervivum ciliosum* ssp.)

Octopoma Gr. 'okto', eight; and Gr. 'poma', cover, lid; for the usually eight valves of the fruit capsules. (*Aizoaceae*)

ocuilensis For the occurrence near Ocuilán, state of México, Mexico. (*Sedum*)

oculatus Lat., eye-shaped, with an eye; (**1**) for to the coloration of the flowers. (*Drosanthemum, Huernia, Phyllobolus*) – (**2**) for the white spot on the corolla lobes. (*Ceropegia*)

odieri For James (?) Odier (fl. 1849), French cactus collector in Paris. (*Eriosyce*)

odontocalyx Gr. 'odous, odontos', tooth; and Gr. 'calyx', calyx. (*Ruschia*)

odontolepis Gr. 'odous, odontos', tooth; and Gr. 'lepis', scale; for the shape of the staminal corona. (*Sarcostemma viminale* ssp.)

Odontophorus Gr. 'odous, odontos', tooth; and Gr. '-phoros', -carrying; for the prominently dentate leaves. (*Aizoaceae*)

odontophorus Gr. 'odous, odontos', tooth; and Gr. '-phoros', -carrying; for the prominent teeth on the branch angles. (*Euphorbia*)

odontophyllus Gr. 'odous, odontos', tooth; and Gr. 'phyllon', leaf; for the toothed leaf margins. (*Phedimus*)

Odontostelma Gr. 'odous, odontos', tooth; and Gr. 'stelma', crown, garland, wreath; for the nature of the corona. (*Asclepiadaceae*)

odoratissimus Superl. of Lat. 'odoratus', pleasantly scented. (*Pelargonium*)

odoratus Lat., pleasantly scented. (*Ceropegia, Hereroa, Stathmostelma*)

odorus Lat., fragrant; (**1**) for the scented flowers. (*Arthrocereus melanurus* ssp.) – (**2**) for the strong cumarine scent produced by the plant tissue. (*Coryphantha*) – (**3**) application obscure. (*Senecio anteuphorbium* var.)

Odosicyos Gr. 'odous, odontos', tooth; and Gr. 'sicyos', cucumber; for the toothed leaf margins and the family placement. (*Cucurbitaceae*)

oehleri For Dr. Oehler (fl. 1907), German (?) plant collector in Namibia and Tanzania. (*Delosperma*)

oenanthemus Gr. 'oinos', wine; and Gr. 'anthemon', flower; for the dark red flowers. (*Gymnocalycium*)

oenotherae For the similarity to the genus *Oenothera* ("Evening Primrose", *Onagraceae*). (*Ipomoea*)

oenotheroides Gr. '-oides', like; for the similarity to the genus *Oenothera* ("Evening Primrose", *Onagraceae*). (*Turbina*)

oertendahlii For I. A. Oertendahl (fl. 1924), head horticulturist at the University Botanical Gardens, Uppsala, Sweden. (*Plectranthus*)

officinalis Lat., used medicinally. (*Aloe, Hoodia*)

officinarum Gen. Pl. of Lat. 'officina', working place, drugstore; for the medicinal use. (*Euphorbia*)

ogadensis For the occurrence in the Ogaden region, Ethiopia. (*Kleinia, Orbea sprengeri* ssp.)

oianthus Gr. 'oion', egg; and Gr. 'anthos', flower; for the corolla shape. (*Brachystelma*)

oishii For M. Oishi (fl. 1971), Japanese collector of the type. (*Hylotelephium sordidum* var.)

okinawensis For the occurrence on the Japanese island of Okinawa (one of the Ryuku Islands). (*Portulaca pilosa* ssp.)

oksapminensis For the type locality Oksapmin, Papua New Guinea. (*Myrmecodia*)

olaboensis From the Madagascan vernacular name "Olaboay" for the plant. (*Adenia*)

oleifolius Lat. 'Olea', olive tree; and Lat. '-folius', -leaved; for the leaf shape. (*Adenium obesum* ssp.)

oleraceus Lat., herbage, pertaining to the kitchen garden (as pot-herb, vegetable or weed); (**1**) for the utilization of the plants. (*Portulaca*) – (**2**) for the cabbage-like leaves. (*Cyphostemma*)

oligocarpus Gr. 'oligos', few; and Gr. 'karpos', fruit. (*Glottiphyllum, Sedum*)

oligocladus Gr. 'oligos', few; and Gr. 'klados', branch; for the habit. (*Euphorbia*)

oligogonus Gr. 'oligos', few; and Gr. 'gonia', tubercle, angle; for the few-ribbed stems. (*Armatocereus*)

oligolepis Gr. 'oligos', few; and Gr. 'lepis', scale; for the scales o f the pericarpel of the flowers. (*Pilosocereus*)

oligophyllus Gr. 'oligos', few; and Gr. 'phyllon', leaf. (*Aloe*)

oligospermus Gr. 'oligos', few; and Gr. 'sperma', seed. (*Portulaca, Sedum, Talinum*)

olivaceus Lat., olive-brown; (**1**) for the general colour of the plants. (*Kalanchoe*) – (**2**) for the leaf colour. (*Echeveria, Haworthia reinwardtii* fa., *Lithops*) – (**3**) for the colour of the tuft of bristles at the leaf tips. (*Trichodiadema*) – (**4**) for the flower colour. (*Stapelia*)

olivifer Lat. 'oliva', olive; and Lat. '-fer, -fera, -ferum', -carrying; for the olive-shaped fruits. (*Rhipsalis*)

olosirawa From the local vernacular name of the taxon in Kenya as noted on the type collection. (*Portulaca*)

olowinskianus For a Mr. Olowinski (fl. 1937), without further data. (*Haageocereus acranthus* ssp.)

omarianus For the occurrence at Sof [= cave] Omar, Ethiopia. (*Euphorbia*)

omasensis For the occurrence near Omas, Dept. Lima, Peru. (*Eriosyce*)

omissus Lat., omitted; because the taxon is based on material collected already 1931, but was only described 1963. (*Brachystelma*)

ommanneyi For Henry T. Ommanney (1849–1936), Englishman working in India and collecting 1901–1902 in the Johannesburg area in RSA. (*Ipomoea*)

oncocladus Gr. 'ogkos, onkos', swelling, swollen; and Gr. 'klados', branch; for the thickened branches. (*Euphorbia leucodendron* ssp.)

onychacanthus Gr. 'onyx, onychos', claw, onyx; and Gr. 'akantha', thorn, spine; for the claw-like spination. (*Melocactus bellavistensis* ssp.)

onychopetalus Gr. 'onyx, onychos', claw, onyx; and Gr. 'petalon', petal. (*Sedum*)

Oophytum Gr. 'oon', egg; and Gr. 'phyton', plant; for the shape of the pairs of fused leaves. (*Aizoaceae*)

opacus Lat., opaque; for the papillae covering the leaves. (*Drosanthemum*)

Operculicarya Lat. 'operculum' small lid; and Gr. 'karya', nut tree; for the operculate nut-like seeds. (*Anacardiaceae*)

Ophionella Dim. of Gr. 'ophis', snake, serpent; for the snake-like sinuous stems. (*Asclepiadaceae*)

ophiophyllus Gr. 'ophis', snake, serpent; and Gr. 'phyllon', leaf; for the tangled mass of leaves resembling a serpent nest. (*Bulbine*)

opimus Lat., fat; for the thick leaves. (*Othonna*)

oppositifolius Lat. 'oppositus', opposite; and Lat. '-folius', -leaved. (*Umbilicus*)

opticus Gr. 'optikos', concerning vision, pertaining to eyes; for the eye-like appearance of the top of the leaf pairs. (*Lithops*)

opulentus Lat., rich, wealthy; for the richly branched inflorescences. (*Mesembryanthemum*)

Opuntia Origin debatable, usually said to derive from Lat. 'Opuntius', a native of the ancient Greek city of Opus; but more probably from Gr. 'opos', fig juice; for the fig-like fruits; or a composition from the Aztec name 'nopalli' (cf. present-day Sp. 'nopal' for *Opuntia* sp.) and Lat. 'pungere', to prick, to sting. (*Cactaceae*)

opuntioides Gr. '-oides', resembling; and for the genus *Opuntia* ("Prickly Pear Cactus", *Cactaceae*); for the similarly segmented branches. (*Euphorbia, Schlumbergera*)

oramicola 'oram-', unresolved; and Lat. '-cola', -dwelling. (*Hoya australis* ssp.)

orangeanus For the occurrence in the former Orange Free State (now Free State), RSA. (*Cynanchum, Sarcostemma viminale* ssp.)

orbatus Lat., orphaned, abandoned; because no wild origin was known for the taxon at the time of description. (*Sedum*)

Orbea Lat. 'orbis', circle; for the thickened corolla part (annulus) surrounding the flower centre. (*Asclepiadaceae*)

Orbeanthus For the genus *Orbea* (*Asclepiadaceae*); and Gr. 'anthos', flower; for the *Orbea*-like flowers. (*Asclepiadaceae*)

orbicularis Lat., orbicular; for the leaf shape found in some forms. (*Crassula*)

orbiculatus Lat., orbicular; (**1**) for the shape of the stem segments. (*Opuntia*) – (**2**) for the leaf shape found in some forms. (*Cotyledon*)

orbiculifolius Lat. 'orbiculus', small circle; and Lat. '-folius', -leaved. (*Euphorbia*)

orcuttii For Charles R. Orcutt (1864–1929), US-American natural history dealer and publisher in San Diego, and very active explorer and collector. (*Dudleya attenuata* ssp., *Escobaria, Mammillaria,* ×*Pacherocactus*)

oreades Lat. (originally Gr.) 'oreas, oreades', Mountain Nymph; for the high-mountain habitat. (*Sedum*)

oreas Lat. (originally Gr.) 'oreas, oreades', Mountain Nymph; for the mountain habitat. (*Melocactus*)

oreganus For the occurrence in the state of Oregon, USA. (*Sedum*)

oregonensis As above. (*Sedum*)

Oreocereus Gr. 'oreios', pertaining to mountains; and *Cereus*, a genus of columnar cacti; for the occurrence at high altitudes. (*Cactaceae*)

oreodoxus Gr. 'oreios', pertaining to mountains; and Gr. 'doxa', beauty, glory; for the attractive flowers and the high-altitude occurrence. (*Matucana*)

oreophilus Gr. 'oreios', pertaining to mountains; and Gr. 'philos', friend; for the occurrence in mountains. (*Echeveria, Pelargonium, Perrierastrum, Rhipsalis floccosa* ssp.)

oresbius From Gr. 'oros', mountain; and Gr. '-bius', living; for the preferred habitat. (*Sarcostemma*)

organensis For the occurrence in the Organ Mts., New Mexico, USA. (*Escobaria*)

orgyalis Gr. 'orgyia', fathom; and Lat. suffix '-alis', pertaining to, i.e. fathom-long, 6 feet long or high; for the size of the plants. (*Kalanchoe*)

oribiensis For the occurrence in the Oribi

Gorge, S KwaZulu-Natal, RSA. (*Plectranthus*)

oricola Gr. 'oros', mountain; and Lat. '-cola', inhabiting; for the occurrence in mountains. (*Opuntia*)

orientalis Lat., eastern; (**1**) for the more E distribution in relation to other infraspecific taxa. (*Aloe marlothii* ssp., *Opuntia megasperma* var., *Orbea semota* ssp., *Sedum eriocarpum* ssp., *Sedum montanum* ssp.) – (**2**) for the occurrence in E Madagascar. (*Aloe*) – (**3**) for the occurrence in the Eastern Cape Prov., RSA. (*Ruschia, Trichodiadema*) – (**4**) for the E-most distribution within Sect. *Adromischus*. (*Adromischus liebenbergii* ssp.)

ormindoi For Paulo Ormindo (fl. 1996), Brazilian botanical artist from Niteroi, Rio de Janeiro. (*Rhipsalis*)

ornatulus Dim. of Lat. 'ornatus', adorned, i.e. ornamental, showy. (*Delosperma*)

ornatus Lat., adorned; (**1**) for the ornamental value. (*Plectranthus*) – (**2**) for the attractive spination. (*Astrophytum*) – (**3**) probably for the reddish margins and keels of the pale green leaves. (*Oscularia*)

ornithobroma Gr. 'ornis, ornithos', bird; and Gr. 'broma', food; because buds and flowers are eaten by small parrots. (*Agave*)

Ornithogalum Lat. 'ornithogale' = Gr. 'ornithogalon', "Bird's Milk", a plant (from Gr. 'ornis, ornithos', bird, and Gr. 'gala', milk); for the egg-shale-coloured flowers of some European taxa; or going back to a Roman allusion of something rare or beautiful "as bird's milk". (*Hyacinthaceae*)

ornithopus Gr. 'ornis, ornithos', bird; and Gr. 'pous', foot; for the shape of the branched nectar glands. (*Euphorbia*)

orobanchoides Gr. '-oides', resembling; and for *Orobanche* ("Broomrape", *Orobanchaceae*); for the similarity of the flowering stems to this genus. (*Monadenium*)

oroensis For the occurrence near Concepción del Oro, Zacatecas, Mexico. (*Agave*)

Orostachys Gr. 'oros', mountain; and Gr. 'stachys', spike; for the occurrence in mountainous regions, and for the inflorescence shape. (*Crassulaceae*)

Oroya For the occurrence near the town of Oroya, Dept. Puno, Peru. (*Cactaceae*)

orpenii For Mr. Redmond Orpen (fl. 1921), of Kleinzee, Northern Cape, RSA, without further data. (*Prepodesma*)

orssichianus For Countess Beatrix Orssich (fl. 1978), Brazilian plant lover in Teresopolis. (*Schlumbergera*)

ortegae For Jesús González Ortega (1876–1936), Mexican botanist. (*Echinocereus*)

Ortegocactus For the Ortega family of San José Lachiguiri, Mexico, who aided with the discovery of the taxon; and Lat. 'cactus', cactus. (*Cactaceae*)

ortgiesii For Karl Eduard Ortgies (1829–1916), German horticulturist. (*Oxalis*)

orthogonus Gr. 'orthos', erect, straight; and Gr. 'gonia', angle; for the straight vertical ribs of the stems. (*Cleistocactus*)

ortholophus Gr. 'orthos', erect, straight; and Gr. 'lophos', crest; for the row of secund flowers. (*Aloe*)

Orthopterum Gr. 'orthos', erect, straight; and Gr. 'pteron', wing; for the erect upper parts of the septa of the fruit capsules. (*Aizoaceae*)

oryzifolius Lat. '-folius', -leaved; and for the similarity of the leaves to species of the genus *Oryza* ("Rice", *Poaceae*). (*Sedum uniflorum* ssp.)

Oscularia Lat. 'osculum' = Dim. of Lat. 'os', mouth; for the gaping leaves of a pair, having dentate margins in some species. (*Aizoaceae*)

ossetiensis For the occurence in Ossetia, Georgia. (*Sempervivum*)

Osteospermum Gr. 'osteon', bone; and Gr. 'sperma', seed; for the bone-hard seeds. (*Asteraceae*)

otallensis For the occurrence at Otallo, Ethiopia. (*Aloe*)

otaviensis For the occurrence in the Otavi Mts., Namibia. (*Pelargonium*)

oteroi For Felipe Otero (fl. 1975), Mexican plant collector. (*Echeveria setosa* var., *Mammillaria, Sedum*)

Othonna Gr., a plant from Syria or Arabia used medicinally; or from Gr. 'othone', linen, used in classical times for a different plant with perforated leaves. (*Asteraceae*)

otjipembanus For the occurrence at Otjipemba in Namibia. (*Euphorbia*)

ottonis For Christoph Friedrich Otto (1783–1856), German botanist and horticulturist at the Botanical Garden Berlin. (*Coryphantha, Parodia*)

Ottosonderia For Dr. Otto W. Sonder (1812–1881), German pharmacist and botanist at RBG Kew, with a large personal herbarium rich in African and Australian plants. (*Aizoaceae*)

otuyensis For the occurrrence near Otuyo, Prov. Linares, Dept. Potosi, Bolivia. (*Corryocactus*)

otzenianus For Max Otzen († c. 1947), German who went to RSA in the 1880s, then to Namibia for the Diamond Company, becoming a member of the board of directors, and succulent plant enthusiast, in 1930 retired to Cape Town, RSA. (*Lampranthus, Lithops*)

outeniquensis For the occurrence in the Outeniqua Mts., S Western Cape, RSA. (*Haworthia*)

ovalifolius Lat. 'ovalis', oval, egg-shaped; and Lat. '-folius', -leaved. (*Aspidoglossum, Moringa*)

ovatifolius Lat. 'ovatus', egg-shaped, ovate; and Lat. '-folius', -leaved. (*Agave, Dudleya cymosa* ssp., *Ledebouria*)

ovatostipulatus Lat. 'ovatus', egg-shaped, ovate; and Lat. 'stipulatus', with stipules. (*Pelargonium stipulaceum* ssp.)

ovatus Lat., egg-shaped, ovate; (**1**) for the shape of the stem segments. (*Maihueniopsis*) – (**2**) for the leaf shape. (*Adenia, Crassula, Dischidia, Myrsiphyllum*) – (**3**) for the fruit shape. (*Tetragonia*)

ovifer Lat. 'ovum', egg; and Lat. '-fer, -fera, -ferum', -carrying; for the shape of the turgid leaves. (*Pachyphytum*)

oviformis Lat. 'ovum', egg; and Lat. '-formis', -shaped; (**1**) for the leaf shape. (*Antimima*) – (**2**) for the shape of the pairs of fused leaves. (*Oophytum*)

ovoideus Lat., egg-shaped; for the leaf form. (*Senecio*)

Oxalis Gr. / Lat. 'oxalis', ancient name for the sorrels; from Gr. 'oxaleios', acidic; for the acrid taste of the leaves, which are rich in the eponymous oxalic acid. (*Oxalidaceae*)

oxycalyptra Gr. 'oxys', sharp, pointed; and Gr. 'kalyptra', covering, woman's hat; for the apiculate tips of the floral bracts. (*Trianthema*)

oxycoccoides Gr. '-oides', resembling; and for the genus *Oxycoccus* (*Ericaceae*). (*Sedum*)

oxycostatus Gr. 'oxys', sharp, pointed; and Lat. 'costatus', ribbed; for the sharply ribbed plant bodies. (*Parodia*)

oxygonus Gr. 'oxys', sharp, pointed; and Gr. 'gonia', corner, margin; for the sharply defined ribs of the plant bodies. (*Echinopsis*)

oxypetalus Gr. 'oxys', sharp, pointed; and Gr. 'petalon', petal. (*Epiphyllum, Sedum*)

oxyriifolius For the genus *Oxyria* ("Mountain Sorrel", *Polygonaceae*); and Lat. '-folius', -leaved. (*Senecio*)

oxysepalus Gr. 'oxys', sharp, pointed; and Lat. 'sepalum', sepal. (*Erepsia*)

oxystegius Gr. 'oxys', sharp, pointed; and Gr. 'stege', protection, covering; for the protection by the sharp persistent peduncles. (*Euphorbia*)

P

paardebergensis For the occurrence on the Paardeberg, Western Cape, RSA. (*Oscularia*)

paarlensis For the occurrence in the Paarl Distr., Western Cape, RSA. (*Lampranthus*)

pacalaensis For the occurrence near Pacala, Dept. Libertad, Peru. (*Haageocereus*)

pacensis For the occurrence near San Luis de la Paz, Guanajuato, Mexico. (*Sedum*)

pachacoensis For the occurrence near Pachaco, San Juan, Argentina. (*Eriosyce strausiana* var.)

pachanoi For Prof. Abelardo Pachano (fl. 1918), Ecuadorian who was assistant to the US-American botanist J. N. Rose on his expedition to South America in 1918. (*Echinopsis*)

pacheco-leonis For Prof. Dr. Antonio Pacheco Leão (1872–1931), Brazilian botanist and 1915–1931 director of the Botanical Garden of Rio de Janeiro. (*Rhipsalis*)

pachona Span. 'pachón, pachona', hairy, shaggy (from Náhuatl 'pacho', woolly, and 'pachtli', hemp); most probably for the irregularly arranged and sometimes twisted spines. (*Opuntia*)

pachyacanthus Gr. 'pachys', thick; and Gr. 'akanthos', spine, thorn. (*Melocactus*)

pachyanthus Gr. 'pachys', thick; and Gr. 'anthos', flower. (*Cyphostemma*)

Pachycarpus Gr. 'pachys', thick; and Gr. 'karpos', fruit; for the mostly very large fruits. (*Asclepiadaceae*)

pachycaulos Gr. 'pachys', thick; and Gr. 'kaulos', stem. (*Aichryson*)

pachycentrus Gr. 'pachys', thick; and Gr. 'kentron', centre; most probably for the massive shortly conical central bud of the rosettes. (*Agave*)

Pachycereus Gr. 'pachys', thick; and *Cereus*, a genus of columnar cacti; for the thick columnar stems. (*Cactaceae*)

pachycladon Gr. 'pachys', thick; and Gr. 'klados', branch, twig. (*Cynanchum*)

pachycladus Gr. 'pachys', thick; and Gr. 'klados', branch, twig. (*Cleistocactus, Euphorbia, Hoya, Pilosocereus, Rhodiola*)

Pachycormus Gr. 'pachys', thick; and Gr. 'kormos', trunk; for the thick trunk and branches. (*Anacardiaceae*)

pachygaster Gr. 'pachys', thick; and Gr. 'gaster', stomach; for the flower shape. (*Aloe*)

pachyglossus Gr. 'pachys', thick; and Gr. 'glossa', tongue; perhaps for the corona segments. (*Schizoglossum bidens* ssp.)

pachyphyllus Gr. 'pachys', thick; and Gr. 'phyllon', leaf. (*Cerochlamys, Graptopetalum, Sedum*)

Pachyphytum Gr. 'pachys', thick; and Gr. 'phyton', plant; for the usually thick leaves. (*Crassulaceae*)

pachyphytus Gr. 'pachys', thick; and Gr. 'phyton', plant; for the thick leaves. (*Dudleya*)

pachypodioides Gr. '-oides', resembling; and for the similarity of the stems with young plants of *Pachypodium geayi* (*Apocynaceae*). (*Euphorbia*)

Pachypodium Gr. 'pachys', thick; and Gr. 'podion', small foot; for the thick caudiciform trunks formed by some species. (*Apocynaceae*)

pachypodius Gr. 'pachys', thick; and Gr. 'podion', small foot; (**1**) for the massive root tubers. (*Brachystelma*) – (**2**) presumably for the short stems. (*Gibbaeum*)

pachypodus Gr. 'pachys', thick; and Gr. 'pous, podos', foot; for the basal thickened stem part. (*Othonna, Talinella*)

pachypterus Gr. 'pachys', thick; and Gr. 'pteron', wing; for the architecture of the stem segments. (*Rhipsalis*)

pachypus Gr. 'pachys', thick; and Gr. 'pous', foot; (**1**) for the tuberous roots. (*Schismocarpus*) – (**2**) for the massively thickened trunk. (*Cyphostemma, Operculicarya*) – (**3**) for the long thickened pericarpel. (*Austrocylindropuntia*)

Pachyrhizus Gr. 'pachys', thick; and Gr. 'rhiza', root; for the thickly tuberous roots. (*Fabaceae*)

pachyrhizus Gr. 'pachys', thick; – (**1**) for the thickly tuberous roots. (*Cereus, Delosperma, Epithelantha micromeris* ssp.)

pachyrrhizus Gr. 'pachys', thick; and Gr. 'rhiza', root. (*Jatropha, Opuntia, Tridentea*)

pachysanthus Gr. 'pachys', thick; and Gr. 'anthos', flower. (*Euphorbia*)

pachystelma Gr. 'pachys', thick; and Gr. 'stelma', crown, garland, wreath; for the massive corona. (*Ceropegia*)

pachystemon Gr. 'pachy-', thick-; and Gr. 'stemon', stamen; for the stout stamina. (*Crassula lanuginosa* var.)

pacificus For the occurrence near the Pacific coast of Baja California, Mexico. (*Echinocereus polyacanthus* ssp.)

padcayensis For the occurrence at Padcaya, Dept. Tarija, Bolivia. (*Rebutia*)

padifolius From [*Prunus*] *padus*, wild cherry; and Lat. '-folius', -leaved. (*Pedilanthus tithymaloides* ssp.)

paediophilus Gr. 'paidion', small child; and Gr. 'philos', friend; for the richly offsetting nature of the plant. (*Gymnocalycium*)

paedogonus Gr. 'paedo-', pertaining to children; and Gr. 'gonos', seed; because the plants are grown in Angola as a fertility charm. (*Aloe*)

paganorum Gen. Pl. of Lat. 'paganus', peasant, farmer; because the taxon is planted in rural villages. (*Euphorbia*)

pageae For Miss Mary M. Page (1867–1925), British-born botanical artist, emigrated to RSA in 1911. (*Conophytum, Crassula*)

pageanus As above. (*Delosperma*)

pailanus For the occurrence in the Sierra de la Paila, Coahuila, Mexico. (*Opuntia, Turbinicarpus mandragora* ssp.)

painteri For Joseph H. Painter (1879–1908), US-American collecting assistant to J. N. Rose in Mexico 1905. (*Villadia*)

pakhuisensis For the occurrence at Pakhuis Pass, Clanwilliam Distr., Western Cape, RSA. (*Lampranthus*)

pakpassensis For the occurrence at Pakpass, Western Cape, RSA. (*Lampranthus*)

palhuayensis For the occurrence in the valley of the Río Palhuaya, Prov. Muñecas, Dept. La Paz, Bolivia. (*Cleistocactus*)

pallens Lat., pale, pale greenish; (**1**) for the leaf colour. (*Crassula, Hereroa, Monanthes, Nananthus, Prenia, Ruschia*) – (**2**) for the flower colour. (*Drosanthemum, Quaqua*) – (**3**) application obscure. (*Galenia*)

pallescens Lat., becoming pale; for the flower colour. (*Hylotelephium*)

pallidiflorus Lat. 'pallidus', pale; and Lat. '-florus', -flowered. (*Rosularia*)

pallidinervius Lat. 'pallidus', pale; and Lat. 'nervus', vein, fibre; for the leaf patterning. (*Zehneria*)

pallidus Lat., pale; (**1**) for the pale green stems, leaves and bracts. (*Tradescantia*) – (**2**) for the pale green leaves. (*Delosperma, Echeveria, Sedum, Yucca*) – (**3**) for the flower colour. (*Aloe prostrata* ssp., *Coryphantha, Crassula alba* var., *Sedum laconicum* ssp.)

palmadora From the local vernacular name "Palmatória" for the taxon in Bahia, Brazil. (*Tacinga*)

palmatus Lat., palmate; for the leaf shape. (*Cucurbita*)

palmensis For the occurrence on La Palma, Canary Islands. (*Aeonium canariense* var., *Aichryson*)

palmeri For A. H. Palmer Esq. (fl. 1873), onetime Colonial Secretary of Queensland, for his services to horticulture. (*Doryanthes*) – (**2**) For Edward Palmer (1831–1911), English-born botanist and plant collector in the USA. (*Agave, Dudleya, Echinocereus, Pedilanthus, Pseudobombax, Sedum, Talinum*)

palmiformis Lat., having the form of a palm. (*Aloe*)

paluster Lat. 'paluster, palustris, palustre', pertaining to swamps; for the preferred habitat. (*Agave, Lampranthus, Raphionacme*)

pamanesiorum For General Fernando Pámanes Escobedo (fl. 1981), Mexican politician and former governor of the state of Zacatecas, for the help provided to A. B. Lau during his travels. (*Echinocereus*)

pamiroalaicus For the occurrence in the Pamiro-Alaj Mts., C Asia. (*Rhodiola*)

pampaninii For Dr. Renato Pampanini (1875–1949), Italian botanist. (*Sedum*)

pampanus For the occurrence on the Pampa de Arrieros, Dept. Arequipa, Peru. (*Echinopsis*)

pampeanus For the occurrence in the Pampa vegetation of Argentina. (*Opuntia*)

panamensis For the occurrence in Panama. (*Pseudorhipsalis amazonica* ssp., *Weberocereus*)

panchganiensis For the occurrence in Panchgani, W Ghat Mts., Maharashtra, India. (*Euphorbia*)

paniculatus Lat., paniculate; (1) for the richly branched plants. (*Leptocereus*) – (2) for the inflorescence. (*Aizoon, Cissus, Cistanthe, Cussonia, Dioscorea sylvatica* var., *Dudleya cymosa* ssp., *Echeveria, Kalanchoe, Pelargonium, Sinocrassula indica* var., *Stapelia, Talinum, Tylecodon*)

paniculiformis Lat. 'panicula', panicle; and Lat. '-formis', -shaped; for the inflorescences. (*Umbilicus*)

papaveroides Gr. '-oides', resembling; and for the genus *Papaver* ("Poppy", *Papaveraceae*); perhaps for the similar leaves. (*Othonna*)

papilionus Lat. 'papilio', butterfly; because the flowering plants observed were full of butterflies. (*Euphorbia*)

papillaris From Lat. 'papilla', papilla; for the occasionally papillate leaves. (*Cotyledon*)

papillatus Lat., papillate; (1) for the stems and leaves. (*Gunniopsis*) – (2) for the stems and leaf margins. (*Disphyma*) – (3) for the papillate leaf surface. (*Antimima*) – (4) for the inflated corolla base, which is papillate inside. (*Ceropegia*) – (5) for the papillate corolla. (*Cynanchum*) – (6) for the petals, which are papillate in the lower half. (*Drosanthemum*) – (7) for the papillate seed testa. (*Parakeelya*)

papillicaulis Lat. 'papilla', papilla; and Lat. 'caulis', stem; for the papillate stems. (*Sedum*)

papillosus Lat., papillose; (1) for the leaves. (*Crassula, Echeveria*) – (2) for the ribs dissolved into tubercles. (*Echinocereus*)

paposanus For the occurrence at Paposo, N Chile. (*Oxalis*)

papschii For Wolfang Papsch (*1946), Austrian cactus hobbyist, *Gymnocalycium* specialist and since 2000 president of the Gesellschaft Österreichischer Kakteenfreunde. (*Gymnocalycium*)

papulifer Lat. 'papula', vesicle, pustule, a relatively large papilla; and Lat. '-fer, -fera, -ferum', -carrying; for the lower leaf face. (*Portulaca*)

papulosus Lat., pustular; (1) for the stems. (*Galenia*) – (2) for the seed testa. (*Cleretum, Portulaca*)

papyracanthus Gr. 'papyros', the Papyrus plant, paper; and Gr. 'akanthos', spine; for the nature of the spines. (*Sclerocactus*)

papyraceus Lat., papery; for the scales enveloping the green leaves. (*Avonia*)

papyrocarpus Gr. 'papyros', the Papyrus plant, paper; and Gr. 'karpos', fruit; for the paper-thin fruit walls. (*Agave*)

paradinei For N. Paradine (fl. 1957), US-American discoverer of the taxon. (*Pediocactus*)

paradisus Lat., paradise; for the occurrence in a paradise-like place in the Trinity Mts., California, USA. (*Sedum obtusatum* ssp.)

paradoxus Lat., strange, paradoxical; (1) because the taxonomic relatiohships were not clear when the taxon was described. (*Haworthia mirabilis* var., *Jatropha, Myrmecodia, Nolana, Orbea*) – (2) for the intermediate position between the genera *Orostachys* and *Hylotelephium* (*Crassulaceae*). (*Orostachys*) – (3) for the appearance of the plants. (*Eriospermum, Rhipsalis, Rhytidocaulon*)

paraguariensis For the occurrence near the city of Paraguarí, Dept. Paraguarí, Paraguay. (*Cleistocactus*)

paraguayensis For the occurrence in Paraguay. (*Amphipetalum, Gymnocalycium*) – (2) For the erroneously presumed occurrence in Paraguay. (*Graptopetalum*)

Parakeelya From "Periculia" or "Parakeelya", the Australian aboriginal name for the seed meal prepared from *P. balonensis* and / or *P. polyandra*. (*Portulacaceae*)

parallelifolius Lat. 'parallelus', parallel; and Lat. '-folius', -leaved; for the strap-shaped leaves with parallel margins. (*Aloe*)

paranganiensis For the occurrence at Hacienda Parangani, Prov. Ayopaya, Dept. Cochabamba, Bolivia. (*Lepismium*)

parapetiensis For the occurrence in the valley of the Río Parapeti, Prov. Cordillera, Dept. Santa Cruz, Bolivia. (*Cleistocactus*)

Parasicyos Gr. 'para', near; and Gr. 'sicyos', cucumber; for the apparent affinity with the genus *Sicyos* (*Cucurbitaceae*). (*Cucurbitaceae*)

parasiticus Lat., parasitic; (**1**) noted as such on an herbarium sheet by Pavón, but this is obviously erroneous. (*Pedilanthus tithymaloides* ssp.) – (**2**) the epiphytically growing plants were erroneously first interpreted as being parasites. (*Dischidiopsis*)

parciramulosus Lat. 'parcus', frugal, scanty, thrifty; and Lat. 'ramulosus', branched. (*Euphorbia*)

parcus Lat., frugal, scanty, thrifty; application obscure. (*Lampranthus*)

paricymus Lat. 'par, paris', paired, equal; and Lat. 'cymus', cyme; for the paired inflorescence. (*Ceropegia*)

paripetalus Lat. 'par, paris', paired, equal; and Lat. 'petalum', petal; for the petals being all of the same length. (*Ruschia*)

parishii For Samuel B. Parish (1838–1928), US-American botanist in California. (*Grusonia*)

parkeri For Charles F. Parker (1820–1883), US-American plant collector. (*Cylindropuntia californica* var.) – (**2**) For David Parker (fl. 1988), English cactus hobbyist in Birmingham, and founder of the English *Echinocereus* Reference Collection. (*Echinocereus*)

parkinsonii For John Parkinson (fl. 1840), English Consul in Mexico. (*Mammillaria*)

parksianus Erroneously for a non-existent "Mrs. Parks", but the "Parks" in the reference of the type specimen sent to Europe referred to the Port Elizabeth Parks & Recreation Department, RSA. (*Haworthia*)

parlatorei For Prof. Dr. Filippo Parlatore (1816–1877), Italian anatomist and botanist in Palermo and Florence. (*Aichryson*)

Parodia For Dr. Domingo Parodi (1823–1890), Italian-born pharmacist and botanist in Montevideo, Paraguay and from 1878 in Buenos Aires, Argentina. (*Cactaceae*)

parrasanus For the occurrence in the Sierra de Parras, Coahuila, Mexico. (*Agave*)

parryi For Dr. Charles C. Parry (1823–1890), English-born US-American botanist, explorer and physician, "King of Colorado botany". (*Agave, Cistanthe, Echinocactus, Nolina*)

parvibracteatus Lat. 'parvus', small; and Lat. 'bracteatus', bracteate. (*Aloe, Cephalophyllum, Eberlanzia*)

parvicapsula Lat. 'parvus', small; and Lat. 'capsula', capsule; for the small fruits in comparison to the allied *Aloe woodii*. (*Aloe*)

parviceps Lat. 'parvus', small; and Lat. '-ceps', -headed; application obscure. (*Euphorbia*)

parvicladus Lat. 'parvus', small; and Gr. 'klados', branch; for the small stem segments. (*Opuntia*)

parvicomus Lat. 'parvus', small; and Lat. 'coma', hair tuft, mane; for the few-leaved rosettes at the stem tips. (*Aloe*)

parvicyathophorus Lat. 'parvus', small; and Gr. 'kyathos', cup, cyathium; and Gr. '-phoros', -carrying; for the small cyathia. (*Euphorbia*)

parvidens Lat. 'parvus', small; and Lat. 'dens', tooth; for the small teeth on the leaf margins. (*Aloe*)

parvidentatus Lat. 'parvus', small; and Lat. 'dentatus', toothed; for the small teeth on the leaf margins. (*Agave*)

parviflorus Lat. 'parvus', small; and Lat. '-florus', -flowered. (*Agave, Brachystelma, Carica, Carpobrotus edulis* ssp., *Cephalophyllum, Cleistocactus, Cyclantheropsis, Delosperma, Duvalia, Hesperaloe, Hoodia, Hoya, Monadenium, Psilocaulon, Quaqua, Ruschia, Sclerocactus, Sesuvium, Talinum, Umbilicus*)

parvifolius Lat. 'parvus', small; and Lat. '-folius', -leaved. (*Acrodon, Adenia penangiana* var., *Aeollanthus, Cyphostemma laza* var., *Drosanthemum, Euphorbia baga* var., *Hoya, Peperomia, Rhinephyllum, Ruschia, Sedum dendroideum* ssp., *Trianthema*)

parvilobus Lat. 'parvus', small; and Lat.

'lobus', lobe; for the smaller corolla lobes in comparison to *Orbea wissmannii*. (*Orbea wissmannii* var.)

parvipetalus Lat. 'parvus', small; and Lat. 'petalum', petal. (*Pelargonium*)

parvipunctus Lat. 'parvus', small; and Lat. 'punctum', spot, dot; for the dotted corolla. (*Tridentea*)

parvisepalus Lat. 'parvus', small; and Lat. 'sepalum', sepal. (*Crassula alba* var., *Sedum*)

parvulus Lat., small, minute; (**1**) for the small stature of the plants. (*Aloe, Brachystelma, Cephalophyllum, Crassula muscosa* var., *Gymnocalycium, Huernia hystrix* var., *Talinum*) − (**2**) for the small flowers. (*Piaranthus, Stapelia*)

parvus Lat., small; (**1**) for the small size of the plants. (*Dudleya abramsii* ssp., *Pseudosedum ferganense* ssp., *Rhodiola bupleuroides* var., *Sedum*) − (**2**) for the relatively small size of the plant. (*Sansevieria*) − (**3**) for the smaller size of leaves, fruits and seeds. (*Adenia olaboensis* var.) − (**4**) for the small leaves. (*Echeveria pringlei* var.)

passiflorus Lat. 'passio', passion; and Lat. '-florus', -flowered; application and possible parallels with the genus *Passiflora* ("Passionflower", *Passifloraceae*) obscure. (*Fevillea*)

patagonicus For the occurrence in Patagonia, S Argentina. (*Austrocactus, Maihuenia*)

patellitectus Lat. 'patella', small dish, plate, patella; and Lat. 'tectus', covered; because the operculum of the fruits has the shape of a dish covering the base of the fruit. (*Trianthema*)

patens Lat., spreading, widely open; (**1**) for the leaf arrangement. (*Argyroderma, Ruschia, Sansevieria*) − (**2**) for the stellate flowers. (*Umbilicus*)

patentispinus Lat. 'patens', spreading, widely open; and Lat. '-spinus', -spined. (*Euphorbia*)

patersoniae For Mrs. Florence M. Paterson (1869–1936), South African amateur naturalist and active plant collector, esp. around Port Elizabeth, Eastern Cape. (*Delosperma*)

patersonii For William Paterson (1755–1810), English naturalist, administrator and collector, travelled in S Africa 1777–1779. (*Monsonia*) − (**2**) For Andrew Paterson (fl. 1978), without further data. (*Aloe*)

patonii For Carlos Patoni (1853–1918), Mexican geographer, naturalist and botanist in the state of Durango. (*Mammillaria mazatlanensis* ssp.)

patriciae For Patricia Halliday (*1930), English botanist at Kew and student of *Kleinia* and *Senecio*. (*Senecio picticaulis* ssp.)

patulifolius Lat. 'patulus', open, spreading widely; and Lat. '-folius', -leaved. (*Ruschia*)

patulus Lat., open, widely spreading; (**1**) for the leaf arrangement. (*Sansevieria cylindrica* var.) − (**2**) for the tepals. (*Lachenalia*) − (**3**) for the flower shape. (*Erepsia, Stomatium, Villadia*)

paucartambensis For the occurrence in the Paucartambo Valley, Cuzco, Peru. (*Oxalis*)

paucicostatus Lat. 'pauci-', few; and Lat. 'costatus', ribbed; for the small number of ribs on the plant body. (*Eriosyce taltalensis* ssp., *Gymnocalycium hyptiacanthum* ssp., *Matucana*)

paucidens Lat. 'pauci-', few; and Lat. 'dens', tooth; for the obscurely toothed leaf margins. (*Stomatium*)

pauciflorus Lat. 'pauci-', few; and Lat. '-florus', -flowered. (*Brachystelma, Caralluma, Dudleya, Hoya, Huernia verekeri* var., *Lampranthus, Ruschia, Stathmostelma*)

paucifolius Lat. 'pauci-', few; and Lat. '-folius', -leaved. (*Antimima, Haworthia angustifolia* var., *Lampranthus*)

paucipetalus Lat. 'pauci-', few; and Lat. 'petalum', petal. (*Ruschia*)

paucispinus Lat. 'pauci-', few; and Lat. '-spinus', -spined. (*Gymnocalycium riojense* ssp., *Melocactus*)

paucistaminatus Lat. 'pauci-', few; and Lat. 'staminatus', provided with stamens. (*Portulaca*)

paucituberculatus Lat. 'pauci-', few; and Lat. 'tuberculatus', tuberculate; for the leaves. (*Aloe compressa* var.)

pauculifolius Lat. 'pauculus', very few; and Lat. '-folius', -leaved; for the commonly solitary leaf. (*Haemanthus*)

paulianae For Mrs. L. Paulian (fl. 1956), wife of R. Paulian, then Deputy Director of the Institut Scientifique de Madagascar. (*Aloe bulbillifera* var.)

paulianii For R. Paulian (fl. 1955), zoologist and Deputy Director of the Institut Scientifique de Madagascar. (*Euphorbia*)

pauper Lat., poor; (**1**) for the poor appearance of the plants with more brown twigs than leaves. (*Drosanthemum*) – (**2**) for the small and insignificant flowers. (*Antimima*)

pavelkae For Petr Pavelka (fl. 2002), Czech molecular biologist and keen student of plants. (*Othonna*)

pavonii For José A. Pavón y Jiménez (1754–1844), Spanish botanist, travelling 1778–1788 with H. Ruiz in S America. (*Matelea*)

paynei For George Payne (fl. 1930), succulent plant collector. (*Haworthia herbacea* var.)

pazoutianus For Ing. Frantisek Pazout (1909–1975), Czech cactus amateur. (*Gymnocalycium*)

peacockiae For Mrs. W. Peacock (fl. 1917), without further data. (*Lampranthus*)

peacockii For John T. Peacock († before 1889), English estate owner and succulent plant collector. (*Agave, Echeveria, ×Gasteraloe*)

pearsonii For Prof. Henry Harold W. Pearson (1870–1916), English botanist naturalized in RSA and founder and first director of the Kirstenbosch National Botanical Gardens. (*Aloe, Argyroderma, Cheiridopsis, Sansevieria, Sarcostemma, Stapelia, Tylecodon*)

pechuelii For Prof. Dr. Eduard Pechuel-Loesche (1840–1913), German geographer travelling repeatedly in Africa. (*Adenia*)

peckii For Major E. A. Peck (fl. 1956, 1962), officer in charge of the Veterinary and Agricultural services in northern Somalia (British Somaliland) before and after World War II, and a keen collector of the native flora. (*Aloe, Caralluma*)

pecten-aboriginum Lat. 'pecten', comb; and Lat. 'aboriginus', native, aboriginal; because the spiny fruits were used as combs by the native people. (*Pachycereus*)

Pectinaria Lat. 'pectinarius', comb- (from Lat. 'pecten', comb); for the comb-like processes of the corona. (*Asclepiadaceae*)

pectinatus Lat., combed; for the arrangement of the spines. (*Echinocereus*)

pectinifer Lat. 'pecten', comb; and Lat. '-fer, -fera, -ferum', -carrying; for the comb-like spine arrangement. (*Mammillaria, Uebelmannia*)

peculiaris Lat., special, particular, peculiar; (**1**) for the peculiar leaves. (*Cheiridopsis, Tylecodon*) – (**2**) for the short and broad flowers. (*Cleistocactus*) – (**3**) application obscure. (*Crassula, Tridentea*)

pedatus Lat., pedate, i.e. palmate with lateral lobes divided; for the leaf shape. (*Telfairia*)

pedemontanus Lat. 'pes, pedis', foot; and Lat. 'montanus', -mountain; for the occurrence at the foot of a mountain. (*Euphorbia*)

pedicellatus Lat. 'pedicellatus', with a pedicel. (*Hoodia, Sedum*)

pedilanthoides Gr. '-oides', resembling; and for the genus *Pedilanthus* (*Euphorbiaceae*); for the superficially similar cyathia. (*Euphorbia*)

Pedilanthus Gr. 'pedilon', slipper, shoe, and Gr. 'anthos', flower; for the shape of the cyathium. (*Euphorbiaceae*)

Pediocactus Gr. 'pedion', plane, field; and Lat. 'cactus', cactus; for the frequent occurrence on level ground. (*Cactaceae*)

pedroi For Prof. A. Gomes Pedro (fl. 1997), Portuguese agronomist. (*Euphorbia*)

peduncularis Lat., pedunculate, with a peduncle; (**1**) for the inflorescence. (*Oxalis*) – (**2**) for the long pedicels (which were mistaken for a peduncle). (*Crassula*)

pedunculatus Lat., pedunculate; (**1**) for the long peduncles of the cyathia. (*Monadenium*) – (**2**) for the long peduncles. (*Stathmostelma, Tetragonia*) – (**3**) for the long pedicels (which were mistaken for peduncles). (*Oscularia, Tromotriche*)

pedunculifer Lat. 'pedunculus', peduncle, inflorescence stalk; and Lat. '-fer, -fera, -ferum', -carrying. (*Agave*)

peeblesianus For Dr. Robert H. Peebles (1900–1955), US-American botanist and one of the world's leading cotton breeders, and especially interested in cacti. (*Pediocactus*)

Peersia For Victor Stanley Peers (1874–1940), Australian civil servant, amateur archaeologist and plant collector, living in RSA from 1899. (*Aizoaceae*)

peersii As above. (*Antimima, Carruanthus, Deilanthe, Delosperma, Glottiphyllum, Lampranthus, Stomatium, Trichodiadema*)

peglerae For Miss Alice M. Pegler (1861–1929), teacher in RSA and keen naturalist and collector. (*Aloe, Delosperma, Stapelia*)

pehlemanniae For Mrs. Inge Pehlemann (fl. 2002), succulent plant enthusiast in Windhoek, Namibia. (*Haworthia nortieri* var.)

pelargoniifolius For the genus *Pelargonium* (*Geraniaceae*); and Lat. '-folius', -leaved; resembling the leaves of some Pelargoniums. (*Jatropha*)

Pelargonium Gr. 'pelargos', stork; for the ripening fruits, which resemble stork's bills. (*Geraniaceae*)

Pelecyphora Lat. 'pelekys', hatchet, axe; and Gr. '-phoros', carrying; for the hatchet-shaped tubercles of the plant body. (*Cactaceae*)

pellacibellus Lat. 'pellax, pellacis', deceptive, full of intrigues; and Lat. 'bellus', beautiful; for the beautiful but unpleasantly odorous flowers. (*Brachystelma*)

pellitus Lat., covered with felt, hairy; (**1**) for the hair-like bladder cells on the leaves. (*Mesembryanthemum*) – (**2**) for the leaves. (*Ipomoea*)

pellucidus Lat., translucent; (**1**) for the windows on the leaves. (*Conophytum*) – (**2**) probably for the colourless leaf-margins. (*Crassula*)

pelona From the local vernacular name "Mescal Pelón"; from Span. 'pelón', bald-headed person; for the absence of leaf marginal teeth. (*Agave*)

peltatus Lat., shield-shaped, peltate; (**1**) for the leaf shape. (*Begonia, Jatropha, Kalanchoe, Pelargonium*) – (**2**) for the leaves, which are only very occasionally peltate, however. (*Uncarina*)

pembanus For the occurrence on Pemba Island, Tanzania. (*Aloe*)

penangianus For the occurrence on Penang Island, Malaysia. (*Adenia*)

pendens Lat., hanging; (**1**) for the growth form. (*Aloe, Cotyledon, Myrmecodia*) – (**2**) for the hanging leaves. (*Bulbine*)

penduliflorus Lat. 'pendulus', hanging down; and Lat. '-florus', -flowered. (*Aloe, Echeveria*)

pendulosus Lat., pendulous; for the pendulous stems. (*Echeveria*)

pendulus Lat., hanging down; (**1**) for the stems. (*Huernia, Senecio*) – (**2**) for the inflorescences. (*Agave*)

penicillatus Lat., like an artist's brush; (**1**) for the arrangement of the spines on each areole of the plant body. (*Parodia*) – (**2**) for the terminal cephalium (*Arrojadoa*) – (**3**) for the hairs on the petal tips. (*Caralluma*)

penicilliger Lat. 'penicillus', (paint) brush; and Lat. '-ger, -gera, -gerum', -carrying, -bearing; for the arrangement of the glochids. (*Opuntia*)

peniculinus Lat. 'peniculus', small tail, brush; application obscure. (*Monsonia*)

peninsulae Lat. 'peninsula', peninsula; for the occurrence on the peninsula of Baja California. (*Ferocactus*)

peninsularis From Lat. 'peninsula', peninsula; for the occurrence on the peninsula of Baja California. (*Mammillaria*)

Peniocereus Gr. 'penia', poverty; or Gr. 'penis', tail, penis; and *Cereus*, a genus of columnar cacti; for the long and slender stems, or for the small spines of some taxa (*Cactaceae*)

pennellii For Dr. Francis W. Pennell (1886–1952), US-American botanist. (*Opuntia*)

pennispinosus Lat. 'penna', feather; and Lat. 'spinosus', spiny; for the "feathery" pubescent spination. (*Mammillaria*)

pensilis Lat., pendent; for the growth form. (*Echinocereus*)

pentaedrophorus Gr. 'penta', five; Gr. 'hedra', plane; and Gr. '-phoros', -carrying; for the five-angled stems. (*Pilosocereus*)

pentagonus Gr. 'penta-', five; and Gr. 'gonia', angle; (**1**) for the five-ribbed branches. (*Euphorbia*) – (**2**) for the pentagonal flower receptacle. (*Erepsia, Phyllobolus splendens* ssp.)

pentalophus Gr. 'penta-', five; and Gr. 'lophos', crest; for the frequently five ribs of the stems. (*Echinocereus*)

pentandrus Gr. 'penta-', five; and Gr. 'aner, andros', man, [botany] stamen; for the five stamens. (*Graptopetalum, Sedella, Tetragonia, Zaleya*)

pentapetalus Gr. 'penta-', five; and Gr. 'petalon', petal; because related taxa have more than five petals. (*Sedum*)

pentapterus Gr. 'penta-', five; and Gr. 'pteron', wing; for the often five-ribbed stem segments. (*Rhipsalis*)

pentastamineus Gr. 'penta-', five; and Lat. 'stamineus', staminal; for the five stamens. (*Sedum*)

pentheri For Arnold Penther (1865–1931), Austrian botanist in Vienna, collecting in S Africa. (*Plectranthus*)

pentlandii For Josef B. Pentland (1797–1873), British Secretary to the Consul in Peru 1827, and British Consul in Bolivia 1836–1839, collecting cacti in the region. (*Cumulopuntia, Echinopsis*)

pentops Gr. 'penta-', five; and Gr. 'ops', eye; for the distinctive cyathial glands. (*Euphorbia*)

Peperomia Gr. 'peperi', pepper; and probably from Gr. 'homos, homoios', similar to; for the similarity to the genus *Piper* (*Piperaceae*). (*Piperaceae*)

peperomioides Gr. '-oides', similar to; and for the genus *Peperomia* (*Piperaceae*); because some taxa have similarly peltate leaves. (*Pilea*)

peploides Gr. '-oides', resembling; and for the former genus *Peplis* (= *Lythrum, Lythraceae*). (*Crassula*)

perangustus Lat. 'per-', very; and Lat. 'angustus', narrow; for the very thin winged stem angles. (*Euphorbia*)

perarmatus Lat. 'per-', very; and Lat. 'armatus', armed; for the strong spination. (*Euphorbia*)

perbellus Lat. 'per-', very; and Lat. 'bellus', beautiful. (*Echinocereus reichenbachii* ssp., *Mammillaria*)

percarneus Lat. 'per-', very; and Lat. 'carneus', flesh-coloured; for the reddish-tinged leaves. (*Aeonium*)

percrassus Lat. 'per-', very, and Lat. 'crassus', thick; for the succulent leaves. (*Aloe*)

perditus Lat., lost; because the taxon had only been found twice at the time of description. (*Brachystelma*)

perdurans Lat. 'per-', very; and Lat. 'durans', enduring, remaining; for the relative longevity of cultivated plants. (*Conophytum depressum* ssp.)

perennis Lat., perennial. (*Portulaca*)

Pereskia For Nicolas-Claude Fabri de Peiresc (1580–1637), French humanist, historian and astronomer. (*Cactaceae*)

pereskiifolius For the genus *Pereskia* (*Cactaceae*); and Lat. '-folius', leaved; for the similarly fleshy leaves. (*Peperomia, Synadenium*)

Pereskiopsis Gr. '-opsis', similar to; and for the genus *Pereskia* (*Cactaceae*). (*Cactaceae*)

perezassoi For A. Pérez Asso (fl. 1993) of the Museo Nacional de Historia Natural, La Habana, Cuba, who first collected the taxon. (*Melocactus*)

perezdelarosae For Jorge A. Pérez de la Rosa (fl. 1985), Mexican forestry engineer and secretary of the botanical institute of the University of Guadalajara. (*Mammillaria*)

perfectior Comp. of Lat. 'perfectus', perfect, complete. (×*Gasteraloe beguinii* nvar.)

perfoliatus Lat., perfoliate; (**1**) for the leaves with their sheathing bases united around the stem. (*Ruschia*) – (**2**) for the densely arranged persistent leaves. (*Crassula*)

perforatus Lat., pierced; for the fused leaves of a pair, which are pierced by the stem axis. (*Antimima, Crassula*)

pergamentaceus Lat., made from parchment; for the persistent dry leaves. (*Hartmanthus*)

periculosus Lat., dangerous, bringing danger; for the pungent leaf tips. (*Yucca*)

perlatus From French 'perlé', with pearls; application unclear. (*Larryleachia*)

permutatus Lat. 'per-', very; and Lat. 'mutatus', changed; because the taxon clearly belongs in another group than a closely similar taxon. (*Parodia*)

pernambucoensis For the occurrence in the state of Pernambuco, Brazil. (*Pilosocereus pachycladus* ssp.)

perotensis For the occurrence at the volcano Cofre de Perote, Veracruz, Mexico. (*Agave horrida* ssp.)

perperus Gr. 'perperos', false, incorrect; because the first specimens were misidentified. (*Euphorbia*)

perplexus Lat., perplex; (**1**) for the perplexing relationships. (*Euphorbia*) – (**2**) for the perplexing data relating to the origin of the material on which the taxon is based. (*Rebutia*)

perpusillus Lat. 'per-', very, and Lat. 'pusillus', small, dwarf. (*Sedum*)

perreptans Lat. 'per-', very; and Lat. 'reptans', creeping; for the growth habit. (*Lampranthus*)

Perrierastrum For the genus *Perriera* (*Simaroubaceae*); and Lat. '-aster', wild, small, inferior; because the plants are like inferior versions of *Perriera* species. (*Lamiaceae*)

perrieri For J. M. Henri A. Perrier de la Bâthie (1873–1958), French botanist, lived in Madagascar 1896–1933. (*Adansonia, Adenia, Aloe, Corallocarpus, Cynanchum, Euphorbia, Laportea, Uncarina, Xerosicyos*)

Perrierosedum For J. M. Henri A. Perrier de la Bâthie (1873–1958), French botanist, lived in Madagascar 1896–1933; and for the similarity to the genus *Sedum* (*Crassulaceae*). (*Crassulaceae*)

perrotii Perhaps for Prof. Dr. Émile C. Perrot (1867–1951), French pharmacist, anatomist and botanist in Paris. (*Sedum*) – (**2**) For B. Perrot (fl. 1901), without further data. (*Sansevieria*)

perryi For Wykeham Perry (fl. 1881), collected 1880 plants on Socotra, without further data. (*Aloe*)

persicus For the occurrence in Persia (modern Iran). (*Rosularia sempervivum* ssp.)

persistens Lat., persistent; for the persistent leaf sheaths. (*Lampranthus*)

persistentifolius Lat. 'persistens', persistent; and Lat. '-folius', -leaved. (*Euphorbia*)

perumbilicatus Lat. 'per-', very; and Lat. 'umbilicatus', with an umbilicus, with a navel; for the deeply sunken stem apex. (*Frailea*)

peruvianus For the occurrence in Peru. (*Echeveria, Echinopsis, Melocactus, Neowerdermannia chilensis* ssp., *Oroya, Peperomia*) – (**2**) For the erroneously supposed occurrence in Peru. (*Nolana*)

perviridis Lat. 'per-', very; and Lat. 'viridis', green; for the leaves. (*Gibbaeum*)

pervittatus Lat. 'per-', very, and Lat. 'vittatus', longitudinally striped; for the obvious striping of the branches. (*Euphorbia*)

peschii For Mr. C. Pesch (fl. 1935) of Omaruru, Hereroland (Namibia), who discovered the taxon. (*Caralluma*)

pestalozzae Probably for Fortunato Pestalozza (fl. 1850), Italian plant collector. (*Rosularia sempervivum* ssp.)

petasitifolius For the genus *Petasites* (*Asteraceae*); and Lat. '-folius', -leaved. (*Begonia*)

peteri For Prof. Dr. Albert [Gustav] Peter (1853–1937), German botanist in München and Göttingen, and traveller in East Africa. (*Kalanchoe, Portulaca*)

pethamensis Because the hybrid was raised by Mr. Rickets, gardener for W. H. Baldock, at Petham, England. (×*Gasteraloe*)

petignatii For Hermann Petignat († 2000), Swiss-born hotel-owner, artist and horticulturist in Madagascar. (*Ceropegia, Cynanchum*)

petiolaris Lat., with a petiole; for the characteristic long petioles. (*Ficus, Plectranthus*)

petiolatus Lat., petiolate; (**1**) for the long petioles. (*Monadenium*) – (**2**) for the basally long-attenuate leaves. (*Agave*)

Petopentia Anagram of the genus name *Pentopetia* (*Asclepiadaceae*), where the taxon was previously classified. (*Asclepiadaceae*)

petraeus Lat., growing among rocks. (*Brachystelma, Euphorbia*)

petrensis From Gr. / Lat. 'petra', rock, pebble; for the occurrence amongst pebbles. (*Gibbaeum*)

petricola Gr. / Lat. 'petra', rock, pebble; and Lat. '-cola', inhabiting; for the habitat. (*Aloe, Euphorbia*)

petrophilus Gr. 'petra', rock, pebble; and Gr. 'philos', friend; for the habitat preference. (*Agave, Aloe, Haworthia variegata* var., *Mammillaria*)

petroselinifolius Lat. '-folius', -leaved; and for the similarity to the genus *Petroselinum* ("Parsley", *Apiaceae*). (*Pelargonium*)

petterssonii For a Mr. Pettersson (fl. 1886), German cactus collector and friend of the German cactus nurserymen Heinrich Hildmann. (*Mammillaria*) – (**2**) For Dr. B. Pettersson (fl. 1949) from the Botanical Museum at Helsingfors. (*Euphorbia*)

peyrierasii For A. Peyrieras (fl. 1976), French zoologist. (*Aloe*)

pflanzii For Dr. Karl Pflanz (fl. 1923), German consul in Villa Montes, Bolivia who at the time shipped cacti to Berlin. (*Gymnocalycium*)

pfrimmeri For Mr. Pfrimmer (fl. 1931), who produced the cross. (×*Gasteraloe*)

phaeacanthus Gr. 'phaios', darkly glistening, grey, brownish; and Gr. 'akantha', spine, thorn; for the spination. (*Brasilicereus, Opuntia*)

phaeodiscus Gr. 'phaios', darkly glistening, grey, brownish; and Lat. 'discus', disc; for the body colour and shape. (*Frailea*)

phariensis For the occurrence near Phari, Tibet. (*Rhodiola purpureoviridis* ssp.)

pharnaceoides Gr. '-oides', resembling; and for the genus *Pharnaceum* (*Molluginaceae*). (*Crassula alata* ssp.)

phascoides Gr. '-oides', resembling; and for the moss genus *Phascum*. (*Crassula*)

phatnospermus Gr. 'phatnoein', hollow out; and Gr. 'sperma', seed; for the patterning of the seed surface. (*Cereus*)

Phedimus From Gr. 'phaidimos', shining; perhaps for the leaves of some species; or for Phedimus (fl. 235), archbishop at Amasea and Metropolitan of Pontus (in present-day Turkey), for the distribution of some taxa. (*Crassulaceae*)

Phiambolia Anagram of *Amphibolia* (*Aizoaceae*), where most of the species were originally classified. (*Aizoaceae*)

philippii For Rudolph [later Rodolfo] A. Philippi (1808–1904), German botanist emigrating to Chile in 1851 as farmer and teacher, finally director of the Museo Nacional. (*Austrocactus, Portulaca*)

philippinensis For the occurrence on the Philippines. (*Sedum parvisepalum* ssp.)

phillipsiae For Lady Dorothea S. F. A. Phillips (1863–1940), patron of arts and science in RSA, commissioned Marloth to write his 'Flora of South Africa'. (*Adromischus*) – (**2**) For Mrs. Lort Phillips, who collected material in Somalia in 1895, accompanied by her friend Miss Edith Cole. (*Euphorbia, Sansevieria*)

phillipsioides Gr. '-oides', resembling; for the similarity to *Euphorbia phillipsiae* (for Mrs. Lort Phillips) with which it was confused. (*Euphorbia*)

phillipsonii For Peter ("Pete") B. Phillipson (*1957), US-American botanist and Madagascar specialist, and assistant curator of the herbarium at the Missouri Botanical Garden, St. Louis. (*Cynanchum*)

phitauianus For the US-American Phi Kappa Tau fraternity of which the US-American sociologist E. Baxter (who described the taxon) was a member. (*Mammillaria*)

phlebopetalus Gr. 'phleps, phlebos', vein; and Gr. 'petalon', leaf; for the prominently veiny leaves. (*Lepidium*)

phoeniceus Lat., purple-red; for the flower colour. (*Conophytum*)

pholidogynus Gr. 'pholis, pholidos', scale; and Gr. 'gyne', female organ, ovary; application obscure. (*Lepidium*)

phosphoreus Lat., phosphorescent; because the latex is sometimes reported to be luminous. (*Euphorbia*)

phylicoides Gr. '-oides', resembling; and for the genus *Phylica* (*Rhamnaceae*); for the similarity to these heather-like shrubs. (*Ruschia*)

phyllacanthus Gr. 'phyllon', leaf; and Gr. 'akantha', thorn, spine; for the strongly flattened (like a leaf) spines of the plant body. (*Stenocactus*)

Phyllanthus Gr. 'phyllon', leaf; and Gr. 'anthos', flower; because the flowers originate from what appear to be leaves (but are phyllocladia). (*Euphorbiaceae*)

phyllanthus Gr. 'phyllon', leaf; and Gr. 'anthos', flower; (**1**) for the flowers that appear from the leaf-like flattened stem parts. (*Epiphyllum*) – (**2**) application unclear. (*Sedum*)

Phyllobolus Gr. 'phyllon', leaf; and Gr. 'bolos', cast, throw; for the deciduous leaves of the type species. (*Aizoaceae*)

phyllopodium Gr. 'phyllon', leaf; and Gr. 'podion', small foot; for the swollen persistent leaf bases (phyllopodia) covering the stem. (*Tylecodon reticulatus* ssp.)

Phylohydrax Anagram of *Hydrophylax* (*Rubiaceae*), where the taxa were formerly classified. (*Rubiaceae*)

phymatocarpus Gr. 'phyma, phymatos', wart, abscess; and Gr. 'karpos', fruit; for the warty fruits. (*Cissus*)

Phytolacca Gr. 'phyton', plant; and MLat. 'lacca', lacquer, varnish; for the dark red fruits, which are used as dye. (*Phytolaccaceae*)

phyturus Gr. 'phyton', plant; and Gr. 'oura', tail; for the densely leafy stems, which resemble animal tails. (*Crassula schimperi* ssp.)

Piaranthus Gr. 'piar', fat; and Gr. 'anthos', flower; for the fleshy corolla. (*Asclepiadaceae*)

piauhyensis For the occurrence in the state of Piauí, Brazil. (*Pilosocereus*)

picardae For L. Picarda (fl. ± 1890), botanical collector in the Caribbean. (*Consolea*)

pickeringii Perhaps for Dr. Charles Pickering (1805–1878), US-American botanist, zoologist and anthropologist, participated in the U.S. Exploring Expedition. (*Parakeelya*)

pickhardii For Mr. R. Pickhard (fl. 1932), without further data. (*Drosanthemum*)

picticaulis Lat. 'pictus', painted; and Lat. 'caulis', stem; for the arrow-head markings on the stems. (*Senecio*)

pictifolius Lat. 'pictus', painted; and Lat. '-folius', -leaved; for the spotted leaves. (*Aloe*)

picturatus Lat., embroidered, painted, coloured; for the green leaves with contrasting translucent areas. (*Haworthia gracilis* var.)

pictus Lat., painted, spotted; (**1**) for the dark blotched and veined sepals. (*Cistanthe*) – (**2**) for the corolla. (*Larryleachia*) – (**3**) application obscure. (*Mammillaria*)

pienaarii For Dr. U. de Villiers Pienaar (*1930), South African histologist and biochemist, long-time Director of Nature Conservation in Pretoria, and succulent plant specialist. (*Malephora*)

Pierrebraunia For Dr. Pierre Braun (*1959), German agronomist and specialist on Brazilian cacti. (*Cactaceae*)

pierrei For Jean B. P. Pierre (1833–1905), French botanist, 1865–1877 director of the Saigon Botanical Gardens, and explorer of Cambodia. (*Stephania*)

piersii For Mr. C. P. Piers (fl. 1900–1931), Government Surveyor, RSA, and field collector. (*Huernia*)

pilcayensis For the occurrence in the Barranca de Pilcaya, Guerrero, Mexico. (*Mammillaria spinosissima* ssp.)

Pilea Lat. 'pileus', felt hat, cap; for the persistent calyx covering the fruit. (*Urticaceae*)

pilifer Lat. 'pilus', hair; and Lat. '-fer, -fera, -ferum', -carrying; (**1**) for the spination. (*Opuntia*) – (**2**) for the leaf margins and tips. (*Haworthia cooperi* var.) – (**3**) for the hairy corolla. (*Hoodia, Rhytidocaulon*) – (**4**) for the hairy-papillate corolla. (*Quaqua*)

pilispinus Lat. 'pilus', hair; and Lat. '-spinus', -spined; for the fine spination. (*Eriosyce taltalensis* ssp., *Mammillaria*)

pillansii For Neville S. Pillans (1884–1964), South African botanist and collector, from 1918 at the Bolus Herbarium. (*Aloe, Apatesia, Arenifera, Cephalophyllum, Cheiridopsis, Duvalia, Erepsia, Euphorbia, Gasteria, Hoodia pilifera* ssp., *Huernia, Nelia, Pelargonium, Quaqua, Rhinephyllum, Schwantesia, Stapelia, Stapeliopsis, Tetragonia*)

pilleifer Lat. 'pil[l]eus', felt hat, felt cap; and Lat. '-fer, -fera, -ferum', -carrying; for the raised felted areoles. (*Browningia*)

pilocarpus Lat. 'pilus', hair; and Gr. 'karpos', fruit; for the bristly fruits. (*Rhipsalis*)

piloshanensis For the occurrence on Pi-lo-shan [Mt.], Yunnan, China. (*Sedum*)

Pilosocereus Lat. 'pilosus', hairy; and *Cereus*, a genus of columnar cacti; for the hairs associated with the spination and/or flowering zones of some taxa. (*Cactaceae*)

pilosulus Dim. of Lat. 'pilosus', hairy, i.e. a

little hairy; (**1**) for the hairy stems and pedicels. (*Delosperma*) – (**2**) for the papillate leaf surface. (*Antimima, Cheiridopsis*) – (**3**) for the hairy leaf surface. (*Gibbaeum*)

pilosus Lat., hairy; (**1**) for the hairy stems. (*Trianthema*) – (**2**) for the stem tubercles drawn out into a filiform appendage. (*Stapelianthus*) – (**3**) for the hairy leaves. (*Echeveria, Miraglossum, Prometheum*) – (**4**) for the usually bristly-hairy radial spines. (*Ferocactus*) – (**5**) for the tufts of hairs in the leaf axils and in the inflorescences. (*Portulaca*)

piltziorum Lat. Gen. Pl. of Piltzius; for Jörg Piltz (fl. 2002) and his wife Brigitte, German *Gymnocalycium* specialists and cactus seed dealers. (*Gymnocalycium riojense* ssp.)

piluliformis Lat. 'pilula', little ball, pill; and Lat. '-formis', -shaped; for the shape of the fused leaf pair. (*Conophytum*)

pinetorum Gen. Pl. of Lat. 'pinetum', pine forest; (**1**) for the occurrence near the deserted Pine City, California, USA. (*Sedum*) – (**2**) for the occurrence in pine forests. (*Echeveria*)

pinguiculus Dim. of Lat. 'pinguis', fat; i.e. somewhat fat; for the thick leaves. (*Sansevieria*)

pinguifolius Lat. 'pinguis', fat; and Lat. '-folius', -leaved. (*Senecio*)

pinguis Lat., fat; for the fleshy leaves. (*Ruschia*)

pinkavae For Prof. Dr. Donald J. Pinkava (*1933), US-American botanist and specialist on N American *Opuntia* species. (*Opuntia*)

pinnatifidus Lat., pinnatifid, pinnately cleft; for the leaves. (*Aethephyllum*)

pinnatus Lat., pinnate, having the leaf blade divided into pairs of lobes (leaflets). (*Kalanchoe, Othonna, Pelargonium*)

piquetbergensis For the occurrence at the Piquetberg, Western Cape, RSA. (*Oscularia*)

piraymirensis For the occurrence at Piraymirí, Prov. Valle Grande, Dept. Santa Cruz, Bolivia. (*Cleistocactus*)

pirottae For Pietro R. Pirotta (1853–1936), Italian botanist, and director of the Rome botanic garden until 1928. (*Aloe*)

piscatorius Lat., concerning the catching of fish; because the latex was used as a fish-poison. (*Euphorbia*)

piscidermis Lat. 'piscis', fish; and Gr. 'derma', skin; for the stem tubercles resembling fish-scales. (*Euphorbia*)

piscodorus Lat. 'piscis', fish; and Lat. 'odorus', smelling; for the scent of the whole plant. (*Antimima*)

pisidicus For the occurrence in the ancient region of Pisidia, SW Turkey. (*Sempervivum*)

pisiformis Lat. 'pisum', pea; and Lat. '-formis', -shaped; for the regularly constricted stems, resembling a row of peas. (*Monilaria*)

pisinus Lat., having the quality of peas; for the pea-green leaves. (*Anacampseros*)

Pistorinia For Jacobo (Jaime or Santiago) Pistorini (fl. 1766–1775), Spanish physician of Italian origin, physician in ordinary of King Carlos III. (*Crassulaceae*)

pittenii Erroneous spelling; for Joost van Putten (fl. 1929), farmer in RSA on whose farm the species was discovered. (*Lampranthus*)

pittieri For Dr. Henri-François Pittier (1857–1950), Swiss botanist and naturalist in Costa Rica and later in Venezuela. (*Echeveria, Epiphyllum, Opuntia, Rhipsalis floccosa* ssp.)

pittonii For Josef C. Pittoni, Ritter von Dannenfeldt (1797–1878), Austrian botanist and administrator. (*Sempervivum*)

pituitosus Lat., full of slime, full of moisture; for the mucilage ducts of the stem tissue. (*Opuntia*)

placentiformis Lat. 'placenta', cake; and Lat. '-formis', -shaped; for the depressed globose plant bodies. (*Discocactus*)

plagianthus Gr. 'plagios', oblique; and Gr. 'anthos', flower; for the laterally produced inflorescences. (*Euphorbia*)

plagiostoma Gr. 'plagios', oblique; and Gr. 'stoma', mouth, opening; for the zygomorphic flowers with oblique mouth. (*Cleistocactus*)

planiceps Lat. 'planus', level, flat; and Lat. '-ceps', -headed; for the growth habit resulting in flat cushions. (*Euphorbia*)

planiflorus Lat. 'planus', level, flat; and Lat. '-florus', -flowered. (*Echidnopsis*)
planifolius Lat. 'planus', level, flat; and Lat. '-folius', -leaved. (*Agave, Crassula, Sedum*)
planus Lat., flat; for the flat calyx. (*Hallianthus*)
platanifolius For the genus *Platanus* ("Plane", *Platanaceae*); and Lat. '-folius', -leaved. (*Cavanillesia*)
platinospinus Lat. 'platinum', platin (see also Span. 'plata', silver); and Lat. '-spinus', -spined; for the grey to silvery spination. (*Haageocereus*)
platyacanthus Gr. 'platys', flat, broad; and Gr. 'akanthos', spine, thorn. (*Echinocactus*)
platycarpus Gr. 'platys', flat, broad; and Gr. 'karpos', fruit. (*Uncarina*)
platycephalus Gr. 'platys', flat, broad; and Gr. 'kephale', head; for the shape of the cyathia. (*Euphorbia*)
platycladus Gr. 'platys', flat, broad; and Gr. 'klados', branch; for the flattly compressed branches. (*Euphorbia*)
platypetalus Gr. 'platys', flat, broad; and Gr. 'petalon', leaf; for the foliage. (*Lepidium*)
platyphyllus Gr. 'platys', flat, broad; and Gr. 'phyllon', leaf. (*Agave, Crassula nudicaulis* var., *Echeveria, Jatropha zeyheri* var., *Rhadamanthus, Rosularia, Thompsonella*)
platyrrhiza Gr. 'platys', flat, broad; and Gr. 'rhiza', root; for the flattish tuber. (*Euphorbia*)
platysepalus Gr. 'platys', flat, broad; and Gr. 'sepalon', sepal. (*Delosperma, Sedum*)
platystylus Gr. 'platys', flat, broad; and Gr. 'stylos', style. (*Sedum*)
platytyreus Gr. 'platys', flat, broad; and probably Gr. 'tyros', cheese; application obscure. (*Myrmecodia*)
plautus Lat., broad, flat; for the broad receptacle of the flowers. (*Lampranthus*)
Plectranthus Gr. 'plektron', cock's spur; and Gr. 'anthos', flower; for the spurred corolla of some species. (*Lamiaceae*)
plegmatoides Gr. 'plegma', basket, wickerwork; and Gr. '-oides', resembling; perhaps for the growth form. (*Crassula*)
pleiocephalus Gr. 'pleios', more, more than usual; and Gr. 'kephale', head; for the offsetting growth habit. (*Parodia werneri* ssp.)
pleiopetalus Gr. 'pleios', more, more than usual; and Gr. 'petalon', petal; for the eight petals, which is more than usually found in related taxa. (*Parakeelya*)
Pleiospilos Gr. 'pleios', more, more than usual; and Gr. 'spilos', fleck, spot; for the conspicuously dotted leaves. (*Aizoaceae*)
pleniflorus Lat. 'plenus', full, filled, "double"; and Lat. '-florus', -flowered; perhaps for the two rows of petals and the numerous staminodes and stamens. (*Lampranthus*)
plenispinus Lat. 'plenus', full, filled, "double"; and Lat. '-spinus', -spined; because spines and stipular spines are of equal length. (*Euphorbia*)
plenus Lat., full, filled, "double"; because the flowers open during the whole day. (*Lampranthus*)
pleurocarpus Gr. 'pleuron', rib; and Gr. 'karpos', fruit. (*Cipocereus minensis* ssp.)
pliabilis Perhaps Latinized from Engl. 'pliable'; for the flexible leaves. (*Beaucarnea*)
plicatilis Lat., foldable; for the fan-like arrangement of the leaves. (*Aloe*)
plicatus Lat., folded; (**1**) for the calyx. (*Nolana*) – (**2**) for the folded petals. (*Sedum*)
Plinthus Gr. 'plinthos', brick; application obscure. (*Aizoaceae*)
plocamoides Gr. '-oides', resembling; and for the genus *Plocama* (*Rubiaceae*). (*Brachystelma*)
plowesii For Darrel C. H. Plowes (*1925), South African-born agricultural officer and naturalist in Zimbabwe, specialist on Stapeliads, who discovered the taxa. (*Aloe, Huernia*)
Plumeria For Charles Plumier (1646–1704), French Franciscan monk, explorer and botanist in the West Indies. (*Apocynaceae*)
plumerioides Gr. '-oides', resembling; and for the genus *Plumeria* ("Frangipani", *Apocynaceae*). (*Euphorbia*)
plumosus Lat., plumose, feathery, downy; for the feathery spines. (*Mammillaria*)
pluricaulis Lat. 'pluri-', many; and Lat. 'caulis', stem. (*Hylotelephium*)
pluricostatus Lat. 'pluri-', many; and Lat. 'costatus', ribbed. (*Coleocephalocereus*)
pluridens Lat. 'pluri-', many; and Lat. 'dens',

tooth; for the leaf margins. (*Aloe*, *Stomatium*)
pluriflorus Lat. 'pluri-', many; and Lat. '-florus', -flowered. (*Haageocereus*)
pluvisilvaticus Lat. 'pluvius', rainy; and Lat. 'silvaticus', forest-, forest-dwelling; for the occurrence in a rain forest. (*Peperomia*)
pocockiae For Dr. Mary A. Pocock (1886–1977), South African algologist, plant collector and painter. (*Lampranthus*)
podagricus Lat., with gouty feet; for the basally swollen stems. (*Jatropha*)
Poellnitzia For Joseph Karl L. A. von Poellnitz (1896–1945), German agriculturist and botanist in Thüringen, strongly interested in succulent plant systematics. (*Aloaceae*)
poellnitzianus As above. (*Haworthia minima* var.)
poeppigii For Eduard F. Poeppig (1798–1868), German botanist, zoologist and explorer, esp. in S America. (*Maihuenia*, *Polyachyrus*)
poeschlii For Josef Pöschl (*1954), Austrian bookbinder and reprography specialist in Wels, for his support of the "Arbeitsgruppe Gymnocalycium". (*Gymnocalycium*)
poincarei For Raymond Poincaré, President of the French Republic, elected 17. 2. 1913 on the day when this taxon was recognized as new. (*Kalanchoe*)
poissonii For Eugène A. Poisson (1871–1910), French traveller and naturalist, government agent in West Africa involved with cotton and palm oil, participating in a commercial expedition to Dahomey (now Benin) in 1901. (*Euphorbia*) – (**2**) For Dr. Henri L. Poisson (1877–1963), French veterinary surgeon and botanist, resident in Madagascar 1916–1954. (*Aloe vaombe* var.)
pojoensis For the occurrence near the city of Pojo, Prov. Carrasco, Dept. Cochabamba, Bolivia. (*Echinopsis*)
Polaskia For Charles and Mary Polaski (fl. 1949), US-American cactus hobbyists from Oklahoma. (*Cactaceae*)
pole-evansii For Dr. Illtyd B. Pole-Evans (1877–1968), Welsh botanist and plant pathologist, lived in RSA from 1905 and travelled widely. (*Dinteranthus*, *Nananthus*)

polianthes Gr. 'polios', grey, whitish; and Gr. 'anthos', flower; and also Gr. 'polion', a strongly scented Labiate; for the whitish and strongly scented flowers. (*Agave*)
polianthiflorus For the former genus *Polianthes* (*Agavaceae*); and Lat. '-florus', -flowered. (*Agave*)
politus Lat., polished (from Lat. 'polire', to polish); (**1**) for the shiny leaves. (*Erepsia*) – (**2**) for the corolla. (*Duvalia*)
pollardii For Mr. Pollard (fl. 1960), security officer for the diamond mines in S Namibia who facilitated botanical journeys in the diamond company's protected territory. (*Ruschia*)
polpodaceus Unknown. (*Crassula muscosa* var.)
poluninianus For Oleg V. Polunin (1914–1985), England-based botanist and plant photographer. (*Ceropegia*)
polyacanthus Gr. 'poly', many; and Gr. 'akantha', thorn, spine; for the spiny plant bodies. (*Echinocereus*, *Euphorbia*, *Opuntia*)
Polyachyrus Gr. 'poly', many; and Gr. 'achyron', chaff; probably for the scabrid bristly pappus. (*Asteraceae*)
polyancistrus Gr. 'poly', many; and Gr. 'ankistron', hook; for the number of hooked central spines. (*Sclerocactus*)
polyandrus Gr. 'poly', many; and Gr. 'aner, andros', man, [botany] stamen; for the numerous stamens. (*Conophytum velutinum* ssp., *Parakeelya*)
polyanthon Gr. 'poly', many; and Gr. 'anthos', flower. (*Lampranthus*)
polyanthus Gr. 'poly', many; and Gr. 'anthos', flower. (*Micranthocereus*)
polycephalus Gr. 'poly', many; and Gr. 'kephale', head; (**1**) for the offsetting growth habit. (*Echinocactus*, *Epithelantha micromeris* ssp.) – (**2**) for the numerous short branches. (*Euphorbia*) – (**3**) for the branched inflorescence. (*Osteospermum*)
polyedrus Gr. 'polyedros', many-sided; for the angled tubercles of the plant body. (*Mammillaria*)
polygaloides Gr. '-oides', resembling; and for the genus *Polygala* ("Milkwort", *Polygalaceae*). (*Talinum*)

polygonoides Gr. '-oides', resembling; and for *Polygonum aviculare* (*Polygonaceae*); for the similar leaves. (*Begonia*)

polygonus Gr., many-angled; for the numerous ribs of stems and branches. (*Euphorbia, Facheiroa squamosa* ssp., *Pilosocereus*)

polylophus Gr. 'poly', many; and Gr. 'lophos', crest; for the numerous ribs of the plant bodies. (*Neobuxbaumia*)

Polymita Gr. 'poly', many; and Gr. 'mitos', thread; for the numerous linear petals. (*Aizoaceae*)

polymorphus Gr. 'poly', many; and Gr. 'morphe', shape; for the variability. (*Trochomeria*)

polypetalus Gr. 'poly', many; and Gr. 'petalon', petal; for the high number of petals in comparison with related taxa. (*Erepsia, Parakeelya*)

polyphyllus Gr. 'poly', many; and Gr. 'phyllon', leaf; for the rosettes with numerous leaves. (*Aloe, Monanthes*)

polythele Gr. 'poly', many; and Gr. 'thele', nipple; for the body tubercles. (*Mammillaria*)

polytomus Gr. 'poly-', many; and Gr. 'tomos', slice, piece; for the whorled branching. (*Senecio*)

polytrichoides Gr. '-oides', resembling; and for the moss genus *Polytrichum*. (*Sedum*)

polzii For Frank Polz (fl. 1986), German cactus and succulent plant enthusiast in München. (*Matucana*)

pomeridianus Lat., pertaining to the afternoon; for the opening time of the flowers. (*Carpanthea, Trichodiadema*)

pomonae For the occurrence near Pomona, Namibia. (*Namibia*)

ponderosus Lat., heavy; (**1**) because the plant becomes heavy when fully grown. (*Euphorbia*) – (**2**) for the large and thick leaves. (*Cheiridopsis*)

pondii For Charles F. Pond (fl. 1889), member of the US-American navy who collected plants on islands off the coast of Baja California. (*Mammillaria*)

pondoensis For the occurrence in Pondoland, KwaZulu-Natal, RSA. (*Delosperma, Plectranthus saccatus* ssp.)

pontii For Johannes W. Pont (1898–1977), Dutch plant physiologist in RSA. (*Delosperma*)

populifolius Lat. '-folius', -leaved; for the similarity of the leaves to those of *Populus* ("Poplar", *Salicaceae*). (*Hylotelephium*)

porifer Lat. 'porus', pore; and Lat. '-fer, -fera, -ferum', -carrying; application obscure. (*Parakeelya*)

porphyranthus Gr. 'porphyreos', purplish-red; and Gr. 'anthos', flower; for the purple nectary glands. (*Euphorbia*)

porphyreus Gr., purplish-red; (**1**) for the stem colour. (*Stoeberia*) – (**2**) for the red outside of the petals. (*Sedum eriocarpum* ssp.)

porphyrocalyx Gr. 'porphyreos', purplish-red; and Gr. 'kalyx', calyx. (*Kalanchoe*)

porphyrogennetos Gr. 'porphyreos', purplish-red; and perhaps from Gr. 'gennaios', noble, excellent. (*Aichryson*)

porphyrostachys Gr. 'porphyreos', purplish-red; and Gr. 'stachys', spike; for the inflorescences. (*Aloe*)

porphyrotrichus Gr. 'porphyreos', purplish-red; and Gr. 'thrix, trichos', hair; for the purplish hairs of the corolla. (*Ceropegia*)

porteri Probably for Prof. Dr. Thomas C. Porter (1822–1901), US-American botanist and presbyterian clergyman, ultimately professor of botany and zoology at Lafayette College, Pennsylvania. (*Pereskiopsis*)

portoricensis For the occurrence on Puerto Rico. (*Agave, Harrisia*)

Portulaca Lat., "Purslane"; probably from Lat. 'portula', small door; for the fruits, which open with a small lid. (*Portulacaceae*)

portulacaceus For the genus *Portulaca* ("Purslane", *Portulacaceae*); and Lat. '-aceus', similar to; for the similar shoots. (*Galenia*)

Portulacaria For the genus *Portulaca* ("Purslane", *Portulacaceae*); and Lat. '-arius', pertaining to; first used as a specific epithet by Linné. (*Portulacaceae*)

portulacastrum For the genus *Portulaca* ("Purslane", *Portulacaceae*); and Lat. '-aster', wild, small, inferior. (*Sesuvium, Trianthema*)

portulacifolius For the genus *Portulaca* ("Purslane", *Portulacaceae*); and Lat. '-folius', -leaved. (*Pereskia, Talinum*)

portulacoides Gr. '-oides', resembling; and for the genus *Portulaca* ("Purslane", *Portulaceae*). (*Tetragonia*)

poselgeri For Dr. Heinrich (Hermann according to other sources) Poselger (1818–1883), German physician, chemist and botanist and succulent plant collector in Berlin, collected 1849–1851 in North America. (*Echinocereus, Mammillaria*)

poselgerianus As above. (*Coryphantha*)

postgenitus Lat. 'post', behind or after; and Lat. 'genitus', produced, born of, arising from; application obscure. (*Aloe brevifolia* var.)

potatorum Gen. Pl. of Lat. 'potator', drunkard; for the possible use in the manufacture of the alcoholic beverages Tequila and Mescal. (*Agave*)

potosianus For the occurrence in the Mexican state of San Luis Potosí. (*Coryphantha*)

potosiensis As above. (*Agave asperrima* ssp.)

potosinus As above. (*Agave, Sedum, Yucca*)

potreranus For the occurrence on Cerro Potrero, Chihuahua, Mexico. (*Agave*)

pottsii For John Potts (fl. 1848), English-born director of the mint at Ciudad Chihuahua, Mexico. (*Ferocactus, Hoya, Mammillaria, Opuntia macrorhiza* var.) – (**2**) For Prof. Dr. George Potts (1877–1948), British botanist, from 1905 at the University of Orange Free State, RSA. (*Delosperma*)

powellii For Mr. H. Powell, 1905–1911 Chief of the Economic Plant Division, Nairobi, Kenya. (*Sansevieria*)

powysii For J. Gilfrid L. Powys (*1938), Kenyan farmer of Welsh descent, explorer and field collector of succulents in Kenya, Tanzania, S Ethiopia and S Sudan. (*Ceropegia arabica* var.)

powysiorum For J. Gilfrid L. Powys (*1938), and his wife Patricia G., Kenyan farmers of British descent, explorers and field collectors of succulents in Kenya, Tanzania, S Ethiopia and S Sudan. (*Aloe*)

praealtus Lat., very tall; for the stature of the plant. (*Sedum dendroideum* ssp.)

Praecereus Lat. 'prae-', before, early; and *Cereus*, a genus of columnar cacti; for the assumed systematic position in relation to *Cereus*. (*Cactaceae*)

praecipitatus Lat., cast down, precipitated; for the pendulous branches. (*Lampranthus*)

praecox Lat., very early; (**1**) for the flowers produced before the leaves. (*Cynanchum, Senecio*) – (**2**) for the early ripening edible fruits. (*Sarcozona*)

praecultus Lat. 'prae-', before, early; and Lat. 'cultus', cultivated, tilled, civilized; application obscure. (*Drosanthemum*)

praelongus Lat., very long; for the long pedicels. (*Brachystelma*)

praerupticola Lat. 'praerupta', steep place, cliff; and Lat. '-cola', -dwelling; for the inaccessible rocky habitat of the taxon. (*Melocactus*)

praesectus Lat., deeply divided; for the shortly fused leaf pairs. (*Conophytum*)

praestans Lat., surpassing, excelling, exceeding. (*Huernia*)

praetermissus Lat., overlooked, missed out; because the taxon was previously overlooked. (*Aloe, Ceropegia, Plectranthus, Stapelia*)

prainii For Sir David Prain (1857–1944), British botanist, in Calcutta (India) 1887–1905, director of RBG Kew 1905–1922. (*Rhodiola*)

prasinopetalus Gr. 'prasinos', leek-green; and Gr. 'petalon', petal; for the green flowers. (*Sedum*)

prasinus Gr., leek-green; (**1**) for the leaf colour. (*Delosperma, Oscularia*) – (**2**) for the flower colour. (*Phyllobolus*)

pratensis Lat., growing in meadows; (**1**) for the preferred habitat. (*Agave*) – (**2**) for the erroneously assumed preferred habitat. (*Aloe*)

pratoalpinus Lat. 'pratum', meadow; and Lat. 'alpinus', alpine; for the habitat. (*Sedum*)

Prenia Gr. 'prenes', hanging forward, drooping, prone; for the decumbent flowering branches of most species. (*Aizoaceae*)

Prepodesma Gr. 'prepo', conspicuous; and Gr. 'desmis', bundle; for the very prominent central bundle of stamens. (*Aizoaceae*)

pretoriensis For the occurrence near Pretoria, RSA. (*Aloe*, *Ceropegia decidua* ssp.)

preussii For Paul R. Preuss (1861–?), German (Prussian) botanist and explorer in Africa, America and Asia. (*Dorstenia cuspidata* var.)

primavernus MLat. 'prima vera', beginning of spring (from Lat. 'ver', spring); because the taxon is the first amongst related taxa to flower in spring. (*Conophytum bolusiae* ssp., *Oscularia*)

primolanatus Lat. 'primus', the first; and Lat. 'lanatus', woolly; for the outstanding white colour of the spination. (*Echinocereus*)

primosii For Mr. Richard Primos (fl. 1928–1936), of Cape Town, Western Cape, RSA, field collector. (*Ruschia*, *Vanheerdea*)

primuliflorus For the genus *Primula* ("Primrose", *Primulaceae*); and Lat. '-florus', -flowered; for the resemblance of the flowers to those of some Primroses. (*Parakeelya*)

primulifolius For the genus *Primula* ("Primrose", *Primulaceae*); and Lat. '-folius', -leaved; for the resemblance of the leaves to those of some Primroses. (*Euphorbia*)

primulinus Lat., like a Primula; for the pale yellow flowers. (*Huernia thuretii* var.)

primuloides Gr. '-oides', resembling; and for the genus *Primula* ("Primrose", *Primulaceae*); perhaps for the superficially similar appearance. (*Rhodiola*)

principis Gen. of Lat. 'princeps', prince; because the genus was named for the Prince of Leuchtenberg. (*Leuchtenbergia*)

pringlei For Cyrus G. Pringle (1838–1911), US-American plant breeder and explorer-collector. (*Agave deserti* ssp., *Echeveria*, *Mammillaria rhodantha* ssp., *Pachycereus*, *Sedum*, *Villadia*) – (2) For Victor L. Pringle (RSA) who first collected the taxon in 1973. (*Haworthia decipiens* var.)

prinslooi For Gerry Prinsloo (fl. 1965), South African grower of succulents. (*Aloe*)

priogonius Gr. 'prion', saw; and Gr. 'gonia', edge; for the saw-toothed ribs. (*Caralluma*)

priscus Lat., old, traditional; perhaps because the subspecies is diploid and thus evolutionarily older than the other subspecies. (*Sedum wrightii* ssp.)

prismaticus Lat., prism-shaped; for the more or less three-angled leaves. (*Tanquana*)

prittwitzii For Hauptmann von Prittwitz (fl. 1907), without further data. (*Kalanchoe*)

proballyanus For Peter R. O. Bally (1895–1980), Swiss botanist, widely travelling in E Africa, and resident in Kenya from the 1930s. (*Euphorbia*)

procerus Lat., tall, slender; (1) for the growth habit. (*Alluaudia*, *Armatocereus*, *Echeveria*, *Parodia*, *Rhodiola integrifolia* ssp.) – (2) for the tall inflorescences. (*Aloe*)

prochazkianus For Jaroslav Procházka (fl. 1995), Czech cactus hobbyist and *Gymnocalycium* specialist in Brno. (*Gymnocalycium valnicekianum* ssp.)

procumbens Lat., procumbent, creeping, prostrate. (*Caralluma*, *Echinocereus pentalophus* ssp., *Galenia*, *Huernia*, *Lampranthus*, *Raphionacme*)

productus Lat., extended, elongated; (1) for the long leaves. (*Agave*) – (2) for the long calyx lobes. (*Lampranthus*) – (3) for the long segments of the corona. (*Schizoglossum bidens* ssp.)

profundus Lat., deep; for the closing bodies, which are deeply placed in the fruit locules. (*Lampranthus*)

profusus Lat., spread out, profuse, abundant; for the abundantly produced flowers and fruits. (*Opuntia galapageia* var.)

prognathus Lat., bulging forwards; for the annulus of the flowers. (*Orbea*)

prolifer Lat., proliferating, producing offsets; (1) for the growth habit. (*Conophytum swanepoelianum* ssp., *Crassula columnaris* ssp., *Cylindropuntia*, *Mammillaria*) – (2) for the branched inflorescences with several many-flowered part-inflorescences. (*Pelargonium*) – (3) for the numerous bulbils produced in the inflorescences. (*Kalanchoe*)

prolificus Lat., prolific, fertile; for the easy propagation from leaves and bracts. (*Echeveria*)

prolongatus Lat., lengthened, extended; for the long branches. (*Antimima*)

Prometheum After the Greek mythological figure Prometheus, who stole the fire from Zeus and was subsequently chained to the

Caucasus for punishment; for the blood-red flowers of the type species which occurs in the Caucasus. (*Crassulaceae*)

prominens Lat., prominent; for the prominent and distinctive size of the plants. (*Avonia*)

prominulus Lat., slightly raised; for the raised dots on the leaf surface. (*Lampranthus*)

promontorii Gen. of Lat. 'promontorium', promontory; (**1**) for the occurrence at the cape of of Baja California, Mexico. (*Agave*) − (**2**) for the occurrence on the Cape Peninsula, RSA. (*Erepsia, Lampranthus*)

promontorium Lat., promontory; for the occurrence on the Cape Peninsula, RSA. (*Crassula capensis* var.)

pronus Lat., leaning forward, prostrate; for the growth habit. (*Euphorbia*)

propagulifer Lat. 'propagulum', bulbil; and Lat. '-fer, -fera, -ferum', -carrying; for the production of bulbils in the inflorescence. (*Aloe*)

propinquus Lat., near, neighbouring; perhaps for the similarity to another taxon. (*Antimima, Gunniopsis, Stathmostelma*)

prorumpens Lat., bursting forth, bursting through; application obscure. (×*Gasteraloe*)

prostratiflorus Lat., 'prostratus', prostrate, creeping; and Lat. 'florus', -flowered. (*Stapelia erectiflora* var.)

prostratus Lat., prostrate, creeping; (**1**) for the growth habit. (*Antimima, Brachystelma, Drosanthemum, Galenia, Leptocereus, Plectranthus*) − (**2**) for the spreading leaves. (*Aloe*)

protectus Lat., defended, covered (from Lat. protegere, to defend, to cover); because the plants grow in the shade of Mesembs. (*Othonna*)

protoamericanus Gr. 'protos', first, original, chief; and Lat. 'americanus', American; for the status as proposed origin of the cultivated *Agave americana*. (*Agave americana* ssp.)

protoparcoides Gr. '-oides', resembling; for the resemblance to the bristly larvae of *Protoparce quinquemaculata* ("Tomato Hornworm", Sphinx Moth family); for the minute prickles near the leaf tips. (*Odontophorus angustifolius* ssp.)

proximus Lat., nearest, next; presumably for the close affinity to *Lampranthus prominulus*. (*Lampranthus*)

pruinatus Lat., pruinose, covered with a waxy bloom; for the leaves. (*Sedum*)

pruinosifolius Lat. 'pruinosus', pruinose, covered with a waxy bloom; and Lat. '-folius', -leaved. (*Peperomia*)

pruinosus Lat., pruinose, covered with a waxy bloom; (**1**) for the stem surface. (*Galenia, Quaqua*) − (**2**) for the leaves. (*Crassula, Sedum spathulifolium* ssp.) − (**3**) for the peduncle and flowers. (*Aloe*)

pruninus Lat., like a plum or prune; for the glaucous plum-coloured leaves. (*Echeveria*)

przewalskii For Nikolai M. Przewalski (1839–1888), Russian soldier, traveller, geographer and naturalist, explorer of Central Asia. (*Sedum*)

psammophilus Gr. 'psamme, psammos', sand; and Gr. 'philos', friend; for the preference of sandy habitats. (*Crassula brevifolia* ssp., *Euphorbia, Kedrostis, Plectranthus*)

Psammophora Gr. 'psamme, psammos', sand; and Gr. '-phoros', -carrying; because sand sticks to the viscid leaf surfaces. (*Aizoaceae*)

Pseudoacanthocereus Gr. 'pseudo-', false; and for the similarity to the genus *Acanthocereus* (*Cactaceae*). (*Cactaceae*)

Pseudobombax Gr. 'pseudo-', false; and for the genus *Bombax* (*Bombacaceae*). (*Bombacaceae*)

pseudoburuanus Gr. 'pseudo-', false; and for the resemblance to *Euphorbia buruana*. (*Euphorbia*)

pseudocactus Gr. 'pseudo-', false; and for the resemblance to *Euphorbia cactus*. (*Euphorbia*)

pseudocampanulatus Gr. 'pseudo-', false; and for the resemblance to *Kalanchoe campanulata*. (*Kalanchoe*)

pseudocruciger Gr. 'pseudo-', false; and for the resemblance to *Mammillaria crucigera*. (*Mammillaria formosa* ssp.)

pseudodeminutus Gr. 'pseudo-', false; and for the resemblance to *Rebutia deminuta*. (*Rebutia*)

pseudoduseimatus Gr. 'pseudo-', false; and for the resemblance to *Euphorbia duseimata*. (*Euphorbia*)

pseudoechinus Gr. 'pseudo-', false; and Lat. 'echinus', sea urchin, hedgehog; for the resemblance to *Coryphantha echinus*. (*Coryphantha*)

pseudofossulatus Gr. 'pseudo-', false; and because of the earlier confusion with *Cleistocactus fossulatus*; Lat. 'fossulatus', with little furrows; for the rib architecture of the stems. (*Oreocereus*)

pseudoglobosus Gr. 'pseudo-', false; (**1**) and for the resemblance to *Adenia globosa*. (*Adenia globosa* ssp.) − (**2**) and for the resemblance to *Euphorbia globosa*. (*Euphorbia*)

pseudohemisphaericus Gr. 'pseudo-', false; and for the resemblance to *Crassula hemisphaerica*. (*Crassula*)

pseudoinsignis Gr. 'pseudo-', false; and for the resemblance to *Discocactus insignis*. (*Discocactus*)

Pseudolithos Gr. 'pseudo-', false; and Gr. 'lithos', stone; for the appearance of the stems. (*Asclepiadaceae*)

pseudomacrochele Gr. 'pseudo-', false; and for the resemblance to *Turbinicarpus macrochele*. (*Turbinicarpus*)

pseudomamillosus Gr. 'pseudo-', false; and for the resemblance to *Echinopsis mamillosa*. (*Echinopsis*)

pseudomelanostele Gr. 'pseudo-', false; and for the resemblance to *Espostoa melanostele*. (*Haageocereus*)

pseudomulticaulis Gr. 'pseudo-', false; and for the resemblance to *Sedum multicaule*. (*Sedum*)

pseudonickelsiae Gr. 'pseudo-', false; and for the resemblance to *Coryphantha nickelsiae*. (*Coryphantha*)

pseudopectinatus Gr. 'pseudo-', false; (**1**) and for the resemblance to *Echinocereus pectinatus*. (*Echinocereus*) − (**2**) and for the resemblance to *Pelecyphora pectinata* (now *Mammillaria pectinifera*). (*Turbinicarpus*)

pseudopulcherrimus Gr. 'pseudo-', false; and for the resemblance to *Frailea pulcherrima*. (*Frailea*)

pseudoracemosus Gr. 'pseudo-', false; and Lat. 'racemosus', racemose; for the appearance of the flowering stems. (*Monadenium*)

pseudoradians Gr. 'pseudo-', false; and for the resemblance to *Coryphantha radians*. (*Coryphantha*)

Pseudorhipsalis Gr. 'pseudo-', false; and for the resemblance to the genus *Rhipsalis* (*Cactaceae*). (*Cactaceae*)

pseudorubroviolaceus Gr. 'pseudo-', false; and for the resemblance to *Aloe rubroviolacea*. (*Aloe*)

pseudoschlichtianus Gr. 'pseudo-', false; and for the resemblance to *Brownanthus schlichtianus*, the latter for Albert W. H. von Schlicht (1817–1893), German-born pharmacist in RSA. (*Brownanthus*)

Pseudosedum Gr. 'pseudo-', false; and for the resemblance to the genus *Sedum* ("Stonecrop"; *Crassulaceae*). (*Crassulaceae*)

pseudotruncatellus Gr. 'pseudo-', false; and for the resemblance to *Mesembryanthemum truncatellum* (now synonymous with *Conophytum truncatum*). (*Lithops*)

pseudotuberosus Gr. 'pseudo-', false; and for the resemblance to *Euphorbia tuberosa*. (*Euphorbia*)

pseudoversicolor Gr. 'pseudo-', false; and for the resemblance to *Haageocereus versicolor*. (*Haageocereus*)

Psilocaulon Gr. 'psilos', bare, naked; and Gr. 'kaulos', stem; for the leafless stems. (*Aizoaceae*)

pteranthus Gr. 'pteron', wing, feather; and Gr. 'anthos', flower; for the feathery appearance caused by the spreading narrow outer perianth segments. (*Selenicereus*)

pteroaspidus Gr. 'pteron', wing; and Gr. 'aspis, aspidos', shield; for the winged shield-shaped swelling from which each leaf arises. (*Myrmecodia*)

Pterocactus Gr. 'pteron', wing; and Lat. 'cactus', cactus; for the uniquely winged seeds. (*Cactaceae*)

pterocladus Gr. 'pteron', wing; and Gr. 'klados', branch; for the winged branch angles. (*Euphorbia*)

Pterodiscus Gr. 'pteron', wing; and Lat. 'discus', disc; for the fruit structure. (*Pedaliaceae*)

pteroneurus Gr. 'pteron', wing; and Gr. 'neuron', nerve, vein; for the stems with ridges extending from each leaf attachment point. (*Euphorbia*)

Pteronia Gr. 'pteron', wing; probably for the seeds, which are wind-dispersed. (*Asteraceae*)

ptychocladus Gr. 'ptyche', fold; and Gr. 'klados', branch; application obscure. (*Oxalis*)

ptychospermus Gr. 'ptyche', fold; and Gr. 'sperma', seed; for the ribbed ("folded") seeds. (*Parakeelya*)

puberulus Lat., puberulous; (**1**) for the stem segments. (*Opuntia*) – (**2**) for the leaf surface with long papillae. (*Dinteranthus*)

pubescens Lat., pubescent, finely hairy, downy; (**1**) for the pubescent stems. (*Cissus quadrangularis* var., *Opuntia*) – (**2**) for the hairy leaves. (*Agave, Brownanthus, Conophytum, Crassula, Galenia, Gibbaeum, Haworthia, Xerosicyos*) – (**3**) for the sometimes hairy leaves. (*Ceropegia*) – (**4**) for the hairy flowers. (*Aloe, Duvalia*) – (**5**) application obscure. (*Kalanchoe, Sedum, Senecio tuberosus* var.)

pubicalyx Lat. 'pubes', pubescence, hairiness; and Lat. 'calyx', calyx. (*Conophytum*)

pubiglans Lat. 'pubes', pubescence, hairiness; and Lat. 'glans', gland; for the hairs on the cyathial glands. (*Euphorbia*)

pubipetalus Lat. 'pubes', pubescence, hairiness; and Lat. 'petalum', petal. (*Delosperma, Pelargonium*)

pubispinus Lat. 'pubes', pubescence, hairiness; and Lat. '-spinus', -spined; for the pubescent spination. (*Sclerocactus*)

pudibundus Lat., modest, bashful; for its insignificant appearance. (*Monadenium*)

pugionacanthus Lat. 'pugio, pugionis', dagger; and Gr. 'akantha', spine, thorn. (*Echinopsis, Gymnocalycium*)

pugioniformis Lat. 'pugio, pugionis', dagger; and Lat. '-formis', -shaped; for the leaves. (*Conicosia*)

pugniformis Lat. 'pugnus', (clenched) fist; and Lat. '-formis', -shaped; perhaps for the shape of the branches. (*Euphorbia*)

pujupatii For Shawintu Pujupat (fl. 1969), a Peruvian boy who accompanied the German plant collector A. B. Lau on his trip in Peru. (*Matucana*)

pulchellus Lat., little and beautiful (Dim. of Lat. 'pulcher', beautiful); (**1**) for the general appearance. (*Cephalophyllum, Cistanthe, Echeveria, Echinocereus, Grusonia, Haworthia, Miraglossum, Orbea, Pedilanthus, Pelargonium, Raphionacme, Sedum, Talinum*) – (**2**) for the beautiful flowers. (*Brachystelma, Drosanthemum, Ruschia*)

pulcher Lat. 'pulcher, pulchra, pulchrum', beautiful. (*Adenia, Aloe cooperi* ssp., *Cephalophyllum, Drosanthemum, Frithia, Gasteria, Quaqua, Rhipsalis, Sansevieria kirkii* var., *Sulcorebutia*)

pulcherrimus Comp. of Lat. 'pulcher', beautiful. (*Aloe*)

pulidonis For Miguel Pulido (fl. 1959) of Mexico City. (*Echeveria*)

pullatus Lat., clothed in dark garments; for the spine colour. (*Cleistocactus acanthurus* ssp.)

pulleineanus For Dr. Robert H. Pulleine (fl. 1955), cactus hobbyist in Adelaide, Australia. (*Coryphantha*)

pulquinensis For the occurrence near Pulquina, Prov. Florida / Prov. Valle Grande, Dept. Santa Cruz, Bolivia. (*Corryocactus*)

pulverulentus Lat., powdered, dusty; (**1**) for the powdery pubescence of the leaves. (*Pelargonium*) – (**2**) for the powdery farina of the leaves. (*Dudleya*) – (**3**) application obscure. (*Drosanthemum*)

pulvinaris Lat., belonging to cushions or pads; for the growth habit. (*Ruschia*)

pulvinatus Lat., cushion-shaped, strongly convex; (**1**) for the growth forming compact cushions. (*Euphorbia, Mesembryanthemum*) – (**2**) for the cushion-like leaves. (*Echeveria*) – (**3**) for the thickened base of the leaves. (*Sedum*) – (**4**) for the thickened corolla. (*Stapelia*)

pulviniger Lat. 'pulvinus', cushion; and Lat. '-ger, -gera, -gerum', -carrying; for the woolly areoles. (*Rhipsalis floccosa* ssp.)

pulvinosus Lat. 'pulvinus', cushion; and Lat. '-osus', full of; for the cushion-forming growth habit. (*Rebutia*)

pumilus Lat., dwarfed, minute; (**1**) for the habit. (*Agave, Antimima, Delonix, Dudleya cymosa* ssp.*, Frailea, Kalanchoe, Marsdenia, Opuntia, Parakeelya, Phyllobolus, Sedella, Sempervivum*) – (**2**) for the small (for the genus) flowers. (*Epiphyllum*)

punae For the occurrence in the Puna vegetation in Andean South America. (*Talinum*)

punctatus Lat., dotted (from Lat. 'punctus', sting, dot); (**1**) for the purplish spots along the leaf margins. (*Aichryson*) – (**2**) for the dots found on the corolla. (*Alsobia, Brachystelma, Pelargonium, Piaranthus*) – (**3**) perhaps for the verrucose corolla. (*Dischidia*)

punctulatus Lat., with small dots; for the raised dots above the idioblasts of the leaf surface. (*Ruschia*)

pungens Lat., pungent, piercing; (**1**) for the spination. (*Cleistocactus*) – (**2**) for the sharp-pointed leaf tips. (*Haworthia*) – (**3**) for the spines developing from sterile branches of the inflorescences. (*Arenifera*) – (**4**) for the spines developing from the pedicels after the fruits have fallen off. (*Ruschia*)

puniceodiscus Lat. 'puniceus', crimson, Phoenician purple; and Lat. 'discus', disc; for the colour of the base of the inside of the flowers. (*Rhipsalis*)

puniceus Lat., crimson, Phoenician purple; for the colour of the involucral bracts. (*Euphorbia*)

punta-caillan For the occurence at Punta Caillan, Cordillera Negra, Peru. (*Austrocylindropuntia*)

puquiensis For the occurrence near Puquio, Prov. Ayacucho, Peru. (*Corryocactus brevistylus* ssp.*, Echinopsis peruviana* ssp.)

purcellii For William F. Purcell (1866–1919), English zoologist emigrating to RSA when still a child. (*Crassula atropurpurea* var.)

purdomii For the botanical collector Purdom (fl. 1916), without further data. (*Sedum*)

purdyi For Carl Purdy (fl. 1936), possibly Carlton E. Purdy (1861–1945), US-American horticulturist, nurseryman and plant collector. (*Sedum spathulifolium* ssp.)

purpurascens Lat., becoming purple; for the inside of the corolla. (*Ceropegia*)

purpuratus Lat., purple-red; for the colour of the lower face of the leaves. (*Plectranthus*)

purpureoalbus Lat. 'purpureus', purple; and Lat. 'albus', white; for the petals. (*Cephalophyllum*)

purpureocroceus Lat. 'purpureus', purple; and Lat. 'croceus', saffron-yellow; for the flower colour. (*Malephora*)

purpureofuscus Lat. 'purpureus', purple; and Lat. 'fuscus', sombre brown; for the flower colour. (*Hoya*)

purpureostylus Lat. 'purpureus', purple; and Lat. 'stylus', style. (*Cerochlamys*)

purpureoviridis Lat. 'purpureus', purple; and Lat. 'viridis', green; for the flowering stems often flushed purplish. (*Rhodiola*)

purpureus Lat., purple; (**1**) for the colour of the plants. (*Tetragonia*) – (**2**) for the purplish leaf margins. (*Aloe*) – (**3**) for the colour of the central spine. (*Echinocereus engelmannii* var.) – (**4**) for the flower colour. (*Cheiridopsis, Coleocephalocereus, Delosperma, Lampranthus, Micranthocereus, Schizoglossum elingue* ssp.*, Sulcorebutia*)

purpusii For Carl A. Purpus (1851–1941), German botanical explorer of Mexico and the W USA, brother of J. A. Purpus. (*Hylocereus, Pilosocereus*) – (**2**) For Joseph Anton Purpus (1860–1932), German horticulturist in St. Petersburg and Darmstadt, brother of C. A. Purpus. (*Fouquieria, Pleiospilos*)

purpusorum For the Purpus brothers, Carl A. Purpus (1851–1941), German botanical explorer in Mexico and the USA, and Joseph A. Purpus (1860–1932), German horticulturist. (*Echeveria*)

pusilliflorus Lat. 'pusillus', very small, minute; and Lat. '-florus', -flowered. (*Cipocereus, Coryphantha*)

pusillus Lat., very small, minute; for the small size of the plants. (*Antimima, Bulbine, Ceropegia, Graptopetalum, Litanthus, Odontophorus, Opuntia, Portulaca, Sedum, Tetragonia, Tylecodon*)

pustulatus Lat., pustulate, covered with blisters or papillae; for the hairy leaves. (*Crassula*)

pustuligemmus Lat. 'pustula', pustule, blister;

and Lat. 'gemma', bud; for the blistered surface of the flower buds. (*Aloe*)

putterillii For Victor A. Putterill (fl. 1917–1926), South African mycologist and plant pathologist, Chief Fruit Inspector in the Western Cape. (*Ruschia*)

puttkamerianus For Mr. Puttkamer (fl. 1914), without further data. (*Hereroa*)

pycnacanthus Gr. 'pyknos', dense, stout, compact; and Gr. 'akantha', thorn, spine; for the spination. (*Coryphantha*, *Opuntia*)

pycnoneuroides Gr. '-oides', like; and for the similary to the genus *Pycnoneurum* (now synonymous with *Cynanchum*, Asclepiadaceae). (*Cynanchum*)

Pygmaeocereus Lat. 'pygmaeus', dwarf; and *Cereus*, a genus of columnar cacti; for the small body size of the plants. (*Cactaceae*)

pygmaeus Lat., dwarf; for the small size of the plants. (*Brachystelma*, *Ceraria*, *Cistanthe*, *Eriosyce taltalensis* var., *Frailea*, *Haworthia*, *Lewisia*, *Moringa*, *Portulaca*, *Rebutia*, *Trichodiadema*, *Tylecodon*)

pyramidalis Lat., pyramidal; for the outline shape of the densely leafy stems. (*Crassula*)

pyramidatus Lat., pyramidal; for the shape of the inflorescence. (*Senecio*)

Pyrenacantha Gr. 'pyren', a stone fruit, pyrene; and Gr. 'akanthos', thorn; for the peg-like protuberances from the inner surface of the fruit penetrating the cotyledons. (Icacinaceae)

pyrifolius Lat. 'pyrus', pear; and Lat. '-folius', -leaved. (*Crassula expansa* ssp.)

pyriformis Lat. 'pyrus', pear; and Lat. '-formis', -shaped; for the shape of the cladodes. (*Opuntia*)

pyromorphus Gr. 'pyr', fire; and Gr. 'morphe', shape; because the plants are shaped by fire. (*Adenia*)

pyrrhacanthus Gr. 'pyrrhos', fire-red; and Gr. 'akantha', spine, thorn; for the spine colour. (*Cumulopuntia*)

Q

qaharensis For the occurrence on Jabal Qahar, Saudi Arabia. (*Aloe*)

qarad From the local vernacular name of the plants in Yemen. (*Euphorbia*)

qoatlhambensis For the occurrence in the Qoatlhamba mountain range in E Lesotho. (*Crassula*)

quadrangularis Lat. 'quadr-', four-; and Lat. 'angularis', angled; (**1**) for the four-angled stems. (*Corryocactus*) – (**2**) for the four-angled branches. (*Cissus, Euphorbia*) – (**3**) for the four-ranked leaf arrangement. (*Crassula montana* ssp.) – (**4**) for the four-angled corolla tube. (*Kalanchoe*)

quadrangulus Lat. 'quadr-', four-; and Lat. 'angulus', angle; for the stems. (*Caralluma*)

quadratiumbonatus Lat. 'quadratus', square, four-sided; and Lat. 'umbonatus', umbonate, with a rounded projection in the middle; for the almost rectangularly tuberculate ribs of the stems. (*Echinopsis*)

quadratus Lat., square, four-sided; (**1**) for the subquadrate nectary glands. (*Euphorbia*) – (**2**) for the square outline of the fruits. (*Drosanthemum*)

quadrialatus Lat. 'quadr-', four-; and Lat. 'alatus', winged; for the four-winged branches. (*Euphorbia*)

quadricentralis Lat. 'quadr-', four-; and Lat. 'centralis', centrally; for the number of central spines. (*Pilosocereus*)

quadricostatus Lat. 'quadr-', four-; and Lat. 'costatus', ribbed; for the four-ribbed stems. (*Leptocereus, Pilosocereus floccosus* ssp.)

quadridens Lat. 'quadr-', four-; and Lat. 'dens', tooth; for the apically cleft corona segments. (*Schizoglossum*)

quadrifidus Lat. 'quadr-', four-; and Lat. '-fidus', divided; (**1**) for the flowers with 4 calyx lobes. (*Carpobrotus*) – (**2**) for the flowers with 4 petals. (*Portulaca, Rhodiola*) – (**3**) for the stamens arranged in 4 bundles. (*Gunniopsis*)

quadrilaterus Lat. 'quadr-', four-; and Lat. 'latus, lateris', side; for the four-sided branches. (*Euphorbia*)

quadripetalus Lat. 'quadr-', four-; and Lat. 'petalum', petal. (*Cistanthe, Sedum*)

quadrisepalus Lat. 'quadr-', four-; and Lat. 'sepalum', sepal. (*Octopoma*)

quadrispinus Lat. 'quadr-', four-; and Lat. '-spinus', -spined; for the four spines (two regular and two stipular) per spine shield. (*Euphorbia*)

quadrivalvis Lat. 'quadr-', four-; and Lat. 'valva', valve; for the often four-valvate fruits. (*Parakeelya*)

quaesitus Lat., sought; for the presumed rarity of the taxon. (*Conophytum*)

quaitensis For the occurrence in the Quaita area, S Yemen. (*Euphorbia*)

Quaqua From the local vernacular name "Qua-Qua" or "Kam-qua-qua" for *Q. hottentottorum* in the Nama language. (*Asclepiadaceae*)

quarcicola Lat. from Fr. / Engl. 'quartz', quartz rocks; and Lat. '-cola', inhabiting. (*Acrodon*)

quarciticola Lat. from Fr. / Engl. 'quartzite', quartzite rocks; and Lat. '-cola', inhabiting. (*Pelargonium*)

quartinianus For Richard Quartin-Dillon († 1841), French botanist and explorer of Ethiopia. (*Kalanchoe*)

quartzicola Lat. from Fr. / Engl. 'quartz', quartz rocks; and Lat. '-cola', inhabiting. (*Bulbine*)

quartziticola Lat. from Fr. / Engl. 'quartzite', quartzite rocks; and Lat. '-cola', inhabiting. (*Aloe capitata* var., *Euphorbia, Senecio*)

quartziticus Lat. from Fr. / Engl. 'quartz', quartz rocks; for the occurrence on quartzite rocks. (*Phyllobolus*)

quarziticus Lat. from Germ. 'Quarz', quartz; for the occurrence on quartzitic rocks. (*Antimima*)

quaternatus Lat. 'quaterna', four each; for the leaves in whorls of four. (*Sedum*)

quehlianus For Leopold Quehl (1849–1922), German cactus hobbyist and founder member of the Deutsche Kakteen-Gesellschaft (DKG). (*Gymnocalycium*)

quercetorum Gen. Pl. of Lat. 'quercus', oak,

i.e. 'of the oaks'; perhaps for an occurrence in oak woodland. (*Pelargonium*)

quercifolius For the genus *Quercus* (oak, Fagaceae), and Lat. '-folius', -leaved; for the leaves, which resemble those of some species of oak. (*Carica*)

queretaroensis For the occurrence in the Mexican state of Querétaro. (*Stenocereus*, *Yucca*)

quevae For Prof. Charles Queva (fl. 1894), French botanist at Dijon University. (*Sedum*)

quevedonis For Miguel A. de Quevedo (1862–1946), Mexican astronomer, meteorologist, later engineer and then conservationist ("Apostle of the Tree"). (*Stenocereus*)

quezaltecus For the occurrence in Dept. Quezaltenango, Guatemala. (*Disocactus*)

Quiabentia From the local vernacular name "Quiabento" for the plants in Brazil. (*Cactaceae*)

quicheensis For the occurrence in Prov. Quiché, Guatemala. (*Furcraea*)

quimilo From the local vernacular name for the plants in N Argentina. (*Opuntia*)

quinarius Lat., five-partite (from Lat. 'quinque', five); for the arrangement of the leaves in five rows. (*Avonia*)

quinatus Lat., in fives; for the five-foliolate leaves. (*Cyphostemma*)

quinquangularis Lat. 'quinque', five; and Lat. 'angularis', angled; for the stems. (*Cissus*)

quinquecostatus Lat. 'quinque', five; and Lat. 'costatus', ribbed; for the five-angled branches. (*Euphorbia*)

quinquelobatus Lat. 'quinque', five; and Lat. 'lobatus', lobed; for the leaf shape. (*Pelargonium*)

quintus Lat., the fifth one; because it was the fifth variety of *H. scabra*. (*Huernia*)

quipa From the local vernacular name "Quipa" for several opuntioid taxa in NE Brazil. (*Tacinga*)

quisqueyanus For the occurrence on the island of Quisqueya, Dominican Republic. (*Pereskia*)

quitensis For the occurrence near Quito, Ecuador. (*Echeveria*, *Opuntia*)

R

rabaiensis For the occurrence at the Rabai Hills, Kenya. (*Aloe*)

Rabiea For Mr. W. A. Rabie (fl. 1930), priest and plant collector in the then Oranje Free State, RSA. (*Aizoaceae*)

rabiei As above. (*Phyllobolus*)

rabiesbergensis For the occurrence at Rabie's Berg, Langeberg area, Western Cape, RSA. (*Lampranthus*)

racemosus Lat., racemose; for the inflorescence. (*Echeveria, Tylecodon*)

radians Lat., radiate, like the spokes of a wheel; (**1**) for the spination. (*Coryphantha*) – (**2**) for the spreading coloured bracts. (*Euphorbia*) – (**3**) for the spreading corolla lobes. (*Echidnopsis*)

radiatus Lat., radiate, like the spokes of a wheel; (**1**) for the arrangement of the corolla lobes. (*Quaqua*) – (**2**) for the arrangement of the corona lobes. (*Cynanchum*) – (**3**) for the stellately spreading follicles. (*Sedum*)

radicans Lat., rooting; for the creeping and rooting stems. (*Aeschynanthus, Ceropegia, Crassula pubescens* ssp., *Prenia, Ruschia, Senecio*)

radicosus Lat., with many roots (from Lat. 'radix', root). (*Rosularia*)

radlii For Florian Radl (fl. 1896), head gardener for the firm of Haage & Schmidt, Erfurt, Germany. (×*Gasteraloe*)

radula Lat., rasp; for the roughly papillate leaves. (*Brunsvigia, Haworthia attenuata* var.)

radulifolius Lat. 'radula', rasp; and Lat. '-folius', -leaved; for the rough leaves. (*Pelargonium*)

raffillii For Charles P. Raffill (1876–1951), without further data. (*Sansevieria*)

ragonesei For Arturo E. Ragonese (*1909), Argentinian botanist. (*Gymnocalycium, Halosicyos, Portulaca*)

rakotozafyi For Armand Rakotozafy (fl. 1984), botanist at the herbarium of Tsimbazaza, Madagascar. (*Euphorbia cremersii* var.)

ramentaceus Lat., covered in chaffy scales; application obscure. (*Sedum*)

ramiglans Lat. 'ramus', branch; and Lat. 'glans', gland; for the branched cyathial glands. (*Euphorbia*)

ramillosus From the Dim. of Lat. 'ramus', branch; and Lat. '-osus', full of; for the similarity of the dense spination to small bundles of dried twigs. (*Coryphantha*)

ramnadensis For the occurrence in the Ramnad Distr. of Madras State, India. (*Jatropha villosa* var.)

ramosissimus Superl. of Lat. 'ramosus', branched. (*Aloe, Brachystelma, Cylindropuntia, Drosanthemum, Euphorbia pillansii* var., *Portulaca*)

ramosus Lat., branched; (**1**) for the growth form. (*Adromischus umbraticola* ssp., *Bulbine, Erepsia, Euphorbia monteiri* ssp., *Haworthia cymbiformis* var., *Pachypodium lamerei* var., *Portulaca, Quaqua*) – (**2**) for the branched cymes. (*Euphorbia heptagona* var.)

ramulosus Lat., with many small branches (from the Dim. of Lat. 'ramus', branch). (*Dicrocaulon, Euphorbia, Pseudorhipsalis*)

rangeanus For Dr. Paul Range (1879–1952), German geologist emigrating to Namibia, and keen naturalist. (*Orbea maculata* ssp., *Tetragonia*)

rapaceus Lat., turnip-like (from Lat. 'rapa', beetroot, rape); for the thick tuber. (*Caulipsolon, Pelargonium*)

Raphionacme Gr. 'r[h]apys, r[h]aphys', beetroot; or Gr. 'rhaphys', needle; and Gr. 'akme', sharpness, cutting edge; application obscure. (*Asclepiadaceae*)

rariflorus Lat. 'rarus', rare; and Lat. '-florus', -flowered; for the well separated flowers in terminal and lateral cymes. (*Ruschia*)

rarus Lat., rare; for the restricted distribution range. (*Marsdenia*)

rastrerus From Span. 'rastrero', crawling (from Lat. 'rastrere', to scrape); for the growth habit. (*Opuntia*)

ratticaudatus MLat. 'rattus', rat; and Lat. 'caudatus', with a tail; for the very long inflorescences. (*Peperomia*)

rattrayi For George Rattray (1872–1941), Scottish teacher and naturalist in RSA. (*Crassula pubescens* ssp.)
ratus Lat., fixed, settled; because the generic position was uncertain before the taxon was described. (*Conophytum*)
Rauhia For Werner Rauh (1913–2000), German botanist in Heidelberg, and specialist on Madagascan succulents. (*Amaryllidaceae*)
rauhianus As above. (*Cynanchum*)
rauhii As above. (*Aloe, Armatocereus, Conophytum uviforme* ssp.*, Pachypodium, Pereskia horrida* ssp., *Weberbauerocereus*)
Rauhocereus For Werner Rauh (1913–2000), German botanist in Heidelberg, and specialist on Madagascan succulents; and *Cereus*, a genus of columnar cacti. (*Cactaceae*)
rauschii For Walter Rausch (*1928), Austrian cactus specialist and traveller in South America. (*Gymnocalycium*)
rawlinsonii For S. I. Rawlinson, collector and grower of succulents in RSA. (*Gasteria*)
raymondii For Dr. Raymond Hamet (1890–1972), French physician and botanist in Paris, and *Crassulaceae* specialist. (*Sedum*)
rayonesensis For the occurrence in the valley of Rayones, Nuevo León, Mexico. (*Echinocereus*)
razafindratsirae For Alfred Razafindratsira (fl. 2001), Madagascan plant collector and owner of a succulent plant nursery. (*Euphorbia*)
razafinjohanii For Monsieur Razafinjohany (fl. 1955), artist at the botanical garden of Tsimbazaza, Madagascar. (*Euphorbia*)
reae For Julio Rea (fl. 1957), Bolivian agronomist. (*Cleistocactus*)
rebmannii For Prof. Norbert Rebmann (fl. 1993, 2002), French university teacher and *Aloe* enthusiast. (*Aloe*)
Rebutia For Pierre Rebut (1830–1898), French vine-grower and owner of a succulent plant nursery in Chazay d'Azergues near Lyon. (*Cactaceae*)
rebutii As above. (×*Gasteraloe*)
rechbergeri For Bruno Rechberger (fl. 2002), Swiss succulent plant enthusiast in Zürich. (*Uncarina leandrii* var.)

rechensis For the occurrence near the village of Ana Rech, Rio Grande do Sul, Brazil. (*Parodia*)
rechingeri For Prof. Karl H. Rechinger (*fil.*) (1906–1998), Austrian botanist at Vienna, and editor of 'Flora Iranica'. (*Othonna, Prometheum*) – (**2**) Unresolved. (*Kalanchoe*)
reclinatus Lat., bent down, reclined; for the spreading decumbent habit. (*Euphorbia*)
reconditus Lat., hidden; for the habitat. (*Conophytum, Eriosyce, Huernia*)
recticaulis Lat. 'rectus', straight; and Lat. 'caulis', stem. (*Rhodiola*)
rectiramus Lat. 'rectus', straight; and Lat. 'ramus', branch. (*Euphorbia*)
rectispinus Lat. 'rectus', straight; and Lat. '-spinus', -spined. (*Echinocereus fendleri* ssp., *Ferocactus emoryi* ssp., *Stenocactus*)
recurvatus Lat., curved backwards; (**1**) for the arrangement of the leaves. (*Beaucarnea*) – (**2**) for the recurved leaves of the plants from the type locality. (*Echeveria*) – (**3**) for the disposition of the scales covering the leaves. (*Avonia*) – (**4**) for the arrangement of the spines. (*Coryphantha, Cumulopuntia*) – (**5**) for the disposition of the petals. (*Brachystelma*) – (**6**) for the disposition of the corolla lobes. (*Ceropegia*)
recurvifolius Lat. 'recurvus', recurved; and Lat. '-folius', -leaved. (*Yucca*)
recurvispinus Lat. 'recurvus', recurved; and Lat. '-spinus', -spined. (*Anthorrhiza*)
recurvus Lat., recurved; (**1**) for the leaf tips. (*Ruschia*) – (**2**) for the calyx lobes. (*Lampranthus*) – (**3**) for the petal tips. (*Villadia*)
redactus Lat., reduced; (**1**) for the smaller size of the plants. (*Aloe ciliaris* var.) – (**2**) for the reduced number of petals and fertile stamens. (*Pelargonium*)
reddii For Dr. V. B. Reddi (fl. 1994) of East London, RSA. (*Haworthia cymbiformis* var.)
redimitus Lat., girdled; for the operculum of the fruits, which are nearly flat and whose edges are prolonged into a number of teeth forming a star-like crown. (*Zaleya*)
redivivus Lat., reviving from the dry state, living again; because herbarium specimens started to grow again. (*Lewisia*)

reduncispinus Lat. 'reduncus', curved or bent backwards; and Lat. '-spinus', -spined; for the spination. (*Coryphantha*)

reduplicatus Lat., reduplicate, double; for the leaves folded lengthwise. (*Tetragonia*)

reenensis For the occurrence at the Van Reenens Pass, KwaZulu-Natal, RSA. (*Aspidonepsis*)

reflexipetalus Lat. 'reflexus', reflexed; and Lat. 'petalum', petal; for the reflexed upper petals. (*Pelargonium*)

reflexispinus Lat. 'reflexus', reflexed; and Lat. '-spinus',-spined; for the spines directed backwards. (*Grusonia*)

reflexus Lat., reflexed; (**1**) for the reflexed tubercles on the stem. (*Monadenium*) – (**2**) for the leaves. (*Lenophyllum*) – (**3**) for the lower and upper lips of the flowers. (*Plectranthus*)

refractus Lat., curved back; application obscure. (*Adenia*)

regalis Lat., royal; for the very attractive flowers. (*Cephalophyllum, Conophytum*)

regelii For Dr. Eduard A. von Regel (1815–1892); German botanist, 1842–1855 head gardener at the Botanical Garden Zürich, then at the Imperial Botanic Gardens St. Petersburg and from 1875 its director. (*Harrisia pomanensis* ssp.)

reginae Gen. of Lat. 'regina', queen; for the superior appearance of the plant. (*Rosularia, Schlumbergera*)

reginae-amaliae Gen. of Lat. 'regina', queen; and for Amalia von Oldenburg (1818–1875), Queen of Greece and wife of King Otto of Greece. (*Sempervivum marmoreum* ssp.)

regis-jubae Gen. of Lat. 'Rex Juba', King Juba of Mauritania, who gave the name *Euphorbia* to the genus. (*Euphorbia*)

regius Lat., king-like; because this was "the most striking species of the whole genus" at the time of description. (*Glottiphyllum*)

regnellii For Dr. Anders F. Regnell (1807–1884), Swedish physician and botanist, settled 1840 in Minas Gerais, Brazil. (*Lepismium houlletianum* fa.)

rehmannii For Dr. Anton Rehmann (1840–1917), Austrian botanist and cartographer collecting in the then Natal and Transvaal (RSA) 1875–1880. (*Aeollanthus, Coccinia, Dioscorea sylvatica* var., *Plectranthus, Plinthus*)

rehneltianus For F. Rehnelt (fl. 1920), German succulent plant enthusiast. (*Hereroa*)

reichenbachii For Friedrich Reichenbach (fl. 1842), German engineer from Dresden, collected cacti in Mexico. (*Echinocereus*)

reinwardtii For Prof. Dr. Caspar G. C. Reinwardt (1773–1854), German botanist and professor of chemistry, pharmacy and natural history in Holland, founder of the Botanical Garden Buitenzorg, Java, collected in RSA in 1816. (*Haworthia*)

reitzii For F. W. Reitz (fl. 1937), without further data. (*Aloe*)

rekoi Nach Dr. Blas P. Reko (1876–1953), German (?) ethnobotanist in Mexico. (*Mammillaria*)

remotus Lat., remote, distant; (**1**) for the distribution. (*Stapelia*) – (**2**) application obscure. (*Brachystelma, Parakeelya*)

reniformis Lat., kidney-shaped; for the leaf shape. (*Cissus, Pelargonium, Sedum*)

renneyi For C. Arnold Renney (*1925), English chartered secretary and succulent plant hobbyist, resident in Nairobi (Kenya) since 1955, who first recognized the taxon as undescribed. (*Monadenium*)

repandus Lat., repand, with slightly uneven and wavy margins; (**1**) for the ribs. (*Cereus*) – (**2**) for the leaves. (*Adenia*)

repens Lat., creeping. (*Aloe andongensis* var., *Callisia, Cissus, Delosperma, Echidnopsis, Huernia volkartii* var., *Opuntia*)

reppenhagenii For Werner Reppenhagen (1911–1996), Austrian cactus collector, cactus nursery owner and *Mammillaria* specialist. (*Ferocactus alamosanus* ssp., *Mammillaria albilanata* ssp.)

reptans Lat., creeping; for the growth form. (*Euphorbia, Lampranthus, Sedum*)

resedolens Lat. 'olens', smelling; and for the genus *Reseda* ("Mignonette", *Resedaceae*); for the pleaseantly scented flowers. (*Stomatium*)

resiliens Lat., springing back; for the corolla lobes, which are more reflexed than in other taxa of the genus. (*Sarcostemma*)

resinifer Lat. 'resina', aromatic resin; and Lat. '-fer, -fera, -ferum', -carrying; for the "Gum Euphorbium" which was used medicinally as a drug. (*Euphorbia*)

restioides Gr. '-oides', resembling; and for the genus *Restio* (*Restionaceae*). (*Aspidoglossum*)

restitutus Lat., re-established; for the confused identification preceeding the publication of this taxon as a distinct species. (*Euphorbia*)

restrictus Lat., restricted; for the small size and the restricted range. (*Euphorbia*)

resurgens Lat., reappairing; for the annually deciduous growth from a tuberous rootstock. (*Phyllobolus*)

reticulatus Lat., reticulate, networked; (**1**) for the patterning of the stems. (*Pterocactus*) – (**2**) for the patterning of the leaves. (*Haworthia*) – (**3**) for the branched persistent inflorescences. (*Tylecodon*) – (**4**) probably for the coloration of the corolla. (*Huernia*) – (**5**) for the patterning of the seed testa. (*Parakeelya*)

retrofractus Lat. 'retro', backwards; and Lat. 'fractus', broken (from Lat. 'frangere', to break); application obscure. (*Othonna*)

retrorsus Lat., turned backwards; (**1**) for the arrangement of the spines on the stems. (*Opuntia anacantha* var.) – (**2**) application obscure. (*Othonna*)

retrospiciens Lat., looking back; for the orientation of the buds and flowers. (*Aloe*)

retrospinus Lat. 'retro-', backwards, and Lat. '-spinus', -spined; for the orientation of the spines. (*Euphorbia*)

retroversus Lat. 'retro', backwards; and Lat. 'versus', turned in the direction of; for the leaves of the second leaf pair that lie flat on the ground. (*Diplosoma*)

retusus Lat., rounded with a slightly notched tip, retuse; (**1**) for the flat-topped rosettes. (*Anacampseros*) – (**2**) for the notched leaf-tip. (*Gasteria carinata* var., *Haworthia*, *Hoya*, *Pedilanthus tithymaloides* ssp., *Sedum obtusatum* ssp., *Sedum*, *Tetragonia*) – (**3**) for the tubercles of the plant bodies. (*Ariocarpus*, *Coryphantha*)

reverchonii For Julien Reverchon (1834–1905), French botanical collector, settled in Texas 1856. (*Yucca*)

revolutifolius Lat. 'revolutus', revolute, rolled back; and Lat. '-folius', -leaved. (*Sesuvium*)

revolutus Lat., revolute, rolled back; (**1**) for the leaves. (*Agave*) – (**2**) for the leaf margins. (*Hoya*) – (**3**) for the corolla lobes. (*Tromotriche*)

rex Lat., King; for the outstanding size and colour of the plants. (*Conophytum herreanthus* ssp.)

reynoldsii For Dr. Gilbert W. Reynolds (1895–1967), Australian optometrist, emigrated to RSA in 1902, *Aloe* specialist. (*Aloe*, *Delosperma*)

rhabdodes Gr. 'rhabdos', rod or wand; perhaps for the rod-like stems when leafless. (*Euphorbia*)

rhabdophyllus Gr. 'rhabdos', rod or wand; and Gr. 'phyllon', leaf; for the shape of the leaves. (*Dracaena*)

rhacodes Gr., torn, ragged, tattered; for the ragged gynostegium head. (*Stathmostelma*)

Rhadamanthus For Rhadamanthus, the son of Zeus and Europa, and brother of Minos. (*Hyacinthaceae*)

Rhinephyllum Gr. 'rhine', file, rasp; and Gr. 'phyllon', leaf; for the rough leaf surface. (*Aizoaceae*)

Rhipsalis From Gr. 'rhips', willow twig, wickerwork; for the tangled thin stems. (*Cactaceae*)

rhizomatosus Lat., rhizomatous; for the growth habit. (*Euphorbia cuneneana* ssp.)

rhizophorus Gr. 'rhiza', root; and Gr. '-phoros', carrying; for the growth habit. (*Monadenium*)

rhodacanthus Gr. 'rhodos' rose-red; and Gr. 'akantha', spine, thorn; (**1**) for the spination. (*Denmoza*) – (**2**) for the red leaf marginal teeth. (*Agave*)

rhodanthus Gr. 'rhodos' rose-red; and Gr. 'anthos', flower. (*Arrojadoa*, *Ferocactus echidne* var., *Mammillaria*, *Rhodiola*)

rhodesiacus For the occurrence in the former Southern Rhodesia (now Zimbabwe). (*Euphorbia confinalis* ssp., *Orbea caudata* ssp.)

rhodesianus As above. (*Aloe*, *Portulaca*, *Sansevieria*)

rhodesicus As above. (*Aspidoglossum, Avonia, Crassula*)

Rhodiola Gr. 'rhodon', rose; and Lat. Dim. suffix '-iola'; the roots have the scent of roses. (*Crassulaceae*)

rhodocarpus Gr. 'rhodos' rose-red; and Gr. 'karpos', fruit. (*Sedum*)

Rhodocodon Gr. 'rhodos' rose-red; and Gr. 'kodon', bell; for the bell-shaped rose-red flowers of the type species. (*Hyacinthaceae*)

rhodotrichus Gr. 'rhodos' rose-red; and Gr. 'thrix, trichos', hair; for the hairs on the flower tube. (*Echinopsis*)

rhombeus Lat., diamond-shaped, rhombic, lozenge-shaped; for the leaf shape. (*Peperomia*)

rhombifolius Lat. 'rhombus', rhombus, lozenge; and Lat. '-folius', -leaved. (*Dischidia, Euphorbia*)

rhomboideus Lat., rhomboid, lozenge-shaped; for the leaf shape. (*Rhombophyllum*)

Rhombophyllum Gr. 'rhombos', rhombus, lozenge; and Gr. 'phyllon', leaf; for the leaf shape. (*Aizoaceae*)

rhombopilosus Lat. 'rhombus', rhombic, lozenge-shaped; and Lat. 'pilosus', hairy; for the leaf shape and indumentum. (*Kalanchoe*)

rhopalophyllus Gr. 'rhopalon', club; and Gr. 'phyllon', leaf; for the club-shaped leaves. (*Bulbine, Fenestraria*)

rhynchocalyptra Gr. 'rhynchos', beaked; and Gr. 'kalyptra', covering, woman's hat; for the persistent stigma on the operculum of the fruit. (*Trianthema*)

Rhytidocaulon Gr. 'rhytidos', wrinkle, fold; and Gr. 'kaulos', stem; for the sculptured stems. (*Asclepiadaceae*)

ricardianus For Richard Grässner (1875–1942), German cactus hobbyist and nurseryman in Perleberg near Berlin. (*Conophytum*)

richardianus For Dr. Richard Barad (1952–1985), US-American physician and tragically deceased son of the US-American succulent plant enthusiast Gerald Barad, who discovered the taxon. (*Rhytidocaulon*)

richardsiae For Mrs. H. Mary Richards (1885–1977), British botanical collector, esp. in Zambia and Tanzania, resident in East Africa 1952–1974. (*Aloe, Euphorbia*)

richardsii For David ("Dave") J. Richards (*1925), Zimbabwean engineer with the Department of Works, City of Salisbury / Harare, succulent plant enthusiast and seed bank organizer for the Aloe, Cactus & Succulent Society of Zimbabwe, and who often accompanied the discoverer of the plant, R. Peckover, on field trips in Zimbabwe. (*Brachystelma*)

Richtersveldia For the occurrence in the Richtersveld, Northern Cape, RSA. (*Asclepiadaceae*)

rigens Lat., rigid; (**1**) for the stems. (*Ruschia*) – (**2**) for the leaves. (*Aloe*)

rigidicaulis Lat. 'rigidus', rigid, stiff; and Lat. 'caulis', stem. (*Ruschia*)

rigidiflorus Lat., rigid, stiff; and Lat. '-florus', -flowered. (*Dudleya*)

rigidissimus Comp. of Lat. 'rigidus', rigid, stiff; for the spination. (*Echinocereus*)

rigidus Lat., rigid, stiff; (**1**) for the branches. (*Cephalophyllum, Galenia, Sceletium*) – (**2**) perhaps for the thick woody stems. (*Dudleya*) – (**3**) for the stiff young branches. (*Ruschia*) – (**4**) for the leaves. (*Beschorneria, Yucca*) – (**5**) application obscure. (*Aizoon*)

rileyi For Laurence A. M. Riley (1888–1928), British botanist, volunteer worker at Kew. (*Opuntia*) – (**2**) For Mr. A. W. Riley (fl. 1959), farmer in the Pietersburg Distr., former Transvaal, RSA. (*Delosperma*)

rinconensis For the occurrence near La Rinconada, Nuevo León, Mexico. (*Thelocactus*)

ringens Lat., gaping; (**1**) for the large gap between the leaves of a pair. (*Argyroderma, Carruanthus*) – (**2**) for the gaping corolla tips. (*Ceropegia*)

Riocreuxia For Alfred Riocreux (1820–1912), French botanical artist. (*Asclepiadaceae*)

riojensis For the occurrence near the city of La Rioja, Prov. La Rioja, Argentina. (*Gymnocalycium*)

riomajensis For the occurrence at the Río Majes, Dept. Arequipa, Peru. (*Armatocereus*)

riosaniensis For the occurrence in the valley

of the Río Saña, Dept. Lambayeque, Peru. (*Rauhocereus*)

rioverdensis For the occurrence near the city of Rioverde, San Luis Potosí, Mexico. (*Turbinicarpus*)

riparius Lat., pertaining to a stream bank (from Lat. 'ripa', shore); erroneously applied as the taxon does not grow at such places. (*Tetradenia*)

ritchiei For Capt. Archie T. A. Ritchie († 1962), British zoologist and army officer, professional soldier for several years, then lived in Kenya from 1920, becoming Chief Game Warden for Kenya 1924–1948, established the Kenya National Parks. (*Monadenium*)

ritteri For Friedrich Ritter (1898–1989), German geologist and self-taught botanist, explorer and traveller esp. in South America, and cactus specialist. (*Aztekium, Cleistocactus, Espostoa, Eulychnia, Matucana, Melocactus violaceus* ssp., *Opuntia, Oreocereus, Parodia, Rebutia*)

ritterianus As above. (*Gymnocalycium*)

rivae For Dr. Domenico Riva (± 1856–1895), Italian botanist, collected in NE Africa during the expedition by Prince Ruspoli in 1892–1894, unable to overcome the shock of Ruspoli's death in Ethiopia, committed suicide after return to Italy. (*Aloe, Euphorbia, Sesamothamnus*)

riviere-de-caraltii For Fernando Riviere de Caralt (1904–1992), Spanish industrialist, grower of succulents and owner of the private botanical garden "Pinya de Rosa". (*Echinopsis*)

rivierei As above. (*Aloe*)

roanianus For Mr. H. Michael Roan (1909–2003), English businessman and succulent plant collector and one of the founders of the National Cactus and Succulent Society, UK. (*Adromischus*)

robbinsorum For James A. Robbins (fl. 1976) and his sons Jimmy and John, who first discovered the taxon. (*Escobaria*)

robecchii For the Italian engineer Robecchi-Bricchetti who collected in Ethiopia and Somalia during three journeys 1889 to 1891. (*Euphorbia*)

robertsianus Perhaps for Austin Roberts (1883–1948), naturalist in RSA. (*Turbina*) – (**2**) For Kate Roberts (fl. 1936), on whose ranch the original material was collected. (*Sedum*)

robinsonii For Prof. Dr. Benjamin L. Robinson (1864–1935), US-American botanist at the Gray Herbarium, Cambridge, and 1899–1935 Asa Gray Professor of systematic botany. (*Opuntia*) – (**2**) For Mr. Robinson (fl. 1984), who brought living plants into cultivation. (*Euphorbia decaryi* var.)

robivelonae For Mme. Adrienne Robivelo (fl. 1994), Madagascan ethnobotanist in Lyon and later Colmar. (*Ceropegia, Euphorbia*)

roborensis For the occurrrence at Roboré, Prov. Chiquitos, Dept. Santa Cruz, Bolivia. (*Opuntia*)

robustior Comp. of Lat. 'robustus', robust; i.e. more robust. (*Pediocactus simpsonii* var.)

robustispinus Lat. 'robustus', robust, firm, hard; and Lat. '-spinus', -spined; for the spination. (*Coryphantha*)

robustus Lat., robust, firm, hard; (**1**) for the general appearance. (*Cheiridopsis, Crassula tetragona* ssp., *Ferocactus, Huernia hislopii* ssp., *Pilosocereus pentaedrophorus* ssp., *Sansevieria*) – (**2**) for the habit. (*Delosperma, Euphorbia richardsiae* ssp., *Gymnocalycium, Kalanchoe, Opuntia, Ruschia, Tetragonia*) – (**3**) for the robust spination. (*Eriosyce curvispina* var., *Euphorbia tenuispinosa* var.)

roczekii For Mr. Bernd Roczek (fl. 2002), German cactus collector. (*Mammillaria saboae* ssp.)

rodentiophilus Lat. 'rodentia', rodents; and Gr. 'philos', friend; because the fruits are eaten by rodents. (*Eriosyce*)

rodolfi For Rodolfo Martínez Gallegos (fl. 2000), son of the Mexican botanist José G. Martínez-Ávalos. (*Echeveria*)

rodwayi For Mr. F. A. Rodway (fl. 1907), without further data. (*Gunniopsis*)

roemeri For Dr. Richard C. Römer (fl. 2002), German cactus collector. (*Mammillaria*)

roemerianus As above. (*Echinocereus adustus* ssp.)

roeoeslianus For Walter Röösli (fl. 2002),

Swiss succulent plant enthusiast in Zürich specializing in Madagascar plants. (*Uncarina*)

roezlii For Benedict Roezl (1824–1885), Bohemian horticulturist and seedsmen, collecting in North and South America between 1854 and 1875. (*Cleistocactus*)

rogersiae For Bertha Rogers (fl. 1928), who collected the type specimen. (*Trichodiadema*)

rogersii For Frederick A. Rogers (1876–1944), English missionary and amateur botanist, lived in RSA from 1904. (*Crassula, Delosperma, Orbea, Sterculia*)

roggeveldii For the occurrence in the Roggeveld in the Northern Cape, RSA. (*Crassula*)

rolandi-bonapartei For Prince Roland Bonaparte (1858–1924), French botanist and fern specialist, second brother of Napoleon I. (*Kalanchoe*)

rolfii For Rolf Rawé (fl. 1970s, 1980s), German succulent plant enthusiast and *Conophytum* specialist in RSA. (*Conophytum obcordellum* ssp.)

ronaldii For Mr. Ronald (fl. 1932), without further data. (*Stomatium*)

rondonianus For Cândido M. da Silva Rondon (1865–1958), Brazilian explorer, army major, and founder of the "Indian Protection Service", and natural history specimen collector. (*Arthrocereus*)

roodiae For Mrs. Petrusa Benjamina Rood (1861–1946), South African housewife and active plant collector. (*Conophytum, Vanheerdea*)

rooneyi For the occurrence at Rooney's Place, Big Bend region, Texas, USA. (*Opuntia*)

roridus Lat., covered with dew droplets; for the droplets produced by extrafloral nectaries on the inflorescences. (*Sansevieria*)

rosae For Rösli Uebelmann (*1921), wife of the Swiss cactus nurseryman Werner Uebelmann. (*Pilosocereus fulvilanatus* ssp.) – (**2**) For Rosa Till (*1926), wife of the Austrian cactus specialist Hans Till. (*Gymnocalycium*)

rosaricus For the occurrence near El Rosario, Baja California, Mexico. (*Cylindropuntia californica* var.)

roseanus For Dr. Joseph N. Rose (1862–1928), US-American botanist in Washington D.C. and eminent specialist on *Cactaceae* and American *Crassulaceae*. (*Acharagma, Agave sobria* ssp., *Kalanchoe*) – (**2**) For Mr. Henri Rose (fl. 1962), curator at the Gardens of the Paris Natural History Museum. (*Euphorbia milii* var.)

rosei For Dr. Joseph N. Rose (1862–1928), US-American botanist in Washington D.C. and eminent specialist on *Cactaceae* and American *Crassulaceae*. (*Agave, Peniocereus, Weberocereus*)

roseiflorus Lat. 'roseus', rose-red; and Lat. '-florus', -flowered. (*Cereus, Cleistocactus, Neoraimondia arequipensis* ssp.)

roseiglandulosus Lat. 'roseus', rose-red; and Lat. 'glandulosus', glandular; for the pubescence of the leaves. (*Cyphostemma*)

roseoalbus Lat. 'roseus', rose-red; and Lat. 'albus', white; for the spine colours. (*Mammillaria*)

roseolus Lat., pink, pale rose; for the petal colour. (*Antimima*)

roseopurpureus Lat. 'roseus', rose-red; and Lat. 'purpureus', purple; for the flower colour. (*Delosperma*)

roseus Lat., rose-like; (**1**) for the rose-like scent of the roots. (*Rhodiola*) – (**2**) for the flower colour. (*Aloe divaricata* var., *Aloe, Cylindropuntia, Hatiora, Huernia, Lampranthus, Leipoldtia, Mitrophyllum, Phyllobolus, Tetragonia*) – (**3**) for the rose-coloured inflorescences. (*Cistanthe, Echeveria*)

rossianus For Georg Ross (1887–1963), German horticulturist and owner of a cactus nursery in Bad Krotzingen. (*Cumulopuntia*)

rossii For Mr. Ross (fl. 1820) of Stoke Newington, London, who provided Haworth with seeds. (*Carpobrotus*) – (**2**) For Georg Ross (1887–1963), German horticulturist and owner of a cactus nursery in Bad Krotzingen. (*Echinopsis pugionacantha* ssp.) – (**3**) For Erich Ross (fl. 1960), sponsor of Madagascar botany in Heidelberg. (*Cynanchum, Euphorbia*)

rostellus Lat. 'rostellum', small beak; (**1**) for the leaf shape. (*Ruschia*) – (**2**) for shape of the pairs of closely appressed leaves. (*Cephalophyllum*)

rosthornianus For the German botanical collector Baron von Rosthorn (fl. 1900). (*Sedum*)

rostratus Lat., beaked; (**1**) for the shape of the flower buds. (*Nolana*) – (**2**) for the fruit shape. (*Momordica, Yucca*) – (**3**) perhaps for the leaf tips. (*Cheiridopsis*)

Rosularia Lat. 'rosula', a small rose; because the leaves are arranged in a rosette. (*Crassulaceae*)

rosularis Lat., with a rosette. (*Sedum longipes* ssp.)

rosulatobulbosus Lat. 'rosulatus', with a rosette; and Lat. 'bulbosus', bulbous; for the axillary bulbils. (*Sedum*)

rosulatus Lat., rosetted, rosulate, with a rosette (from Lat. 'rosula', a small rose). (*Aloinopsis, Pachypodium, Rosularia*)

rosynensis For the occurrence on the Rosyntjieberg Mts., Northern Cape, RSA. (*Conophytum taylorianum* ssp.)

rotundatus Lat., rounded; for the globose flowers. (*Rhodocodon*)

rotundifolius Lat. 'rotundus', round; and Lat. '-folius', -leaved. (*Cissus, Kalanchoe, Monadenium letestuanum* var., *Monadenium pudibundum* var., *Peperomia, Pereskiopsis, Portulaca, Solenostemon*)

rourkei For Dr. John P. Rourke (*1942), South African botanist and curator of the Compton Herbarium, Kirstenbosch. (*Dorotheanthus*)

rouxii For C. H. D. Roux (fl. 1935), collected plants (esp. Mesembs) for L. Bolus. (*Rhinephyllum, Stomatium*)

rowlandii For Lt. Col. Rowland Jones, 1949 in charge of the northern section of Kruger National Park, RSA. (*Euphorbia*)

rowleyanus For Gordon D. Rowley (*1921), English botanist at Reading University and widely known succulent plant specialist and author. (*Senecio*)

roxburghianus For Dr. William Roxburgh (1751–1815), British botanist and physician with the East India Company. (*Sansevieria*)

royenii For Dr. Adriaan van Royen (1704–1779), Dutch botanist and physician in Leiden. (*Pilosocereus*)

royleanus For John F. Royle (1798–1858), British botanist and physician, collecting extensively in India. (*Euphorbia*)

ruamahanga For the occurrence at the Ruamahanga River, New Zealand. (*Crassula*)

rubellus Lat., reddish; (**1**) for the red colouring of all plant parts. (*Monadenium*) – (**2**) for the red leaf colour. (*Crassula atropurpurea* var.) – (**3**) for the colour of the cyathophylls. (*Euphorbia*) – (**4**) for the corolla. (*Brachystelma, Kalanchoe*)

rubens Lat., becoming red; (**1**) for the leaf colour. (*Sedum*) – (**2**) for the flower colour. (*Dudleya*)

ruber Lat., red; (**1**) for the leaves. (*Crassula setulosa* var., *Gunniopsis*) – (**2**) for the reddish leaves and inflorescences. (*Synadenium compactum* var.) – (**3**) for the flower colour. (*Astridia, Huernia*)

rubescens Lat., becoming red, reddening; for the colour of the stem segments. (*Consolea*)

rubiginosus Lat., rusty-red; for the flower colour. (*Pelargonium, Schizoglossum, Stapelia*)

rubineus Lat., ruby-red; (**1**) for the reddish leaves. (*Crassula sieberiana* ssp.) – (**2**) for the colour of the base of the perianth segments. (*Selenicereus*)

rubispinus Incorrectly formed compound from Lat. 'rubineus', ruby-red; and Lat. '-spinus', -spined. (*Echinocereus rigidissimus* ssp.)

rubricaulis Lat. 'ruber, rubra, rubrum', red; and Lat. 'caulis', stem. (*Crassula, Portulaca, Ruschia*)

rubriflorus Lat. 'ruber, rubra, rubrum', red; and Lat. '-florus', -flowered. (*Conophytum ricardianum* ssp., *Cypselea, Poellnitzia*)

rubrimarginatus Lat. 'ruber, rubra, rubrum', red; and Lat. 'marginatus', margined; (**1**) for the leaf margins. (*Echeveria*) – (**2**) for the colouring of the nectary glands. (*Euphorbia*)

rubriseminalis Lat. 'ruber, rubra, rubrum', red; and Lat. 'semen', seed. (*Euphorbia*)

rubrispinosus Lat. 'ruber, rubra, rubrum', red; and Lat. 'spinosus', spiny, thorny. (*Euphorbia*)

rubrocoronatus Lat. 'ruber, rubra, rubrum', red; and Lat. 'coronatus', crowned; for the often purple-red filaments. (*Epiphyllum*)

rubrogemmius Lat. 'ruber, rubra, rubrum', red; and Lat. 'gemma', bud; for the colour of the flower buds. (*Opuntia viridirubra* ssp.)

rubroglandulosus Lat. 'ruber, rubra, rubrum', red; and Lat. 'glandulosus', glandular; for the glandular hairs covering various plant parts. (*Cyphostemma*)

rubrograndis Lat. 'ruber, rubra, rubrum', red; and Lat. 'grandis', large; for the large red flowers. (*Mammillaria melanocentra* ssp.)

rubrolineatus Lat. 'ruber, rubra, rubrum', red; and Lat. 'lineatus', striped; (**1**) for the markings on top of the fused leaf pair. (*Conophytum*) – (**2**) for the flower colour. (*Aeonium arboreum* var., *Aloinopsis*)

rubroluteus Lat. 'ruber, rubra, rubrum', red; and Lat. 'luteus', saffron-yellow; (**1**) for the flower colour. (*Echidnopsis*) – (**2**) for the colour of the filaments. (*Lampranthus*)

rubropunctatus Lat. 'ruber, rubra, rubrum', red; and Lat. 'punctatus', dotted; for the red gland-dots on calyx and vegetative parts. (*Plectranthus*)

rubrostipus Lat. 'ruber, rubra, rubrum', red; and Lat. 'stipes', trunk. (*Adansonia*)

rubrotinctus Lat. 'ruber, rubra, rubrum', red; and Lat. 'tinctus', coloured, dyed; for the colour of leaves and sepals. (*Sedum*)

rubrovenosus Lat. 'ruber, rubra, rubrum', red; and Lat. 'venosus', veined; for the reddish striations on the leaves. (*Tylecodon*)

rubroviolaceus Lat. 'ruber, rubra, rubrum', red; and Lat. 'violaceus', violet; for the colour of the dry leaves. (*Aloe*)

rudatisii For Hans Rudatis (fl. 1937), German naturalist. (*Ceropegia*)

rudibuenekeri For Rudi W. Büneker (fl. 1987), Brazilian cactus collector of German descent, son of R. H. Büneker. (*Parodia*)

rudis Lat. 'rudis', untidy, wild; (**1**) perhaps for the habitat. (*Crassula tetragona* ssp.) – (**2**) for the withered remains of the previous year's leaf sheaths. (*Cheiridopsis*) – (**3**) for its irregular branching. (*Euphorbia*)

rudolfii For Dr. [Friedrich Richard] Rudolf Schlechter (1872–1925), German traveller, collector and botanist in Berlin and specialist (amongst many other groups) on orchids. (*Crassula, Euphorbia*)

ruedebuschii For Mr. Rüdebusch (fl. 1927), farmer on Farm Vahldorn, S of Warmbad, Namibia, on whose farm the German botanist Kurt Dinter was guest during several visits. (*Schwantesia*)

rufescens Lat., becoming reddish; for the purplish lower face of the leaves. (*Anacampseros*)

ruffingianus For Dr. E. Ruffing (fl. 1999), German physician working for several years in Prov. Toliara, Madagascar. (*Aloe*)

ruficeps Lat. 'rufus', reddish; and Lat. '-ceps', headed; for the reddish cephalia. (*Espostoa*)

rufidus Lat., becoming reddish; for the colour of the glochids. (*Opuntia*)

rufus Lat., reddish; for the flower colour. (*Stapelia*)

rugosiflorus Lat. 'rugosus', rugose, wrinkled; and Lat. '-florus', -flowered; for the wrinkled surface of the nectary glands. (*Euphorbia*)

rugosifolius Lat. 'rugosus', rugose, wrinkled; and Lat. '-folius', -leaved; for the leaf surface. (*Aloe*)

rugospermus Lat. 'rugosus', rugose, wrinkled; and Gr. 'sperma', seed. (*Talinum*)

rugosquamosus Lat. 'ruga', wrinkle; and Lat. 'squamosus', scaly; for the upper leaf surface. (*Aloe compressa* var.)

rugosus Lat., wrinkled; (**1**) for the roughness of the dry stems. (*Sedum multicaule* ssp.) – (**2**) for the leaf surfaces. (*Conophytum*) – (**3**) for the rough upper leaf surface. (*Monadenium*)

runyonii For Robert Runyon (1881–1968), US-American photographer, tradesman, politician and amateur botanist in Brownsville, Texas, and co-author of the book "Texas Cacti" (1930). (*Coryphantha macromeris* ssp., *Echeveria*)

rupestris Lat. rock-; for the often rocky habitat of the plants. (*Aloe, Brachychiton, Copiapoa, Crassula, Lampranthus, Sedum, Umbilicus*)

rupicola Lat. 'rupes', rock, cliffs; and Lat. '-cola', inhabiting; for the occurrence amongst rocks. (*Aloe, Armatocereus, Bulb-*

ine, Ceropegia, Crassula sarcocaulis ssp., *Cyphostemma, Hoya australis* ssp., *Lasiocereus, Lewisia columbiana* var., *Nolana, Ruschia, Sedum, Sempervivum, Trichodiadema, Yucca*)

rupigenus Lat. 'rupes', rock, cliffs; and Lat. 'genus', birth, origin; for the preferred habitat. (*Octopoma*)

rupis-arcuatae Lat. 'rupes', rock, cliffs; and Lat. 'arcuatus', arcuate, curved; for the occurrence at a rock formation called "Bogenfels" ("curved rock"), Namibia. (*Amphibolia*)

ruprechtii For Franz J. Ruprecht (1814–1870), Austrian-Bohemian botanist in Prague and St. Petersburg. (*Hylotelephium telephium* ssp.)

ruralis Lat., rural; for the origin. (*Ruschia*)

rusapensis For the occurrence near Rusape, Zimbabwe. (*Glossostelma*)

rusbyi For Prof. Dr. Henry H. Rusby (1855–1940), US-American botanist, physician and plant explorer. (*Graptopetalum*)

Ruschia For Ernst J. Rusch (1867–1957), German farmer and businessmen in Namibia. (*Aizoaceae*)

Ruschianthemum For Ernst J. Rusch (1867–1957), German farmer and businessmen in Namibia; and Gr. 'anthemon', flower. (*Aizoaceae*)

Ruschianthus For Ernst J. Rusch (1867–1957), German farmer and businessmen in Namibia; and Gr. 'anthos', flower. (*Aizoaceae*)

ruschianus For Ernst J. Rusch (1867–1957), German farmer and businessmen in Namibia. (*Ruschia, Tromotriche*) – (**2**) For Augusto Ruschi (fl. 1980), Santa Teresa, Brazil, who helped the Dutch cactus specialist A. F. H. Buining during one of the trips to Brazil. (*Pilosocereus brasiliensis* ssp.)

ruschii For Ernst J. Rusch (1867–1957), German farmer and businessmen in Namibia. (*Avonia, Conophytum wettsteinii* ssp., *Hoodia*)

ruschiorum Gen. Pl.; for the family of Ernst J. Rusch (1867–1957), German farmer and businessmen in Namibia. (*Lithops*)

ruscifolius Lat. '-folius', -leaved; for the similarity tothe leaves of *Ruscus* (*Ruscaceae*). (*Dischidia*)

ruspolianus For Prince Eugenio Ruspoli (1866–1893), Italian nobleman, explorer and plant collector in NE Africa, killed by an elephant in Ethiopia. (*Aloe*)

ruspolii As above. (*Pterodiscus*)

russanthus Lat. 'russus', reddish; and Gr. 'anthos', flower; for the flower colour. (*Echinocereus*)

russellianus For Lord John Russell (1792–1878), Duke of Bedford, English politician and plant enthusiast. (*Schlumbergera*)

russellii For Paul G. Russell (1889–1963), US-American botanist, travelled 1915 in Brazil together with the US-American botanist J. N. Rose. (*Rhipsalis*)

rustii For Johann C. Rust (1855–1921), German farmer and merchant, emigrated to RSA in 1879, moving to Namibia in 1900. (*Lampranthus*)

rutenbergianus For Dietrich C. Rutenberg (1851–1878), German plant collector and traveller, murdered in Madagascar on an expedition. (*Pachypodium*)

ruthenicus For the occurrence in Ruthenia (S European Russia). (*Sempervivum*)

rutilans Lat., reddish-orange; (**1**) for the colour of the stems. (*Cyphostemma*) – (**2**) for the colour of the spination. (*Parodia*) – (**3**) for the cyathium colour. (*Euphorbia nubigena* var.)

rutteniae For Dr. (Mrs.) C. J. Rutten-Pekelharing, Dutch geologist whose 1930-expedition was accompanied by the Dutch botanist W. Hummelinck. (*Agave*)

ruwenzoriensis For the occurrence on the Ruwenzori mountain range, Uganda. (*Sedum*)

rycroftianus For Prof. Dr. Hedley Brian Rycroft (1918–1990), South African botanist, Harold Pearson Professor of Botany at Cape Town University, and third director of the National Botanical Gardens Kirstenbosch. (*Haworthia mucronata* var.)

ryderae For Mrs. Eleanore F. Ryder (née Fisher-Rowe) († 1958), English plant enthusiast, collected in RSA during visits in the 1920s und 1930s. (*Stomatium, Trichodiadema*)

rzedowskianus For Prof. Dr. Jerzy Rzedowski Rotter (*1926), Poland-born botanist and plant geographer, emigrating to Mexico in 1946, and his wife Graciela Calderón de Rzedowski (*1931), Mexican botanist. (*Portulaca*) – (**2**) For Prof. Dr. Jerzy Rzedowski Rotter (*1926), Poland-born botanist and plant geographer, emigrating to Mexico in 1946. (*Agave*)

S

sabaeus Probably commemorating the State of "Saba" (Sheba), Arabia. (*Aloe*)

saboae For Kathryn ("Kitty") Sabo (*1917), US-American cactus collector in Woodland Hills, California, president of the American Cactus and Succulent Society 1980–1981. (*Mammillaria*)

saboureaui For Pierre Saboureau (fl. 1947–1960), French Nature Conservation officer in Madagascar. (*Senecio*)

sabulicola Lat. 'sabulum', sand; and Lat. '-cola', inhabiting; for the preferred habitat. (*Ruschia*)

sabulosus Lat., sandy; for the preferred habitat. (*Apatesia*, *Kleinia*)

saccatus Lat., like a bag; (**1**) for the broadened corolla base. (*Plectranthus*) – (**2**) for the shape of the corona segments. (*Schizoglossum*) – (**3**) application obscure. (*Pterodiscus*)

sacculatus Lat., like a small bag; probably for the flower shape. (*Orbea*)

sacer Lat. 'sacer, sacra, sacrum', sacred; application obscure. (*Rhodiola chrysanthemifolia* ssp.)

sacharosa From the local vernacular name "sacharosa" of the plants used in Argentina (from Quechua 'sacha', false; and Lat. / Span. 'rosa', rose). (*Pereskia*)

saddianus For Prof. Nagib Saddi (fl. 1983), Brazilian botanist at the Universidade Federal de Mato Grosso. (*Cereus*)

saginatus Lat. 'sagina', good forage, fattening forage; (**1**) for the thick stems. (*Senecio mweroensis* ssp.) – (**2**) for the turgid leaves. (*Amphibolia*)

saginoides Gr. '-oides', resembling; and for the genus *Sagina* ("Pearlwort", Caryophyllaceae). (*Crassula*)

sagittarius Lat., pertaining to arrows; because the latex was used as a source of arrow-poison. (*Euphorbia avasmontana* var.)

sagittatus Lat., arrow-like; for the shape of the processes of the corona. (*Dischidia*)

sagittipetalus Lat. 'sagitta', arrow; and Lat. 'petalum', petal; for the petal shape. (*Sedum*)

saglionis For Joseph Saglio (fl. 1847), French cactus amateur in Strasbourg (or Paris according to some sources). (*Gymnocalycium*)

saint-pieanus For Paul Saint-Pie (fl. 1957), French cactus horticulturist at the Côte d'Azur. (*Parodia*)

sakalava For the occurrence in the region of the Sakalava tribe, Madagascar. (*Uncarina*)

sakalavensis As above. (*Cyphostemma*)

sakarahaensis For the occurrence in the Sakaraha Forest, Madagascar. (*Euphorbia*)

salazarii For António de Oliveira Salazar (1889–1970), Portuguese politician and State President 1932–1968. (*Kalanchoe*)

salicifolius For the genus *Salix* ("Willow", Salicaceae); and Lat. '-folius', -leaved; for the similar leaf shape. (*Ceropegia*)

salicola Lat. 'sal, salis', salt; and Lat. '-cola', inhabiting; for the occurrence in a salty habitat. (*Drosanthemum*, *Lampranthus*, *Lithops*)

salicornioides Gr. '-oides', resembling; and for the genus *Salicornia* ("Marsh Samphire", "Glasswort", Chenopodiaceae); for the similar stems. (*Hatiora*, *Psilocaulon*)

salignus Lat., pertaining to willows; presumably for the long narrow leaves as found in some species of willows. (*Tetragonia*)

salinensis For the occurrence near Salinas Victoria, Nuevo León, Mexico. (*Coryphantha*)

salmianus For Fürst Joseph Salm-Reifferscheid-Dyck (1773–1861), German (Prussian) botanist, botanical artist and horticulturist and succulent plant collector. (*Agave*, *Opuntia*)

salmii As above. (*Glottiphyllum*)

salmoniflorus Lat. 'salmoneus', salmon-pink, carrot-yellow, orange; and Lat. '-florus', -flowered. (*Monsonia*)

salsoloides Gr. '-oides', resembling; and for the genus *Salsola* (Chenopodiaceae). (*Cistanthe*, *Trianthema*)

saltensis For the occurrence in Prov. Salta, Argentina. (*Echinopsis*)

salteri For Terence M. Salter (1883–1969),

Royal Navy Captain who settled in the Cape, RSA, after retirement in 1931, specialist on *Oxalis*. (*Lampranthus*)

salvadorensis For the occurrence in El Salvador. (*Opuntia, Sedum*) – (**2**) For the occurrence near the city of Salvador, Bahia, Brazil. (*Melocactus, Pilosocereus catingicola* ssp.)

samaipatanus For the occurrence near the city of Samaipata, Prov. Florida, Dept. Santa Cruz, Bolivia. (*Cleistocactus*)

Samaipaticereus For the occurrence near the city of Samaipata, Prov. Florida, Dept. Santa Cruz, Bolivia; and *Cereus*, a genus of columnar cactus. (*Cactaceae*)

samalanus For the occurrence in the valley of Samalá, Guatemala. (*Furcraea*)

sambiranensis For the occurrence in the Sambirano Distr., Madagascar. (*Sansevieria*)

samburuensis For the occurrence in the region inhabited by the Samburu people in Kenya. (*Euphorbia*)

samius Lat., pertaining to the Greek Island of Samos; for the occurrence on Samos. (*Sedum*)

samnensis For the occurrence at Samne, Dept. Cajamarca, Peru. (*Cleistocactus fieldianus* ssp.)

san-angelensis For the occurrence on the Pedregal de San Angel, México, Mexico. (*Mammillaria haageana* ssp.)

sanae For Mrs. Sana Jardine (fl. 1897), without further data. (*Hoya australis* ssp.)

sanchez-mejoradae For Hernando Sánchez-Mejorada (1926–1988), Mexican amateur botanist and cactus specialist. (*Echeveria halbingeri* var., *Mammillaria*)

sanctae-martae For the occurrence near Santa Marta, N Colombia. (*Portulaca*)

sancti-spirituensis For the occurrence in Prov. Sancti Spíritu, Cuba. (*Agave brittoniana* ssp.)

sanctulus Dim. of Lat. 'sanctus', sacred; because the taxon was discovered near a local sanctuary. (*Kalanchoe*)

sandbergensis For the occurrence at Sandberge near Komaggas, Northern Cape, RSA. (*Ruschia*)

sandbergii For George Sandberg (fl. 1975), employee of the White Sands Missile Range, New Mexico, USA, who first discovered the taxon. (*Escobaria*)

sandersonii For John Sanderson (1820–1881), Scottish journalist, trader and amateur botanist emigrating to RSA in 1850. (*Brachystelma, Ceropegia*)

sanfelipensis For the occurrence in the San Felipe Desert, Baja California, Mexico. (*Cylindropuntia*)

sanguineus Lat., blood-red; (**1**) because the whole plants turn red when older. (*Trianthema*) – (**2**) for the leaf colour. (*Conophytum roodiae* ssp.) – (**3**) for the flower colour. (*Opuntia*) – (**4**) for the red pollen. (*Plectranthus*)

sanguiniflorus Lat. 'sanguis, sanguinis', blood; and Lat. '-florus', -flowered; for the flower colour. (*Echinopsis*)

sanluisensis For the occurrence in the state of San Luis Potosí, Mexico. (*Ariocarpus agavoides* ssp.)

Sansevieria For Count Pietro Antonio Sanseverino, Italian patron of horticulture in Naples around 1785. (*Dracaenaceae*)

santa-maria For the occurrence at the Bahia Santa Maria, Isla Magdalena, Baja California, Mexico. (*Ferocactus*)

santa-rita For the occurrence in the Santa Rita Mts., Arizona, USA. (*Opuntia*)

santacruzensis For the occurrence in the city of Santa Cruz, Dept. Santa Cruz, Bolivia. (*Cleistocactus baumannii* ssp.)

santaensis For the occurrence in the valley of the Río Santa, Dept. Ancash, Peru. (*Echinopsis*)

santamaria For the occurrence near Bahia Santa Maria, Isla Magdalena off the coast of Baja California, Mexico. (*Cylindropuntia*)

santamarinae For J. Santamarina Guerra (fl. 1992), Cuban official concerned with the Government's environmental program. (*Leptocereus*)

santapaui For Rev. Fr. Dr. Hermenegild Santapau (1903–1970), Spanish-born jesuit priest and botanist, onetime director of the Botanical Survey of India. (*Ceropegia, Euphorbia*)

santarosa For the occurrence in the Sierra

Hermosa de Santa Rosa and the fact that it was first collected near Santa Rosa, Coahuila, Mexico. (*Coryphantha ramillosa* ssp.)

Saphesia Gr. 'saphes', distinct; for the unique characters shown by the plants, making them distinct from all other known members of the family. (*Aizoaceae*)

sapinii For A. Sapin (fl. 1908), chief of a scientific mission established by the Compagnie du Kasai. (*Euphorbia*)

sarcocaulis Gr. 'sarx, sarkos', flesh; and Gr. 'kaulon', stem. (*Crassula*)

sarcodes Gr. 'sarx, sarkos', flesh; for the succulent stems. (*Euphorbia*)

sarcophyllus Gr. 'sarx, sarkos', flesh; and Gr. 'phyllon', leaf. (*Galenia, Neoalsomitra, Tetragonia*)

Sarcopilea Gr. 'sarx, sarkos', flesh; and for the genus *Pilea* (*Urticaceae*), to reflect the generic relationship. (*Urticaceae*)

Sarcorrhiza Gr. 'sarx, sarkos', flesh; and Gr. 'rhiza', root; for the root tubers. (*Asclepiadaceae*)

sarcospathula Gr. 'sarx, sarkos', flesh; and Lat. 'spat[h]ula', spatula; for the fleshy spatulate leaflets. (*Cyphostemma*)

Sarcostemma Gr. 'sarx, sarkos', flesh; and Gr. 'stemma', garland, wreath; for the fleshy tangling stems. (*Asclepiadaceae*)

sarcostemmoides Gr. '-oides', resembling; and for *Sarcostemma australe* (*Asclepiadaceae*); for the similar growth form. (*Euphorbia*)

Sarcozona Gr. 'sarx, sarkos', flesh; and Gr. 'zone', ring, girdle; for the complete sheath of bracteoles enveloping the flower base. (*Aizoaceae*)

sardienii For Tommy Sardien (*1932), longtime curator for South African succulents at Kirstenbosch National Botanical Garden. (*Ornithogalum*)

sargentii For Mr. Sargent (fl. 1912), who supplied the type plant to the Missouri Botanical Garden. (*Agave vivipara* var.)

sarkariae For Dr. Jagdish Singh Sarkaria (fl. 1978–2002), Indian Sikh medical doctor and succulent plant enthusiast in Chandigarh, India. (*Caralluma*)

sarmentosus Lat., sarmentose, producing long runners. (*Aizoon, Crassula, Ruschia, Sedum*)

sartorii For Carl C. W. Sartorius (1796–1872), German interested in botany, emigrated to Mexico and settled 1830 in Veracruz, where he was visited by naturalists such as Galeotti, Karwinski etc. (*Mammillaria*)

satumensis For the occurrence on the Satsuma Peninsula, Japan, and presumably correctable to 'satsumensis'. (*Sedum*)

saturatus Lat., saturated; (**1**) for the rich 'saturated' petal colour. (*Antimima, Delosperma, Erepsia, Phyllobolus*) – (**2**) for the rich 'saturated' filament colour. (*Lampranthus*)

saudi-arabicus For the occurrence in Saudi Arabia. (*Huernia*)

sauerae For Miss Mary Sauer (fl. 1933), without further data. (*Lampranthus*)

saueri For Paul Sauer (fl. 1928), brother-in-law of H. W. Viereck, German emigrator to Mexico. (*Turbinicarpus*) – (**2**) For Niko Sauer (fl. 2002), South African mathematician and amateur botanist. (*Conophytum pellucidum* ssp.)

saundersiae For Lady Katherine Saunders (1824–1901), British collector and artist, emigrated to RSA in 1854. (*Aloe*)

saundersii For "Mr. Charles Saunders" (fl. 1892, the discoverer, without further data. (*Pachypodium lealii* ssp.) – (**2**) For William W. Saunders (1809–1879), English botanist, horticulturist and entomologist. (*Aeonium*)

saxatilis Lat., dwelling or found among rocks. (*Ceropegia, Stapeliopsis, Tacinga*)

saxetanus From Lat. 'saxum', rock; for the preferred stony habitat. (*Conophytum*)

saxicola Lat. 'saxum', rock; and Lat. '-cola', inhabiting; for the rocky habitat. (*Antimima, Commiphora, Delosperma, Opuntia, Praecereus, Psammophora*)

saxifragoides Gr. '-oides', resembling; and for the genus *Saxifraga* (*Saxifragaceae*). (*Graptopetalum, Portulaca, Rhodiola*)

saxifragus Lat. 'saxum', rock; and Lat. 'frangere', to break; for the occurrence in rock crevices. (*Crassula*)

saxorum Gen. Pl. of Lat. 'saxum', rock; i.e. of the rocks, for the habitat. (*Euphorbia, Streptocarpus*)

saxosus Lat., full of rocks; for the habitat. (*Dudleya*)

sayulensis For the occurrence near Sayula, Jalisco, Mexico. (*Echeveria*)

scaber Lat. 'scaber, scabra, scabrum', rough; (**1**) for the rough leaves. (*Agave, Crassula, Haworthia heidelbergensis* var., *Haworthia, Ruschia, Zehneria*) – (**2**) for the scabrous calyx base. (*Lampranthus*) – (**3**) for the papillose inside of the inflated corolla base. (*Ceropegia*)

scabridus Lat., somewhat rough; for the roughly papillate stems. (*Rhodiola coccinea* ssp.)

scabrifolius Lat. 'scaber, scabra, scabrum', rough; and Lat. '-folius', -leaved; for the leaf surface. (*Aloe*)

scabripes Lat. 'scaber, scabra, scabrum', rough; and Lat. 'pes', foot; for the roughened pedicels. (*Delosperma*)

scabrispinus Lat. 'scaber, scabra, scabrum', rough; and Lat. '-spinus', -spined; for the rough spination. (*Haworthia arachnoidea* var.)

scabrocostatus Lat. 'scaber, scabra, scabrum', rough; and Lat. 'costatus', ribbed; for the roughened longitudinal ribs of the leaves. (*Ornithogalum*)

scandens Lat., climbing; for the climbing habit. (*Heliophila, Lepidium, Tylecodon*)

scaphirostris From Gr. 'skaphe', boat; and Lat. 'rostrum', beak; for the shape of the tubercle tips. (*Ariocarpus*)

scapiger Lat. 'scapus', scape; and Lat. '-ger, -gera, -gerum', -carrying, bearing; (**1**) for the long-pedunculate inflorescences. (*Kalanchoe*) – (**2**) for the long-pedicellate flowers. (*Bergeranthus*)

scaposus Lat., having a well-developed scape or leafless peduncle. (*Agave, Nolana, Senecio*)

scarlatinus Lat., scarlet; for the red cyathia. (*Euphorbia*)

Sceletium Lat. 'scletus', skeleton (from Gr. 'skeletos', dried up, withered); for the persistent vascular skeletons of the dry dead leaves. (*Aizoaceae*)

schaeferianus For Dr. Fritz Schäfer († 1911), physician and plant collector from Lüderitz, Namibia. (*Tylecodon*)

schaffneri For Wilhelm (later José Guillermo) Schaffner (1830?–1882), German plant collector and pharmacist, from 1856 in Mexico. (*Echeveria*)

schaijesii For Michel Schaijes (fl. 1986), Belgian botanist. (*Monadenium*)

schatzlii For Stefan Schatzl (1922–2001), Austrian horticulturist, cactus collector and former curator of the cactus collection at the Botanical Garden Linz, Austria. (*Melocactus*)

scheeri For Frederick Scheer (1792–1868), German merchant, living in Kew (England) for much of his life, and interested in many plant groups incl. cacti. (*Coryphantha robustispina* ssp., *Echeveria, Echinocereus, Opuntia, Sclerocactus*)

scheffleri For Georg Scheffler († 1910), missionary who collected plants in Tanzania 1899, and in the Kibwezi Distr., Kenya, 1906–1910. (*Adenia lanceolata* ssp.)

schellenbergii For Dr. Gustav A. L. D. Schellenberg (1882–1963), German botanist in Zürich, Berlin, Kiel and Göttingen, later publisher of a newspaper in Wiesbaden, Germany. (*Aizoon*)

schelpei For Prof. Edmund A. C. L. E. Schelpe (1924–1985), South African botanist at Cape Town University. (*Aloe*)

schenckii For Dr. Adolf Schenck (1847–1936), German geographer and geologist, visited Namibia and RSA 1884–1886. (*Brownanthus ciliatus* ssp., *Tetragonia*) – (**2**) For Dr. Heinrich Schenck (1860–1927), German botanist, travelling in Brazil and Mexico, 1896–1927 director of the Botanical Garden Darmstadt. (*Myrtillocactus*)

schereri For Egon Scherer (fl. 1976), German cactus collector. (*Echinocereus*)

schickendantzii For Friedrick Schickendantz (1837–1896), German chemist emigrating to Argentina in 1861, working for a mining company, then for a sugar refinery, and founder of the Lillo Herbarium in Tucumán. (*Echinopsis, Gymnocalycium, Opuntia*)

schidiger Gr. 'schidia', brushed linen (from Gr. 'schizein', split, divide); and Lat. '-ger, -gera, -gerum', carrying; for the fibres on the leaf margins. (*Agave filifera* ssp., *Yucca*)

schiedeanus For Christian J. W. Schiede (1798–1836), German gardener and later botanist, traveller and collector who spent several years in Mexico. (*Columnea, Mammillaria*)

schielianus For Wolfgang Schiel (1904–1978), German cactus hobbyist in Freiburg. (*Echinopsis*)

schiffneri For Prof. Dr. Victor F. Schiffner (1862–1944), Austrian botanist. (*Sinningia*)

schilinzkyanus For Guido von Schilinzky (1823–1898), Baltic naturalist and and Privy Councillor in St. Petersburg. (*Frailea*)

schillianus For Prof. Dr. Rainer Schill (fl. 2002), German botanist at Heidelberg University, published 1973 on *Lomatophyllum*. (*Aloe*)

schimperi For Georg [H.] W. Schimper (1804–1878), German-born botanist and plant collector, collecting widely in Egypt, Arabia and Ethiopia, married a daughter of Ras Ubies and settled in the then Abyssinia in 1837, adopting Abyssinian nationality. (*Corallocarpus, Crassula, Delosperma, Euphorbia*)

schimperianus As above. (*Kalanchoe*)

schinzii For Dr. Hans Schinz (1858–1941), Swiss botanist and long-time director of the Zürich Botanical Garden. (*Brachystelma, Euphorbia, Stapelia*)

Schismocarpus Gr. 'schisma', split, cleft; and Gr. 'karpos', fruit; for the five-cleft free portion of the fruit capsules. (*Loasaceae*)

schistophilus Gr. 'schistos', split, here: schist rock; and Gr. 'philos', friend; for the preferred habitat. (*Aloe compressa* var.)

schistorhizus Gr. 'schistos', split; and Gr. 'rhiza', root; for the shape of the thickened roots. (*Parakeelya*)

schizacanthus Gr. 'schizein', to split; and Gr. 'akantha', spine, thorn; for the forked spine tips. (*Euphorbia*)

Schizobasis Gr. 'schizein', to split; and Gr. 'basis', base; for the fruit capsules. (*Hyacinthaceae*)

schizoglossoides Gr. '-oides', resembling; and for the genus *Schizoglossum* (*Asclepiadaceae*). (*Brachystelma*)

Schizoglossum Gr. 'schizein', to split; and Gr. 'glossa', tongue; for the often bifid corona lobes. (*Asclepiadaceae*)

schizolepis Gr. 'schizein', to split; and Gr. 'lepis', scale; for the deeply lobed nectary scales. (*Sedum*)

schizopetalus Gr. 'schizein', to split; and Gr. 'petalon', petal; for the petal shape. (*Pelargonium*)

schizophyllus Gr. 'schizein', to split; and Gr. 'phyllon', leaf; for the deeply divided leaves. (*Kalanchoe*)

schlechtendalii For Prof. Dr. Diederich F. L. von Schlechtendal (1794–1866), German botanist in Berlin and Halle, long-time editor of the journal Linnaea. (*Bursera*)

Schlechteranthus For Maximilian (Max) Schlechter (1874–1960), German trader and plant collector, lived in Namibia and RSA from 1896, brother of the German botanist Rudolf Schlechter; and Gr. 'anthos', flower. (*Aizoaceae*)

Schlechterella For Dr. [Friedrich Richard] Rudolf Schlechter (1872–1925), German traveller, collector and botanist in Berlin and specialist (amongst many other groups) for orchids. (*Asclepiadaceae*)

schlechteri As above. (*Antimima, Cleretum papulosum* ssp., *Dorstenia hildebrandtii* var., *Jatropha, Lampranthus, Myrmecodia, Nelia*) – (**2**) For Maximilian (Max) Schlechter (1874–1960), German trader and plant collector, lived in Namibia and RSA from 1896, brother of the German botanist Rudolf Schlechter. (*Cheiridopsis, Conophytum*)

schliebenii For Hans-Joachim E. Schlieben (1902–1975), German botanist working in E and S Africa. (*Adenia, Portulaca*)

Schlumbergera For Frédéric M. Schlumberger (fl. 1840), French nurserymen and cactus collector in the Normandy. (*Cactaceae*)

schmidianus For Heinz Schmid (fl. 1994), Swiss cactus enthusiast from Biberist. (*Gymnocalycium catamarcense* ssp.)

schmidtii For Johann A. Schmidt (1823–1905) who first collected the taxon in 1850.

(*Umbilicus*) – (**2**) For E. Schmidt (fl. 1886), who imported the taxon from Africa. (*Crassula*)
schmiedickeanus For Karl Schmiedicke (1870–1926), German cactus hobbyist in Berlin. (*Turbinicarpus*)
schmitzii For André Schmitz, Belgian collector stationed in Zaïre 1938–1959. (*Euphorbia*)
schmollii For Ferdinand Schmoll († 1950), German artist (painter) and later cactus collector and important exporter in Cadereyta, Querétaro, Mexico. (*Echinocereus, Mammillaria haageana* ssp., *Thelocactus leucacanthus* ssp.)
schneiderianus For Camillo K. Schneider (1876–1951), German botanist, garden architect, horticultural journalist and author in Berlin, and friend of the German botanist A. Berger. (*Eberlanzia, Huernia*)
schoelleri For Max Schoeller (fl. 1894), German ethnologist travelling widely in Africa. (*Aloe*)
schoenii For Mr. E. Schön (fl. 1956), German in Arequipa who supported the German botanist W. Rauh on his travels in Peru. (*Echinopsis*)
schoenlandianus For Dr. Selmar Schönland [also Schonland] (1860–1940), German-born botanist, emigrated 1889 to RSA and became director of the Albany Museum in Grahamstown. (*Brachystelma, Drosanthemum*)
schoenlandii As above. (*Euphorbia, Orostachys, Rhinephyllum*)
schollii For Georg Scholl (fl. 1786–1800), German gardener at Schönbrunn, Vienna, collecting in RSA 1786–1799. (*Ruschia*)
schomeri For Mr. Menko Schomerus (fl. 1966), mine-owner in Ampanihy, Madagascar. (*Aloe*)
schonlandii For Dr. Selmar Schönland [also Schonland] (1860–1940), German-born botanist, emigrated 1889 to RSA and became director of the Albany Museum in Grahamstown. (*Adromischus cristatus* var.)
schooneesii For Mr. D. H. Schoonees (fl. 1931), teacher at Steytlerville, RSA. (*Aloinopsis*)

schottii For Arthur Schott (1814–1875), German naturalist, explorer and plant collector, working in various government functions in Washington and surveyor of the Mexican Boundary Survey 1851–1864. (*Agave, Grusonia, Pachycereus, Yucca*)
schrankii For Prof. Dr. Franz von Paula von Schrank (1747–1835), German botanist, entomologist and Jesuit teacher, and first director of the Munich Botanical Garden 1809–1832. (*Disocactus*)
schreiteri For R. Schreiter (fl. 1927), Argentinian botanist. (*Echinopsis*)
Schreiteria As above. (*Portulacaceae*)
schroederianus For Dr. J. Schroeder [Schröder?] (fl. 1941), a friend of the German-Uruguayan botanist Cornelius Osten. (*Gymnocalycium hyptiacanthum* ssp.)
schubei For Dr. Theodor Schube (1860–1934), German botanist and high school teacher. (*Monadenium*)
schuldtianus For Mr. Schuldt, 1936 owner of the horticultural establishment of Albert Schenkel, Hamburg, Germany. (*Adromischus*)
schultzei For Dr. Leonhard S. Schultze (1872–1955), German zoologist, anthropologist, geographer and philologist in Jena, Kiel and Marburg. (*Brachystelma, Leipoldtia*)
schumannianus For Dr. Karl M. Schumann (1851–1904), German botanist in Berlin with a strong interest in cacti and other succulents. (*Parodia*)
schumannii As above. (*Mammillaria*)
Schwantesia For Prof. M. H. Gustav Schwantes (1881–1960), German professor of prehistory at Kiel University, and Mesemb specialist. (*Aizoaceae*)
schwantesii As above. (*Gibbaeum, Lithops, Titanopsis*)
schwartzii For Dr. Herman Schwartz (fl. 2002), US-American physician, succulent plant enthusiast and owner of Strawberry Press. (*Senecio mweroensis* fa.)
schwarzii For Fritz Schwarz (fl. 1940, 1955), cactus collector of German descent in Mexico. (*Echinocereus adustus* ssp., *Ferocactus, Mammillaria, Thelocactus bicolor* ssp., *Turbinicarpus schmiedickeanus* ssp.)

schwebsianus For Willy Schwebs (1876–1934), German cactus horticulturist near Dresden. (*Parodia*)

schweinfurthii For Dr. Georg A. Schweinfurth (1836–1925), German botanist, geographer and explorer of NE Africa and Arabia. (*Aloe, Jatropha, Orbea*)

scimitariformis Engl. 'scimitar', scimitar (from Fr. 'cimeterre' and Ital. 'scimitarra', both perhaps ultimately of Persian origin); and Lat. '-formis', -shaped; for the shape of the leaf cross section. (*Sansevieria*)

scitulus Lat., pretty, neat. (*Euphorbia, Stapelia*)

sciurus Lat., Red Squirrel; perhaps for the colour and texture of the spination resembling a squirrel tail. (*Echinocereus*)

Sclerocactus Gr. 'skleros', hard, dry, cruel; and Lat. 'cactus', cactus; for the fierce hooked spines. (*Cactaceae*)

sclerocarpus Gr. 'skleros', hard, dry, cruel; and Gr. 'karpos', fruit; for the hard fruits. (*Jasminocereus thouarsii* var.)

scobinifolius Lat. 'scobina', rasp; and Lat. '-folius', -leaved; for the rough leaf surface. (*Aloe*)

scopa Lat., thin twig, Pl. 'scopae', faggot, broom; for the spination. (*Parodia*)

scoparius Lat., pertaining to a broom; for the spination. (*Neobuxbaumia*)

scopatus Lat., brushlike, covered in bristles (Lat. 'scopae', faggot, broom). (*Anacampseros*)

Scopelogena Gr. 'skopelos', mountain top, cliff, rock; and Gr. 'genos', birth, origin; for the occurrence in rocky habitats. (*Aizoaceae*)

scopulicola Lat. 'scopulus', mountain top, cliff, rocks; and Lat. '-cola', -dwelling; for the preferred habitat. (*Echinopsis*)

scopulinus Lat., like a little broom (from the Dim. of Lat. 'scopa', broom); for the dense leaf arrangement. (*Sedum*)

scopulophilus Gr. 'skopelos', mountain top, cliff, rock; and Gr. 'philos', friend; for the preferred habitat. (*Leptocereus*)

scopulorum Gen. Pl. of Lat. 'scopulus', mountain top, cliff, rocks; for the preferred habitat. (*Echinocereus*)

scorpioides Lat., scorpioid; for the shape of the peduncle. (*Aloe*)

scottii Probably for George F. Scott-Elliot (1862–1934), British botanist. (*Senecio longiflorus* ssp.)

scrippsianus For Edward W. Scripps (1854–1926), US-American newspaper publisher and investor, founder of the Scripps Institution for Oceanography of the University of California. (*Mammillaria*)

sculptilis Lat., sculptured; for the appearance of the plant. (×*Gasteraloe*)

scutatus Lat., shield-shaped; for the hard obovate or rectangular leaf sheaths. (*Monilaria*)

scutellatus Lat., shield-, scutellate (from Lat. 'scutella', bowl); for the shape of the leaf rudiments on the stems. (*Echidnopsis*)

scutellifolius Lat. 'scutellum', small shield; and Lat. '-folius', -leaved. (*Peperomia*)

scyphadenus Gr. 'skyphos', cup; and Gr. 'aden', gland; for the cup-shaped nectary glands. (*Euphorbia*)

sebaeoides Gr. '-oides', resembling; and for the genus *Sebaea* (*Gentianaceae*). (*Crassula*)

sebsebei For Prof. Dr. Sebsebe Demissew (*1953), Ethiopian botanist at Addis Ababa University, and leader of the Ethiopian Flora Project. (*Euphorbia*)

secundiflorus Lat. 'secundus', secund; and Lat. '-florus', -flowered. (*Aloe*)

secundus Lat., secund, turned towards the same side; for the inflorescence architecture. (*Echeveria, Galenia, Rhadamanthus*)

securiger Lat., carrying an axe or hatchet; for the shape of the stem segments. (*Opuntia*)

Sedella Dim. of *Sedum*; for the diminutive size of the plants and their relationships. (*Crassulaceae*)

sediflorus For the genus *Sedum* ("Stonecrop", *Crassulaceae*); and Lat. '-florus', -flowered; for the similar flowers. (*Crassula*)

sedifolius For the genus *Sedum* ("Stonecrop", *Crassulaceae*); and Lat. '-folius', -leaved. (*Aeonium, Bulbine, Crassula exilis* ssp., *Medinilla, Nolana, Othonna, Portulaca, Trianthema*)

sediformis For the genus *Sedum* ("Stonecrop", *Crassulaceae*); and Lat. '-formis', -shaped. (*Sedum, Talinum*)

sedoides For the genus *Sedum* ("Stonecrop", *Crassulaceae*); and Gr. '-oides', resembling. (*Eberlanzia, Portulaca, Sedum*)

Sedum Lat. 'sedum', Houseleek, Stonecrop, i.e. the Lat. vernacular name of several *Crassulaceae*, with unresolved origin. (*Crassulaceae*)

seelemannii For Monsieur Seelemann (fl. 1916), physician and friend of the French botanist and physician Raymond Hamet. (*Sedum*)

seemannianus For Dr. Berthold Seemann (1825–1871), German botanist, publisher and traveller, died as director of a gold mining company in Nicaragua. (*Agave*)

seibanicus For the occurrence on the Kor [Kaur] Seiban in the Hadhramaut, Yemen. (*Echidnopsis, Euphorbia*)

seineri For Franz Seiner (1874–1940), Austrian journalist, traveller and collector. (*Jatropha*)

sejunctus Lat., separated; for the isolated habitat. (*Euphorbia grandicornis* ssp.)

sekiteiensis For the occurrence at Sekitei, Taiwan. (*Sedum*)

sekukuniensis For the occurrence in Sekukuniland in present-day Mpumalanga, RSA. (*Euphorbia*)

selebicus For the occurrence in Celebes (= Sulawesi). (*Myrmephytum*)

Selenicereus Gr. 'selene', moon; and *Cereus*, a genus of columnar cacti; for the flowers opening at night. (*Cactaceae*)

seleri For Dr. Georg E. Seler (1849–1922), German (Prussian) plant collector, archaeologist and ethnologist in Berlin and travelling in C and S America. (*Peperomia*)

selloanus Perhaps corrupted for Friedrich Sellow (1789–1831), German botanical explorer in Brazil. (*Furcraea*)

sellowii For Friedrich Sellow (1789–1831), German botanical explorer in Brazil. (*Parodia, Sinningia*)

selskianus For Ilarion S. Selsky (fl. 1858), Secretary of the Siberian branch of the Russian Geographical Society in Irkutsk. (*Phedimus*)

semenaliundatus Lat. 'semenalis', seed-; and Lat. 'undatus', wavy, undate; for the undulate margins of the seeds. (*Bulbine*)

semenovii For Peter P. von Semenov (1827–1914), traveller in C Asia. (*Rhodiola*)

semibarbatus Lat. 'semi-', half, and Lat. 'barbatus', bearded; for the hairs on the filaments. (*Bulbine*)

semidentatus Lat. 'semi-', half; and Lat. 'dentatus', toothed; for the leaves with few teeth. (*Ruschia*)

semiensis For the occurrence in the Semien Mts., Ethiopia. (*Afrovivella*)

semiglobosus Lat. 'semi-', half; and Lat. 'globosus', globose; (**1**) for the shape of the flower receptacle. (*Drosanthemum*) – (**2**) for the bell-shaped fruit base. (*Ruschia*)

semilunatus Lat. 'semi-', half; and Lat. 'lunatus', crescent-shaped; application obscure. (*Sedum*)

seminudus Lat. 'semi-', half; and Lat. 'nudus', naked; for the corolla. (*Duvalia sulcata* ssp.)

semiteres Lat. 'semi-', half; and Lat. 'teres', terete; for the leaves. (*Dudleya, Sedum*)

semitubiflorus Lat. 'semi-', half; Lat. 'tubus', tube; and Lat. '-florus', -flowered. (*Orbea*)

semivestitus Lat. 'semi-', half; and Lat. 'vestitus', clothed; (**1**) for the glabrous or only slightly pubescent leaves and the strongly pubescent inflorescences. (*Echeveria*) – (**2**) for the partly glabrous and partly hairy leaves. (*Conophytum*)

semivivus Lat. 'semi-', half, and Lat. 'vivus', living; for the leaves, which appear half dead during the dry season. (*Haworthia*)

semotus Lat., distant, far removed; for the distribution range. (*Orbea*)

semperflorens Lat. 'semper', always; and Lat. '-florens', -flowering; for the long flowering period. (*Euphorbia*)

sempervivi Gen. of Lat. 'sempervivum', an evergreen plant, ever living, houseleek (from Lat. 'semper', always; and Lat. 'vivus', living). (*Mammillaria*)

sempervivoides Gr. '-oides', resembling; and for the genus *Sempervivum* ("Houseleek", *Crassulaceae*). (*Prometheum*)

Sempervivum Lat., an evergreen plant, ever living, houseleek (from Lat. 'semper', always, and Lat. 'vivus', living). (*Crassulaceae*)

sempervivum As above. (*Rosularia*, *Senecio*)
senarius Lat. 'senex', old man; and Lat. suffix '-arius', having, pertaining to; either for the dark grey old stems, or for the pubescence of the herbaceous plant parts. (*Ruschia*)
Senecio Lat. 'senex, senecis', old man; for the white-haired pappus. (*Asteraceae*)
senecioides Gr. '-oides', resembling; and for the similarity of the leaves with those of many species of the genus *Senecio* (*Asteraceae*). (*Pelargonium*)
senegambicus For the occurrence in Senegambia (traditional name for the W tropical African region embracing Senegal, Gambia, Guinea and Guinea Bissau). (*Sansevieria*)
senilis Lat., old; (**1**) for the numerous white hairs giving the plants an old appearance. (*Cephalocereus*) – (**2**) for the white to greyish hair-like to bristly spination. (*Eriosyce*, *Espostoa*, *Mammillaria*)
sennii For Lorenzo Senni (1879–1954), Italian forester and botanist, repeatedly collecting in Eritrea and Somalia. (*Zaleya*)
sepalosus Lat., with well developed sepals. (*Oxalis*)
sepicola Lat. 'sepes' (also 'saepes'), fence; and Lat. '-cola', -dwelling. (*Matelea*)
sepium Gen. Pl. of Lat. 'sepes' (also 'saepes'), fence; perhaps because it was found growing in a fence. (*Ceropegia*, *Cleistocactus*)
septemfidus Lat. 'septem', seven; and Lat. '-fidus', -divided; for the seven-lobed leaves. (*Jatropha macrorhiza* var.)
septentrionalis Lat., northern; (**1**) for the distribution in comparison to related taxa. (*Beschorneria*, *Echinocereus cinerascens* ssp., *Ferocactus macrodiscus* ssp., *Orbea halipedicola* ssp., *Rhipsalis paradoxa* ssp.) – (**2**) for the distribution in N Kenya. (*Euphorbia*)
septifragus Lat. 'septum', septum; and Lat. 'frangere', to break; for the mode of fruit opening. (*Gunniopsis*)
sepultus Lat., buried; because the plant bodies shrink below ground-level during the dry season. (*Euphorbia*)
serendipitus Lat., a lucky find; for the chance discovery of the taxon. (*Euphorbia*)

seretii For Felix Seret (fl. 1905–1909), Belgian forestry officer and plant collector in the then Congo (formerly Zaïre, today Democratic Republic of Congo). (*Aloe*, *Euphorbia*)
sericeus Lat., silky with appressed glossy hairs; for the leaves. (*Crassula*, *Dolichos*, *Plinthus*)
sericifer Lat. 'sericum', cloth made from silk; and Lat. '-fer, -fera, -ferum', -carrying; for the spination. (*Cereus fernambucensis* ssp.)
sericifolius Lat. 'sericum', cloth made from silk; and Lat. '-folius', -leaved. (*Pelargonium*)
serotinus Lat., late coming, happening late; for the flowers, opening at dusk. (*Aridaria*)
serpens Lat., creeping; for the growth form. (*Cleistocactus*, *Hoya*, *Lampranthus*, *Senecio*)
serpentinicus For the occurrence on serpentine rocks. (*Prometheum*)
serpentinus Lat., snake-like; for the growth form. (*Ceropegia stapeliiformis* ssp., *Peniocereus*)
serpentisulcatus Lat. 'serpentinus', snake-like; and Lat. 'sulcatus', furrowed; for the undulating furrows between the ribs of the plant bodies.. (*Copiapoa*)
serpyllaceus From Lat. 'serpyllum', Wild Thyme; for the similarly small leaves. (*Pilea*)
serratifolius Lat. 'serratus', serrate, saw-edged; and Lat. '-folius', -leaved; for the serrate leaf margins. (*Dasylirion*)
serratus Lat., serrate, saw-edged; (**1**) for the serrate leaf margins. (*Kalanchoe*, *Lewisia cantelovii* var., *Rhodiola*, *Rosularia*) – (**2**) for the serrate leaf tip margins. (*Sinocrassula indica* var.) – (**3**) for the serrate keels of the leaves. (*Circandra*) – (**4**) for the spines on the leaf margins. (*Haworthia*)
serriyensis For the occurrence at Serriya, Yemen. (*Aloe*)
serrulatus Lat., finely serrulate; for the leaf margins. (*Codonanthe*, *Dracaena*, *Ruschia*)
Sesamothamnus From the herbaceous genus *Sesamum* ("Sesame", *Pedaliaceae*); and Gr. 'thamnos', shrub; for the relationships and the growth form. (*Pedaliaceae*)

sessilicymulus Lat. 'sessilis', sessile, stalkless; and Lat. 'cymulus', small cyme; for the inflorescence architecture. (*Crassula capitella* ssp.)

sessiliflorus Lat. 'sessilis', sessile; and Lat. '-florus', -flowering; (**1**) for the sessile flowers. (*Cynanchum, Echeveria, Momordica, Nolana*) − (**2**) for the sessile cyathia. (*Euphorbia*)

sessilifolius Lat. 'sessilis', sessile; and Lat. '-folius', -leaved. (*Coccinia*)

sessilis Lat., sessile, stalkless; for the flowers. (*Ruschia, Sesuvium*)

sesuvioides Gr. '-oides', resembling; and for the genus *Sesuvium* (*Aizoaceae*). (*Sesuvium*)

Sesuvium Lat. 'sesuvium', "Houseleek" (*Sempervivum, Crassulaceae*); most probably for the Houseleek-like succulent leaves. (*Aizoaceae*)

setaceus Lat., bristly; for the bristly pericarpel and fruit. (*Selenicereus*)

setatus Lat., bristly; for the white leaf spination. (*Haworthia arachnoidea* var.)

setchellii For Prof. Dr. William A. Setchell (1864−1943), US-American botanist (algologist), plant geographer and botanical historian. (*Dudleya*)

setifer Lat. 'saeta' / 'seta', bristle, stiff hair; and Lat. '-fer, -fera, -ferum', -carrying; for the stipules. (*Jatropha schlechteri* ssp.)

setispinus Lat. 'saeta / seta', bristle, bristly hair; and Lat. '-spinus', -spined; (**1**) for the bristle-like spines. (*Euphorbia, Thelocactus*) − (**2**) for the slender radial spines. (*Mammillaria*)

setosiflorus Lat. 'setosus', bristly-hairy; and Lat. '-florus', -flowered; for the hairy pericarpel and flower tube. (*Eriosyce heinrichiana* var.)

setosus Lat., bristly-hairy; (**1**) for the hairiness of the plants. (*Brachystelma*) − (**2**) for the setose leaves. (*Echeveria*)

setulifer Lat. 'setula', small bristle; and Lat. '-fer, -fera, -ferum', -carrying; (**1**) for the spines on the leaf margins. (*Haworthia cymbiformis* var.) − (**2**) for the tuft of bristles on the leaf tips. (*Trichodiadema*)

setulosus Lat., minutely setose (from Lat. 'seta', bristle); for the leaves. (*Crassula*)

sexangularis Lat. 'sex', six; and Lat. 'angularis', angled; (**1**) for the stems. (*Kalanchoe*) − (**2**) for the arrangement of the leaves in six rows. (*Sedum*)

sexfolius Lat. 'sex', six; and Lat. '-folius', -leaved; for the leaves in whorls of six. (*Rhodiola chrysanthemifolia* ssp.)

Seyrigia For André Seyrig (1897−1945), French miner, colonist and amateur entomologist, from 1928 in Madagascar. (*Cucurbitaceae*)

shadensis For the occurrence on Jabal Shada, Hijaz Prov., Saudi Arabia. (*Aloe*)

shaferi For John A. Shafer (1863−1918), US-American botanist, pharmacist and plant collector, esp. in the West Indies, collected 1916−1917 cacti for the US-American botanists N. L. Britton and J. N. Rose in South America. (*Agave, Austrocylindropuntia, Rhipsalis baccifera* ssp.)

shandii For Mr. John Shand (fl. ± 1920), magistrate in Ladismith, Western Cape, RSA. (*Gibbaeum*)

sharpei For H. B. Sharpe (fl. 1937), who collected the type, without further data. (*Echidnopsis*)

sharpii For Peter Sharp (fl. 1989), English cactophile who emigrated to the USA, and who presumably collected the taxon first in 1971. (*Echinocereus pulchellus* ssp.)

shavianus After Shaw's Garden, a popular name for the Missouri Botanical Garden in St. Louis, USA, founded by Henry Shaw. (*Echeveria*)

shawii For Henry Shaw (1800−1889), English-born merchant and benefactor in St. Louis, USA, founder of the Missouri Botanical Garden in St. Louis. (*Agave*)

shebae For the occurrence in the Mt. Sheba Nature Reserve, Mpumalanga, RSA. (*Aspidonepsis*)

shebeliensis For the occurrence near the Shebele River, SE Ethiopia. (*Monadenium*)

sheilae For Mrs. Iris Sheila Collenette (*1927), intrepid English plant collector and photographer, esp. in Saudi Arabia. (*Aloe, Rhytidocaulon*)

sheldonii For Charles Sheldon (1867−1928), US-American naturalist, once owner of a successful mining business in Mexico. (*Mammillaria*)

shepherdii For John Shepherd (1764–1836), English botanist and first curator of the Liverpool Botanic Garden. (*Hoya*)

sherriffii For George Sherriff (1898–1967), British explorer of Tibet and Bhutan. (*Rhodiola*)

shigatsensis For the occurrence at Shigatse, Tibet. (*Sedum*)

shitaiensis For the occurrence at Shitai in Anhui Prov., China. (*Sedum*)

shrevei For Dr. Forrest Shreve (1878–1950), US-American botanist and ecologist working esp. in the Sonoran Desert. (*Agave*, *Fouquieria*)

siamicus For the occurrence in Siam (former name of Thailand). (*Hoya*)

sibbettii For Mr. Cecil J. Sibbett († 1967), South African naturalist, Chairman of the Council for the Botanical Society of South Africa. (*Neohenricia*)

sibthorpiifolius Lat. '-folius', -leaved; and for the similarity to *Sibthorpia europaea* ("Cornish Moneywort", Scrophulariaceae). (*Pelargonium*)

sicariguensis For the occurrence near Sicarigua, Lará, Venezuela. (*Pseudoacanthocereus*)

sichotensis Most probably for the occurrence in the Sikhot Alin Mts. in the Russian Far East. (*Phedimus*)

sidoides Gr. '-oides', resembling; for the similarity of the leaves to those of *Sida rhombifolia* ("Queensland Hemp", Malvaceae). (*Pelargonium*)

sieberianus For Franz W. Sieber (1789–1844), Bohemian botanist, traveller and plant collector. (*Crassula*)

sieboldii For Dr. Philipp F. von Siebold (1796–1866), German physician, natural scientist and orientalist famous for his travels in Japan. (*Hylotelephium*)

sigridiae For Prof. Dr. Sigrid Liede (*1957), German botanist and Asclepiadaceae specialist in Münster and later Bayreuth. (*Cynanchum*)

sikokianus For the occurrence on Shikoku Island, Japan. (*Meterostachys*, *Phedimus*)

silenifolius For the genus *Silene* ("Campion", "Catchfly", Caryophyllaceae); and Lat. '-folius', -leaved. (*Euphorbia*)

sileri For Mr. A. M. Siler (fl. 1883), US-American cactus collector, without further details. (*Agave*)

silicicola Lat. 'silicis', silica; and Lat. '-cola', -inhabiting; for the preferred habitat. (*Aloe*)

sillamontanus Lat. 'montanus', relating to mountains; for the occurrence on Cerro [= mountain] de la Silla, Nuevo León, Mexico. (*Tradescantia*)

silvaticus Lat., forest-, forest-dwelling. (*Echinopsis mamillosa* ssp.)

silvestrii For Filippo [Philippo] Silvestri (1873–1949), Italian entomologist and botanist, 1898–1899 at the Museum of Buenos Aires, and a friend and supporter of the Argentinian botanist C. Spegazzini. (*Echinopsis*)

silvestris Lat., growing wild, growing in forests; application obscure. (*Sedum sarmentosum* var., *Tunilla*)

silvicola Lat. 'silva' ['sylva'], forest; and Lat. '-cola', inhabiting. (*Aloe capitata* var.)

simii For Dr. T. R. Sim (1858–1938), Scottish forestry botanist, emigrated to RSA in 1889. (*Aloe*)

similirameus Lat. 'similis', similar; and Lat. 'rameus', branch; for the similarities of the branches to those of *Euphorbia graciliramea*. (*Euphorbia*)

similis Lat., like, similar; for the similarity to other species. (*Huernia*, *Stapelia*, *Tylecodon*)

simoneae For Simone Petignat (fl. 1993), wife of Herman Petignat, hotel owner and plant enthusiast in Madagascar. (*Ceropegia*)

simonianus For Wilhelm Simon (1909–1989), German cactus hobbyist, 1952–1955 president of the Deutsche Kakteen-Gesellschaft DKG. (*Rebutia*)

simplex Lat., simple, undivided; (**1**) for the unbranched growth. (*Brachystelma*) – (**2**) for the solitary rosettes. (*Agave deserti* ssp.) – (**3**) for the undivided leaves. (*Ipomoea*) – (**4**) for the simple inflorescences of solitary cyathia. (*Monadenium*)

simplicifolius Lat. 'simplex, simplicis', simple; and Lat. '-folius', -leaved. (*Adenia fruticosa* ssp.)

simpsonii For Captain J. H. Simpson (fl. 1876), US-American topographical engineer and leader of the exploration across the Great Basin of Utah by the US Army. (*Pediocactus*) – (**2**) For Charles T. Simpson (fl. 1920), naturalist, long resident in Florida. (*Harrisia*) – (**3**) For Mr. & Mrs. Simpson (fl. 1922), station managers of the railway station of Halenberg, Namibia, in whose garden the plants were first discovered. (*Juttadinteria*)

simsii For Dr. John Sims (1749–1831), British physician and botanist. (*Aeonium*)

simulans Lat., imitating, resembling; (**1**) for having a growth form like that of *Erepsia* (Aizoaceae). (*Lampranthus*) – (**2**) probably for the similarity to some other taxon. (*Antimima, Crassula, Echeveria, Eriosyce heinrichiana* ssp.) – (**3**) probably for the resemblance of the leaves to pieces of rock, or for the similarity to a closely related taxon. (*Pleiospilos*)

sinaicus For the occurrence in the Sinai region, Egypt. (*Caralluma*)

sinanus For the occurrence at Debre Sina, Ethiopia. (*Aloe*)

sinclairii For Dr. Andrew Sinclair (± 1796–1861), British physician and plant collector, Colonial Secretary in New Zealand 1844–1856. (*Crassula*)

sinensis Lat., Chinese; for the occurrence in China. (*Chirita, Sedum tosaense* ssp.)

singularis Lat., alone, singular, solitary; (**1**) for the unique characteristics and the normally solitary leaf. (*Tylecodon*) – (**2**) perhaps because the taxon is only known from the type locality. (*Schizoglossum*) – (**3**) application obscure. (*Dischidia*)

singuliflorus Lat. 'singulus', one to each, single; and Lat. '-florus', -flowered. (*Agave*)

singulus Lat., one to each, single; for the solitary flowers. (*Ruschia*)

sinkatanus For the occurrence at Sinkat, Sudan. (*Aloe*)

Sinningia For Wilhelm Sinning (1792–1874), head gardener at the University of Bonn Botanical Garden. (*Gesneriaceae*)

sino-alpinus Lat. 'sinensis' ['sino-'], Chinese; and Lat. 'alpinus', alpine; for the occurrence at high ('alpine') altitudes in China. (*Rhodiola cretinii* ssp.)

Sinocrassula Lat. 'sinensis' ['sino-'], Chinese; for the occurrence of some taxa in China, and for the flowers with a single whorl of stamens as in the genus *Crassula* (Crassulaceae). (*Crassulaceae*)

sinoerectus Lat. 'sinensis' ['sino-'], Chinese; and Lat. 'erectus', erect; for the growth form. (*Ceropegia*)

sinoglacialis Lat. 'sinensis' ['sino-'], Chinese; and Lat. 'glacialis', of glaciers; for the occurrence at high altitudes in China. (*Sedum*)

sinuatus Lat., waved, sinuate; (**1**) for the ribs of the plant bodies. (*Ferocactus hamatacanthus* ssp.) – (**2**) for the lobed leaves. (*Rhodiola*) – (**3**) for the undulate leaf margins. (*Fockea*) – (**4**) for the leaflets. (*Cussonia paniculata* var.)

sinuosus Lat., sinuous, curving; for the conspicuously flexuose prostrate shoots. (*Phyllobolus*)

sinus-simiorum Lat. 'sinus', recess, bay; and Lat. 'simiorum', of the monkeys; for the type locality Monkey Bay, Malawi. (*Sansevieria*)

sipolisii For Abbé Michel Marie Sipolis (fl. 1856), Director of the Seminary of Diamantina, Minas Gerais, Brazil. (*Euphorbia*)

sisalanus For the Sisal fibre, which is manufactured from the leaves of this and other species of *Agave*; originally named for the town of Sisal in Yucatán (Mexico) from where the fibre was exported. (*Agave*)

skiatophytoides Gr. '-oides', resembling; and for the genus *Skiatophytum* (Aizoaceae). (*Caryotophora*)

Skiatophytum Gr. 'skias, skiatos', shade; and Gr. 'phyton', plant; for the habitat preference. (*Aizoaceae*)

skinneri For Don B. Skinner (fl. 1972?), US-American succulent plant collector in Los Angeles. (*Echeveria*)

skottsbergii For Carl J. F. Skottsberg (1880–1963), Swedish botanist and explorer. (*Echinopsis*)

sladenianus For the Percy Sladen Memorial Expeditions, during one of which the type was collected. (*Aloe, Prenia*)

sladenii For William Percy Sladen († 1900), British naturalist whose funds in the Percy Sladen Memorial Trust furthered botanical exploration in RSA. (*Crassula*)

smallii For Dr. John K. Small (1869–1938), US-American botanist at the New York Botanical Garden. (*Pedilanthus tithymaloides* ssp., *Portulaca*, *Sedum*)

smaragdiflorus Lat. 'smaragdinus', emerald-green; and Lat. '-florus', -flowered; for the green-tipped flowers. (*Cleistocactus*)

smaragdinus Lat., emerald-green; for the leaf colour. (×*Gasteraloe*)

Smicrostigma Gr. 'smikros', little; und Gr. 'stigma', stigma; for the supposedly small stigmas. (*Aizoaceae*)

smithianus For Major Cornelius C. Smith of the US Army, who assisted the US-American botanist J. N. Rose 1916 during some excursions in N Venezuela. (*Praecereus euchlorus* ssp.)

smithii For Prof. Christen Smith (1785–1816), Norwegian botanist and physician. (*Aeonium*) – (**2**) For William Wright Smith (1875–1956), British botanist in Edinburgh and long-time director of the Edinburgh Botanical Garden. (*Rhodiola*) – (**3**) For Gerald G. Smith (1892–1976), engineer, amateur botanist and student of *Haworthia* in RSA. (*Ceropegia radicans* ssp.) – (**4**) For Christo A. Smith (1898–1956), South African botanist at the Division of Botany, Pretoria, RSA, later an agricultural journalist. (*Malephora*) – (**5**) For Prof. Dr. Albert C. Smith (*1906), US-American botanist. (*Melocactus*)

smorenskaduensis For the occurrence on the farm Smor[g]enskadu near Springbok, Northern Cape, RSA. (*Conophytum*)

smrzianus For Oskar Smrz (1885–1938), Czech gardener and cactus hobbyist, author of an important Czech cactus book published 1929. (*Echinopsis*)

smytheae For Mrs. D. Smythe (fl. 1926), without further data. (*Delosperma*)

sneedii For J. R. Sneed (fl. 1923), who first found the taxon in Texas, USA. (*Escobaria*)

soanieranensis For the Soanierana Section of the Tsimbazaza Botanical Garden, Antananarivo, Madagascar, where the plant was found growing. (*Euphorbia*)

sobolifer Lat. 'soboles', branch, offspring; and Lat. '-fer, -fera, -ferum', -carrying; (**1**) for the offsetting nature of the plants. (*Aloe secundiflora* var.) – (**2**) for the bulbilliferous inflorescences. (*Agave*) – (**3**) perhaps for the inflorescences with several flowers. (*Ceropegia*)

sobrinus Lat., cousin; application obscure, perhaps for the similarity to some other taxon, and not further explained. (*Antimima*)

sobrius Lat., sober; because the plant cannot be used to produce the alcoholic beverage Mescal. (*Agave*)

sociabilis Lat., having the ability to occur socially; unresolved, perhaps for the occurrence in the same region as some related taxa. (*Eriosyce*)

socialis Lat., social; for the clustering habit. (*Aloe*, *Crassula*, *Ledebouria*)

sociorum Gen. Pl. of Lat. 'socius', companion; for the travel companions of the South African botanist L. Bolus. (*Lampranthus*)

socotranus For the occurrence on the island of Socotra. (*Adenium obesum* ssp., *Caralluma*, *Dendrosicyos*, *Dorstenia*, *Echidnopsis*, *Sarcostemma*)

Socotrella For the occurrence on Socotra; and Lat. Dim. suffix '-ellus', indicating small size. (*Asclepiadaceae*)

soederstromianus For Ludovic Söderstrom (fl. 1919), Quito, Ecuador, working for an US-American telecommunication company, without further data. (*Opuntia*)

soehrensii For Prof. Johannes Söhrens († 1934), German botanist and later director of the Botanical Garden at Santiago de Chile. (*Tunilla*)

sofiensis For the occurrence in the basin of the Sofia River, Madagascar. (*Pachypodium rutenbergianum* var.)

solaris Lat., pertaining to the sun; for the occurrence above the usually fog-covered part of the Chilean coast. (*Copiapoa*)

solenophorus Gr. 'solen', pipe, tube; and Gr. '-phoros', -bearing; for the tubular flowers. (*Caralluma*)

Solenostemon Gr. 'solen', pipe, tube; and Gr. 'stemon', stamen; for the filaments, which are basally fused to form a tube. (*Lamiaceae*)

solidus Lat., solid; probably for the staminodes and stamens collected into a bundle. (*Antimima*)

solieri For Antoine J. J. Solier (1792–1851), French botanist, entomologist and soldier. (*Crassula*)

solisioides Gr. '-oides', resembling; and for the former genus *Solisia* (*Cactaceae*). (*Mammillaria*)

solitarius Lat., solitary; for the solitary flowers. (*Ruschia*)

somalensis For the occurrence in Somalia. (*Adenium obesum* ssp., *Ceropegia*, *Duvalia sulcata* ssp., *Orbea baldratii* ssp.)

somalicus As above. (*Caralluma*, *Huernia*, *Kalanchoe bentii* ssp., *Portulaca*)

somaliensis As above. (*Aloe*, *Calyptrotheca*, *Cyanotis*)

somenii For Dr. H. Somen (fl. 1916), physician and friend of the French botanist and physician Raymond Hamet. (*Sedum*)

songweanus For the occurrence in the Songwe valley, SW Tanzania. (*Euphorbia*)

sonorae For the occurrence in the state of Sonora, Mexico. (*Ibervillea*)

sonorensis As above. (*Mammillaria*)

sootepensis For the occurrence at the Sootep Mountain, Thailand. (*Ceropegia*)

sordidus Lat., dingy, dirty-looking; (**1**) for the dark-coloured leaves. (*Haworthia*, *Hylotelephium*) – (**2**) for the flower buds, which were described as pale dingy greenish. (*Sansevieria*)

sorgerae For Dr. Friederike Sorger (fl. 1984), Austrian botanist and specalist in the Turkish flora. (*Sedum*)

sororius Lat., sisterly (from Lat. 'soror', sister); for the close relationship with another taxon. (*Pleiospilos compactus* ssp.)

sosnowskyi Perhaps for Dmitrii I. Sosnowsky (1885–1952), Russian botanist. (*Sempervivum*)

soutpansbergensis For the occurrence on the Soutpansberg, Northern Prov., RSA. (*Aloe*)

spachianus For E. Spach (fl. 1839), "aide-naturaliste" at the "Jardin du Roi", Paris, France. (*Echinopsis*)

sparsiflorus Lat. 'sparsus', scattered, sparse; and Lat. '-florus', -flowered. (*Lampranthus*)

spartarius For the similarity of the stems to those of the genus *Spartium* ("Spanish Broom", *Fabaceae*). (*Euphorbia*)

spartioides Gr. '-oides', resembling; and for the genus *Spartium* ("Spanish Broom", *Fabaceae*); for the photosynthetic peduncles. (*Absolmsia*)

spathaceus Lat., spathe-like; for the deeply boat-shaped bracts. (*Tradescantia*)

spathulatus Lat., spatulate; for the leaf shape. (*Aeonium*, *Aloinopsis*, *Crassula*, *Glossostelma*, *Kalanchoe*, *Nolana*, *Pereskiopsis*, *Thompsonella*)

spathulifolius Lat. 'spat[h]ulatus', spatulate; and Lat. '-folius', -leaved. (*Sedum eriocarpum* ssp., *Sedum*)

spathulisepalum Lat. 'spat[h]ulatus', spatulate; and Lat. 'sepalum', sepal. (*Sedum*)

speciosus Lat., beautiful; (**1**) for the flowers. (*Aeschynanthus*, *Aloe*, *Astridia*, *Caralluma*, *Ceropegia*, *Cheiridopsis*, *Cyanotis*, *Disocactus*, *Drosanthemum*, *Monsonia*, *Orbea*, *Pterodiscus*, *Schwantesia*) – (**2**) for the red cymes contrasting with the blue-green branches. (*Euphorbia*)

specksii For Ernst Specks (fl. 1999), German succulent plant collector and nursery owner. (*Euphorbia*)

spectabilis Lat., notable, showy; (**1**) for the general appearance. (*Lampranthus*, *Plectranthus*, *Stathmostelma*) – (**2**) for the spectacular inflorescences. (*Hylotelephium*, *Monadenium*) – (**3**) for the showy flowers. (*Echeveria*)

spegazzinianus For Prof. Dr. Carlos Spegazzini (1858–1926), emigrated 1879 from Italy to Argentina, pharmacist and later botanist and mycologist in Buenos Aires (first as assistant in the pharmacy of Dr. Parodi), specialist on Argentinian cacti. (*Rebutia*)

spegazzinii As above. (*Cereus*, *Gymnocalycium*)

spektakelensis For the occurrence at Spektakel, Northern Cape, RSA. (*Othonna retrorsa* var.)

spergularinus Lat. '-inus', indicating resemblance; and for the genus *Spergularia* (Caryophyllaceae). (*Parakeelya*)

sphacelatus Lat., with brown or blackish speckling; for the dark spine tips. (*Mammillaria*)

sphaericus Gr. 'sphairikos' = Lat. 'sphaericus', globose; (**1**) for the shape of the plant body. (*Mammillaria*) – (**2**) for the shape of the stem segments. (*Cumulopuntia*)

sphaerocarpus Gr. 'sphaira', globe; and Gr. 'karpos', fruit. (*Tetragonia*)

sphaerocephalus Gr. 'sphaira', globe; and Gr. 'kephale', head; for the head-like round-topped inflorescences. (*Crassula southii* ssp.)

sphaerophyllus Gr. 'sphaira', globe; and Gr. 'phyllon', leaf; for the leaf shape. (*Parakeelya*)

sphalmanthoides Gr. '-oides', resembling; and for the genus *Sphalmanthus* (Aizoaceae). (*Delosperma*)

sphenophyllus Gr. 'sphen, sphenos', wedge; and Gr. 'phyllon', leaf. (*Adromischus*)

Sphyrospermum Gr. 'sphyros', ankle; and Gr. 'sperma', seed; for the centrally bulging seeds resembling the bulging ankle. (Ericaceae)

spicatus Lat., with a spike (from Lat. 'spica', spike); (**1**) for the inflorescences. (*Agave, Aloe, Euphorbia, Plectranthus*) – (**2**) for the spike-like appearance of the inflorescences. (*Tetragonia*) – (**3**) for the spike-like inflorescence tips. (*Jatropha*)

spiculatus Lat., with points, with warts; for the raised warts on the leaves. (*Neohenricia*)

spinescens Lat., spinescent, becoming spiny; (**1**) for the spiny persistent midribs of the leaves. (*Talinum*) – (**2**) for the stout spines below the leaves. (*Monadenium*) – (**3**) for the spines developing on the inflorescence. (*Arenifera*)

spineus Lat., spiny; for the spine-tipped branches. (*Euphorbia*)

spinibarbis Lat. 'spina', spine, thorn; and Lat. 'barba', beard; for the dense spination. (*Eriosyce aurata* var., *Opuntia sulphurea* ssp.)

spiniflorus Lat. 'spina', spine, thorn; and Lat. '-florus', -flowered; (**1**) for the spine-tipped scales on the floral tube. (*Acanthocalycium*) – (**2**) for the spines on the pericarpel of the flowers. (*Austrocactus*)

spiniformis Lat. 'spina', spine, thorn; and Lat. '-formis', -shaped; for the leaf shape. (*Lampranthus*)

spinigemmatus Lat. 'spina', spine, thorn; und Lat. 'gemmatus', having buds; for the spine-covered flower buds. (*Echinocereus*)

spinosibacca Lat. 'spinosus', spiny; and Lat. 'bacca', berry; for the spiny fruits. (*Opuntia*)

spinosior Comp. of Lat. 'spinosus', spiny; (**1**) for the numerous spines per areole. (*Cylindropuntia, Sclerocactus*) – (**2**) for the prickles near the leaf tips. (*Aloe glauca* var.)

spinosissimus Superl. of Lat. 'spinosus', spiny; for the very spiny nature of the plants. (*Arthrocereus, Consolea, Mammillaria, Rebutia*)

spinosus Lat., spiny; (**1**) for the spiny stems. (*Adenia, Euphorbia ambroseae* var., *Momordica, Monsonia*) – (**2**) for the persistent spiny stipules. (*Pelargonium*) – (**3**) for the spiny leaf tips. (*Orostachys*) – (**4**) for the spinescent inflorescences. (*Ruschia*)

spinulifer Lat. 'spinula', little spine (Dim. of Lat. 'spina', thorn, spine); and Lat. '-fer, -fera, -ferum', -carrying; (**1**) for the spination. (*Opuntia*) – (**2**) according to the protologue for the spinescent stipules, but no stipules are reported for this taxon. (*Phyllobolus*)

spinulosus Lat., with small spines (Lat. 'spinula' = Dim. of Lat. 'spina', spine); (**1**) for the spination of the stems. (*Selenicereus*) – (**2**) for the spiny stipules on the stem tubercles. (*Monadenium*)

spiralis Lat., spiralled; (**1**) for the spiralling leaves. (*Albuca*) – (**2**) for the spirally arranged leaves. (*Astroloba*) – (**3**) for the spirally arranged ribs of the plant body. (*Ferocactus latispinus* ssp.) – (**4**) for the spirally arranged spine-shields. (*Euphorbia*) – (**5**) for the spirally twisted corolla lobes. (*Ceropegia*)

spirostichus Lat. 'spira', spiral; and Gr. 'stichos', row; for the spirals of leaves and stipules. (*Euphorbia decaryi* var.)

spissus Lat., thick, crowded, dense; (**1**) for the compact growth form. (*Cephalophyllum*) – (**2**) for the numerous lateral short shoots. (*Dicrocaulon*)

splendens Lat., brilliant; (**1**) for the appearance of the plant. (*Haworthia magnifica* var., *Phyllobolus*) – (**2**) for the flower colour. (*Aloe, Drosanthemum, Euphorbia milii* var., *Fouquieria*) – (**3**) application obscure. (*Raphionacme*)

spongiosus Lat., spongy; (**1**) because carpel base and receptacle become spongy when the seeds are mature. (*Crassula pellucida* ssp.) – (**2**) for the light and spongy fruit capsule. (*Jordaaniella*)

sponsaliorum Gen. Pl. of Lat. 'sponsalia', engagement; for the offsetting growth form. (*Conophytum bachelorum* ssp.)

spraguei For Dr. Thomas A. Sprague (1877–1958), British botanist at the Royal Botanic Gardens Kew. (*Opuntia*)

sprengeri For Carl Ludwig Sprenger (1846–1917), German horticulturalist, Director of Achilleion Garden on Corfu, Greece. (*Orbea*)

springbokvlakensis For the occurrence on the Springbokvlakte, Little Karoo, Eastern Cape, RSA. (*Haworthia*)

sprucei For Dr. Richard Spruce (1817–1893), British botanist and plant collector widely travelling in S America 1849–1864. (*Echeveria quitensis* var., *Peperomia*)

spurius Lat., false, spurious; application obscure. (*Cotyledon orbiculata* var., *Phedimus*)

Squamellaria Lat. 'squamella', a little scale, a little plate (Dim. of Lat. 'squama', scale); for the four fringed plates inside the corolla tube. (*Rubiaceae*)

squamosus Lat., scaly, provided with scales; for the numerous scales on the pericarpel and receptacle of the flowers. (*Facheiroa*)

squamulatus Lat., with small scales (from Lat. 'squamula', small scale); for the leaf rudiments along the stems. (*Echidnopsis*)

squamulosus Lat., with small scales (from Lat. 'squamula', small scale); (**1**) for the scaly stems and leaves. (*Galenia*) – (**2**) for the large papery scales covering the pericarpels and receptacles of the flowers. (*Neobuxbaumia*) – (**3**) for the comparatively obvious nectary scales. (*Villadia*)

squarrosus Lat., spreading, recurved; (**1**) for the growth form. (*Corryocactus, Senecio polytomus* ssp.) – (**2**) for the tubercle-teeth, which are disposed at right angles to the branch axis. (*Euphorbia*) – (**3**) for the branches, which spread at right angles. (*Euphorbia beharensis* var.) – (**4**) for the rough leaf surface in the sense of "rough with scales, tips of bracts, etc., projecting outwards" (Stearn 1992). (*Aloe*)

stagnensis Lat. 'stagnum', pool, waterhole; for the occurrence at Ross's Waterhole, Australia. (*Parakeelya*)

stahlii For Prof. Dr. [Christian] Ernst Stahl (1848–1919), Alsatian botanist, travelling in Mexico in 1894. (*Sedum*)

staintonii Probably for Adam Stainton (fl. 1969, 1988), British plant collector and botanical author. (*Caralluma, Sempervivum*)

stalagmifer Lat. 'stalagmium', drop-shaped ear-ring; and Lat. '-fer, -fera, -ferum', -carrying; for the shape of the hairs on the petal tips. (*Caralluma*)

stamineus Lat., staminal; because only male plants are known. (*Rhodiola*)

staminodiosus Lat., full of staminodes; for the numerous staminodes. (*Cephalophyllum, Lampranthus, Ruschia*)

staminosus Lat., with prominent stamens. (*Rauhia*)

stampferi For Josef Stampfer (fl. 1979), Austrian forester working for some time in Durango, Mexico. (*Mammillaria longiflora* ssp.)

standleyanus For Paul C. Standley (1884–1963), American botanist and botanical explorer of C and W North America. (*Mammillaria, Matelea*)

standleyi As above. (*Stenocereus*)

stanfordiae For Kate Canova Stanford († 1952), English-born governess in RSA, started a plant nursery in the 1930s and collected many plants. (*Hereroa, Lampranthus*)

stanleyi For Victor Stanley Peers (1874–1940), Australian civil servant, amateur archaeologist and plant collector, living in RSA from 1899. (*Chasmatophyllum*)

Stapelia For Jan Bode van Stapel (±1602–1636), Dutch physician (Genaust 1983). (*Asclepiadaceae*)

Stapelianthus For the genus *Stapelia* (*Asclepiadaceae*); and Gr. 'anthos', flower. (*Asclepiadaceae*)

stapeliiformis For the genus *Stapelia* (*Asclepiadaceae*); and Lat. '-formis', -shaped; for the succulent stems. (*Ceropegia*, *Senecio*)

stapelioides Gr. '-oides', resembling; and for the genus *Stapelia* (*Asclepiadaceae*); (**1**) for the succulent stems. (*Euphorbia*, *Monadenium*) – (**2**) for the flower shape (referring to species now classified as *Orbea*). (*Huernia*)

Stapeliopsis Gr. '-opsis', similar to; and for the genus *Stapelia* (*Asclepiadaceae*). (*Asclepiadaceae*)

stapfii For Dr. Otto Stapf (1857–1933), Austrian botanist, 1890 becoming curator of the Royal Botanic Gardens Kew. (*Euphorbia*, *Rhodiola*)

starkianus For Prof. Peter Stark (fl. 1934), without further data. (*Haworthia scabra* var.)

Stathmostelma Gr. 'stathmos', plumb-line; and Gr. 'stelma', crown, garland, wreath; for the straight appendages of the inner corona segments. (*Asclepiadaceae*)

stayneri For Frank J. Stayner (1907–1981), horticulturist and curator of the Karoo Botanic Garden Worcester 1959–1969. (*Antimima*, *Braunsia*, *Lampranthus*, *Machairophyllum*, *Stapeliopsis saxatilis* ssp., *Trichodiadema*)

Stayneria As above. (*Aizoaceae*)

stebbinsii For Dr. George L. Stebbins jr. (1906–2000), US-American botanist in Davis, California. (*Lewisia*)

steenbergensis For the occurrence on Mt. Steenberg, Western Cape, RSA. (*Oscularia*)

steenbokensis For the occurrence on a farm called Steenbok, Northern Cape, RSA. (*Polymita*)

steffanieanus For Mrs. Steffanie Paulsen (fl. 1999), German horticulturist responsable for the Madagascar collection at the Heidelberg Botanical Garden. (*Aloe*)

Steganotaenia Gr. 'stegano-', covered, roofed (from Gr. 'stege', roof); and Gr. 'taenia', ribbon, band; application obscure. (*Apiaceae*)

steinbachii For José Steinbach († 1930 or 1931), cactus collector. (*Sulcorebutia*)

steineckeanus For Karl Steinecke (fl. 1936, 1951), German succulent plant nurseryman in Ludwigsburg near Stuttgart, and one of the early opponents against imported succulent plants. (*Lithops*)

steinmannii For Prof. Dr. Johann H. C. G. Gustav Steinmann (1856–1929), German palaeontologist and geologist. (*Rebutia*)

stellariifolius Lat. '-folius', -leaved; and for the genus *Stellaria* (*Caryophyllaceae*); for the similarity to *Stellaria media*. (*Sedum*)

stellatus Lat., star-shaped; (**1**) for the radiating branches. (*Euphorbia*) – (**2**) for the radiating ribs. (*Gymnocalycium*) – (**3**) for the disposition of the spines in each areole. (*Stenocereus*) – (**4**) for the star-shaped clusters of spiny stipules. (*Monadenium*) – (**5**) for the flower shape. (*Brachystelma*, *Echidnopsis virchowii* var., *Phedimus*)

stelliformis Lat. 'stella', star; and Lat. '-formis', -shaped; probably for the flowers. (*Sedum*)

stellispinus Lat. 'stella', star; and Lat. '-spinus', -spined; for the branched spiny peduncles. (*Euphorbia*)

stellulato-tuberculatus Lat. 'stellulatus', with starlets; Lat. 'tuberculatus', tuberculate; for the patterning of the seed testa. (*Portulaca*)

stellulifer Dim. of Lat. 'stella', star; and Lat. '-fer, -fera, -ferum', -carrying; for the stellately-headed mucilage glands on the leaves. (*Uncarina*)

stenandrus Gr. 'stenos', narrow, slender; and Gr. 'aner, andros', man, [botany] stamen. (*Conophytum obcordellum* ssp., *Delosperma*, *Mesembryanthemum*)

stenanthus Gr. 'stenos', narrow, slender; and Gr. 'anthos', flower; (**1**) for the narrow corolla lobes. (*Ceropegia*) – (**2**) for the slender tubular flowers. (*Pereskia*)

stenarthrus Gr. 'stenos', narrow, slender; and Gr. 'arthron', join, segment; for the shape of the stem segments. (*Opuntia*)

Stenocactus Gr. 'stenos', narrow, slender; and Lat. 'cactus', cactus; for the very thin ribs of the plant bodies. (*Cactaceae*)

stenocaulis Gr. 'stenos', narrow, slender; and Gr. 'kaulos', stem. (*Tylecodon*)

Stenocereus Gr. 'stenos', narrow, slender; and *Cereus*, a genus of columnar cacti. (*Cactaceae*)

stenocladus Gr. 'stenos', narrow, slender; and Gr. 'klados', branch; for the short spine-tipped branchlets. (*Euphorbia*)

stenodactylus Gr. 'stenos', narrow, slender; and Gr. 'daktylos', finger; for the narrow leaflets. (*Adenia*)

stenoglossus Gr. 'stenos', narrow, slender; and Gr. 'glossa', tongue; for the narrow corona segments. (*Schizoglossum*)

stenogonus Gr. 'stenos', narrow, slender; and Gr. 'gonia', corner, margin; for the narrow ribs. (*Cereus*)

stenolobus Gr. 'stenos', narrow, slender; and Gr. 'lobos', lobe; for the corolla lobes. (*Ceropegia*)

stenopetalus Gr. 'stenos', narrow, slender; and Gr. 'petalon', petal. (*Lampranthus, Machairophyllum, Moringa, Opuntia, Sedum, Sempervivum*)

stenophyllus Gr. 'stenos', narrow, slender; and Gr. 'phyllon', leaf. (*Brachystelma, Ceropegia, Hereroa, Marlothistella, Nolana, Sedum*)

stenopleurus Gr. 'stenos', narrow, slender; and Gr. 'pleuron', rib; for the shape of the ribs of the plant bodies. (*Gymnocalycium*)

stenopterus Gr. 'stenos', narrow, slender; and Gr. 'pteron', wing; for the shape of the stems. (*Hylocereus*)

stenosiphon Gr. 'stenos', narrow, slender; and Gr. 'siphon', tube; for the flower tube. (*Kalanchoe*)

stenostachyus Gr. 'stenos', narrow, slender; and Gr. 'stachys', spike; for the inflorescences. (*Orostachys*)

Stenostelma Gr. 'stenos', narrow, slender; and Gr. 'stelma', crown, garland, wreath; application obscure. (*Asclepiadaceae*)

stentiae For Mrs. Sydney M. Stent (1875–1942), botanist in RSA and grass specialist. (*Ceropegia*)

stenus Gr. 'stenos', narrow, slender; for the filiform stems. (*Lampranthus*)

Stephania Gr. 'stephanos', wreath, crown; for the connate stamens (Genaust 1983) [and not for Christian F. Stephan (1757–1814), Saxonian botanist and physician (Jackson 1990)]. (*Menispermaceae*)

stephanii For Prof. [Christian] Friedrich Stephan (1757–1814), German (Saxonian) botanist and physician, from 1792 in St. Petersburg. (*Rhodiola*) – (**2**) For Paul Stephan (fl. 1927–1933), German gardener in charge of succulents at the Botanical Garden Hamburg, and a friend of the German Mesemb specialist G. Schwantes. (*Conophytum, Lampranthus*) – (**3**) For Stephan Martínez (fl. 1991), Mexican who first collected the taxon. (*Jatropha*)

Stephanocereus Gr. 'stephanos' wreath, crown; and *Cereus*, a genus of columnar cacti; for the ring-like cephalia. (*Cactaceae*)

Sterculia For Sterculius, Roman deity of dung (from Lat. 'stercus', dung); for the unpleasantly smelling flowers of some taxa. (*Sterculiaceae*)

sterilis Lat., sterile; because only sterile fruits are formed. (*Yucca harrimaniae* var.)

sternens Lat., becoming laid out, becoming spread out; for the mat-forming growth. (*Lampranthus*)

sterrophyllus Gr. 'sterros', firm, stiff; and Gr. 'phyllon', leaf. (*Myrmecodia*)

Stetsonia For Francis L. Stetson (1846–1920), US-American lawyer and businessman in New York, plant lover and supporter of the New York Botanical Garden, and involved with the negotiations concerning the Panama Canal. (*Cactaceae*)

steudneri For Dr. H. Steudner (1832–1863), botanist and explorer in NE Africa. (*Aloe*)

stevenianus For Christian von Steven (1781–1863), Finnish botanist frequently working in Russia and the Caucasus and inspector of the Russian silk industry. (*Phedimus*)

stevensii For Dr. Peter F. Stevens (*1944), British botanist. (*Anthorrhiza*)

steyermarkii For Dr. Julian A. Steyermark (1909–1988), US-American botanist noted for his work in the New World tropics, and one of the world's most prolific collectors. (*Echeveria*)

steytlerae Probably for Miss J. W. Steytler (fl. 1928–1940), secretary of the National Botanic Gardens, Kirstenbosch, who appears to have first collected the taxon. (*Delosperma, Erepsia*)

stictanthus Gr. 'stiktos', dotted, colourful; and Gr. 'anthos', flower; for the spotted corolla. (*Ceropegia rupicola* var.)

stictatus Gr. 'stiktos', dotted, colourful; for the spotted leaves. (*Agave*)

Stictocardia Gr. 'stiktos', dotted, colourful; and Gr. 'kardia', heart; application obscure. (*Convolvulaceae*)

stimulosus Lat., well provided with stings, spurred; perhaps for the fimbriate leaf margins or the spurred leaves. (*Sedum*)

stipitaceus Lat., stipitate, provided with a short stalk; for the shortly pedunculate inflorescences. (*Sarcostemma viminale* ssp.)

stipitatus Lat., stipitate, provided with a short stalk; application obscure. (*Coryphantha clavata* ssp.)

stipulaceus Lat., provided with stipules; (**1**) for the conspicuous stipules. (*Pelargonium*) – (**2**) application obscure, since there are no stipules. (*Lampranthus*)

stiriacus For the occurrence in Styria, Austria. (*Sempervivum montanum* ssp.)

stiriifer Lat. 'stiria', icicle; and Lat. '-fer, -fera, -ferum', -carrying; for the icicle-like long papillae on the leaf epidermis. (*Conophytum devium* ssp.)

stockingeri For Francisco "Xico" Stockinger (*1919), Brazilian sculptor, painter and cactus hobbyist in Rio Grande do Sul. (*Parodia*)

stocksii For John E. Stocks (1822–1854), British botanist collecting in Asia. (*Sarcostemma*)

Stoeberia For Mr. E. Stöber (fl. 1927), German-born teacher and botanical explorer at Lüderitzbucht, Namibia. (*Aizoaceae*)

stokoei For Thomas P. Stokoe (1868–1959), British mountaineer, artist and plant collector, living in RSA from 1911. (*Antimima, Drosanthemum, Esterhuysenia*)

stolonifer Lat., stoloniferous; for the growth habit. (*Dudleya, Echeveria, Echinocereus, Euphorbia, Monadenium, Phedimus, Sarcostemma*)

Stomatium Gr. 'stoma, stomatos', mouth; for the spreading dentate leaves of a pair, which resemble an open mouth. (*Aizoaceae*)

Stomatostemma Gr. 'stoma, stomatos', mouth; and Gr. 'stemma', garland, wreath; for the coronal outgrowths at the mouth of the corolla tube. (*Asclepiadaceae*)

stormiae For Marian Storm (fl. 1939), who collected the type, without further data. (*Euphorbia radians* var.)

stramineus Lat., straw-yellow; (**1**) for the spination. (*Echinocereus, Melocactus*) – (**2**) for the flower colour. (*Aridaria noctiflora* ssp.)

strangulatus Lat., throttled, constricted and widened again; for the segmented branches. (*Euphorbia*)

stratiotes Gr., soldier, warrior; perhaps for the armed leaves. (*Furcraea*)

strausianus For Mr. L. Straus (1862–1934), German merchant and cactus hobbyist in Bruchsal near Baden and co-founder of the Deutsche Kakteen-Gesellschaft DKG. (*Eriosyce*)

strausii As above. (*Cleistocactus*)

streptacanthus Gr. 'streptos', twisted; and Gr. 'akanthos', spine, thorn. (*Opuntia*)

streptanthus Gr. 'streptos', twisted; and Gr. 'anthos', flower; application obscure. (*Kalanchoe*)

Streptocarpus Gr. 'streptos', twisted; and Gr. 'karpos', fruit; for the twisted capsules. (*Gesneriaceae*)

streyi For Rudolf G. Strey (*1907), German-born farmer and botanist in RSA. (*Crassula*)

striatulus Lat., with small striae; for the markings on the leaf sheaths. (*Aloe*)

striatus Lat., striate, marked with usually parallel fine lines; (**1**) for the longitudinally striate stems. (*Peniocereus, Tylecodon*) – (**2**) for the horizontal stripes on the branches. (*Euphorbia horrida* var.) – (**3**) for the leaves. (*Agave, Aloe, Bulbine, Portulaca*) – (**4**) for the central stripes on the petals. (*Drosanthemum*) – (**5**) for the red-striped corolla tube. (*Ceropegia*)

stricticaulis Lat. 'strictus', very straight, strictly upright; and Lat. 'caulis', stem; for the erect short shoots. (*Disphyma australe* ssp.)

strictiflorus Lat. 'strictus', very straight, strictly upright; and Lat. '-florus', -flowered; for the narrow straight inflorescences. (*Echeveria*)

strictifolius Lat. 'strictus', very straight, strictly upright; and Lat. '-folius', -leaved. (*Drosanthemum*)

strictus Lat., very straight, strictly upright; (**1**) for the growth habit. (*Adenia, Mesembryanthemum, Opuntia, Ruschia, Sceletium*) – (**2**) for the orientation of the leaves. (*Beaucarnea, Villadia, Yucca glauca* var.) – (**3**) for the leaf structure. (*Agave*) – (**4**) application obscure. (*Cistanthe*)

strigil Lat., scraping iron; for the closely set areoles resulting in a densely spined appearance of the plants. (*Opuntia*)

strigillosus Lat., strigillose, with short appressed hairs. (*Nematanthus*)

striglianus For Franz Strigl (*1937), Austrian cactus hobbyist in Kufstein, founder member of the Austrian "Arbeitsgruppe Gymnocalycium". (*Gymnocalycium*)

strigosus Lat., bristly-hairy (from Lat. 'striga', straight rigid short bristle-like appressed hair); (**1**) for the hairs on the stems. (*Crassula, Plectranthus*) – (**2**) for the spination of the stems. (*Echinopsis*) – (**3**) for the hairy leaves. (*Euphorbia*)

stringens Lat. 'stringere', hurt, injure, string together; perhaps for the armature of the leaves. (*Agave*)

strobiliformis Lat. 'strobilus', cone of gymnosperms; and Lat. '-formis', -shaped; for the appearance of the plant body. (*Pelecyphora*)

Strombocactus Gr. 'strombos', top, spindle, cone of conifers; and Lat. 'cactus', cactus; for the body shape. (*Cactaceae*)

strophiolatus Lat., provided with a strophiola; for the seed morphology. (*Parakeelya*)

strubeniae For Miss Edith Struben († 1936), a keen gardener and artist in South Africa, council member of the Botanical Society of South Africa. (*Ruschia*)

strumosus Lat., having swellings like cushions, with swollen cervical glands; for the thickened tuberous root. (*Trichodiadema*)

stuckertii For Theodore Stuckert (1852–1932), Swiss apothecary from Basel, emigrating to Argentina, owner of a pharmacy in Córdoba in 1885, then professor at the agricultural faculty in Córdoba, from 1913+ in Geneva. (*Gymnocalycium*)

stuckyi For a Mr. Stucky (fl. 1861), without further data. (*Sansevieria*)

stuemeri For Ernst [Ernesto] Stüemer (fl. 1930s), Argentinian field collector and plant trader. (*Parodia*)

stuessyi For Prof. Dr. Tod F. Stuessy (*1943), US-American botanist at Ohio State University and from 1997 at Vienna University, Austria. (*Aeonium*)

stygianus For the steam-vessel 'Styx', from which the British botanist Hewett C. Watson (1804–1881) landed 1843 on the Azores for botanizing. (*Euphorbia*)

stylosus Lat., with a long or persistent style. (*Adenia firingalavensis* var., *Arenifera*)

suarezensis For the occurrence in the region of Diégo Suarez (= Antsiranana), Madagascar. (*Adansonia, Aloe, Kalanchoe*)

suaveolens Lat., pleasantly fragrant (from Lat. 'suavis', sweet; and Lat. 'olens', scented); (**1**) for the scent of the leaves. (*Aeollanthus*) – (**2**) for the sweet-scented flowers. (*Ruschia, Sedum, Stomatium*)

suavis Lat., sweet; for the scented flowers. (*Glottiphyllum*)

suavissimus Comp. of Lat. 'suavis', sweet; for the delicate ("sweet") pink flower colour. (*Lampranthus*)

subacaulis Lat. 'sub-', almost, more or less; and Lat. 'acaulis', stemless; for the almost stemless habit. (*Aeollanthus, Crassula*)

subacutissimus Lat. 'sub-', almost, more or less; and Superl. of Lat. 'acutus', pointed; because the *Aloe* specialist G. W. Reynolds (1895–1967) assumed that the taxon is similar to *Aloe acutissima*. (*Aloe*)

subaequalis Lat. 'sub-', almost, more or less; and Lat. 'aequalis', equal; for the almost equal calyx lobes. (*Lampranthus*)

subalbus Lat. 'sub-', almost, more or less; and Lat. 'albus', white; for the flowers that are only white at the base. (*Argyroderma*)

subalpinus Lat. 'sub-', almost, more or less; and Lat. 'alpinus', alpine; for the subalpine habitat. (*Echeveria, Sedum erici-magnusii* var., *Sedum lanceolatum* ssp.)

subaphyllus Lat. 'sub-', almost, more or less, and Lat. 'aphyllus', leafless. (*Ceropegia, Cissus, Crassula*)

subcapitatus Lat. 'sub-', almost, more or less; and Lat. 'capitatus', capitate; for the inflorescences. (*Hylotelephium*)

subcarnosus Lat. 'sub-', almost, more or less; and Lat. 'carnosus', fleshy; perhaps for the slightly fleshy leaves. (*Galenia*)

subcerulatus Lat. 'sub-', almost, more or less; and Lat. 'cerulatus', a little waxy; for the similarity to *Agave cerulata*. (*Agave cerulata* ssp.)

subclausus Lat. 'sub-', almost, more or less; and Lat. 'clausus', closed; probably for the leaves of a pair, which are closely appressed to each other. (*Drosanthemum*)

subclavatus Lat. 'sub-', almost, more or less; and Lat. 'clavatus', club-shaped, clavate; for the pedicels. (*Delosperma*)

subcompressus Lat. 'sub-', almost, more or less; and Lat. 'compressus', compressed; for the semiterete and somewhat compressed leaves. (*Drosanthemum*)

subcorymbosus Lat. 'sub-', almost, more or less; and Lat. 'corymbosus', corymbose; for the inflorescence. (*Echeveria*)

subcylindricus Lat. 'sub-', almost, more or less; and Lat. 'cylindricus', cylindrical; for the shape of the stem segments. (*Tacinga inamoena* ssp.)

subdenudatus Lat. 'sub-', almost, more or less; and Lat. 'denudatus', denuded, stripped, worn-off; for the insignificant and short spination of the plant bodies. (*Echinopsis*)

subdistichus Lat. 'sub-', almost, more or less; and Lat. 'distichus', distichous, two-ranked; for the leaf arrangement. (*Adromischus*)

subductus Lat., pulled up; for the more porrect central spines. (*Mammillaria laui* ssp.)

suberectus Lat. 'sub-', almost, more or less, and Lat. 'erectus', erect, upright; for the leaf orientation. (*Haworthia turgida* var.)

suberosus Lat., corky; (**1**) for the corky bark of old stems. (*Stephania*) – (**2**) for the corkiness of old stems. (*Sarcostemma viminale* ssp.)

subfastigiatus Lat. 'sub-', almost, more or less; and Lat. 'fastigiatus', fastigiate, clustered; for the somewhat clustered branches. (*Euphorbia carunculifera* ssp.)

subfenestratus Lat. 'sub-', almost, more or less; and Lat. 'fenestratus', windowed; for the poorly developed windows on the leaves. (*Conophytum*)

subgaleatus Lat. 'sub-', almost, more or less; and Lat. 'galeatus', helmet-shaped; for the petal shape. (*Sedum*)

subgibbosus Lat. 'sub-', almost, more or less; and Lat. 'gibbosus', gibbous, tuberculate; for the ribs that are ± dissolved into tubercles. (*Eriosyce*)

subglaucus Lat. 'sub-', almost, more or less; and Lat. 'glaucus', glaucous. (*Haworthia chloracantha* var.)

subglobosus Lat. 'sub-', almost, more or less; and Lat. 'globosus', globose; for the leaf shape. (*Octopoma*)

subincanus Lat. 'sub-', almost, more or less; and Lat. 'incanus', hoary, white; for the leaf colour. (*Conophytum uviforme* ssp., *Delosperma*)

subinermis Lat. 'sub-', almost, more or less; and Lat. 'inermis', unarmed; (**1**) for the absent or short spines. (*Acanthocereus*) – (**2**) for the short spines. (*Echinocereus*)

subinteger Lat. 'sub-', almost, more or less; Lat. 'integer, integra, integrum', entire; for the almost untoothed leaves. (*Faucaria*)

sublaxus Lat. 'sub-', almost, more or less; and Lat. 'laxus', lax; for the branching. (*Lampranthus*)

sublineatus Lat. 'sub-', almost, more or less; and Lat. 'lineatus', lineate, lined; for the leaves. (*Haworthia mirabilis* var.)

sublobatus Lat. 'sub-', almost, more or less; and Lat. 'lobatus', lobed; for the leaf shape. (*Jatropha pelargoniifolia* var.)

submammillaris Lat. 'sub-', almost, more or less; and for the similarity to *Euphorbia mammillaris*. (*Euphorbia*)

submammulosus Lat. 'sub-', almost; and from the Dim. of Lat. 'mamma', breast, tubercle; i.e. full of small tubercles; for the tuberculate plant bodies and the similarity to *Echinocactus* (*Parodia*) *mammulosus*. (*Parodia mammulosa* ssp.)

subnodosus Lat. 'sub-', almost, more or less; and Lat. 'nodosus', with nodes; for the inconspicuous nodes of the stems. (*Psilocaulon*)

subnudus Lat. 'sub-', almost, more or less; and Lat. 'nudus', naked; for the leaves, which are tomentose when young but become glabrous later. (*Anacampseros*)

suboppositus Lat. 'sub-', almost, more or less; and Lat. 'oppositus', opposite; for the leaf arrangement. (*Rhodiola*)

subpaniculatus Lat. 'sub-', almost, more or less; and Lat. 'paniculatus', paniculate; for the well-branched inflorescences. (*Ruschia*)

subpetiolatus Lat. 'sub-', somewhat, more or less; and Lat. 'petiolatus', petiolate. (*Delosperma*)

subplanus Lat. 'sub-', almost, more or less; and Lat. 'planus', plane, flat; (**1**) for the almost flat rosettes. (*Aeonium canariense* var.) − (**2**) for the almost flat-topped ovary. (*Drosanthemum*)

subregularis Lat. 'sub-', almost, more or less; and Lat. 'regularis', regular; for the flower shape. (*Haworthia reticulata* var.)

subrigidus Lat. 'sub-', almost, more or less; and Lat. 'rigidus', rigid; for the branches of the inflorescences. (*Echeveria*)

subrosulatus Lat. 'sub-', almost, more or less; and Lat. 'rosulatus', rosulate; for the crowded rosulate leaves. (*Kalanchoe*)

subrotundus Lat. 'sub-', almost, more or less; and Lat. 'rotundus', round; for the semiglobose receptacle of the flowers. (*Lampranthus*)

subsalsus Lat. 'sub-', almost, more or less; and Lat. 'salsus', salted; for the occurrence at the foot of the Pedra de Sal (Port. 'sal', salt). (*Euphorbia*)

subscandens Lat. 'sub-', almost, more or less; and Lat. 'scandens', climbing. (*Euphorbia, Rhytidocaulon*)

subsessilifolius Lat. 'sub-', almost, more or less; Lat. 'sessilis', sessile; and Lat. '-folius', -leaved. (*Adenia*)

subsessilis Lat. 'sub-', almost, more or less; und Lat. 'sessilis', sessile; for the cyathia. (*Euphorbia heptagona* var.)

subsimilis Lat. 'sub-', almost, more or less; and Lat. 'similis', similar; for the close similarity to another taxon. (*Pilosocereus*)

subsimplex Lat. 'sub-', almost, more or less; and Lat. 'simplex', simple; (**1**) because the plants are only sometimes suckering. (*Agave*) − (**2**) for the sometimes almost entire leaves. (*Jatropha zeyheri* var.)

subsphaericus Lat. 'sub-', almost, more or less; and Lat. 'sphaericus', spherical, globose; for the leaf shape. (*Ruschia*)

subsphaerocarpus Lat. 'sub-', almost, more or less; Gr. 'sphaira', globe; and Gr. 'karpos', fruit. (*Opuntia*)

subspicatus Lat. 'sub-', almost, more or less; and Lat. 'spicatus', spike-like; for the inflorescences. (*Sansevieria*)

subspinosus Lat. 'sub-', almost, more or less; and Lat. 'spinosus', spiny, prickly; for the persistent spinescent pedicels. (*Drosanthemum*)

substerilis Lat. 'sub', almost; and Lat. 'sterilis', sterile; because the only plant originally found in the wild flowered but never produced fruits according to a local peasant. (*Calymmanthium*)

subteres Lat. 'sub-', almost; and Lat. 'teres', terete; for the leaf shape. (*Ruschia*)

subterraneus Lat., subterranean, underground; (**1**) for the large tuberous root. (*Turbinicarpus mandragora* ssp.) − (**2**) for the almost completely underground plant bodies. (*Maihueniopsis, Parodia*) − (**3**) for the underground stems. (*Orbea*)

subtilis Lat., fine, delicate; (**1**) for the nature of the whole plant. (*Cynanchum, Sedum*) − (**2**) for the small flowers. (*Sansevieria*)

subtilispinus Lat. 'subtilis', fine, delicate; and Lat. '-spinus', -spined. (*Haageocereus*)

subtruncatus Lat. 'sub-', almost, more or

less; and Lat. 'truncatus', truncate; (**1**) for the shape of the leaf tips. (*Lampranthus, Mesembryanthemum*) – (**2**) for the shape of the base of the receptacle. (*Antimima*)

subulatoides Gr. '-oides', resembling; and for the similarity to *Mesembryanthemum* (now *Acrodon*) *subulatum*. (*Cephalophyllum*)

subulatus Lat., subulate, awl-shaped; (**1**) for the elongate stems. (*Caralluma*) – (**2**) for the leaf shape. (*Acrodon, Asclepias, Austrocylindropuntia, Crassula, Sedum*)

subviridigriseus Lat. 'sub-', almost, more or less; Lat. 'viridis', green; and Lat. 'griseus', grey; for the colour of the plant bodies. (*Discocactus bahiensis* ssp.)

subviridis Lat. 'sub-', almost, more or less; and Lat. 'viridis', green; for the leaf colour. (*Adromischus*)

succineus Lat., amber-coloured; for the colour of the radial spines. (*Parodia scopa* ssp.)

succotrinus For the presumed but fictitious occurrence on Socotra; or a contraction of Lat. 'succus', sap; and Lat. 'citrinus', lemon-yellow; for the colour of the dried leaf exudate. (*Aloe*)

succulentus Lat., succulent (from Lat. 'suc[c]us', sap); (**1**) for the succulent stems. (*Apodanthera, Pachypodium*) – (**2**) for the succulent leaves. (*Amphibolia, Bulbine, Oxalis, Pteronia*) – (**3**) application obscure. (*Thorncroftia*)

succumbens Lat., liable to succumb, liable to fail; because the plants first found were almost dead and cuttings did not root. (*Schwantesia*)

sucrensis For the occurrence near the city of Sucre, Prov. Oropeza, Dept. Chuquisaca, Bolivia. (*Echinopsis*)

sudanicus For the occurrence in the former French Sudan (now mostly Burkina Faso, Mali and Niger). (*Euphorbia*)

suffrutescens Lat., slightly woody, becoming woody at base, becoming subshrub-like. (*Portulaca*)

suffruticosus As above. (*Chlorophytum, Phyllobolus, Sansevieria, Turbina*)

suffultus Lat., subtended, supported; (**1**) for the weak branches, which are supported by other vegetation. (*Euphorbia, Tylecodon*) – (**2**) for the slender long peduncles, which need support by shrubs. (*Aloe*)

sulcatus Lat., furrowed, grooved; (**1**) for the low ribs of the stem segments. (*Rhipsalis*) – (**2**) for the body tubercles. (*Coryphantha*) – (**3**) for the leaves. (*Sansevieria*) – (**4**) for the grooved sides of the fused leaf pair. (*Conophytum ectypum* ssp.) – (**5**) for the grooves on the flower base. (*Delosperma*) – (**6**) for the grooved corolla lobes. (*Duvalia*)

sulcicalyx Lat. 'sulcus', furrow, groove; and Lat. 'calyx', calyx (here referring to the involucre of phyllaries). (*Senecio*)

sulcifer Lat. 'sulcus', furrow, groove; and Lat. '-fer, -fera, -ferum', -carrying; for the furrows crossing the ribs of the plant body. (*Cleistocactus*)

sulcolanatus Lat. 'sulcus', furrow, groove; and Lat. 'lanatus', woolly; for the wool-filled furrows and axils of the body tubercles. (*Coryphantha*)

Sulcorebutia Lat. 'sulcus', furrow, groove; and for the genus *Rebutia* (*Cactaceae*); for the grooves separating the tubercles of the plant bodies. (*Cactaceae*)

sulphureus Lat., like sulphur, sulphur-yellow; for the flower colour. (*Opuntia, Stenocactus, Tylecodon*)

sumati From the local Samburu vernacular name of the plants in Kenya. (*Euphorbia*)

sundaensis For the occurrence on the Lesser Sunda Islands (Indonesia). (*Portulaca pilosa* ssp.)

superans Lat., overtopping, rising above; perhaps for the upturned branches. (*Euphorbia, Oscularia*)

superbus Lat., superb, very beautiful; (**1**) for the stately appearance. (*Ceropegia arabica* var., *Espostoa, Miraglossum*) – (**2**) for the leaf colouration. (*Graptopetalum pentandrum* ssp.)

supertextus Lat., covered, woven; for the dense spination. (*Mammillaria*)

suppositus Lat., placed underneath; because the bracts enclosing the pedicel appear to support the calyx. (*Zeuktophyllum*)

suppressus Lat., suppressed; presumably for the dwarf habit and the very short branches. (*Euphorbia*)

suprafoliatus Lat. 'supra', above; and Lat. 'foliatus', -leaved; for the leaf arrangement of young plant with distichous leaves one above the other like the pages of an open book. (*Aloe*)

surculosus Lat., surculose, suckering, offsetting; for the growth habit. (*Mammillaria*, *Sedum*)

suricatinus For *Suricata suricata*, one of the African meerkats; probably for the gaping denticulate leaves of a pair, resembling an open mouth. (*Stomatium*)

surrectus Lat., erect; (**1**) for the stems. (*Stapelia*) – (**2**) for the leaves. (*Glottiphyllum*)

susan-holmesiae For Mrs. Susan Carter Holmes (*1933), English botanist at RBG Kew, and specialist on *Euphorbia* and *Aloe* in Tropical Africa. (*Euphorbia*)

susannae For Mlle. Suzanne Leschot (fl. 1910), an acquaintance of the French physician and botanist Raymond Hamet. (*Sedum*) – (**2**) For Susanna Muir (fl. 1929), wife of the Scottish physician and naturalist John Muir, Riversdale, RSA. (*Euphorbia*) – (**3**) For Suzanne Lavranos (fl. 1962), Pretoria, RSA, former wife of the succulent plant collector John Lavranos. (*Crassula*)

sutherlandii For Dr. Peter C. Sutherland (1822–1900), born in England, Surveyor-General of Natal (1855–1887), occasionally collecting plants for Kew. (*Begonia*, *Delosperma*)

suttoniae For Miss Sutton (fl. 1966), without further data. (*Delosperma*)

suzannae For Mlle. Suzanne Decary (fl. 1921), daughter of the French administrator, botanist and plant collector Raymond Decary and his wife Hélène. (*Aloe*)

suzannae-marnierae For Madame Suzanne Marnier-Lapostolle (fl. 2002), wife of Julien Marnier-Lapostolle, who founded the private botanical garden "Les Cèdres" in S France. (*Euphorbia*)

swanepoelianus For Jac Swanepoel (fl. 1971), owner of San Marino Nursery, RSA. (*Conophytum*)

swanepoelii As above. (*Quaqua*)

swartbergensis For the occurrence on the Swartberg, near Prince Albert, Western Cape, RSA. (*Lampranthus*)

swartkopensis For the occurrence in the Swartkop Distr., Eastern Cape, RSA. (*Lampranthus*)

swarupa For Dr. Kundil Swarupanandan (*1952), Indian botanist at the Kerala Forest Research Institute and specialist for *Asclepiadaceae*. (*Brachystelma*)

swazicus For the occurrence in Swaziland, S Africa. (*Adenium obesum* ssp., *Brachystelma*)

swaziensis As above. (*Crassula*)

swaziorum Gen. Pl., for the occurrence in Swaziland, i.e. the region inhabited by the Swazi, S Africa. (*Ceropegia*)

swobodae For Heinz Swoboda (1941–1997), Austrian cactus hobbyist and plant collector in North and South America and China. (*Turbinicarpus*)

swynnertonii For Charles F. M. Swynnerton (1877–1939), English zoologist and naturalist, worked on Tsetse control, studied fauna and flora generally, collecting in Moçambique, Tanzania and Zimbabwe, died in an air crash in Tanzania. (*Aloe*, *Plectranthus*)

sylvaticus Lat., forest-, forest-dwelling. (*Dioscorea*)

sylvestris Lat., growing wild, growing in forests; for the preferred habitat. (*Leptocereus*)

sylvicola Lat. 'sylva', forest; and Lat. '-cola', -dwelling. (*Cissus*, *Raphionacme*)

symmetricus Lat., symmetrical; for the regular shape of the plant bodies. (*Euphorbia obesa* ssp.)

Synadenium Gr. 'syn-', together; and Gr. 'aden, adenos', gland; for the nectary glands, which form a continuous rim around the involucre. (*Euphorbiaceae*)

Synaptophyllum Gr. 'synaptos', united; and Gr. 'phyllon', leaf; for the leaves connate in pairs. (*Aizoaceae*)

synsepalus Gr. 'syn-', together; and Gr. 'sepalon', sepal; for the partly united sepals. (*Kalanchoe*)

T

taboraensis For the occurrence in the Tabora Distr., Tanzania. (*Euphorbia*)

tabularis Lat., plate-like, flat; (**1**) for the depressed-globose plant body. (*Parodia*) – (**2**) for the flat rosettes. (*Crassula*)

tabularius Lat. 'tabula', board shelf, table; because a drawing from the archives of the Bolus Herbarium was used for the protologue. (*Brachystelma*)

tabuliformis Lat. 'tabula', board, shelf, table; and Lat. '-formis', -shaped; for the flat rosettes. (*Aeonium*)

tacaquirensis For the occurrence near Tacaquira, Dept. Chuquisaca, Bolivia. (*Echinopsis*)

Tacinga Anagram of "caatinga", which denominates the shrubby vegetation in the dry NE of Brazil, where the genus is native. (*Cactaceae*)

tacnaensis For the occurrence near Tacna, Dept. Tacna, Peru. (*Oreocereus*)

tacuaralensis For the occurrence near Tacuaral, Prov. Chiquitos, Dept. Santa Cruz, Bolivia. (*Cereus*)

taetra Lat. 'taeter, taetra, taetrum', offensive, ugly; for the unusually strong spination. (*Harrisia*)

taitensis For the occurrence in the Taita [Teita] region, Kenya. (*Calyptrotheca*)

taiticus For the occurrence at Taita Hills, Kenya. (*Orbea*)

takesimensis For the occurrence on the Island of Take-Shima (Tok-to) situated between Korea and Japan. (*Phedimus*)

Talinella For the similarity to the genus *Talinum* (*Portulacaceae*), but the Dim. is hardly justified since the plants grow much larger. (*Portulacaceae*)

talinoides Gr. '-oides', resembling; and for the genus *Talinum* (*Portulacaceae*). (*Senecio*)

Talinum Unresolved, either from the vernacular name of one of the species in Senegal (improbable), or from Lat. 'telinum', a costly ointment prepared from *Trigonella* (Genaust 1983), probably for the similar usage. (*Portulacaceae*)

taltalensis For the occurrence near Taltal, N Chile. (*Eriosyce*)

tamaranae For the occurrence on the Canary Islands, which in the local Guanche language were called Tamaran. (*Dracaena*)

tamaulipensis For the occurrence in the state of Tamaulipas, Mexico. (*Sedum*)

tanaensis For the occurrence near the Tana River, Kenya. (*Euphorbia*)

tananarivae For the occurrence in the Antananarivo region, Madagascar. (*Euphorbia milii* var.)

tanganyikensis For the occurrence in the former Tanganyika (now Tanzania). (*Huernia*)

tanguticus For the occurrence in Tangut (an old name for Gansu and Quinghai Prov.), China. (*Rhodiola*)

taningaensis For the occurrence near Taninga, Prov. Córdoba, Argentina. (*Gymnocalycium*)

tanjorensis For the occurrence in the Tanjore Distr., Madras, India. (*Jatropha*)

Tanquana For the occurrence in the Tanqua Karoo, RSA. (*Aizoaceae*)

tanquanus As above. (*Euphorbia gentilis* ssp.)

tantillus Lat., such a trifle, so little; for the small plant size. (*Conophytum*)

taohoensis For the occurrence in the Taoho region, Kansu Prov., China. (*Rhodiola himalensis* ssp.)

tapecuanus For the occurrence at Tapecua, Prov. O'Connor, Dept. Tarija, Bolivia. (*Echinopsis obrepanda* ssp.)

Tapinanthus Gr. 'tapeinos', humble, common; and Gr. 'anthos', flower; for the small size of the flowers. (*Loranthaceae*)

tapona For the fancied resemblance of the fruit to a bottle-stopper (Span. 'tapón'). (*Opuntia*)

taprobanicus For the occurrence in Sri Lanka; from 'Taprobane', the name that the ancient Greeks and Roman cartographers used for Sri Lanka. (*Ceropegia*)

tapscottii For Sydney Tapscott (fl. 1930), keen collector and photographer of succulents in S Africa. (*Orbea*)

taquimbalensis For the occurrence near Taquimbala, Dept. Cochabamba, Bolivia. (*Echinopsis tacaquirensis* ssp.)

tarapacanus For the occurrence in Prov. Tarapacá, N Chile. (*Nolana*) – (**2**) For the occurrence in the old Prov. Tarapacá, Chile, which in the mid-19. century extended to present-day Prov. Catamarca, Argentina. (*Maihueniopsis*)

taratensis For the occurrence near Tarata, Prov. Tarata, Dept. Cochabamba, Bolivia. (*Echinopsis, Parodia*)

tardieuanus For Dr. Marie L. Tardieu-Blot (*1902), French pharmacist, physician and botanist (pteridologist) in Paris. (*Euphorbia*)

tardissimus Superl. of Lat. 'tardus', late; for the very late opening time of the flowers. (*Ruschia*)

tardus Lat., late; for the flowers opening only in the afternoon. (*Drosanthemum*)

tarijensis For the occurrence near the city of Tarija, Prov. Cercado, Dept. Tarija, Bolivia. (*Cleistocactus, Echinopsis, Sulcorebutia*)

tarkaensis For the occurrence near Tarkastad, Eastern Cape, RSA. (*Aloe broomii* var.)

tarmaensis For the occurrence near Tarma, Dept. Junín, Peru. (*Echinopsis*)

taruensis For the occurrence at Taru, Kenya. (*Euphorbia*)

tashiroi For Yasusada Tashiro (1856–1928), botanist in Taiwan. (*Kalanchoe*)

tatarinowii For Alexander A. Tatarinow (1817–1886), Russian botanist. (*Hylotelephium*)

Tavaresia For José Tavares de Macedo (fl. ± 1850), superior official in the Portuguese Ministry of Marine and the Colonies, and amateur botanist. (*Asclepiadaceae*)

taylorianus For Edward Taylor (1848–1928), British grower of succulent plants, esp. Mesembs. (*Conophytum*)

taylorii As above. (*Corpuscularia*) – (**2**) For Norman Taylor (1883–1967), British-born US-American botanist and botanical explorer. (*Harrisia, Opuntia*)

tayloriorum For Bob and Suzanne Taylor (fl. 1975), US-American cactus hobbyists in El Cajon, California. (*Mammillaria*)

tayopensis For the legendary lost gold mine of Tayopa that is supposed to have existed near the type locality of the taxon in E Sonora, Mexico, and probably also in allusion to the yellow flower colour. (*Echinocereus stolonifer* ssp.)

techinensis For the occurrence at Te Chin, Yunnan, China. (*Sinocrassula*)

tectorum Gen. Pl. of Lat. 'tectum', roof; because plants were frequently planted on roofs to ward off lightning. (*Sempervivum*)

tectus Lat., hidden, covered, concealed; (**1**) for the leaves covered in papillae. (*Crassula*) – (**2**) for the staminodes that cover the stamens. (*Ruschia*) – (**3**) perhaps for the ± imbricate leaves. (*Agave*)

tegelbergianus For Gilbert H. Tegelberg jr. (1924–1997), US-American nurseryman and promotor of the cactus and succulent hobby. (*Mammillaria albilanata* ssp.)

tegelerianus For Wilhelm Tegeler (fl. 1936), councillor of Hamburg. (*Echinopsis*)

tegens Lat., covering, concealing; for the cushion-forming growth habit and the fact that the cushions are covered with flowers at flowering time. (*Lampranthus*)

tehuacanus For the occurrence in the Tehuacán region in Puebla, Mexico. (*Opuntia, Pedilanthus*)

tehuantepecanus For the occurrence on the Isthmus of Tehuantepec, Mexico. (*Callisia, Opuntia*)

tehuaztlensis For the occurrence near Tehuaztepec, state of México, Mexico. (*Sedum*)

teissieri For Marc Teissier (fl. 2002), French horticulturist and curator of the private botanical garden "Les Cèdres" near Nice, France. (*Aloe*)

teixeirae For the Angolan botanist Eng. J. Brito Teixeira (fl. 1974) of the Divisão de Botânica e Fitogeografia at Nova Lisboa, Angola. (*Euphorbia*)

teke From the local vernacular name of the plants in NE Zaïre. (*Euphorbia*)

telephiastrum Lat. 'Telephium', orpine [*Hylotelephium telephium*]; and Lat. Dim. suffix '-astrum', wild, small, inferior. (*Anacampseros*)

telephioides Gr. '-oides', resembling; and for 'Telephium', the pre-Linnean name for *Hylotelephium telephium* (Crassulaceae). (*Hylotelephium*)

telephium Pre-Linnean name for *Hylotele-*

phium telephium (*Crassulaceae*). (*Hylotelephium*)
Telfairia For Charles Telfair (1778–1833), Irish surgeon, naturalist and plant collector on Mauritius. (*Cucurbitaceae*)
tenax Lat., tough; for the branches. (*Euphorbia*)
tenebricus Lat., dark, gloomy; for the dark-coloured plant bodies. (*Eriosyce*)
tenellus Lat., very delicate; (**1**) for the delicately small plants. (*Brachystelma, Pelargonium, Sedum*) – (**2**) for the slender branches. (*Ruschia*)
tener Lat. 'tener, tenera, tenerum', delicate, tender; (**1**) for the general appearance. (*Haworthia gracilis* var.) – (**2**) application unclear (or incorrect form of Lat. 'tenere', to cling, for the twining stems). (*Tinospora*)
tentaculatus Lat., with tentacles; for the long narrow corolla lobes. (*Ceropegia multiflora* ssp., *Quaqua*)
tenuicaulis Lat. 'tenuis', thin, slender; and Lat. 'caulis', stem. (*Crassula, Euphorbia lophogona* var., *Pelargonium*)
tenuicylindricus Lat. 'tenuis', thin, slender; and Lat. 'cylindricus', cylindrical; for the slender cylindrical plant body. (*Parodia*)
tenuiflorus Lat. 'tenuis', thin, slender; and Lat. '-florus', -flowered. (*Opuntia, Phyllobolus, Pseudobombax ellipticum* var.)
tenuifolius Lat. 'tenuis', thin, slender; and Lat. '-folius', -leaved. (*Agave, Gunniopsis, Hereroa, Hesperaloe, Lampranthus, Sedum amplexicaule* ssp.)
tenuior Comp. of Lat. 'tenuis', slender. (*Aloe, Euphorbia gillettii* ssp.)
tenuipedicellatus Lat. 'tenuis', thin, slender; and Lat. 'pedicellatus', pedicellate. (*Crassula*)
tenuipes Lat. 'tenuis', thin, slender; and Lat. 'pes', foot; application obscure. (*Hoya australis* ssp.)
tenuiradiatus Lat. 'tenuis', thin, slender; and Lat. 'radiatus', with rays; for the slender inflorescence appendages. (*Dorstenia*)
tenuis Lat., thin, slender; (**1**) for the smaller overall size. (*Haworthia coarctata* var.) – (**2**) for the slender stems. (*Ceropegia linearis* ssp., *Haageocereus, Sedum oreganum* ssp., *Tylecodon*) – (**3**) for the narrowed petiole. (*Echeveria*) – (**4**) for the slender corolla lobes. (*Brachystelma*) – (**5**) for the narrowly filiform staminodes. (*Lampranthus*)
tenuisectus Lat. 'tenuis', thin, slender; and Lat. 'sectus', divided; for the narrow leaf divisions. (*Ibervillea*)
tenuiserpens Lat. 'tenuis', thin, slender; and Lat. 'serpens', creeping; for the slender creeping stems. (*Cleistocactus*)
tenuispinosus Lat. 'tenuis', thin, slender; and Lat. 'spinosus', spiny, thorny. (*Euphorbia*)
tenuispinus Lat. 'tenuis', thin, slender; and Lat. '-spinus', -spined. (*Euphorbia milii* var.)
tenuissimus Lat., thinnest, narrowest (Superl. of Lat. 'tenuis', thin, slender); (**1**) for the small plant bodies and delicate spines. (*Copiapoa humilis* ssp.) – (**2**) for the slender stems. (*Corallocarpus*) – (**3**) for the leaves. (*Talinum*)
tenuistylus Lat. 'tenuis', thin, slender; and Lat. 'stylus', style. (*Yucca*)
tepalcatepecanus For the occurrence in the drainage of the Río Tepalcatepec, Michoacán, Mexico. (*Peniocereus*)
tepamo From the local vernacular name of the plants in Michoacán, Mexico. (*Pachycereus*)
tepexicensis Nahuátl 'tepexic', from the cliff, cliff; for the preferred habitat. (*Mammillaria*)
tephracanthus Gr. 'tephros', ash-grey; and Gr. 'akantha', thorn, spine; for the spine colour. (*Harrisia*)
Tephrocactus Gr. 'tephros', ash-grey; and Lat. 'cactus', cactus; for the spine colour of some taxa. (*Cactaceae*)
tepoxtlanus For the occurrence near Tepoxtlán, Morelos, Mexico. (*Mammillaria spinosissima* ssp.)
tequilanus For the town of Tequila (Jalisco, Mexico), the centre of tequila production. (*Agave*)
terebinthinus Lat., pertaining to *Pistacia terebinthus* ("Terebinth Tree", *Anacardiaceae*); for the turpentine-scent of the leaves. (*Peperomia*)
teres Lat., terete, oblong and round in cross-section; for the stem segments. (*Rhipsalis*)

teretifolius Lat. 'teres, teretis', terete, oblong and round in cross-section; and Lat. '-folius', -leaved. (*Acrosanthes, Calamophyllum, Echeveria, Hereroa, Portulaca, Talinum*)

teretiusculus Dim. of Lat. 'teres, teretis', terete, i. e. a little terete, somewhat terete; for the leaves. (*Calamophyllum*)

ternato-multifidus Lat. 'ternatus', in threes; Lat. 'multi-', many; and Lat. '-fidus', -cleft; for the ternate leaves with deeply trifid leaflets. (*Cyphostemma*)

ternatus Lat., in threes; (**1**) for the whorled leaves. (*Sedum*) – (**2**) for the 3-foliolate leaves. (*Cyphostemma*)

ternifolius Lat. 'terni', three each; and Lat. '-folius', -leaved; for the 3-foliolate leaves. (*Pelargonium*)

terrae-canyonae For the occurrence in the Canyon Lands National Park, Utah, USA (Lat. 'terrae', of the land). (*Sclerocactus parviflorus* ssp.)

terrae-reginae Lat. translation of Queensland, where the taxon is native. (*Portulaca*)

terscheckii For Mr. Terscheck (fl. 1837), court gardener at the garden "Japanese Palais" at Dresden. (*Echinopsis*)

terweemeanus For Mr. Ter Weeme (fl. 1930), Dutch cactus hobbyist in Neede, customer in the De Laet nursery. (*Gymnocalycium*)

tesajo From the local vernacular name of the plants in Baja California, Mexico. (*Cylindropuntia*)

tescorum Gen. Pl. of Lat. 'tesca', desert, wild region; for the occurrence in desert regions. (*Euphorbia*)

tessellatus Lat., tessellate, checkered; (**1**) for the stem surface. (*Cleistocactus fieldianus* ssp., *Notechidnopsis*) – (**2**) for the leaf patterning. (*Haworthia venosa* ssp.)

tessmannii For G. Tessmann (*1884), German explorer and plant collector. (*Nematanthus*)

testaceus Lat., brick-red; for the flower colour. (*Delosperma, Ruschia*)

testicularis Lat., testiculate, with testicles; for the pairs of rounded leaves. (*Argyroderma*)

testudo Lat., tortoise, turtle; for the growth with segmented stems appressed to tree trunks, likened to a series of green tortoises following each other. (*Selenicereus*)

tetensis For the occurrence at Tete, Moçambique. (*Plectranthus*)

tetetzo From the local vernacular name of the plants in Mexico. (*Neobuxbaumia*)

Tetilla Dim. of Span. 'teta', nipple, udder; for the thickened clavate petiole bases; and perhaps the local vernacular name of the plant. (*Saxifragaceae*)

tetracanthoides Gr. '-oides', resembling; and for the similarity to *Euphorbia tetracantha*. (*Euphorbia*)

tetracanthus Gr. 'tetra-', four-; and Gr. 'akantha', spine, thorn; (**1**) for the four spines per areole. (*Cylindropuntia*) – (**2**) for the four spines (two regular, two stipular) per spine shield. (*Euphorbia*)

tetractinus Gr. 'tetra-', four-; and Gr. 'aktis, aktinos', ray; for the four-merous flowers. (*Sedum*)

Tetradenia Gr. 'tetra', four; and Gr. 'aden', gland; perhaps for the four-lobed ovary. (*Lamiaceae*)

Tetragonia Gr. 'tetra-', four-; and Gr. 'gonia', angle, corner; for the four-angled fruits of many species. (*Aizoaceae*)

tetragonus Gr. 'tetra-', four-; and Gr. 'gonia', angle, corner; (**1**) for the four-angled branches. (*Euphorbia, Pelargonium, Prenia*) – (**2**) for the often four-angled stems. (*Acanthocereus*) – (**3**) for the leaf arrangement. (*Crassula*) – (**4**) for the almost square shape of the fused leaf pair when viewed from above. (*Conophytum angelicae* ssp.)

tetramerus Gr., four-merous; for the flowers. (*Crassula sieberiana* ssp., *Kalanchoe*)

tetrancistrus Gr. 'tetra', four; and Gr. 'ankistron', hook; for the number of hooked central spines. (*Mammillaria*)

tetraphyllus Gr. 'tetra', four; and Gr. 'phyllon', leaf. (*Bulbine, Kalanchoe, Peperomia*)

tetrasepalus Gr. 'tetra', four; and Gr. 'sepalon', sepal. (*Octopoma*)

tetrastichus Gr. 'tetra', four; and Gr. 'stichos', row; for leaf arrangement. (*Cephalophyllum*)

tewoldei For Tewolde-Berhan Gebre-Egziabher (fl. 1997), Ethiopian botanist and one of the joint leaders of the Ethiopian Flora Project. (*Aloe*)

texanus For the occurrence in Texas, USA. (*Lenophyllum*, *Mammillaria prolifera* ssp., *Nolina*)

texensis As above. (*Echinocactus*)

thailandicus For the occurrence in Thailand. (*Hoya*)

thalassoscopicus Gr. 'thalassa', the sea; and Gr. 'skopos', a watcher; for the habitat of the taxon facing the sea. (*Plectranthus*)

theartii For Major Ian Theart (fl. 1997), Infantry Major in RSA and succulent plant student. (*Argyroderma*)

thekii From the Fijian name 'theke theke nkau' for tuber-forming ant-plants, literally meaning 'testicles of the trees'. (*Squamellaria*)

thelegonoides Gr. '-oides', resembling; and for *Cereus* (*Trichocereus* / *Echinopsis*) *thelegonus*. (*Echinopsis*)

thelegonus Gr. 'thele', tubercle; and Gr. 'gonia', edge; for the tuberculate ribs of the stems. (*Echinopsis*)

Thelocactus Gr. 'thele', tubercle; and Lat. 'cactus', cactus; for the tuberculate plant bodies. (*Cactaceae*)

theresae For Theresa Bock (fl. 1967), wife of John Bock, US-American cactus collector in Sharon, Pennsylvania. (*Mammillaria*)

thermarum Gen. of Lat. 'thermae', thermal waters, spa; for the occurrence at Warmbad, Western Cape, RSA. (*Oscularia*)

thinophilus Gr. 'thinos', sand dune; and Gr. 'philos', friend; for the habitat. (*Euphorbia*, *Nolana*)

thionanthus Gr. 'theion', sulphur; and Gr. 'anthos', flower; for the flower colour. (*Acanthocalycium*)

tholicola Lat. 'tholus', dome; and Lat. '-cola', -dwelling; for the habitat on rock-domes. (*Euphorbia*)

thomasae For Señora Dora Thomas (fl. 1915), owner of Hotel Thomas in Guatemala where the taxon was found cultivated. (*Agave*)

thomasiae For Vicky Thomas (fl. 2003), South African botanical artist. (*Bulbine*)

thomasianus For Fritz Thomas (fl. 1893), accountant at the "Reichsbank" and German cactus hobbyist. (*Epiphyllum*)

Thompsonella For Prof. Charles H. Thompson (1870–1931), US-American botanist at the Missouri Botanical Garden and later at the Massachusetts College. (*Crassulaceae*)

thompsoniae For Dr. (Mrs.) Sheila Clifford Thompson (fl. ± 1930) of Haenertsburg, Northern Prov., RSA, mother of Louis Clifford Thompson. (*Aloe*)

thompsonii For Robert (Bob) Thompson (fl. 1985), US-American collector and botanical enthusiast at the US Forest Service in Price, Utah. (*Talinum*)

thompsoniorum For Louis Clifford Thompson (1920–1997) and his wife Eva Horstmeier, farmers and mountain enthusiast at Haenertsburg, Northern Cape, RSA, son of Dr. Sheila Clifford Thompson. (*Tavaresia*)

thomsonii For Dr. Thomas Thomson (1817–1878), British physician and botanist, 1854–1861 superintendent of the Calcutta Botanic Garden. (*Hoya*)

thornberi For Prof. John J. Thornber (1872–1962), US-American botanist at the University of Arizona. (*Cylindropuntia acanthocarpa* var., *Mammillaria*)

Thorncroftia For George Thorncroft (1874–1934), English trader and plant collector, emigrated to RSA in 1882. (*Lamiaceae*)

thorncroftii As above. (*Aloe*, *Thorncroftia*)

thouarsianus For Louis Marie Aubert du Petit Thouars (1758–1831), French botanist exploring in Madagascar before 1802. (*Euphorbia*)

thouarsii For Abel Du Petit-Thouars, French mariner and captain of the "Vénus" that 1836–1839 made a world tour and 1838 visited the Galápagos Islands. (*Jasminocereus*)

thraskii For a Mr. Thrask (fl. 1880), without further data. (*Aloe*)

thudichumii For Jacques Thudichum (1893–?), Swiss-born horticulturalist emigrating to RSA and curator of the Karoo Botanic Garden 1945–1958. (*Deilanthe*, *Drosanthemum*, *Huernia*, *Tromotriche*)

thunbergianus For Prof. Dr. Carl P. Thunberg (1743–1828), Swedish botanist and physician at Uppsala, collected at the Cape 1772–1774. (*Crassula*)

thunbergii As above. (*Brachystelma praelon*-

gum ssp., *Gasteria carinata* var., *Malephora*, *Sarcostemma viminale* ssp.)

thurberi For George Thurber (1821–1890), US-American botanical collector participating in the "Boundary Survey". (*Cylindropuntia, Stenocereus*)

thuretii For Gustave Adolphe Thuret (1817–1875), who established a botanical garden on the French Riviera. (*Huernia*)

thyrsiflorus Lat. 'thyrsus', thyrse; and Lat. '-florus', -flowered; for the inflorescences. (*Crassula capitella* ssp., *Cussonia, Kalanchoe, Orostachys*)

tianmushanensis For the occurrence in the Tianmushan Distr., Zhejiang, China. (*Sedum*)

tibeticus For the occurrence in Tibet. (*Rhodiola*)

tiburonensis For the occurrence on Isla Tiburón in the Gulf of California off the coast of Baja California, Mexico. (*Ferocactus*)

ticnamarensis For the occurrence near Ticnamar, Dept. Tarapacá, N Chile. (*Cumulopuntia*)

tidmarshii For Edwin Tidmarsh (1831–1915), British horticulturist and curator of the Grahamstown Botanical Garden, RSA. (*Aloe ciliaris* var.)

tiegelianus For Ernst Tiegel († 1936), German cactus hobbyist and *Mammillaria* specialist in Duisburg. (*Echinopsis*)

tigrinus Lat., tiger-like; for the strongly toothed leaves that resemble tiger jaws. (*Faucaria*)

tihamanus For the occurrence in the Tihama, i. e. on the arid Red Sea coastal plains in SW Saudi Arabia and Yemen. (*Ceropegia*)

tilcarensis For the occurence near Tilcara, Prov. Jujuy, Argentina. (*Gymnocalycium saglionis* ssp., *Parodia*)

tillaea For the former genus *Tillaea* (for Michelangelo Tilli [1653–1740], Italian botanist), where the taxon was originally described. (*Crassula*)

tillianus For Hans Till (*1920), Austrian horticulturist, cactus hobbyist and *Gymnocalycium* specialist, one of the founders and at a time chairman of the Austrian "Arbeitsgruppe Gymnocalycium". (*Gymnocalycium*)

tingoensis For the occurrrence near Tingo, Dept. Arequipa, Peru. (*Portulaca*)

Tinospora Gr. 'teinos', stretched; and Gr. 'spora', seed; for the elongate fruits (Jackson 1990). (*Menispermaceae*)

tiraquensis For the occurrence near Tiraque, Prov. Carrasco, Dept. Cochabamba, Bolivia. (*Sulcorebutia*)

tirucalli From the local Indian Malayalam name of the plant, from 'tiru', good; and 'kalli', any euphorbia; for the medical use. (*Euphorbia*)

tisserantii For the missionary and plant collector P. C. Tisserant (fl. 1950). (*Adenia*)

Titanopsis Gr. 'titanos', chalk, gypsum; and Gr. '-opsis', resembling; because the exposed leaf tips resemble the surrounding calcareous stones. (*Aizoaceae*)

titanopsis For the resemblance to species of the genus *Titanopsis* (*Aizoaceae*). (*Crassula ausensis* ssp.)

titanopsoides Gr. '-oides', resembling; and for the genus *Titanopsis* (*Aizoaceae*). (*Eriospermum*)

titanotus Gr. 'titanos', chalk, gypsum; for the alabaster-white leaves. (*Agave*)

tithymaloides Gr. '-oides', resembling; and for the former genus *Tithymalus* (*Euphorbiaceae*). (*Pedilanthus*)

tlalocii For the Aztek deity Tlaloc, God of the Rain; in allusion to the Mexican origin. (*Mammillaria crucigera* ssp.)

tobarensis For the occurrence near Tobar, Durango, Mexico. (*Echeveria*)

tobuschii For Hermann Tobusch (fl. 1952), US-American watchmaker and jeweller in Chicago, skilled cultivator of rare and unusual succulents. (*Sclerocactus brevihamatus* ssp.)

tocopillanus For the occurrence near Tocopilla, N Chile. (*Copiapoa*)

toftiae For Catherine A. Toft (fl. 1975), US-American botanist. (*Yucca angustissima* var.)

togoensis For the occurrence in Togo. (*Brachystelma*)

toliari For the occurrence near Toliara, Madagascar. (*Cynanchum*)

tolimanensis For the occurrence in the Barranca de Tolimán, Hidalgo, Mexico. (*Echeveria*)

tolucensis For the occurrence near Toluca, México, Mexico. (*Echeveria*)

tomasi Unknown. (*Conophytum*)

tomentellus Dim. of Lat. 'tomentum', felty matter; (**1**) for the minutely pubescent stem segments. (*Opuntia*) – (**2**) for the minute pubescence on the leaves. (*Jatropha seineri* var., *Pedilanthus*)

tomentosus Lat., covered in matted hairs, felted; (**1**) for the general appearance. (*Anacampseros filamentosa* ssp., *Cotyledon, Didelta carnosa* var., *Kalanchoe, Plectranthus hadiensis* var.) – (**2**) for the tomentose stem segments. (*Opuntia*) – (**3**) for the hairy young shoots and rachis. (*Delonix*) – (**4**) for the hairy leaves. (*Crassula*) – (**5**) for the tomentose lower leaf face. (*Gerrardanthus*) – (**6**) for the dense cobwebby tomentum of the rosettes. (*Sempervivum arachnoideum* ssp.) – (**7**) for the hairy flowers. (*Aloe*) – (**8**) for the hairy corolla lobes. (*Ceropegia*)

tominensis For the occurrence near Tomina, Dept. Chuquisaca, Bolivia. (*Cleistocactus*)

tonalensis For the occurrence at Puente de Tonala, Oaxaca, Mexico. (*Mammillaria*)

tonduzii For Adolphe Tonduz (1862–1921), Swiss botanist and plant collector, 1889–1920 in Costa Rica, 1920–1921 in Guatemala. (*Weberocereus*)

tongaensis For the occurrence in the Tongoland along the coast in N KwaZulu-Natal, RSA. (*Pelargonium, Plectranthus purpuratus* ssp.)

toonensis For the occurrence near Matjestoon (abbreviated to Toon), Western Cape, RSA. (*Haworthia heidelbergensis* var.)

torataensis For the occurrence near Torata, Dept. Moquegua, Peru. (*Weberbauerocereus*)

tormentorii Lat. 'tormentum', catapult, gun, cannon; from the type locality Gunner's Quoin. (*Aloe*)

tororoanus For the occurrence at Tororo Rock, Uganda. (*Aloe*)

torquatus Lat., twisted; application obscure. (*Sedum tsiangii* var.)

torrei For Antonio Rocha da Torre (*1904), Portuguese botanist. (*Aloe, Monadenium*)

torreyi For Prof. Dr. John Torrey (1796–1873), US-American botanist, chemist and physician. (*Yucca*)

torsivus Lat., spirally twisted; for the upper leaf parts. (*Bulbine*)

tortilis Lat., twisted; (**1**) for the branches. (*Euphorbia*) – (**2**) for the inflorescences. (*Trachyandra*)

tortiramus Lat. 'tortus', twisted; and Lat. 'ramus', branch. (*Euphorbia cactus* var., *Euphorbia*)

tortispinus Lat. 'tortus', twisted; and Lat. '-spinus', -spined. (*Cumulopuntia*)

tortistylus Lat. 'tortus', twisted; and Lat. 'stylus', style. (*Euphorbia*)

tortulispinus Lat. 'tortula', pretzel (Dim. of Lat. 'tortus', contorted, twisted); and Lat. '-spinus', -spined; for the spination. (*Ferocactus cylindraceus* ssp.)

tortuosus Lat., bent or twisted in various directions; (**1**) for the growth habit. (*Aichryson, Harrisia, Kleinia, Sceletium*) – (**2**) for the flowering branches. (*Sedum*)

tortus Lat., twisted; (**1**) for the twisted appearance of the stems of the holotype (not apparent in living material of other collections). (*Euphorbia*) – (**2**) for the leaves. (*Bulbine*) – (**3**) for the contorted corolla tips. (*Rhytidocaulon*)

torulosus Lat., torulose, cylindrical with bulges or contractions at intervals; for the stems. (*Crassula ericoides* ssp., *Pelargonium, Sedum, Tylecodon*)

tosaensis For the occurrence in the Tosa Distr., Japan. (*Sedum*)

totolapensis For the occurrence near Totolapan on the Isthmus of Tehuantepec, Oaxaca, Mexico. (*Cephalocereus*)

totoralensis For the occurrence near Totoral Bajo, Chile. (*Eriosyce crispa* ssp.)

totorensis For the occurrence near Totora, Prov. Totora, Dept. Cochabamba, Bolivia. (*Echinopsis tarijensis* ssp.)

toumeyanus For Prof. Dr. James W. Toumey (1865–1932), US-American forestry botanist. (*Agave*)

tovarii For Prof. Dr. Óscar Tovar Serpa (*1923), Peruvian botanist. (*Cistanthe*)

townsendianus For Dr. Charles H. Townsend (1859–1944), US-American naturalist, 1902–1937 director of the New York Aquarium, leader of the expedition of 1911 to Baja California. (*Ferocactus*)

toxotis Gr. 'toxotis', of the bowman; for Joseph Archer (1871–1954), Englishman, emigrated to RSA in 1890 as railway worker, became station master at Matjiesfontein, and succulent plant collector, 1921–1939 curator of Karoo Garden, Whitehill, RSA. (*Senecio*)

Trachyandra Gr. 'trachys', rough; and Gr. 'aner, andros', male; for the rough filaments. (*Asphodelaceae*)

Trachycalymma Gr. 'trachys', rough; and Gr. 'calymma', covering; perhaps for the ± coarse hairiness of the plants. (*Asclepiadaceae*)

trachyticola Engl. / Fr. 'trachyte', trachyte rock; and Lat. '-cola', -inhabiting; for the habitat. (*Aloe*)

Tradescantia For John Tradescant (± 1570–1638), gardener to Charles I of England. (*Commelinaceae*)

tradescantioides Gr. '-oides', resembling; and for the genus *Tradescantia* (*Commelinaceae*). (*Delosperma*)

transcaucasicus For the occurrence in the Transcaucasus, Georgia. (*Sempervivum*)

transiens Lat., passing over into; for the intermediate nature of the taxon. (*Haworthia cymbiformis* var.)

transitensis For the occurrence near El Transito in the valley of the Río Huasco E of Vallenar, Chile. (*Eriosyce kunzei* var.)

transmontanus Lat. 'trans-', across; und Lat. 'montanus', mountain-; i.e. across the mountains; for the distribution range. (*Peniocereus greggii* var.)

transvaalensis For the occurrence in the former Transvaal (now Gauteng and N-ward adjacent provinces), RSA. (*Crassula lanceolata* ssp., *Euphorbia*, *Huernia*, *Ipomoea*)

traskiae For Mrs. Blanche Trask († 1916), US-American naturalist in California. (*Dudleya*)

treculeanus For Auguste A. Trécul (1818–1896), French botanist and pharmacist, travelled in North America 1848–1850, later at the Muséum d'Histoire Naturelle in Paris. (*Yucca*)

treleasei For Prof. Dr. William Trelease (1857–1945), US-American botanist, Engelmann Professor at the Shaw School of Botany, and Agavaceae specialist. (*Agave schottii* var., *Opuntia basilaris* var., *Sedum*, *Stenocereus*)

triacanthus Gr. 'tri-', three-; and Gr. 'akantha', thorn, spine; for the spines arranged in threes. (*Opuntia*)

triactinus Gr. 'tri-', three-; and Gr. 'aktis, aktinos', ray; for the normally 3 carpels. (*Sedum*)

triaculeatus Lat. 'tri-', three-; and Lat. 'aculeatus', spiny; for the single spine accompanied by two stipular spines. (*Euphorbia*)

triandrus Gr. 'tri-', three-; and Gr. 'aner, andros', man, [botany] stamen; for the three fertile stamens. (*Pelargonium*)

triangularis Lat., triangular; (**1**) for the three-angled branches. (*Euphorbia*, *Hylocereus*) – (**2**) for the leaf shape. (*Agave*) – (**3**) for the peduncle. (*Talinum*)

Trianthema Gr. 'tri-', three-; and Gr. 'anthemon', flower; for the groups of three flowers in some species. (*Aizoaceae*)

trianthemoides Gr. '-oides', resembling; and for the genus *Trianthema* (*Aizoaceae*). (*Sesuvium*)

trianthinus Gr. 'tri-', three-; and Gr. 'anthos', flower; perhaps for the three stages of anthesis (suggested by V. Reiter). (*Echeveria*)

tribblei For Derek V. Tribble (*1952), English computer software applications engineer, enthusiast of South African leaf succulents, who discovered this and other new Tylecodons on one of his many field trips to RSA. (*Tylecodon*)

tribracteatus Lat. 'tri-', three-; and Lat. 'bracteatus', bracteate; for the three-flowered inflorescences and the fact that all pedicels are bracteate. (*Ruschia*)

Tribulocarpus From the genus *Tribulus* (*Zygophyllaceae*); and Gr. 'karpos', fruit; for the similarly spiny fruits. (*Aizoaceae*)

tricae Lat. 'tricae', a tangle of difficulties, nonsense; for the ambiguous relationships of the plants. (*Selenicereus*)

tricarpus Gr. 'tri-', three-; and Gr. 'karpos', fruit. (*Sedum*)

trichadenia Gr. 'trichos', hair; and Gr. 'aden', gland; for the slender hair-like processes of the nectary glands. (*Euphorbia*)

trichanthus Gr. 'trichos', hair; and Gr. 'anthos', flower; for the thread-like corolla lobes. (*Ceropegia*)

Trichodiadema Gr. 'trichos', hair; and Gr. 'diadema', crown; for the tuft of hairs or bristles on the leaf tips. (*Aizoaceae*)

trichophorus Gr. 'trichos', hair; and Gr. '-phoros', carrying; for the curly hair-like spination of the stems. (*Weberocereus*)

trichosanthus Gr. 'trichos', hair; and Gr. 'anthos', flower; for the hairy perianth. (*Aloe*)

trichospermus Gr. 'trichos', hair; and Gr. 'sperma', seed; for the long papillae of the seeds. (*Sedum*)

trichosus Lat., full of hair, hairy; for the densely hair-covered pericarpel. (*Echinopsis*)

trichotomus Gr., three-parted, branched in threes; application obscure. (*Phyllobolus*)

trichromus Gr. 'tri-', three; and Gr. 'chromos', colour; for the three-coloured petals. (*Sedum*)

tricolorus Lat., three-coloured; for the flowers with yellow and red petals, reddish filaments and brown anthers. (*Cephalophyllum*)

tridentatus Lat. 'tri-', three-; and Lat. 'dentatus', toothed; (**1**) for the corona segments. (*Schizoglossum atropurpureum* ssp.) – (**2**) for the processes on the cyathial glands. (*Euphorbia*) – (**3**) application obscure. (*Stathmostelma fornicatum* ssp.)

Tridentea Lat. 'tri-', three-; and Lat. 'dens, dentis', tooth; for the frequently three-toothed intrastaminal corona segments. (*Asclepiadaceae*)

triebneri For Wilhelm Triebner (1883–1957), German horticulturist, went to the then German Southwest-Africa (now Namibia) in 1904 for military service, stayed there as gardener and farmer, established a succulent plant nursery near Windhoek in 1930, collected plants for Jacobsen, von Poellnitz and others. (*Hoodia, Schwantesia*)

triebnerianus As above. (*Haworthia mirabilis* var.)

trifarius MLat., triple, three-ranked; for the three rows of teeth on the leaves. (*Stomatium*)

trifasciatus Lat. 'tri-', three-; and Lat. 'fasciatus', banded; for the leaf markings. (*Sansevieria*)

trifidus Lat. 'tri-', three-; and Lat. '-fidus', -divided; for the segmented leaves. (*Pelargonium*)

triflorus Lat. 'tri-', three; and Lat. '-florus', -flowered. (*Adromischus, Ruschia*)

trifoliolatus Lat. 'tri-', three; and Lat. 'foliolatus', -foliolate. (*Adenia fruticosa* ssp., *Pelargonium*)

triglochidiatus Gr. 'tri-', three; Gr. 'glochis, glochidos', arrow-head, arista; for the frequently triple main spines. (*Echinocereus*)

trigonanthus Gr. 'trigonos', triangular; and Gr. 'anthos', flower. (*Aloe*)

trigonus Gr. 'trigonos', triangular; (**1**) for the stems. (*Euphorbia, Hylocereus, Rhipsalis*) – (**2**) for the leaf shape. (*Cerochlamys*) – (**3**) for the tubercles of the plant body. (*Ariocarpus retusus* ssp.)

trigynus Gr. 'tri-', three; and Gr. 'gyne', female organ, carpel, ovary; application obscure and perhaps an error of observation. (*Adromischus*)

trilobatus Lat. 'tri-', three; and Lat. 'lobatus', lobed; for the usual leaf shape. (*Coccinia*)

trinervis Lat. 'tri-', three; and Lat. '-nervis', -nerved; for the prominent veins of the leaves. (*Monadenium*)

tripalmatus Lat. 'tri-', three; and Lat. 'palmatus', palmately compound; for the leaf shape. (*Pelargonium*)

tripartitus Lat. 'tri-', three; and Lat. 'partitus', partite; i.e. with three parts, for the three-foliolate leaves. (*Zygosicyos*)

triphyllus Lat. 'tri-', three; and Lat. '-phyllus', -leaved; (**1**) for the leaves which are sometimes in whorls of three. (*Lewisia*) – (**2**) for the leaves with three leaflets. (*Pelargonium*)

Tripogandra Gr. 'tri-', three; Gr. 'pogon', beard; and Gr. 'aner, andros', man, anther; because the type species has three bearded and three glabrous stamens. (*Commelinaceae*)

tripolium Gr. / Lat. 'tripolion', a plant growing on rocky cliffs, also used as epithet for many small herbaceous plants, esp. also for *Aster tripolium* ("Sea Aster", *Asteraceae*); perhaps for the superficially similar flowers. (*Skiatophytum*)

tripugionacanthus From Lat. 'tri-', three; Lat. 'pugio' (Gen. 'pugionis'), dagger; and Gr. 'akantha', spine, thorn; for the central spines. (*Coryphantha*)

triqueter Lat.,, three-edged; for the leaves. (*Antimima, Trianthema*)

trisectus Lat. 'tri-', three; and Lat. 'sectus', cut, incised; for the three-lobed or -foliolate leaves. (*Adenia*)

tristis Lat., sad, dull-coloured; for the flower colour. (*Pelargonium*)

tristriatus Lat. 'tri-', three; and Lat. 'striatus', striped; application obscure. (*Sedum*)

tritelii For a Mr. Tritel (fl. 1913), without further data. (*Sedum*)

triticiformis Lat. 'triticum', wheat; and Lat. '-formis', -shaped; application obscure. (*Ruschia cradockensis* ssp.)

Trochomeria Gr. 'trochos', wheel, spreading wheel-like; and Gr. 'meros', part; possibly for the spreading corolla lobes. (*Cucurbitaceae*)

Trochomeriopsis Gr. '-opsis', similar to; and for the genus *Trochomeria* (*Cucurbitaceae*). (*Cucurbitaceae*)

trollii For Prof. Dr. Wilhelm Troll (1897–1978), German botanist and authority on plant morphology. (*Didierea, Peperomia*) – (**2**) For Prof. Dr. Carl Troll (1899–1975), German geographer and botanist in Munich and later in Bonn, Germany, younger brother of the German botanist Wilhelm Troll. (*Oreocereus*)

Tromotriche Gr. 'tromos', trembling; and Gr. 'thrix, trichos', hair; for the vibratile corolla hairs of some taxa. (*Asclepiadaceae*)

tropaeolifolius For the genus *Tropaeolum* ("Garden Nasturtium", *Tropaeolaceae*); and Lat. '-folius', -leaved. (*Dorstenia barnimiana* var., *Jatropha, Senecio oxyriifolius* ssp., *Umbilicus*)

trullipetalus Lat. 'trulla', bricklayer's trowel; and Lat. 'petalum', petal; for the petal shape. (*Sedum*)

truncatus Lat., truncate; (**1**) for the growth form resulting in flat-topped cushions. (*Euphorbia clavarioides* var.) – (**2**) for the truncate stem segments. (*Schlumbergera*) – (**3**) for the distinctly truncate leaves. (*Agave parryi* var., *Bulbine, Conophytum, Euphorbia beharensis* var., *Haworthia cooperi* var., *Haworthia*) – (**4**) for the truncate intrastaminal parts of the corona. (*Tridentea parvipuncta* ssp.) – (**5**) application obscure. (*Dischidia*)

trunciformis Lat. 'truncus', trunk; and Lat. '-formis', -shaped. (*Euphorbia breviarticulata* var.)

truteri For J. Truter (fl. 1961), farmer at Brakfontein, Eastern Cape, RSA. (*Delosperma, Ruschia*)

tsangii For Peter Tsang (fl. 1988?), without further data. (*Hoya*)

tsavoensis For the occurrence in the Tsavo National Park, Kenya. (*Euphorbia heterochroma* ssp.)

tsiangii For Ying Tsiang (1898–1982), Chinese botanist. (*Sedum*)

tsimbazazae For the occurrence in the Tsimbazaza Botanical Garden, Antananarivo, Madagascar. (*Euphorbia, Euphorbia viguieri* var.)

tsinghaicus For the occurrence in the Tsinghai Prov. (today Quinghai), China. (*Sedum*)

tsomoensis For the occurrence at Tsomo, Transkei, RSA. (*Stapelia*)

tsugaruensis For the occurrence in Prov. Tsugaru, Honshu, Japan. (*Hylotelephium ussuriense* var.)

tuberculatoides Gr. '-oides', resembling; and for *Euphorbia tuberculata*. (*Euphorbia*)

tuberculatus Lat., tuberculate; (**1**) for the tuberculate stems. (*Euphorbia, Kleinia*) – (**2**) for the papillose stems. (*Sedum*) – (**3**) for the tuberculate ribs of the plant body. (*Matucana, Parodia, Pilosocereus*) – (**4**) for the papillate-tuberculate corolla. (*Caralluma*) – (**5**) for the tuberculate seeds. (*Portulaca*)

tuberculifer Lat. 'tuberculus', small tuber (Dim. of Lat. 'tuber', tuber, swelling); and Lat. '-fer, -fera, -ferum', -carrying; for the conspicuous closing bodies of the fruit capsules. (*Drosanthemum*)

tuberculosus Lat., tuberculate; (**1**) for the tuberculate plant bodies. (*Escobaria*) – (**2**) for the tuberculate leaves. (*Faucaria*) – (**3**) for the large closing bodies of the fruit capsules. (*Antimima*)

tuberellus Lat., small tuber (Dim. of Lat. 'tuber', tuber, swelling); for the root tubers. (*Crassula*)

tuberifer Lat. 'tuber', tuber, swelling; and Lat. '-fer, -fera, -ferum', -carrying; (**1**) for the tuberous rootstock. (*Adenia, Euphorbia cylindrifolia* ssp.) – (**2**) for the root tubers. (*Sedum*)

tuberisulcatus Lat. 'tuber', tuber, swelling; and Lat. 'sulcatus', furrowed; for the deep furrow separating the tubercles of the ribs. (*Eriosyce curvispina* ssp.)

tuberosus Lat., tuberous (from Lat. 'tuber', tuber, swelling); (**1**) for the caudex. (*Ceratosanthes, Cissus, Impatiens, Myrmecodia, Othonna, Senecio, Tylecodon*) – (**2**) for the large root tubers. (*Brachystelma, Euphorbia, Jatropha, Matelea, Mestoklema, Pachyrhizus, Pterocactus*) – (**3**) for the rhizomes. (*Furcraea, Oxalis, Sedum*) – (**4**) for the bulbous plant base. (*Kalanchoe*)

tubiflorus Lat. 'tubus', tube; and Lat. '-florus', -flowered. (*Beschorneria, Echinopsis*)

tubiformis Lat. 'tubus', tube; and Lat. '-formis', -shaped; for the flowers. (*Orbea*)

tubiglans Lat. 'tubus', tube; and Lat. 'glans', gland; for the tubular nectar glands. (*Euphorbia*)

tubulatus Lat., provided with a tube; for the rather deep flower tube. (*Agave*)

tuckeyanus For Captain James K. Tuckey (1776–1816), English naval officer who led the British expedition to explore the Zaïre River in 1816. (*Euphorbia*)

tucumanensis For the occurrence in Prov. Tucumán, Argentina. (*Rhipsalis floccosa* ssp.)

tugelensis For the occurrence by the Tugela River, KwaZulu-Natal, RSA. (*Euphorbia*)

tugenensis For the occurrence in the Tugen Hills, Kenya. (*Aloe*)

tugwelliae For Mrs. Anna M. Tugwell (fl. 1914–1929), plant collector in RSA and an "old friend" of the South African Mesemb specialist Louisa Bolus. (*Bijlia, Cylindrophyllum*)

tulbaghensis For the occurrence at Tulbagh, Western Cape, RSA. (*Lampranthus*)

tulearensis For the occurrence near Tuléar (Toliara), Madagascar. (*Euphorbia milii* var., *Euphorbia*)

tulensis For the occurrence near Tula, San Luis Potosí, Mexico. (*Echinocereus cinerascens* ssp., *Thelocactus*)

tulhuayacensis Presumably for the occurrence at a place called Tulhuayaca or similarly, Peru. (*Echinopsis*)

Tumamoca For Tumamoc Hill, the locality of the Desert Laboratory of the Carnegie Institution near Tucson, Arizona, USA. (*Cucurbitaceae*)

tumidulus Dim. of Lat. 'tumidus', swollen; for the slightly bulging leaf sheaths. (*Ruschia*)

tumidus Lat., swollen; (**1**) for the globose stem segments and the short broad fruits. (*Cumulopuntia*) – (**2**) probably for the globose fruits. (*Parakeelya*)

tuna Span. 'tuna', Prickly Pear (taken from the Arabic word for fig); for the fruits resembling figs. (*Opuntia*)

tunariensis For the occurrence on Cerro Tunari, Prov. Cercado, Dept. Cochabamba, Bolivia. (*Echinopsis*)

tunicatus Lat., tunicate, having a coat or envelope; for the loose-fitting sheaths covering the spines. (*Cylindropuntia*)

Tunilla Dim. of Span. 'tuna', Prickly Pear, in general species of the genus *Opuntia* (used for plant and fruits); for the small size of this *Opuntia*-relative. (*Cactaceae*)

tunilla Dim. of Sp. 'tuna', Prickly Pear, in general species of the genus *Opuntia* (used for plant and fruits); for the small oblong fruits; or perhaps from the local vernacular name "tunilla" (frequently used for different cacti). (*Weberocereus*)

tupizensis For the occurrence near Tupiza, Dept. Potosí, Bolivia. (*Cleistocactus*)

turbidus Lat., confused, disordered, disturbed; perhaps for the variable spination. (*Haageocereus pseudomelanostele* ssp.)

Turbina Lat., spinning object, spinning top; probably for the fruits. (*Convolvulaceae*)

turbinatus Lat., top-shaped; (**1**) for the shape

of the plant bodies. (*Parodia*) – (**2**) for the receptacle shape. (*Cheiridopsis*) – (**3**) for the fruit shape. (*Opuntia*) – (**4**) application obscure. (*Lampranthus*)

Turbinicarpus Lat. 'turbo, turbinis', top, spindle; and Gr. 'karpos', fruit; for the fruit shape. (*Cactaceae*)

turbiniformis Lat., top-shaped; (**1**) for the obconical plant body. (*Euphorbia*) – (**2**) for the shape of the fused leaf pair. (*Conophytum auriflorum* ssp., *Lithops*)

turecekianus For Victor Turecek (fl. 1995), US-American collector of Argentinian cacti in Los Angeles. (*Parodia*)

turgidifolius Lat. 'turgidus', turgid, swollen; and Lat. '-folius', -leaved. (*Trianthema*)

turgidus Lat., turgid, swollen; for the leaves. (*Echeveria*, *Haworthia*)

turicanus To honour the efforts of the City of Zürich (Lat. 'Turicum Helvetiorum'), Switzerland, for maintaining the Municipal Succulent Plant Collection since 1931. (*Uncarina*)

turkanensis For the occurrence in the Turkana Distr., Kenya. (*Aloe*, *Euphorbia*)

turneri For H. J. Allen Turner (1876–1953), British taxidermist at the Coryndon Museum, Nairobi, lived in Kenya from 1908, and active field collector who found many new species of animals and plants. (*Caralluma*)

turnerianus For Mr. V. A. Turner (fl. 1963), farmer in the Vanrhynsdorp region, RSA, on whose farm the type of the taxon was collected. (*Antimima*)

turriculus Dim. of Lat. 'turris', tower, turret; perhaps for the large solitary erect flowers. (*Ceropegia*)

turriger Lat. 'turris', tower, turret; and Lat. '-ger, -gera, -gerum', -carrying, bearing; for the shape of the free leaf lobes. (*Conophytum*)

turumiquirensis For the occurrence on Cerro Turumiquire, Sucre, Venezuela. (*Echeveria bicolor* var.)

tuyensis For the occurrence near Capilla Tuya, Paraguay. (*Frailea cataphracta* ssp.)

tweediae For Mrs. E. Marjorie Tweedie (fl. 1942), British artist and collector, resident in Kenya from ± 1918 onwards. (*Aloe*)

tweedyi For Frank Tweedy (1854–1937), US-American topographic engineer and amateur botanist. (*Cistanthe*)

Tylecodon Anagram of the genus name *Cotyledon* (*Crassulaceae*), where the species were formerly placed. (*Crassulaceae*)

Tylosema Gr. 'tylos', swelling; and Gr. 'sema', mark, distinguishing mark; perhaps for the tuberous rootstocks, which distinguish the genus from *Bauhinia*. (*Fabaceae*)

tymphaeum For the occurrence on Mt. Timfi, Greece. (*Prometheum*)

U

ubomboensis For the occurrence in the Ubombo Mts., KwaZulu-Natal, RSA. (*Haworthia limifolia* var., *Orbea*)

Uebelmannia For Mr. Werner Uebelmann (*1921), Swiss cactus horticulturist and expert on Brazilian cacti. (*Cactaceae*)

uebelmannianus As above. (*Gymnocalycium*)

uhligianus For Dr. C. Uhlig, German naturalist who collected 1901–1904 in Maasailand, Kenya and Tanzania, and in the Kilimanjaro Distr., Tanzania. (*Euphorbia*)

uhlii For Charles H. Uhl (*1918), US-American botanist at Cornell University and specialist in the cytology of *Crassulaceae*. (*Echeveria*)

uitenhagensis For the occurrence near Uitenhage, Eastern Cape, RSA. (*Delosperma, Malephora, Ruschia*)

ukambensis For the occurrence in the former Ukambani Distr. (now divided into Kitui Distr. and Machakos Distr.), Kenya. (*Aloe, Caralluma turneri* ssp.)

ulei For Ernst H. G. Ule (1854–1915), German botanist and botanical explorer of Brazil. (*Facheiroa, Pilosocereus*)

ulricae Unknown. (*Sedum*)

umadeave Unknown. (*Eriosyce*)

umbella Lat., parasol, sunshade, [botany] umbel; for the parasol-like fused leaves of a pair. (*Crassula*)

umbellatus Lat., umbellate; (**1**) for the umbellate inflorescence. (*Caralluma, Cistanthe, Crassula*) – (**2**) for the well-branched inflorescence. (*Ruschia*)

umbilicatus Lat., provided with a navel, navel-like; for the concave peltate leaves. (*Peperomia*)

Umbilicus Lat., navel; for the central navel-like depression of the peltate leaves. (*Crassulaceae*)

umbonatus Lat., with a navel-like tubercle; for the rounded spine shields. (*Euphorbia*)

umbracula Lat. 'umbraculum', sunshade, parasol; for the flower shape. (*Orbea*)

umbraticola Lat. 'umbra', shade; and Lat. '-cola', inhabiting. (*Adromischus, Ceropegia, Crassula, Portulaca*)

umbrosus Lat., full of shade, dark; for the preference to grow in shaded rock crevices. (*Cheiridopsis*)

umdausensis For the occurrence at Umdaus, Northern Cape, RSA. (*Cheiridopsis, Tromotriche*)

umfoloziensis For the occurrence in the Umfolozi River valley, KwaZulu-Natal, RSA. (*Aloe, Euphorbia*)

Uncarina Lat. 'uncus', hook, barb; for the barbed spines on the fruits. (*Pedaliaceae*)

uncinatus Lat., uncinate, barbed, hooked; (**1**) for the hooked central spines. (*Coryphantha robustispina* ssp., *Mammillaria, Sclerocactus*) – (**2**) for the recurved leaf tips. (*Delosperma, Ruschia*) – (**3**) for the hooked tips of the corona segments. (*Aspidoglossum*)

uncinulatus Dim. of Lat. 'uncinatus', uncinate, barbed, hooked; for the hooked hairs on the leaves and the hook-like teeth of the lobe margins. (*Jatropha*)

uncus Lat., hook, barb; for the recurved leaf tips. (*Lampranthus*)

undatus Lat., wavy, undate; for the margin of the ribs of the stems. (*Hylocereus*)

underwoodii For Dr. Lucien M. Underwood (1853–1907), US-American botanist. (*Agave*)

undulatifolius Lat. 'undulatus', undulate, wavy; and Lat. '-folius', -leaved; for the undulate leaf margins. (*Crassula arborescens* ssp., *Euphorbia*)

undulatus Lat., undulate, wavy; (**1**) for the nature of the stem segments. (*Opuntia*) – (**2**) for the leaves. (*Aeonium, Apodanthera, Pterodiscus*) – (**3**) for the leaf margins. (*Furcraea*) – (**4**) for the margins of the petals. (*Pelargonium*) – (**5**) for the margins of the seeds. (*Bulbine*)

undulosus From Lat. 'undula', small wave, i.e. undulate; for the ribs. (*Dendrocereus*)

unguentarius Lat. 'unguen, unguinis', ointment, pomade; or Lat. 'unguentarius', trader in pomades; because the roots are used locally as ingredient for a pomade. (*Plectranthus*)

unguiculatus Lat. 'unguiculatus', clawed, with a nail; for the leaf tip. (*Echeveria*)

unguispinus Lat. 'unguis', claw, nail; and Lat. '-spinus', -spined; for the spination. (*Cumulopuntia, Echinomastus, Epithelantha micromeris* ssp.)

unicornis Lat. 'uni-', one-; and Lat. '-cornis', horned; (**1**) for the solitary spine. (*Euphorbia*) – (**2**) for the single central spine. (*Coryphantha*) – (**3**) for the single inner corona lobe. (*Stapelia*)

unicostatus Lat. 'uni-', one-; and Lat. 'costatus', ribbed; for the prominent midrib of the leaves. (*Jatropha*)

unidens Lat. 'uni-', one-; and Lat. 'dens, dentis', tooth; for the single tooth on the leaf keel. (*Ruschia*)

uniflorus Lat. 'uni-', one- and Lat. '-florus', -flowered. (*Delosperma, Jordaaniella, Kalanchoe, Lampranthus, Leipoldtia, Parakeelya, Sedum*)

unifoliatus Lat. 'uni-', one-; and Lat. 'foliatus', leafy (from Lat. 'folium', leaf); for the single leaf. (*Ornithogalum*)

uniondalensis For the occurrence at Uniondale, Eastern Cape, RSA. (*Marlothistella*)

unispinus Lat. 'uni-', one-; and Lat. '-spinus', -spined; for the single spines. (*Euphorbia*)

uralensis For the occurrence in the Ural Mts., Russia. (*Hylotelephium*)

urbanianus For Prof. Dr. Ignatz Urban (1848–1931), German botanist at Berlin and specialist in the flora of the West Indies. (*Erepsia, Opuntia, Selenicereus*)

urbicus For the occurence near the town of Urbicum (= present-day La Laguna) on Tenerife, Canary Islands. (*Aeonium*)

urbionensis For the occurrence on the Pico de Urbión, Soria, Spain. (*Sempervivum cantabricum* ssp.)

urceolatus Lat., urn-shaped (from Lat. 'urceola', small jar); for the flower shape. (*Echidnopsis, Huernia*)

Urginea For the Beni Urgen tribe in Algeria where the type was collected (Jackson 1990). (*Hyacinthaceae*)

urniflorus Lat. 'urna', urn; and Lat. '-florus', -flowered; for the corolla shape. (*Stapeliopsis*)

ursi For Dr. Urs Eggli (*1959), Swiss botanist and succulent plant specialist in Zürich. (*Sedum*)

uruguayanus For the ocurrence in Uruguay. (*Cereus hildmannianus* ssp.)

uruguayensis As above. (*Gymnocalycium*)

urvillei For Jules S. C. D. d'Urville (1790–1842), French (?) botanist. (*Sedum*)

usambarensis For the occurrence in the Usambara Mts., Tanzania. (*Kalanchoe*)

ussanguensis For the occurrence in the Ussangu Region, Tanzania. (*Euphorbia cooperi* var.)

ussuriensis For the occurrence in the Ussuri region, E Siberia, Russia. (*Hylotelephium*)

ustulatus Lat., burnt, ashen; for the colour of the scales covering the leaves. (*Avonia*)

utahensis For the occurrence in Utah, USA. (*Agave, Yucca elata* var.)

utcubambensis For the occurrence near Utcubamba, Dept. Amazonas, Peru. (*Echeveria*) – (**2**) For the occurrence in the valley of the Río Utcubamba, Dept. Amazonas, Peru. (*Espostoa*)

uter From the vernacular name "odre" for the plants in Angola. (*Cyphostemma*)

utilis Lat., useful; (**1**) for the popular use as firewood. (*Stoeberia*) – (**2**) for the rubber produced from the tubers. (*Raphionacme*)

utkilio Unknown, perhaps from a local vernacular name of the plants in Argentina. (*Opuntia anacantha* var.)

uvifolius Lat. 'uva', grape; and Lat. '-folius', -leaved; for the globosely thickened leaves. (*Sesuvium*)

uviformis Lat. 'uva', grape; and Lat. '-formis', -shaped; for the appearance of the clustering plants. (*Conophytum*)

uwanda Kiswahili, a plain; for the occurrence on the Engusoro Plain, Tanzania. (*Cyphostemma*)

uyupampensis For the occurrence near Uyupampa, Dept. Arequipa, Peru. (*Echinopsis*)

uzmuk From the local vernacular name of the plants in Yemen. (*Euphorbia*)

V

vaalputsianus For the occurrence at Vaalputs, Western Cape, RSA. (*Euphorbia*)
vacillans Lat., variable, changeable. (*Aloe*)
vadensis Because the cross was made at the Institute of horticultural plant breeding at Wageningen (Lat. Vada), Holland. (*Kalanchoe*)
vaduliae For Vadulia Thühellengonn (fl. 1991), which is the pseudonym of a South African amateur botanist. (*Caralluma*)
vagans Lat., wandering; (**1**) for the straggling stems. (*Selenicereus*) – (**2**) for the straggling scapes. (*Bulbine*)
vaginatus Lat., sheathed; for the sheathing leaf bases. (*Crassula, Jacobsenia, Ruschia*)
vagus Lat., uncertain, without particular direction; for the arrangement of the flowers. (*Orbea lutea* ssp.)
vahrmeijeri For Mr. Johannes Vahrmeijer (*1942), Dutch-born economic botanist settling in RSA in 1950. (*Brachystelma*)
vaillantii Probably for Sébastien Vaillant (1669–1722), French botanist and physician. (*Crassula*)
vajravelui For Dr. E. Vajravelu (fl. 1993) of the Botanical Survey of India. (*Euphorbia*)
valdezianus Nach Mrs. L. Valdez (fl. 1930), wife of Arthur Möller and sister-in-law of the Swiss cactus hobbyist Dr. Heinrich Möller. (*Turbinicarpus*)
valentinii For Dr. J. Valentin (fl. 1897), Argentinian botanical collector. (*Pterocactus*)
validulus Dim. of Lat. 'validus', strong, robust-growing, i.e. rather strong or robust. (*Talinum*)
validus Lat., strong, robust; for the general appearance. (*Aspidoglossum, Cereus, Euphorbia meloformis* ssp., *Orbea, Ruschia, Selenicereus, Yucca*)
vallaris Lat., of walls; (**1**) for the habitat. (*Aloe*) – (**2**) for the occurrence on cliffs of an escarpment. (*Euphorbia*)
vallegrandensis For the occurrence near the town of Valle Grande, Dept. Santa Cruz, Bolivia. (*Echinopsis huotii* ssp.)

vallenarensis For the occurrence near Vallenar, N-C Chile. (*Eriosyce subgibbosa* ssp.)
vallis-gratiae For the occurrence at Genadental, Namibia (Germ., "Valley of Grace or Favour" = Lat. 'vallis', valley; and Lat. 'gratia', favour, grace). (*Lampranthus*)
vallis-mariae For the occurrence on the farm Marienthal (Germ., "Valley of Maria"; Lat. 'vallis', valley), Namibia. (*Lithops*)
valnicekianus For Dr. J. Valnicek († 1967), Czech cactus hobbyist. (*Gymnocalycium*)
valvatus Lat., valvate; for the valvate petals (which is exceptional for *Echeveria*). (*Echeveria*)
valverdensis For the occurrence near Valverde, Hierro, Canary Islands. (*Aeonium*)
vanbalenii For Mr. Jan C. Van Balen (1894–1956), Dutch horticulturist, emigrated to RSA in 1919. (*Aloe*)
vanbredae For Philip A. B. van Breda (fl. 1956–1964), officer in charge of the Veld Reserve, Worcester, Western Cape, RSA. (*Ruschia*)
vanderbergiae For Miss M. van der Berg (fl. 1935), without further data. (*Ruschia*)
vanderietiae For Mrs. Van de Riet (fl. 1932), who collected the type, without further data. (*Monsonia*)
vandermerwei For Dr. Frederick Z. van der Merwe (1894–1968), South African medical inspector of schools, specialist in *Aloe* and *Scilla*. (*Aloe, Delosperma, Euphorbia*) – (**2**) For N. J. S. van der Merwe (fl. 1929), without further data. (*Drosanthemum*)
vanderystii For Hyacinthe J. R. Vanderyst (1860–1934), Belgian agronomical engineer, botanist and missionary in the then Belgian Congo. (*Ceropegia*)
Vanheerdea For Pieter van Heerde (1893–1979), South African teacher and school principal in Springbok, Northern Cape, and active field collector. (*Aizoaceae*)
vanheerdei As above. (*Astridia, Conophytum, Lampranthus, Namaquanthus, Peersia, Ruschia*)
vanlessenii For Michael D. van Lessen (fl. 1960ies), highly decorated (Military Cross) major of the British Army serving 1962–

1965 in Aden, Yemen, who discovered the taxon. (*Sarcostemma*)

vanrensburgii For Mr. A. D. van Rensburg (fl. 1953), without further data. (*Braunsia, Prenia*)

Vanzijlia For Dorothy van Zijl (fl. 1922), South African plant collector. (*Aizoaceae*)

vanzijliae As above. (*Lampranthus*)

vanzylii For Gert H. van Zyl (fl. 1930–1932), postmaster at Pofadder, Northern Cape, RSA. (*Antimima, Conophytum calculus* ssp., *Dinteranthus, Ihlenfeldtia*)

vaombe From the local vernacular name of the plants in Madagascar. (*Aloe*)

vaotsanda As above. (*Aloe*)

vargasianus For Dr. Julio César Vargas Calderón (1907–1960), Bolivian botanist. (*Cereus*)

variabilis Lat., variable; (**1**) for the variable leaf shape. (*Jatropha*) – (**2**) for the variable flower colour. (*Lampranthus*) – (**3**) application obscure. (*Avonia*)

varians Lat., varying; (**1**) for the variation in leaf shape. (*Sansevieria*) – (**2**) for the number of flowers, which are never solitary. (*Antimima*) – (**3**) because several characters are at variance with those of related taxa. (*Sceletium*)

variantissimus Comp. of Lat. 'varians', varying, i.e. the most variable; for the variable shape of the branch segments. (*Euphorbia seretii* ssp.)

varicolor Lat. 'varius', various; and Lat. 'color', colour; for the variable spine colour. (*Oreocereus*)

varieaculeatus Lat. 'varius', various; and Lat. 'aculeatus', spiny, thorny; for the variable spination. (*Mammillaria*)

variegatus Lat., mottled, variegated; (**1**) for the variegated stems. (*Ceropegia*) – (**2**) for the variable spine colours. (*Echinocereus engelmannii* var.) – (**3**) for the leaf markings. (*Agave, Aloe, Haworthia*) – (**4**) for the spotted flowers. (*Orbea*) – (**5**) for the darker veined petals. (*Dudleya*)

varifolius Lat. 'varius', various; and Lat. '-folius', -leaved; for the variable leaf shape. (*Jatropha*)

varispinus Lat. 'varius', various; and Lat.

'-spinus', -spined; for the variable spine length. (*Cleistocactus*)

vaseyi For Dr. George Vasey (1822–1893), English-born US-American physician, botanist and from 1872 at the United States Department of Agriculture and curator of the United States National Herbarium. (*Opuntia*)

vasquezii For Roberto Vásquez (*1942), Bolivian cactus specialist. (*Echinopsis*)

vastus Lat., deserted, empty, vast, immense; application obscure. (*Ficus*)

vatteri For Ernesto Vatter (1900–1970), cactus enthusiast and plant collector in Santos Lugares, Prov. Córdoba, Argentina. (*Echinopsis, Gymnocalycium ochoterenae* ssp.)

vaupelianus For Dr. Friedrich Vaupel (1876–1927), German botanist in Berlin and Cactaceae specialist. (*Coryphantha, Stenocactus*)

veenianus For Mr. L. J. van Veen (*1923), Dutch cactus horticulturist in Honselersdijk and friend of the Dutch cactus specialist Dirk van Vliet. (*Parodia rutilans* ssp.)

velox Lat., swift, quick; for the rapid growth. (*Cheiridopsis*)

velutinus Lat., velvety; (**1**) for the pubescent stems and leaves. (*Kalanchoe*) – (**2**) for the pubescent stem segments. (*Opuntia*) – (**3**) for the pubescence of young shoots and the leaf rachis. (*Delonix*) – (**4**) for the leaf surface. (*Astridia, Conophytum, Cotyledon, Crassula sericea* var., *Delosperma, Gibbaeum, Jatropha, Pelargonium reniforme* ssp., *Raphionacme*) – (**5**) for the corolla. (*Duvalia*)

venenatus Lat., poisonous, toxic. (*Adenia, Euphorbia*)

venenificus Lat., poisonous. (*Euphorbia*)

venezuelensis For the occurrence in Venezuela. (*Crassula*)

venosus Lat., veined; (**1**) for the leaf patterning. (*Haworthia*) – (**2**) probably for the veined papery stipules. (*Begonia*)

venteri For Stefanus Venter (*1953), South African botanist. (*Plectranthus*) – (**2**) For F. Venter (fl. 2000), curator of the herbarium at the University of the North, Sovenga, RSA. (*Euphorbia*)

ventricosus Lat., thick-bellied, swollen (esp. one-sided) (from Lat. 'venter, ventris', belly); (**1**) for the bulging sheaths of the bract pair on the pedicel. (*Antimima*) – (**2**) for the centrally bulging corolla tube. (*Tylecodon*)

venustus Lat., beautiful, graceful; to honour Grace Violet Britten (fl. 1996) for her interest in the genus *Haworthia*. (*Haworthia cooperi* var.)

verdiensis For the occurrence in the region of the Verde and East Verde Rivers, Arizona, USA. (*Yucca elata* var.)

verdoorniae For Dr. Inez C. Verdoorn (1896–1989), South African botanist at the Botanical Research Institute, Pretoria, RSA. (*Chasmatophyllum*)

verecundus Lat., modest; (**1**) for the general appearance of the plants. (*Aloe, Delosperma*) – (**2**) perhaps for the solitary flowers. (*Lampranthus*)

verekeri For Mr. L. S. A. Vereker (fl. 1933), of what is now Harare, Zimbabwe, succulent plant collector. (*Aloe chabaudii* var., *Huernia*)

verityi For Dr. David S. Verity (*1930), US-American botanist and entomologist. (*Dudleya*)

vernalis Lat., vernal; (**1**) for the spring-green leaf colour. (*Lampranthus*) – (**2**) for the spring flowering season. (*Aloe reitzii* var.)

vernicolor Lat. 'vernus', pertaining to spring; and Lat. 'color', colour; (**1**) for the leaf colour like new-grown spring leaves. (*Oscularia*) – (**2**) probably for the fresh green leaf colour. (*Delosperma*)

verrucosus Lat., with warts; (**1**) for the warted stems. (*Cynanchum*) – (**2**) for the stem and leaf surfaces. (*Sesuvium, Tetragonia*) – (**3**) for the warty leaf surface. (*Gasteria carinata* var.) – (**4**) for the warty leaf tips. (*Conophytum*) – (**5**) for the corolla. (*Orbea*) – (**6**) for the warty ovary. (*Cheiridopsis*)

verruculatus Lat., with small warts or protuberances; for the leaves. (*Scopelogena*)

verruculosus Lat., covered with small warts; (**1**) for the wrinkled appearance of the skin of the stems. (*Euphorbia*) – (**2**) for the papillate young stems and leaves. (*Antimima*) – (**3**) for the epidermis. (*Lithops*)

versadensis For the occurrence near Versada, Oaxaca, Mexico. (*Sedum*)

verschaffeltii For Ambroise C. A. Verschaffelt (1825–1886), famous Belgian horticulturist at Gand [Gent], founder of the periodical 'Illustration Horticole' in 1854. (*Austrocylindropuntia*)

versicolor Lat., variously coloured; (**1**) for the spine coloration. (*Cylindropuntia, Haageocereus*) – (**2**) for the variable flower colour. (*Delosperma, Ruschia, Sedum*) – (**3**) perhaps for the flowers. (*Aloe, Sempervivum*)

versicolores Lat. 'versicolor', variously coloured; for the colour change in the nectar glands as the cyathia mature. (*Euphorbia*)

verticillacanthus Lat. 'verticillus', verticil, intertwined material; and Gr. 'akantha', thorn, spine; for the more or less tortuous spination. (*Sulcorebutia*)

verticillaris Lat., verticillate; for the arrangement of the appendages of the corona. (*Miraglossum*)

verticillatus Lat., verticillate; (**1**) for the branching pattern. (*Quiabentia*) – (**2**) for the leaf arrangement. (*Ceropegia, Hylotelephium, Peperomia*) – (**3**) for the arrangement of the flowers. (*Hoya, Plectranthus*)

vertongenii For Dr. Herman Vertongen (fl. 1995), Belgian cactus enthusiast. (*Eriosyce*)

veruculoides Gr. '-oides', resembling; and for *Mesembryanthemum verruculatum* (now *Scopelogena verruculata*; Aizoaceae). (*Malephora*)

verus Lat., true; because this is the true *Aloe* of commerce. (*Aloe*)

veseyi For L. Desmond E. F. Vesey-Fitzgerald (1909 or 1910–1974), British entomologist, worked on biological control of insects in many tropical countries, including Kenya, Tanzania and Zambia. (*Aloe*)

vespertinus Lat., belonging to the evening; (**1**) for the flowers, opening in the afternoon and closing in the evening. (*Aridaria, Bergeranthus, Drosanthemum*) – (**2**) for the flowers opening only in the evening. (*Anacampseros*) – (**3**) perhaps for the supposed flowering period (*Yucca baccata* var.)

vestitus Lat., clothed; (**1**) for the hair-covered stems. (*Austrocylindropuntia, Tetragonia*) –

(2) for the stems densely covered with leaves. (*Crassula*) – (3) for the long-hairy corolla. (*Duvalia*)

vetovalidus Lat. 'vetus', old; and Lat. 'validus', strong, robust-growing; to avoid a homonym. (*Ruschia*)

vetulus Lat., oldish; (1) perhaps for the greyish-white spination. (*Mammillaria*) – (2) for the shaggy-hairy flowers. (*Stapelia*)

vexans Lat. 'vexare', torture, annoy, ill-treat; perhaps for the leaf armature. (*Agave datylio* var.)

viatorum Gen. Pl. of Lat. 'viator', traveller; for the occurrence at roadsides. (*Lampranthus*)

vibratilis Lat., vibratile, easily moving; for the vibratile hairs on the corolla. (*Orbea*)

viciifolius Lat. '-folius', -leaved; and for the similarity to species of the genus *Vicia* ("Vetch", *Fabaceae*). (*Pelargonium*)

vicinus Lat., neighbouring; perhaps for the similarity to other species. (*Agave*)

victoriae-reginae Lat. 'regina', queen; for the British Queen Victoria, who reigned from 1837–1901. (*Agave*)

victorianus For the occurrence on Mt. Victoria, Myanmar. (*Sedum*)

victoriensis For the occurrence near Ciudad Victoria, Tamaulipas, Mexico. (*Ferocactus echidne* var.)

victoris For Victor Stanley Peers (1874–1940), Australian civil servant, amateur archaeologist and plant collector, living in RSA from 1899. (*Ruschia*)

vidalii Very probably for Sebastian Vidal y Soler (1842–1889), Spanish botanist working several years in the Philippines. (*Dischidia*)

viduiflorus Lat. 'viduus', without, deprived of; and Lat. '-florus', -flowered; because the taxon was not known to have flowered. (*Euphorbia*)

viereckii For Hans-Wilhelm Viereck (1903–1946), German plant enthusiast and cactus collector, settling in Mexico 1920–1938. (*Echinocereus*, *Mammillaria picta* ssp., *Turbinicarpus*)

vietnamensis For the occurrence in Vietnam. (*Sedum*)

vignei For Mr. Vigne (fl. 1936), who collected the type. (*Raphionacme*)

viguieri For Prof. René Viguier (1880–1931), French botanist in Paris, later in Caen and 1919–1931 director of Caen Municipal Garden, who collected 1912 in Madagascar with H. Humbert. (*Aloe*, *Euphorbia*, *Kalanchoe*, *Rosularia adenotricha* ssp.)

vilaboensis For the occurrence near the city of Vila Boa (= the present-day Goiás), Goiás, Brazil. (*Pilosocereus*)

vilanandrensis For the occurrence at Vilanandro, Madagascar. (*Euphorbia viguieri* var.)

vilis Lat., without value, low, contemptuous. (*Grusonia*)

Villadia For Dr. Manuel M. Villada (1841–1924), Mexican physician and naturalist. (*Crassulaceae*)

villadioides Gr. '-oides', resembling; and for the genus *Villadia* (*Crassulaceae*). (*Sedum versadense* var.)

villardii For "Reb" Villard (fl. 1975), who first collected the taxon. (*Escobaria*)

villetiae For Mrs. Villet, wife of Dr. A. C. T. Villet (fl. 1936–1956), keen collector of succulents in Worcester, RSA. (*Stapelia*)

villetii For Dr. A. C. T. Villet (fl. 1936–1956), keen collector of succulents in Worcester, RSA. (*Lithops*, *Stomatium*)

villicumensis For the occurrence in the Sierra Villicum, San Juan, Argentina. (*Eriosyce*)

villiersii For Mr. H. L. de Villiers (fl. 1932–1959), without further data. (*Erepsia*, *Lampranthus*)

villipetiolus Lat. 'villus', shaggy hair; and Lat. 'petiolus', petiole. (*Peperomia*)

villosus Lat., villous, shaggy with fairly long soft hairs; (1) for the general appearance. (*Aichryson*, *Portulaca pilosa* ssp.) – (2) for the hairy stems and leaves. (*Brachystelma*, *Jatropha*, *Mesembryanthemum*, *Sedum*) – (3) for the dense soft spination. (*Eriosyce*) – (4) for the densely villous flowers. (*Apodanthera*)

vilmorinianus For Maurice L. de Vilmorin (1849–1918), French botanist, dendrologist and sylviculturist, son of L. de Vilmorin. (*Agave*)

viminalis Lat., bearing shoots suitable for

plaiting and wicker-work; for the leafless branches (*Sarcostemma*)

vinaceus Lat., wine-like; for the flower colour. (*Delosperma, Pelargonium*)

vincifolius For the genus *Vinca* ("Periwinkle", *Apocynaceae*); and Lat. '-folius', -leaved; for the similar leaves. (*Ceropegia*)

vinicolor Lat. 'vinum', wine; and Lat. 'color', colour; for the wine-red leaves. (*Sedum*)

violaceus Lat., violet; (**1**) for the violet hue of the young spines. (*Melocactus*) – (**2**) for the colour of the bracts. (*Pereskia grandifolia* ssp.) – (**3**) for the colour of the petals. (*Lampranthus*)

violaciflorus Lat. 'violaceus', violet; and Lat. '-florus', -flowered. (*Conophytum, Micranthocereus*)

violiflorus For the genus *Viola* ("Violet", *Violaceae*); and Lat. '-florus', -flowered; for the flowers resembling those of white violets. (*Pelargonium*)

viperinus Lat., snake-like; for the slender elongate plant bodies. (*Mammillaria sphacelata* ssp., *Peniocereus*)

virchowii For Geheimrat R. Virchow (1821–1902), German pathologist, anthropologist and politician. (*Echidnopsis*)

virellus Lat., a little bit greenish, somewhat greenish, for the similarity to *Haworthia gracilis* var. *viridis*. (*Haworthia decipiens* var.)

virens Lat., becoming green; (**1**) for the bright green leaves. (*Delosperma*) – (**2**) perhaps for the greenish corona. (*Schizoglossum atropurpureum* ssp.) – (**3**) application obscure. (*Ruschia*)

virescens Lat., becoming green; (**1**) probably for the leaf colour. (*Carpobrotus*) – (**2**) for the flower colour. (*Tridentea*)

virgatus Lat., twiggy, rod-like; (**1**) for the long thin branches. (*Aizoon, Aspidoglossum, Commiphora, Crassula subaphylla* var., *Monadenium, Ruschia, Tetragonia, Villadia*) – (**2**) for the erect slender primary stems. (*Lampranthus*)

virgineus Lat., virgin; for the occurrence in the Barranco de la Virgen (Gran Canaria) (Span. 'virgen', virgin). (*Aeonium canariense* var.)

virginicus For the occurrence in Virginia, USA. (*Agave*)

viridescens Lat., becoming green; for the greenish flowers. (*Ferocactus, Hylotelephium*)

viridicatus Lat. 'viridis', green; and Lat. '-atus', having the possession; for the leaf colour. (*Conophytum truncatum* ssp.)

viridiflavus Lat. 'viridis', green; and Lat. 'flavus', yellow; for the yellow-green sepals. (*Sedum lutzii* var., *Sinocrassula indica* var.)

viridiflorus Lat. 'viridis', green; and Lat. '-florus', -flowered. (*Aloe, Cylindropuntia, Echinocereus, Marsdenia, Phyllobolus, Tylecodon*)

viridifolius Lat. 'viridis', green; and Lat. '-folius', -leaved. (*Euphorbia cremersii* fa., *Ruschia*)

viridiruber Lat. 'viridis', green; and Lat. 'ruber, rubra, rubrum', red; for the green stem segments with reddish markings. (*Opuntia*)

viridis Lat., green; (**1**) for the body colour. (*Browningia, Melocactus pachyacanthus* ssp.) – (**2**) for the spine colour. (*Euphorbia atrispina* var., *Euphorbia enopla* var., *Euphorbia heptagona* var.) – (**3**) for the leaf colour. (*Haworthia gracilis* var., *Haworthia marumiana* var., *Lithops, Pachyphytum, Smicrostigma, Stomatium*) – (**4**) probably for the insignificant greenish flowers. (*Crassula, Hylotelephium*)

viridissimus Superl. of Lat. 'viridis', green; i.e. the greenest; for the bright green leaves. (*Echeveria*)

virosus Lat., poisonous; for the latex. (*Euphorbia*)

viscatus Lat., sticky; for the notably sticky leaves. (*Aeonium lindleyi* var.)

viscidus Lat., viscid, glutinous; for the viscid leaves. (*Dudleya*)

viscosus Lat., sticky; for the scabrid leaves. (*Haworthia*)

Viscum Perhaps from Lat. 'viscum', birdlime; for the very sticky interior of the berries. (*Viscaceae*)

vitelliniflorus Lat. 'vitellinus', yellow like egg-yolk; und Lat. '-florus', -flowered. (*Opuntia*)

vitellinus Lat., yellow like egg-yolk; for the flower colour. (*Hoya*)

vitreopapillus Lat. 'vitreus', glassy, transparent; and Lat. 'papilla', papilla; for the glassy papillae of the leaves. (*Conophytum obscurum* ssp.)

vitreus Lat., glassy, transparent; for the leaves. (*Bulbine*)

vittatifolius Lat. 'vittatus', longitudinally striped; and Lat. '-folius', -leaved. (*Bulbine*)

vittatus Lat., longitudinally striped; (**1**) for the branch variegation. (*Euphorbia*) – (**2**) for the longitudinally striped petals. (*Nananthus*)

vituensis For the erroneously presumed occurrence in the Witu region in Kenya, around 1870–1890 the centre of a German-owned farming project, and the starting point for the expedition during which the taxon was found. (*Aloe*)

viviparus Lat., viviparous; (**1**) for the bulbils in the leaf axils. (*Dorstenia*) – (**2**) for the bulbils in the inflorescences. (*Agave, Hylotelephium*) – (**3**) for the proliferating fruits producing roots and new shoots after having fallen to the ground. (*Cylindropuntia*) – (**4**) application obscure. (*Escobaria*)

vizcainoensis For the occurrence in the Sierra Vicaíno, Baja California, Mexico. (*Agave*)

Vlokia For Jan H. J. Vlok (*1957), Environmental Advisor for the Cape Department of Nature Conservation, RSA, and active plant collector. (*Aizoaceae*)

vlokii As above. (*Gasteria, Haworthia*)

voburnensis For the Duke of Bedford's cactus collection at Woburn (fl. 1845), England. (*Mammillaria*)

vogtherrianus For Hans Vogtherr (fl. 1932), German cactus hobbyist in Berlin. (*Coryphantha*)

vogtsii For Mr. Lewis R. Vogts (fl. 1930), South African administrator and successful cultivator of succulent plants in his garden near Pretoria, RSA. (*Aloe, Delosperma*)

volkartii For George Volkart († before 1937), from Winterthur (Switzerland), friend of the Swiss botanist John Gossweiler. (*Huernia*)

volkensii For Prof. Dr. Georg L. A. Volkens (1855–1917), German botanist in Berlin, explorer of the Kilimanjaro 1892–1894. (*Adenia, Aloe, Crassula, Sansevieria, Synadenium*)

volkii For Prof. Dr. Otto H. Volk (1903–2000), German botanist at Würzburg University, with a strong interest in medicinal plants. (*Lithops pseudotruncatella* ssp.)

volkmanniae For Miss Margareta Volkmann (fl. 1928), owner of the Fam Auros, Namibia. (*Euphorbia*)

vollianus For Otto Voll (1884–1958), German (?) botanist in Brazil. (*Echinopsis*)

volubilis Lat., twining (from Lat. 'volvere', to twine). (*Bowiea, Brachystelma, Ceropegia, Parakeelya*)

vomeriformis Lat. 'vomer', ploughshare; and Lat. '-formis', -shaped; application obscure. (*Stathmostelma angustatum* ssp.)

vorwerkii For Wilhelm Vorwerk (1873–1936), horticulturist, later inspector and finally technical director at the Botanical Garden Berlin. (*Neowerdermannia*)

vossii For Mr. Harold Voss (fl. 1936), without further data. (*Aloe*)

vredenburgensis For the occurrence at Vredenburg, Malmesbury Distr., Western Cape, RSA. (*Oscularia*)

vryheidensis For the occurrence at Vryheid, KwaZulu-Natal, RSA. (*Aloe*)

vulcanensis For the occurrence at Volcán, Jujuy, Argentina. (*Grahamia*)

vulcanii For the occurrence on volcanic soils. (*Euphorbia milii* var.)

vulcanorum Gen. Pl. of Lat. 'vulcanus', volcano; for the occurrence on volcanic lava soils. (*Euphorbia*)

vulpes Lat., fox; for the fox-brown spination. (*Haageocereus*)

vulpis-cauda Lat. 'vulpes', fox; and Lat. 'cauda', tail; for the hanging stems with dense reddish-brown spination. (*Cleistocactus*)

vulvaria For the scent of the plant like that of *Chenopodium vulvaria* ("Stinking Goosefoot", Chenopodiaceae). (*Ruschia*)

W

wagenknechtii For Rodolfo Wagenknecht (fl. 1957), cactus collector of German descent in Chile. (*Eriosyce subgibbosa* ssp., *Maihueniopsis*)

wagnerianus For Hermann Wagner (fl. 1932), cactus collector in Ludwigsburg, Germany. (*Mammillaria*)

wakefieldii For Thomas Wakefield, English missionary and amateur naturalist and geographer in Kenya for over 20 years from 1862. (*Euphorbia*)

waldheimii For Dr. Alexander A. Fischer von Waldheim (1839–1920), Russian botanist and director of the Imperial Botanical Garden at St. Petersburg. (*Kalanchoe*)

walgateae For Marion M. Walgate (later Mrs. Macnae) (*1914), English-born botanist, emigrated with her parents to RSA 1920. (*Lampranthus*)

wallichianus For Dr. Nathaniel Wallich (born Nathan Wulff [Wolff]) (1786–1854), Danish physician and botanist, at Calcutta Botanical Garden 1815–1846, collecting widely in Asia (incl. India) and also in RSA. (*Rhodiola*)

wallichii As above. (*Ceropegia*, *Tylecodon*)

wallisii For Gustav Wallis (1830–1878), German gardener and botanical explorer, repeatedly collecting in C and S America, died in Ecuador. (*Agave*)

wallowensis For the occurrence in Wallowa County, Oregon, USA. (*Lewisia columbiana* var.)

walpoleanus For Frederick A. Walpole (1861–1904), US-American botanical artist who painted *Crassulaceae* for J. N. Rose. (*Echeveria*)

walteri For Walter Rausch (*1928), Austrian lithographer, cactus specialist and traveller in South America. (*Echinopsis*) – (**2**) For Dr. Walter Till (*1956), Austrian botanist and Bromeliad specialist at Vienna University. (*Gymnocalycium*)

waltheri For Eric Walther (1892–1959), German-born US-American, emigrated to USA in 1909, from 1917 horticulturist at the Golden Gate Park, California, and eventually director of Strybing Arboretum and Botanical Garden, after retirement in 1957 Botany Research Associate at California Academy of Sciences, specializing in New World *Crassulaceae*. (*Echeveria*)

waltoniae For Miss A. Walton (fl. 1923), without further data. (*Orthopterum*)

wangii For Chi Wu Wang (*1913), Chinese botanist and collector. (*Sedum*)

warasii For Eddie Waras (fl. 1977), plant collector in São Paulo, Brazil. (*Parodia*)

waringiae For Mrs. Gerold (née Waring) (fl. 1998), wife of the plant trader Raymond Gerold in Madagascar. (*Euphorbia*)

warnockii For Prof. Dr. Barton H. Warnock (*1911), US-American botanist at the Sul Ross State College, Alpine, Texas, USA. (*Echinomastus*)

warszewiczianus For Josef Warszewicz, Ritter von Rawicz (1812–1866), Lithuanian-born gardener in Berlin, later independent plant collector in C and S America, 1854–1866 curator of the Cracow Botanical Garden. (*Callisia*)

waterbergensis For the occurrence in the Waterberg area, Northern Prov., RSA. (*Delosperma*, *Euphorbia*)

watermeyeri For Mr. E. B. Watermeyer (1915–1929), collector of succulents in the Vanrhynsdorp area, RSA. (*Antimima*, *Crassula atropurpurea* var., *Lampranthus*)

watsonii For J. M. Watson (fl. 1963), without further data. (*Echidnopsis*)

weberbaueri For Prof. Augusto Weberbauer (1871–1948), German / Polish botanist originally from Breslau, since 1901 mostly living and working in Peru. (*Cistanthe*, *Euphorbia*, *Matucana*, *Sedum*, *Weberbauerocereus*)

Weberbauerocereus For Prof. Augusto Weberbauer (1871–1948), German / Polish botanist originally from Breslau, since 1901 mostly living and working in Peru; and *Cereus*, a genus of columnar cacti. (*Cactaceae*)

weberi For Dr. Frédéric Albert C. Weber

(1830–1903), French military surgeon and amateur botanist strongly interested in cacti. (*Agave*, *Pachycereus*, *Tephrocactus*)

weberianus As above. (*Pereskia*)

Weberocereus For Dr. Frédéric Albert C. Weber (1830–1903), French military surgeon and amateur botanist strongly interested in cacti; and *Cereus*, a genus of columnar cacti. (*Cactaceae*)

websterianus For Gertrude D. Webster (1872–1947); US-American philanthropist and one of the chief patrons of the Desert Botanical Garden, Phoenix, Arizona, USA. (*Echinocereus*)

weigangianus For Mr. Weigang (fl. 1923), without further data. (*Leipoldtia*)

weinbergii For Frank Weinberg († 1941), US-American nurseryman in Long Island (fl. 1906) and California, of German descent. (*Echinocereus pulchellus* ssp., *Lenophyllum*)

Weingartia For Wilhelm Weingart (1856–1936), German manufacturer and amateur botanist. (*Cactaceae*)

weingartianus As above. (*Leptocereus*, *Mammillaria*)

welwitschii For Dr. Friedrich M. J. Welwitsch (1806–1872), Austrian physician and naturalist, travelled widely in Angola 1853–1861 (*Adenia*, *Corallocarpus*, *Ipomoea*, *Kalanchoe*, *Odontostelma*, *Raphionacme*, *Stathmostelma*)

wenchuanensis For the occurrence at Wenchuan, Szechuan Prov., China. (*Sedum*)

wendtii For Dr. Thomas L. Wendt (*1950), US-American botanist at Louisiana State University. (*Agave*)

wenigeri For Delbert ("Del") K. Weniger (1923–1999), US-American biology and ecology teacher in San Antonio, Texas, and specialist on Texan cacti. (*Echinocereus pectinatus* ssp.)

wercklei For Carlos (Karl) Wercklé (1860–1924), French-born botanist and horticulturist in the US and from 1902 onwards in Costa Riva as private horticulturist. (*Agave*, *Selenicereus*)

werdermannianus For Prof. Dr. Erich Werdermann (1892–1959), German botanist in Berlin, specialist on cacti and former director of the Botanischer Garten und Museum Berlin. (*Parodia*)

werdermannii As above. (*Coryphantha*, *Echinopsis*, *Pachyphytum*, *Portulaca*)

werneri For Werner Uebelmann (*1921), Swiss horticulturist and expert on Brazilian cacti. (*Tacinga*) – (2) For Werner Triebner (fl. c. 1950), South African farmer and son of Wilhelm Triebner. (*Lithops*)

wessnerianus For Willi Wessner (1904–1983), German merchant and later owner of a well-known cactus-nursery in Muggensturm, Germany. (*Rebutia*)

westii For James West (1886–1939; alias name for Egon Viktor Moritz Karl Maria von Ratibor und Corney, Prinz zu Hohenloe-Schillingsfürst), well-known Californian horticulturist, who accompanied Goodspeed on the first Andes Expedition. (*Echeveria*, *Echinopsis maximiliana* ssp., *Weingartia*)

wethamae For Mrs. Boddam Wetham (fl. 1928), without further data. (*Delosperma*)

wetmorei For Alexander Wetmore (1886–1978), US-American ornithologist who brought plants from Argentina to the US-American botanists N. L. Britton and J. N. Rose. (*Opuntia*)

wettsteinii For Richard Wettstein von Westersheim (1863–1931), Austrian botanist in Vienna. (*Conophytum*, *Nematanthus*)

wheeleri For George M. Wheeler (1842–1905), lieutnant of the US Army and leader of the geographical and geological survey expedition in the W USA. (*Dasylirion*)

whellanii For James A. Whellan (1915–1995), English-born botanist, entomologist and naturalist in Zimbabwe, later in Malawi and Malaysia. (*Euphorbia*)

whipplei For Amiel W. Whipple (1817–1863), Lieutenant of the US Army, 1853–1854 topographical engineer on the US exploration and survey for a railroad route from the Mississippi River to the Pacific Ocean. (*Cylindropuntia*, *Hesperoyucca*, *Sclerocactus*)

whitcombei For R. P. Whitcombe of Salalah, Oman, who first collected the taxon in 1989. (*Aloe*)

Whiteheadia For Rev. Henry Whitehead (1817–1884), Anglican missionary from England who collected in Namaqualand (RSA). (*Hyacinthaceae*)

whiteheadii As above. (*Crassula*)

whitei For Orlando E. White (fl. 1925), curator of plant breeding at the Brooklyn Botanic Garden, USA. (*Echeveria*)

Whitesloanea For Alain C. White (1880–1951) and Boyd L. Sloane (1886–1955), US-American authors of important books on Euphorbias and Asclepiads. (*Asclepiadaceae*)

whitesloaneanus As above. (*Huernia*)

whytei For Alexander Whyte (fl. 1897), Head of the Scientific Department in Zomba, Malawi, and active plant collector. (*Pelargonium*)

wiesei For Mr. T. G. (Buys) Wiese (*1923); succulent plant grower, farmer and owner of the farm Quaggaskop, Western Cape, RSA. (*Bulbine*)

wiesingeri For Mr. Wiesinger (fl. 1932), cactus collector at Waldshut, Bavaria, Germany, who discovered the taxon. (*Mammillaria*)

wightianus For Dr. Robert Wight (1796–1872), British surgeon and botanist, working for many years in India. (*Adenia, Portulaca*)

wilcoxii For Timothy E. Wilcox (fl. 1892), US-American Brigadier General and enthusiastic student of plants. (*Mammillaria wrightii* ssp.) – (**2**) For Dr. Glover B. Wilcox (fl. 1909), who provided Mexican cacti to the US-American botanists N. L. Britton and J. N. Rose. (*Opuntia*)

wilczekianus For Dr. Ernst Wilczek (1867–1948), Swiss botanist in Zürich and Lausanne. (*Sedum*)

wildemanius For Émile A. J. De Wildeman (1866–1947), Belgian botanist and pioneer student of the Congolese flora. (*Stathmostelma*)

wildii For Prof. Hiram Wild (1917–1982), British botanist, emigrated to Zimbabwe 1945, Director of the National Herbarium, Harare. (*Aloe, Commiphora, Euphorbia, Kalanchoe*)

willdenowii For Prof. Dr. Carl L. Willdenow (1765–1812), German botanist in Berlin. (*Ruschia*)

williamsii Most probably for Reverend Theodore Williams (fl. 1841), owner of a cactus collection at Hendon Vicarage, England; or (more improbably) for Mr. C. H. Williams (fl. 1845), English traveller in Brazil (Bahia). (*Lophophora*)

williamsonii For Dr. Graham Williamson (*1932), Zimbabwean / South African dental surgeon in Zimbabwe, Zambia, Malawi and Namibia, and after retirement botanist in RSA, from 1996 Research Associate at the Bolus Herbarium, Cape Town, throughout life an active field student of succulents and orchids. (*Euphorbia*)

willowmorensis For the occurrence near Willowmore, Eastern Cape, RSA. (*Chasmatophyllum, Hereroa, Ophionella*)

wilmaniae For Maria Wilman (1867–1957), botanist and geologist in RSA, first director of the McGregor Museum in Kimberley, RSA. (*Delosperma, Ebracteola, Euphorbia, Hereroa*)

wilmotianus For Mr. C. Wilmot (fl. 1939), without further data. (*Dinteranthus*)

wilmsii For Friedrich Wilms jr. (1848–1919), German pharmacist, botanist and plant collector, working in RSA 1882–1896. (*Adenia*)

wilsonii For Dr. Ernest H. Wilson (1876–1930), British-born US-American botanist, plant collector and traveller esp. in Asia. (*Sedum*) – (**2**) For Mr. John G. Wilson (*1927), British agricultural officer and ecologist with the Uganda Department of Agriculture 1953–1968, and "who contributed a great deal to our knowledge of Ugandan succulents", later living in Kenya. (*Aloe, Orbea*)

windsorii For the occurrence at the rock formation named Windsor Castle, N Madagascar. (*Pachypodium baronii* var.)

winkleri For Jim Winkler (fl. 1960), US-American who discovered the taxon while on a camping trip with his mother, Mrs. Agnes Winkler. (*Pediocactus*) – (**2**) For

Werner Winkler (fl. 1972), German cactus collector in Bonn and friend of the Dutch cactus enthusiast Dirk van Vliet. (*Parodia mueller-melchersii* ssp.)

winterae For Hildegard Winter (1893–1975), sister of the German cactus specialist F. Ritter, who sold the seeds collected by Ritter. (*Mammillaria*)

winteri For the seed selling business run by Hildegard Winter (1893–1975), sister of the German cactus specialist F. Ritter, selling the seeds collected by Ritter. (*Cleistocactus*)

winterianus For Hildegard Winter (1893–1975), sister of the German cactus specialist F. Ritter, who sold the seeds collected by Ritter. (*Weberbauerocereus*)

wislizeni For Dr. Friedrich A. Wislizenus (1810–1889), Germany-born physician and traveller settling 1839 in St. Louis, physician and botanist partner to George Engelmann. (*Ferocactus*)

wissmannii For Prof. Dr. Hermann von Wissmann (fl. 1927), German geographer, travelled in Yemen and collected plants. (*Orbea*)

wittebergensis For the occurrence in the Witteberg Mts., Western Cape, RSA. (*Drosanthemum, Haworthia*)

wittii For Mr. N. Witt (fl. 1900), German merchant in Amazonian Brazil. (*Selenicereus*)

witzenbergensis For the occurrence at the Witzenberg, Western Cape, RSA. (*Huernia*)

wiumii For E. J. F. Wium (fl. 1967), without further data. (*Delosperma*)

wocomahi From the local vernacular name of the plants with the Warihio Indians in NW Mexico. (*Agave*)

wohlschlageri For Michael Wohlschlager (*1936), Austrian cactus collector in Moosbrunn who travelled widely in Mexico. (*Coryphantha*)

wolfgang-krahnii For Wolfgang Krahn (fl. 1960, 2003), German cactus collector, travelled in Peru in the 1960s together with the US-American botanist Paul C. Hutchison. (*Peperomia*)

wolfii For Dr. Carl B. Wolf (1905–1974), US-American botanist at the Rancho Santa Ana Botanical Garden in California. (*Cylindropuntia*)

wollastonii For A. F. R. Wollaston (fl. 1908), British botanist and collector in East Africa. (*Aloe*)

woodburniae For Mrs. M. Woodburn (fl. 1925), without further data. (*Lampranthus*)

woodii For Dr. John Medley Wood (1827–1915), British botanist and director of the Botanical Garden in what was then Natal, RSA. (*Aspidoglossum, Ceropegia linearis* ssp., *Cotyledon, Euphorbia, Orbea, Plectranthus, Stictocardia*) – (**2**) For John R. I. Wood (*1944), British Inspector of Schools in Yemen and active amateur botanist. (*Aloe*)

woodsii For Robert S. Woods (fl. 1934), US-American cactus enthusiast in Azusa, California. (*Mammillaria hahniana* ssp.)

Wooleya For Major C. H. F. Wooley (fl. 1937, 1960), South African plant collector supplying succulents to Kirstenbosch. (*Aizoaceae*)

woolleyi For Major C. H. F. Wooley [erroneously written as 'Woolley'] (fl. 1937, 1960), South African plant collector supplying succulents to Kirstenbosch. (*Haworthia venosa* ssp.)

woollianus For Mr. Woolley of Barberton, former Transvaal, RSA. (*Aloe chortolirioides* var.)

wootonii For Prof. Elmer O. Wooton (1865–1945), US-American botanist, chiefly at the United States Department of Agriculture. (*Opuntia*)

worcesterae For the occurrence near Worcester, Western Cape, RSA. (*Drosanthemum*) – (**2**) For the erroneously presumed occurrence near Worcester, Western Cape, RSA. (*Pelargonium*)

wordsworthiae For Mrs. R. Wordsworth (fl. 1899?), who collected the type specimen, without further data. (*Lampranthus*)

woronowii For Georg J. N. Woronow (1874–1931), Russian botanist and director of the herbarium at Tbilissi, Georgia. (*Sedum*)

wrefordii For Mr. Herbert Wreford-Smith (1890–1962), variously transporter, farmer, prospector, cattle buyer and naturalist in Kenya and Uganda, lived in Kenya from 1908. (*Aloe*)

wrightiae For Mrs. Dorde Wright (fl. 1961) of Salt Lake City, Utah, USA, who discovered the taxon. (*Sclerocactus*)

wrightii For Charles Wright (1811–1885), US-American botanist and important collector esp. in the S USA, Mexico, C America and Cuba. (*Leptocereus*, *Mammillaria*, *Sclerocactus uncinatus* ssp., *Sedum*)

wulfenii For Franz Xaver Freiherr von Wulfen (1728–1805), Austrian Jesuit, teacher and botanist. (*Sempervivum*)

wurdackii For John J. Wurdack (1921–1998), US-American botanist in New York and later at the Smithsonian Institution, Washington D.C., and specialist on the neotropical flora. (*Echeveria*)

X

xaltianguensis For the occurrence near the village of Xaltianguis, Guerrero, Mexico. (*Mammillaria*)

xanthadenia Gr. 'xanthos', yellow; and Gr. 'aden', gland; for the yellow cyathial glands. (*Euphorbia mahafalensis* var.)

xanthocarpus Gr. 'xanthos', yellow; and Gr. 'karpos', fruit; for the fruit colour. (*Rebutia*)

xanthochlorus Gr. 'xanthos', yellow; and Gr. 'chloros', green; for the colour of the nectary glands. (*Euphorbia attastoma* var.)

xanthosphaericus Gr. 'xanthos', yellow; and Gr. 'sphaira', sphere, globe; for the yellow flowers in globose inflorescences. (*Aspidoglossum*)

xanti For János (John) Xantus de Vesey (1825–1894), Hungarian lawyer and natural history collector in Asia, emigrating 1851 to the USA and becoming scientific collector for the Smithsonian Institution, collected 1862–1864 in Baja California. (*Euphorbia*)

xeranthemoides Gr. '-oides', resembling; and for the genus *Xeranthemum* (*Asteraceae*); for the spine-tipped outer perianth segments, for the similarity to the flower heads of *Xeranthemum*. (*Echinocactus polycephalus* ssp.)

xerophilus Gr. 'xeros', dry; and Gr. 'philos', friend; for the preferred habitat. (*Plectranthus*)

xerophyton Gr. 'xeros', dry; and Gr. 'phyton', plant; for the dry habitat of the taxon. (*Pelargonium*)

Xerosicyos Gr. 'xeros', dry; and Gr. 'sicyos', cucumber; for the xerophytic nature and the family placement of the plants. (*Cucurbitaceae*)

xiphiophyllus Gr. 'xiphion', small sword; and Gr. 'phyllon', leaf; for the leaf shape. (*Haworthia arachnoidea* var.)

xochipalensis For the occurrence near Xochipala, Guerrero, Mexico. (*Thompsonella*)

xylacanthus Gr. 'xylon', wood; and Gr. 'akantha', thorn, spine; for the robust ('woody') spines. (*Euphorbia*)

xylodes Gr., woody; for the woody stems. (*Lepidium*)

xylonacanthus Gr. 'xylon', wood; and Gr. 'akantha', spine, thorn; for the firm leaf marginal teeth and end spines. (*Agave*)

xylophylloides Gr. 'xylon', wood; Gr. 'phyllon', leaf; and Gr. '-oides', similar to; for the photosynthetic woody leaflike branches. (*Euphorbia*)

xylorhizus Gr. 'xylon', wood; and Gr. 'rhiza', root. (*Cleistocactus*)

xysmalobioides Gr. '-oides', resembling; and for the genus *Xysmalobium* (*Asclepiadaceae*). (*Glossostelma*)

Xysmalobium Gr. 'xysma', cleft or scraped place; and Gr. 'lobion', small lobe; for the frequently cleft corona lobes. (*Asclepiadaceae*)

Y

yanganucensis For the occurrence in the Quebrada Yanganuco [Llanganuco], Dept. Ancash, Peru. (*Austrocylindropuntia*)

yaquensis For the occurrence at the Río Yaqui, Sonora, Mexico. (*Mammillaria thornberi* ssp.)

yattanus For the occurrence along the foot of the Yatta Plateau, Kenya. (*Monadenium*)

yavellanus For the occurrence at Yavello, Ethiopia. (*Aloe*)

Yavia For the occurrence in Dept. Yavi, Prov. Jujuy, N Argentina. (*Cactaceae*)

yemenensis For the occurrence in the Yemen. (*Ceropegia, Echidnopsis, Kalanchoe*)

yemenicus As above. (*Aloe*)

yildizianus For Dr. Bayram Yildiz (*1946), Turkish botanist. (*Sedum*)

yorubanus For the occurrence in the region of the Yoruba tribe, Nigeria. (*Ceropegia*)

yosemitensis For the occurrence in Yosemite National Park, California, USA. (*Sedum spathulifolium* ssp.)

youngae For Marie S. Young (1872–1919), US-American teacher and botanist who first found the taxon. (*Talinum*)

ysabelae For Ysabel Wright (fl. 1931), US-American cactus collector, probably in Texas. (*Turbinicarpus*)

yucatanensis For the occurrence in the state of Yucatán, Mexico. (*Mammillaria columbiana* ssp.)

Yucca Name first used 1557 in a German travelogue and probably derived from a name used on Hispaniola through Span. 'yuca', which is, however, used for the edible root tubers of Cassava, and that was perhaps erroneously applied to *Yucca* because of the edible flowers of some species. (*Agavaceae*)

yuccoides Gr. '-oides', resembling; and for the genus *Yucca* (*Agavaceae*). (*Beschorneria*)

yunckeri For Prof. Dr. Truman G. Yuncker (1891–1964), US-American botanist at the De Pau University, Indiana. (*Stenocereus*)

yungasensis For the occurrence in the Yungas vegetation of E Bolivia. (*Echinopsis bridgesii* ssp.)

Yungasocereus For the occurrence in the Bolivian Yungas vegetation; and *Cereus*, a genus of columnar cacti. (*Cactaceae*)

yunnanensis For the occurrence in Yunnan Prov., China. (*Kalanchoe, Rhodiola, Sinocrassula*)

yuquina For the occurrence at Yuquina near Culpina, Prov. Sud-Cinti, Dept. Chuquisaca, Bolivia. (*Echinopsis*)

yvesii For a Mr. Yves (fl. 1910), without further data. (*Sedum*)

Z

za From the Malagasy vernacular name "za" of the plants. (*Adansonia*)

zacanus For the yacht "Zaca" that was used for exploration in the Galápagos Islands when this taxon was found. (*Opuntia echios* var.)

Zaleya Perhaps from an Indian vernacular name (Jackson 1990). (*Aizoaceae*)

zambesiacus For the occurrence in Zambia. (*Ceropegia*)

zambesiensis For the occurrence near the Zambesi River. (*Adenia*)

zambicus For the occurrence in Zambia. (*Jatropha schweinfurthii* ssp.)

zambiensis As above. (*Euphorbia griseola* ssp.)

zamiifolius For the genus *Zamia* (*Zamiaceae*); and Lat. '-folius', -leaved. (*Zamioculcas*)

Zamioculcas For the genus *Zamia* (*Zamiaceae*); and from Arab. 'qolqas', 'kulkas', the name of the Taro plant (*Colocasia*); for the leaves, which resemble those of *Zamia*, and the relationship with *Colocasia*. (*Araceae*)

zamudioi For Sergio Zamudio (fl. 1999), Mexican biologist. (*Opuntia*)

zanaharensis For the Zanahara Section of the Tsimbazaza Botanical Garden, Antananarivo, Madagascar, where the taxon was found growing. (*Euphorbia*)

zangalensis For the occurrence near Zangal, Dept. Cajamarca, Peru. (*Haageocereus*)

zantnerianus For Alfred Zantner († 1953), German succulent plant enthusiast. (*Haworthia*)

zanzibaricus For the presumed occurrence on the island of Zanzibar. (*Dorstenia, Sansevieria*)

zaragosae For the occurrence near the town of Zaragosa, Nuevo León, Mexico. (*Turbinicarpus mandragora* ssp.)

zarcensis For the occurrence near La Zarca, Durango, Mexico. (*Agave asperrima* ssp.)

zebra From Portuguese 'zebro, zebra', zebra, wild donkey; for the zebra-like striping (cross-bands) of the leaves. (*Agave*)

zebrinus Bot. Lat., striped (from Portuguese 'zebro, zebra', zebra, wild donkey); (**1**) for the striped leaves. (*Aneilema, Haworthia reinwardtii* fa., *Tradescantia*) – (**2**) for the cross-banded leaf bases. (*Aloe, Ornithogalum*) – (**3**) for the striped corolla. (*Huernia*)

zeederbergii For Mr. Zeederberg (fl. 1934), without further data. (*Delosperma*)

zegarrae For Ing. German Zegarra Caero (fl. 1958), promotor of cactus explorations in Bolivia. (*Gymnocalycium*)

zehntneri For Leo Zehntner (1864–1961), Swiss botanist, 1912–1916 head of a forest research station in Juazeiro, Bahia, Brazil, where he made several botanical excursions to study cacti etc. (*Discocactus, Melocactus, Pilosocereus gounellei* ssp., *Quiabentia*)

zeleborii For Johann Zelebor (fl. 1850s), Austrian zoologist in Vienna, participated as zoological collector in the Austrian circumnavigation of the world 1857–1859. (*Sempervivum*)

zephyranthoides Gr. '-oides', resembling; and for the genus *Zephyranthes* (*Amaryllidaceae*); for the similarly attractive and large flowers. (*Mammillaria*)

Zeuktophyllum Gr. 'zeuktos', yoked; and Gr. 'phyllon', leaf; for the closely set leaf pairs. (*Aizoaceae*)

zeyheri For Karl [Carl] L. P. Zeyher (1799–1858), German naturalist and botanical explorer in RSA from 1822. (*Adromischus cristatus* var., *Aizoon, Ceropegia, Erythrina, Jatropha, Lampranthus, Raphionacme*)

zeylanicus Lat., from Ceylon (former name of Sri Lanka). (*Sansevieria*)

zilzianus For Prof. Dr. Julian Zilz († 1930?), physician and cactus collector in Vienna, Austria, 1930 president of the Vienna Branch of the Austrian Cactus Society. (*Escobaria*)

zimmermannii For [Philipp Wilhelm] Albrecht Zimmermann (1860–1931), German botanist, 1902–1920 botanist at the Coffee Culture Experiment Station at Amani (E Africa). (*Cyphostemma*)

zinniiflorus For the genus *Zinnia* (*Asteraceae*); and Lat. '-florus', -flowered. (*Pereskia*)

zoeae For Zoë Harris (fl. 1935), without further data. (*Delosperma*)

zokuriensis For the occurrence on Zokurisan (Mt.), Korea. (*Phedimus*)

zombensis For the occurrence at Mt. Zomba, Malawi. (*Crassula*)

zombitsiensis For the occurrence in the Zombitsy Forest, Prov. Toliara, Madagascar. (*Aloe*)

zonalis Lat., zoned; for the colour pattern of the leaves. (*Pelargonium antidysentericum* ssp.)

zopilotensis For the occurrence in the Cañón del Zopilote, Guerrero, Mexico. (*Peniocereus*)

zoutpansbergensis For the occurrence in the Soutpansberg [Zoutpansberg] Ranges, Northern Prov., RSA. (*Delosperma, Euphorbia*)

zublerae For Ruth Zubler (fl. 1987), Swiss cactus enthusiast in Breisach near Basel. (*Mammillaria*)

zuluensis For the occurrence in the then Zululand (now part of KwaZulu-Natal), RSA. (*Euphorbia franksiae* var., *Plectranthus*)

zygophylloides Gr. '-oides', resembling; and for the genus *Zygophyllum* (*Zygophyllaceae*). (*Drosanthemum, Gunniopsis*)

Zygosicyos Gr. 'zygon', yoke; and Gr. 'sicyos', cucumber; for the zygomorphic flowers. (*Cucurbitaceae*)